ROUTLEDGE LIBRARY EDITIONS:
GEOLOGY

Volume 21

INTRODUCTION TO GEOMORPHOLOGY

INTRODUCTION TO GEOMORPHOLOGY

ALISTAIR F. PITTY

Routledge
Taylor & Francis Group

LONDON AND NEW YORK

First published in 1971 by Methuen & Co Ltd

This edition first published in 2020
by Routledge
2 Park Square, Milton Park, Abingdon, Oxon OX14 4RN

and by Routledge
52 Vanderbilt Avenue, New York, NY 10017

Routledge is an imprint of the Taylor & Francis Group, an informa business

© 1971 Alistair F. Pitty

British Library Cataloguing in Publication Data
A catalogue record for this book is available from the British Library

ISBN: 978-0-367-18559-6 (Set)
ISBN: 978-0-429-19681-2 (Set) (ebk)
ISBN: 978-0-367-22024-2 (Volume 21) (hbk)
ISBN: 978-0-367-22079-2 (Volume 21) (pbk)
ISBN: 978-0-429-27313-1 (Volume 21) (ebk)

Publisher's Note
The publisher has gone to great lengths to ensure the quality of this reprint but points out that some imperfections in the original copies may be apparent.

Disclaimer
The publisher has made every effort to trace copyright holders and would welcome correspondence from those they have been unable to trace.

Alistair F. Pitty

Introduction to Geomorphology

METHUEN & CO LTD

First published 1971 by
Methuen & Co Ltd, 11 New Fetter Lane, London EC4P 4EE
Reprinted 1973 and 1977

© *1971 Alistair F. Pitty*

Filmset in Photon Times 11 on 12½ pt by
Richard Clay (The Chaucer Press), Ltd, Bungay, Suffolk
and printed in Great Britain by
Fletcher & Son Ltd, Norwich, Norfolk

ISBN 0 416 29760 9

Distributed in the USA by
HARPER & ROW PUBLISHERS, INC.
BARNES & NOBLE IMPORT DIVISION

Contents

Acknowledgements

A great deal of the effort behind the presentation of this book is that of the technical and secretarial staff in the Department of Geography in the University of Hull, especially that of Mr Keith Scurr and Miss Wendy A. Wilkinson who drew the figures. Mr R. R. Dean arranged the schedule for drawing at a time when illustrations for a wide range of other publications were being requested. Mr A. Key assisted with the drawing at a final stage. Mr J. B. Fisher and Mr S. Moran prepared photographic prints of the drawings. The typing was done with great care by Miss C. Hayward, who typed most of the text, and by Mrs B. Smith, assisted by Miss H. Coburn, who typed the appendix and the captions.

At the outset, when an attempt to span the breadth of geomorphology seemed scarcely possible I was particularly grateful to the following for their assistance in compiling information; when demands on the Map Library were not too heavy, the Curator, Miss A. M. Ferrar, arranged for her assistant, Miss B. Myers, to help with compilation; one of my tutorial groups, comprising Misses J. A. Cavanagh, S. V. Lloyd, C. A. Luland, M. D. V. Raybould, and Mr G. D. Sanderson; and Mrs P. A. Pitty, who also prepared abstracts used in parts of Chapter II. Research students who, by the thoroughness and enthusiasm of their assistance as demonstrators, helped to make my teaching commitments less exacting at the time when I was writing the book include Mr I. Reid (practicals), Mr D. W. John (tutorials), and Mr K. A. Falconer (field work). Previously Dr A. C. Imeson had assisted on these occasions. I am grateful to the publishers for their help and kind editorial assistance and advice. I am also grateful to many other members, past and present, of the University of Hull, for their continued interest in the progress of the work, particularly to Dr J. R. Dennett, Mr M. J. Hutchinson, Mr M. J. Burgess, Mr F. G. Gray, Miss P. Zalasinski, and Miss A. Holloway, and to many more geomorphologists than those listed in the bibliography and text, who, by their absorbing studies, have made a wide reading of the subject a most rewarding experience.

In addition to those who have helped in the immediate task of preparing this book, I would like also to acknowledge those who, at earlier stages, encouraged me to seek the rewards of studying geography in general and landforms in particular; my parents for encouraging me to admire and explore landscapes and for teaching me to use, value, and to draw maps; my sixth-form geography master, Mr Ian H. Watts, for showing me how to start learning for oneself, and the Department of Geology in the University of Manchester, particularly Dr F. M. Broadhurst, for most informative extra-mural courses; as an undergraduate the formation of first notions about geomorphology were influenced by the teaching .of my tutor, Dr J. M. Houston, by Dr R. P. Beckinsale, by Miss M. Marshall's special option on the history of the cycle of erosion concept, and by my post-graduate supervisor, Dr M. M. Sweeting.

I hope that I can make these acknowledgements without associating any of the above with the shortcomings of this book.

The author and publisher would like to thank the following for permission to reproduce copyright figures and tables.

Journals/Editors

American Journal of Science for figs: I.4.E (Vol. 260), I.7 (Vol. 260), II.6 (Vol. 265), II.10.A (Vol. 256), III.1 (Vol. 257), III.17 (Vol. 266), IV. 15.B (Vol. 268), IV.36.A (Vol. 263), IV. 36.C (Vol. 258), IV. 43.B (Vol. 264), IV.49.B (Vol. 262), V.1 (Vol. 263), V.3.C (Vol. 258), V.5 (Vol. 261), V.7 (Vol. 265), V.27 (Vol. 255); *Australian Geographical Studies* for figs. I.3.C, IV.31.A; *The Canadian Geographer* for fig. IV.24.C; *Economic Geology* for figs. IV.42, V.13; *Geografski Glasnik* for fig. V.18; *Geological Magazine* for figs. I.5.A, I.5.B, I.6.A, IV.30.A, IV.45.D; *Nature* for figs. III.1, IV.26.B; *Rassegna Speleologica Italiana* for fig. V.17.A; *Science*, Volume 132, 1960, fig. 5. Copyright 1960 by the American Association for the Advancement of Science, for fig. II.5.A; *Scientific American*, August 1960, page 88 for fig. IV.34.J.

Publishers

Archiv für Wissenschaftliche Geographie for figs. IV.45.E, V.6, V.12.D; Blackwell Scientific Publications for figs. III.10, III.15.A; Clarendon Press for figs. I.8, I.9, III.12.B, III.29, IV.9.D, IV.26.C, IV.39.B, IV.44.B, V.4, V.30, Table III.2, Table IV.2, Table V.3; Elsevier, Amsterdam, for figs. I.12, IV.13.A, IV.13.B, IV.18.D, IV.38; Eyrolles, Paris, for fig. V.25.C from *Plages et cotes de sable* by J. Larras; figs. I.4.B, IV.28.A, IV.34.A–D, IV.36.B adapted from *Fluvial Processes in Geomorphology* by Luna B. Leopold, M. Gordon Wolman, and John P. Miller. W. H. Freeman and Company. Copyright 1964; Generalstabens Litografiska Anstalt for figs.

II.15.A, II.15.B, III.8.B, III.13, IV, IV.1, IV.5.A, IV.10, IV.13.D, IV.15.A, IV.24.D, IV.24.F, IV.30.B, IV.47.A, IV.47.D, IV.51, IV.53, V.3.A, V.3.B, Table IV.1, Table IV.9, Table App. 2; Gebrüder Borntraeger for figs. I.4.A, II.15.E, II.15.F, IV.7.A, IV.7.B, IV.18.A, IV.33.A, IV.33.B, fig. App. 1.C, fig. App. 1.D, Table App. 1; Heinemann Educational Books Ltd for fig. IV.6.D redrawn from the original diagram by Mr A. Bartlett of the Geography Department of the University of Sydney; Liverpool University Press for fig. V.22; Longman Group Ltd for fig. V.11.A; fig. IV.19 from McGraw-Hill *Encyclopedia of Science and Technology* 1962, used with permission of McGraw-Hill Book Company; Masson et Cie for figs. III.6.D, III.9.A, III.9.B, IV.26.D, IV.26.E, IV.39.A, V.24, Table 1.1; Methuen and Co for figs. I.6.B, III.3, IV.18.B; Thomas Nelson and Sons Ltd for fig. IV.43.A, part of IV.6.D; Oliver and Boyd for fig. V.31.E from *Morphology of the Earth* by L. C. King, and Table IV.8 from *Principles of Lithogenesis*, I by N. M. Strakhov; Panstowowe Wydawnictwo Naukowe for fig. IV.8.C (from *Przeglad Geograficzny*) and fig. IV.9.A (from *Geographia Polonica*); Princeton University Press for fig. II.1; *Redakcja Biuletyn Peryglacjalny* for figs. IV.8.A, IV.8.B, and IV.39.C; University of Chicago Press for the following figs. taken from *Journal of Geology*: I.2 (and Marie Morisawa), I.4.C (and H. J. Finkel), II.7 (and R. F. Flint), III.6.B (and E. D. Sneed and R. L. Folk), III.6.C (and P. Kuenen), III.12.D (and D. C. Rhoads), III.18 (and J. S. Olson), IV.11.C (and W. O. Kupsch), IV.15.D (and J. C. Harms, *et al.*), IV.20 and IV.40 (and V. P. Zenkovitch), IV.45.C (and M. Demorest), IV.50 (and P. McKenzie), V.21 (and E. Buffington, D. G. Moore), fig. App. 1, Table IV.11 (and C. B. Beaty), Table V.2 (and R. Le B. Hooke); Universitets Forlaget for figs. IV.3.A, IV.12.B, IV.12.C, IV.49.C, V.31.B, Table IV.9; Van Nostrand Reinhold Co for figs. II.12, IV.3.B, IV.7.C, Table IV.12, and Table IV.13; John Wiley and Sons Inc, for fig. IV.39.E; Wiley-Interscience for fig. V.19.A.

Learned Societies

American Association of Petroleum Geologists for figs. II.8.B–E, IV.23.A, IV.23.B, V.11.B, V.11.C; American Geophysical Union (and W. B. Langbein and S. A. Schumm) for fig. I.6.D, and from *Soviet Hydrology*, figs. IV.36.D, IV.44.C, IV.52; figs. IV.18.C and V.12.A reproduced by permission from the *Annals of the Association of American Geographers*; Cave Research Group of Great Britain for fig. IV.26.A; Drzavna zalozbu Slovenije, Ljubljana for fig. V.8; Field Studies Council for fig. IV.6.B; Geografisko drustvo Slovenje for figs. III.11 and V.18; fig. IV.47.C redrawn from an illustration in the *Journal of Glaciology* by permission of the Glaciological Society; The Geological Society of America for figs. I.3.B, II.4, II.5.B, II.5.C, II.9, II.10.B,

II.10.C, II.11, III.6.E, III.7.A, III.8.A, III.16.A, IV.2, IV.5, IV.14, IV.21, IV.25, IV.27.B, IV.28.E, IV.28.F, IV.34.F–H, IV.43.C, IV.46, IV.48.B, IV.53, V.14, V.23, Table V.1; The Geological Society of London for figs. I.5.C, I.5.D, III.8.A, III.12.E, IV.24.E, V.26, V.31.A; Geologiska Foreningens I Stockholm Forhandlingar for figs. IV.24A; The Geologists Association for figs. I.10, IV.6.D, IV.27.A, V.25.B; Houston Geological Society for figs. IV.17 (and E. H. Rainwater) IV.28.G (and A. S. Naidu); Institute of British Geographers for figs. II.14, IV.8.D, V.15, V.25.A; International Association of Scientific Hydrology for Table IV.7; Libraire des Comptes rendus de l'Academie des Sciences, for Tables IV.4 and IV.5; Macaulay Institute for Soil Research for fig. IV.48.C; New Zealand Geographical Society for fig. IV.34.E; Royal Geographical Society for the following taken from *Geographical Journal*: IV.12.A (and W. H. Ward), IV.13.C (and C. A. Lewis, G. M. Lass), IV.39.D (and M. S. Longuet-Higgs, D. W. Parkin), V.12.B, V.12.C (and J. C. Doornkamp, P. H. Temple), V.31.C (and D. E. Sugden, B. S. John); Royal Scottish Geographical Society for fig. IV.3.C; The Royal Society for figs. II.2, II.3; The Royal Society of New South Wales for fig. IV.31.C; Schweizerische Geologische Gesellschaft for fig. V.31.D; Society of Economic Paleontologists and Mineralogists for the following from the *Journal of Sedimentary Petrology*: II.8.A (and M. Gipson), IV.11.A (and B. J. Bluck), IV.11.B (and S. Sengupta), IV.22 (and E. D. McKee, G. C. Tibbits), IV.23.C (and E. D. McKee and G. C. Tibbits), IV.32, IV.45.B (and B. J. Bluck), IV.37 (and W. F. Tanner), IV.41 (and D. H. Yaalon), IV.45.A (and J. P. Allen), IV.49.A (and C. E. Weaver), V.13 (and P. F. Williams, B. R. Rust); Societe Geologique de France for fig. III.16.B; Soil Science Society of America for figs. I.6.C, I.13, III.12.A, III.12.C, IV.48.A, IV.55.B, Table III.1, Table III.3, Table IV.10, Table App. 2; The Yorkshire Geological Society (and R. Agar) for fig. V.10.

Universities

Bonn University Institute of Geography for figs. IV.6.A, IV.31.D; Florida State University for figs. III.18, IV.27.C, IV.28.D, V.2; University of Hull for figs. III.4.B, III.15.B, IV.9.C, fig. App. 1; Louisiana State University, Coastal Institute (and F. A. Welder) for fig. IV.35; University of Lyon for fig. IV.7.D; University of Strasbourg for figs. III.7.B, IV.44.A.

Miscellaneous

Figs. II.15.C. Based on Crown copyright Geological Survey diagram. Reproduced by permission of the Controller, H.M. Stationery Office; Centre National de la Recherche Scientifique, Paris for figs. IV.9.B, IV.16 and IV.44.D; La Commission géologique de Finlande for figs. IV.24.B and V.20;

Department of Mines and Technical Surveys, Ottawa for figs. I.4.D, III.4.C, IV.47.B from *Geographical Bulletin* reproduced with permission of Information Canada; Service des Etudes Scientifique for fig. IV.9; U.S. Department of Agriculture for fig. IV.55.A; Yugoslav Ministry of Defence for fig. IV.6.C; Professor R. Coque for figs. II.12, II.13, III.14, V.28, V.29; Professor A. Jahn for figs. IV.33.C, IV.33.D.

Preface

In the last twenty-five years there has been considerable progress in all branches of science. On a modest scale the part-time science of geomorphology has achieved a great deal during this time, and the geographers, geologists, hydrologists, and specialists from other fields who all contribute to the study of geomorphology have become aware of the comparative suddenness of change within the context of landform studies. Commenting from a geologist's point of view on the activity in the actual dating of Pleistocene deposits, F. W. Shotton (1967) found it difficult to realize that almost all this work has been done in less than two decades. On the geographical side, O. H. K. Spate (1960) observed that many had been startled by the suddenness of the rise of quantification. Since 1945 many new techniques in addition to radio-carbon dating have been utilized. Vertical air photographs gave penetrating insights into tectonics and led to a demand for certain revisions of traditional geomorphological concepts. In 1947 the first systematic offshore acoustic study was made and since 1955 the profiling of continuous depth surveys of the bottom topography offshore has become widespread and detailed. Information from little-known areas has come in. For instance, in the past twenty years much research on tropical soils has modified profoundly an understanding of weathering processes. Periglacial features, from being picturesque curiosities have become accepted as a fundamental phenomenon recognizable in areas covering more than a third of the earth's land surface. After 1957–8 the first reliable estimate of the volume of the Antarctic ice-sheet was the product of the co-operative investigations of the International Geophysical year. It was as recently as 1960 that the various glacial stages on the Arctic borders were worked out, and it is only in the last seven years that a body of measurements of postglacial submergence and rebound have emerged. Different approaches to landform study have developed, partly because workers in related sciences have found the study of geomorphological problems as rewarding as their own. In 1945 the engineer R. E. Horton discussed in detail his hydrophysical approach to drainage basin study which was considered in Europe by P. Pinchemel and taken up with vigour in the United States by A. N. Strahler and his students as a possible alternative to W. M. Davis's approach. In 1943

J. Tricart began his geomorphological investigations but with the deep feeling
that insufficiencies of Davisian geomorphology left him unarmed for the task.
Today W. M. Davis, still the father of geomorphology for many has become
an Aunt Sally for others. Criticisms and attacks on Davis have taken a variety
of forms, including L. B. Leopold and W. B. Langbein's attempt to run Davis
down with a thermodynamic engine. In 1959 C. A. M. King's book on coasts
and beaches appeared, emphasizing the engineers' use of mathematical theory,
scale models and wave tank experiments to further the study of wave action
and shore processes. It has become increasingly apparent that interest in
landform study stems from more than one source. For those at one extreme,
landforms are essentially a sporting challenge, at the other, a problem in
theoretical physics. Also apparent in much of geomorphological work since
1945 are changes in emphasis. Within this time there has been a rapid
development in the study of slopes, and studies developing from F.
Hjulström's thesis (1935) on the morphological activity of rivers concentrate
on investigations of geomorphological processes. A traditional concern with
erosional forms increasingly includes the study of mass movements and
depositional features. Recently a new factor, government interests and financial support from a variety of sources, has benefited geomorphology. For
instance, government-sponsored surveys in Canada and Australia have not
only brought back factual information from little-known areas but have also
provided some geomorphologists with the opportunity of full-time employment
in the study of landforms and with the stimulus of working in close association with scientists in related disciplines. Another recent factor has been the
intensification of interest in practical problems. The tackling of coastal engineering problems or the military problems of crossing rough terrain has
provided the geomorphologist with useful information. The investigation of
many other practical problems from which geomorphology has benefited
include the possibility of influencing climate artificially which has heightened
interest in glaciers which store three-quarters of the Earth's fresh water. The
surge of interest and major advances in the study of the transport of solutes
and sediment in streams are partly due to increased public awareness of
problems of water supply and water-quality management. In the last two
decades the range of literature available to the geomorphologist has increased.
After the lapse of two geomorphological journals during the Second World
War, the *Revue de géomorphologie dynamique* first appeared in 1950 and the
Zeitschrift für Geomorphologie reappeared in 1957. More specialized journals, the *Journal of Glaciology* in 1947 and the *Biuletyn Peryglacjalny* in
1954, have appeared. A most significant recent trend for geomorphologists
depending on sources written in English has been the degree to which journals
published in many countries include an increasingly large number of articles
or long abstracts in English. Also translation services have made available a

huge store of information on hydrology, soils, and geography compiled by scientists in the Soviet Union.

Scientific progress has had several repercussions within geomorphology. The main one, accentuated by the very smallness of the number of scientists who can devote all their work to landform study, is the tendency for a multitude of specialized fields of interest to emerge. For instance, from the geological point of view, the word geomorphology has been prefaced by words like structural or morphotectonic. Studies orientated towards 'process geomorphology' may include prefixes such as 'climatic' or more specifically 'periglacial' or 'tropical'. Studies of past conditions or of relict features absorb the attention of Pleistocene, stratigraphical, or historical geomorphologists; and some petroleum geologists are increasingly attracted to the study of paleogeomorphology. Approaches to geomorphology have also been described by a variety of terms, sometimes as a slogan by some or as an implied criticism by others, and include the adjectives quantitative, qualitative, descriptive, dynamic, mathematical, theoretical, geological, geographical, regional, systematic, and, not least, Davisian. Also, as conflicting conclusions of geomorphologists working in different areas may reflect their differing field experiences and schools of thought, geomorphology may be linked, for example, with Poland or with Scandinavia. The variety of these diverse themes raises the questions of their validity, their definition, of where they might lead, and to what extent and in what way they are complementary, opposed, or merely tortologous.

Optimistically, it is hoped that the first part of this book will serve as a guide to the reader in placing his further reading, including the subsequent sections of the present book, in the context of the rapidly developing and expanding field of geomorphology. An optimistic approach in attempting to present a balanced sketch of geomorphology is justified because the very reasons of pace and variety which militate against the feasibility of this task make it equally important that some coherent attempt should be made to search for and to illustrate the underlying themes of landform study. Although the intention is to present a guide-book rather than a textbook, it is hoped that sufficient detail is included for the reader to become familiar with some examples of the findings of more recent work. However, with one exception it is not intended deliberately to promote certain themes nor systematically to oppose more traditional studies. It is felt that the pace of progress makes the uncritical adoption of innovations as inadvisable as the dogmatic adherence to outdated ideas. The sole conscious deliberation has been to illustrate, if only with a restricted number of examples, the nature of the actual, observed facts of landform study in preference to summarizing the stockpile of hypotheses which have yet to be tested against facts yet to be observed.

Definitions, nature and basic postulates

A. Definitions

Geomorphology is the study of landforms. This obvious statement is difficult to extend without stating a partial view of the subject. Traditionally the study was essentially that of the origin and evolution of landforms. Many would still agree with N. M. Fenneman that 'the study and interpretation of the records left by erosion constitute the larger part of the science of geomorphology', and see their work, as S. W. Wooldridge did, as answerable for the last brief chapter of the geological record. The historical element in geomorphology is evident from the need to interpret still observable traces of events that once took place on the earth's surface. On the other hand, workers like L. B. Leopold, M. G. Wolman, and J. P. Miller (1964), while observing that 'much of geomorphology is stratigraphic geology', have made the study of contemporary processes a conspicuous part of their own approach to the subject. Much recent work tends increasingly to reflect the definition of F. Hjulström, that geomorphology is the science of landforms and of land-forming processes. Yet another view sees geomorphology not merely as a study of landforms and processes but stresses the importance of the inter-actions at the contact surface between the terrestrial parts of the lithosphere and the liquid and gaseous envelopes which surround it, the ever-changing inter-adjustments between different forces of nature, and the changes in appearance of the land surface as the dynamic balance between these forces is modified.

B. The Nature of Geomorphology

1. Description and interpretation

Any science starts with observations of natural phenomena. Basic descrip-tions in geomorphology are of three kinds, being observational, classificatory, or explanatory–descriptive in nature. The traditions of German geomorphology emphasize the importance of observation, illustrate the desire to provide as detailed descriptions as possible of relief forms, and date back to the work of von Richthofen in the 1880s. Secondly, some landforms have sufficient symmetry and distinctiveness of form to be classified as one of a certain type. Thus a term like river meander conveys immediately to the reader a large amount of descriptive information. A third approach is that of genetic description, which aims at further economy by describing in a word not only information about the appearance of a form but also implies an interpretation of its origin, e.g. peneplain. Geomorphology is unusual in the degree to which this descriptive–explanatory or genetic approach has been employed. Since 1945 there have been changes in the ways in which observations are recorded, with the growing recognition that some degree of schematization is necessary. Increasingly there have been attempts to economize on lengthy verbal descriptions, to sharpen the precision of observation with the development of specialized mapping techniques, and to search for ways of reducing observations of landform to numbers. These attempts have included the description of complicated terrains as well as of simpler symmetrical landforms. There has also been a movement away from genetic description in an attempt to separate clearly description from interpretation.

A. Cailleux and J. Tricart (1956) put an initial problem, that of scale, into perspective by defining a series of size orders. A notion of scale not only provides a framework in which to categorize observations but also influences the methods used to collect the observations and which of the related specialisms, from geophysics to soil science, will have most bearing on the interpretation of the results (Table I.1). The actual size of the scale orders is less important than the stress laid on the significance of scale. Within the size range of features studied in geomorphology little in detail is known beyond the general fact that some forms may vary with size, whereas others may not. For instance, river meanders have the same dimensions in plan regardless of scale. By contrast, in small-scale sand forms the coarsest material collects on the crests, whereas the reverse is invariably the case for large-scale dunes.

Landform maps have been prepared on a variety of scales, over some larger areas where no detailed contour map is available, or from smaller areas to record aspects of the geometry of the ground surface, such as discontinuities, which a contour map does not show. Fig. I.1.A illustrates morphological mapping techniques used by R. A. G. Savigear. A similar approach can be used in coastal studies, as the symbols used by M. A. Arber (1949) in describing cliff profiles show (fig. I.1.B). For mapping purposes, coasts might be divided between cliffed and non-cliffed, with both categories then subdivided according to whether their plan is regular or irregular in outline.

Table I.1 Classification of geomorphological features (*after Tricart, 1965*).

Order	Units of earth's surface in km²	Characteristics of units, with examples	Equivalent climatic units	Basic mechanisms controlling the relief	Time-span of persistence
I	10^7	continents, ocean basins	large zonal systems controlled by astronomical factors	differentiation of earth's crust between sial and sima	10^9 years
II	10^6	large structural entities (Scandinavian Shield, Tethys, Congo basin)	broad climatic types (influence of geographical factors on astronomical factors)	crustal movements, as in the formation of geosynclines. Climatic influence on dissection	10^8 years
III	10^4	main structural units (Paris basin, Jura, Massif Central)	subdivisions of the broad climatic types, but with little significance for erosion	tectonic units having a link with paleogeography; erosion rates influenced by lithology	10^7 years
IV	10^2	basic tectonic units; mountain massifs, horsts, fault troughs	regional climates influenced predominantly by geographical factors, especially in mountainous areas	influenced predominantly by tectonic factors; secondarily by lithology	10^7 years

limit of isostatic adjustments

Order	Units of earth's surface in km²	Characteristics of units, with examples	Equivalent climatic units	Basic mechanisms controlling the relief	Time-span of persistence
V	10	tectonic irregularities, anticlines, synclines, hills, valleys	local climate, influenced by pattern of relief; adret, ubac, altitudinal effects	predominance of lithology and static aspects of structure	10^6–10^7 years
VI	10^{-2}	landforms; ridges, terraces, cirques, moraines, debris, etc.	mesoclimate, directly linked to the landform, e.g. nivation hollow	predominance of processes, influenced by lithology	10^4 years
VII	10^{-6}	microforms; solifluction lobes, polygonal soils, nebka, badland gullies	microclimate, directly linked with the form, e.g. lapies (karren)	predominance of processes, influenced by lithology	10^2 years
VIII	10^{-8}	microscopic, e.g. details of solution and polishing	micro-environment	related to processes and to rock texture	

Further subdivisions could be based on local conditions, such as bay shape, rock type, and gradient.

Description by the measurement of landforms is for the most part difficult because of the extremely complex details of form. Furthermore, where a series of measurements is involved it is not easy to summarize these com-

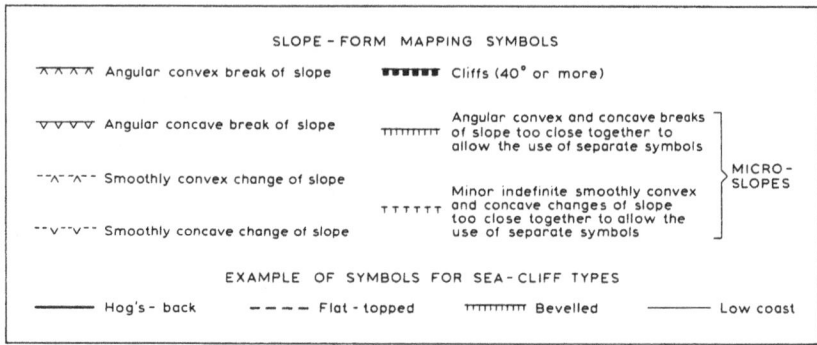

Figure I.1 Examples of landform mapping symbols. (*Above*) Some slope-form mapping symbols (*from Savigear, 1965*). (*Below*) Sea-cliff types (*from Arber, 1949*).

pactly in diagnostic indices, and because of discontinuities and irregularities, equations are largely inapplicable. Efforts to describe landform in numbers are usually based on ratios between various dimensions of the form under consideration. For instance, R. E. Horton (1932) suggested a simple ratio of length of drainage basin, L, to its width, W, length being the longest dimen-

Table I.2 Indices of drainage basin shapes (*from Morisawa, 1958*).

Drainage basin	A	B	C	D
E	0·82	0·86	—	0·59
C	0·58	0·64	0·47	0·45
L/W	1·32	0·97	0·50	2·17
area (sq miles)	25·7	18·7	86·0	10·6

sion from the mouth to the opposite side, with width measured normal to the length. S. A. Schumm suggested the index $E = d/L_m$, where E is an elongation ratio between d, the diameter of a circle with the same area as the drainage basin, and L_m, the maximum length of the basin parallel to the main river. A similar index, suggested by J. P. Miller, is $C = A_b/A_c$, where C is a circularity ratio between basin area, A_b, and the area of a circle A_c, having the same perimeter. Table I.2 lists calculations of these indices for the basin outlines shown in fig. I.2. However, it is apparent that the measurement of basin length is problematic when the long axis of the basin trends away from the line between basin mouth and the opposite side of the basin and that several differently shaped basins might have similar indices due to irregularities in their outline. The circularity ratio, for instance, is a description of increased irregularity of the basin outline rather than an index of basin shape. The description of landforms is full of dilemmas like these, and the efficiency and

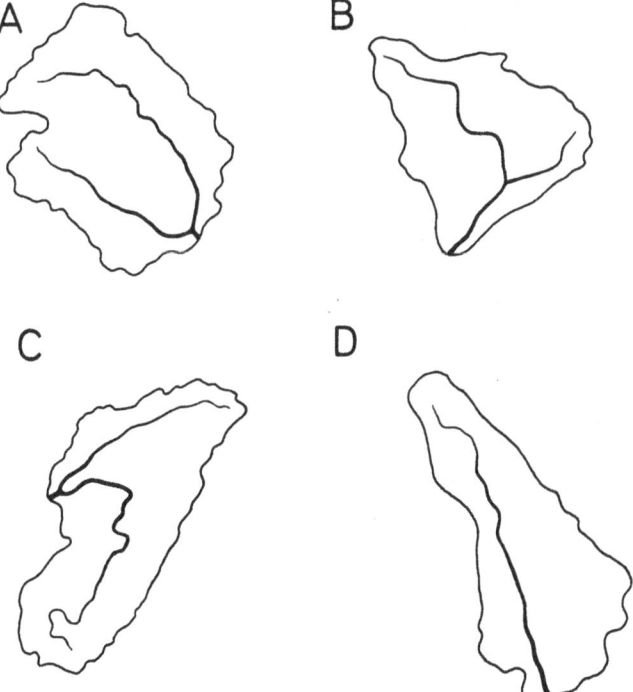

Figure I.2 Various drainage basin shapes (*from Morisawa, 1958*); these examples are from the Appalachian plateau.

exact meaning of every measurement and index must be carefully and critically evaluated. Figs. I.3 and I.4 illustrate the definitions on which some other measurements of landforms have been based. The terrain unit (fig. I.3) used by military engineers could have as much geomorphological significance as the hydrologists' drainage basin and can be described by ratios similar to those used for basin shape. The range of examples in fig. I.4 starts with the definition of depth of a weathering recess, the simplicity of the index belying the infrequency with which such observations have been recorded. By contrast the unparalleled regularity of the river meander has been measured frequently. However, even those forms which are striking in their geometric

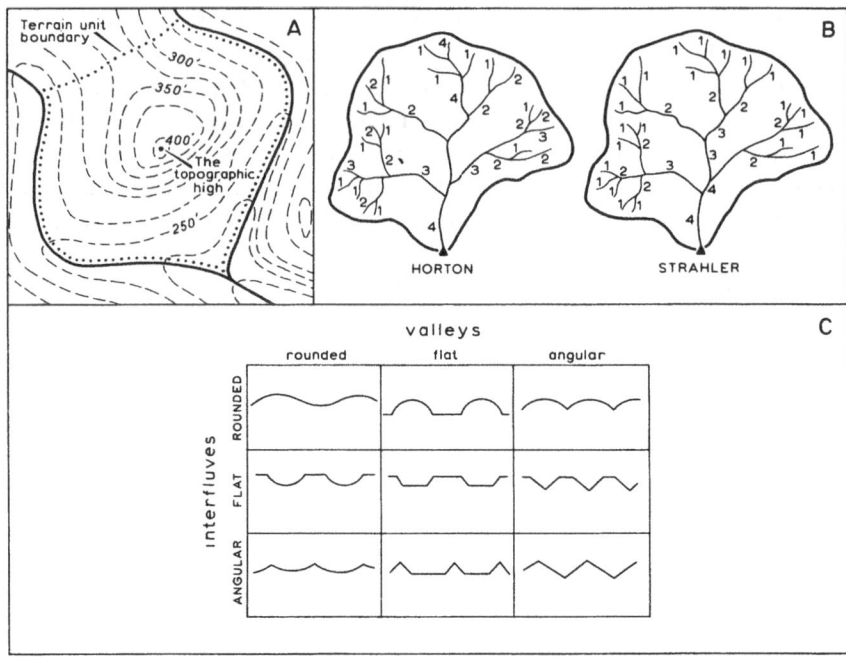

Figure I.3 Basic definitions of some descriptive schemes for categorizing land
surface components.

A. The military engineers' terrain unit (*after Krenkel and Hoadley, 1963,*
Waterways Exp. Sta. Rep. No. 4–86).

B. Stream-ordering systems for describing drainage composition (*after K. L. Bow-
den and J. R. Wallis, 1964,* Bull. geol. Soc. Am., *Vol. 75*).

C. A classification of land surface profiles based on interfluve and valley profiles
(*from Ollier, 1967*).

regularity may show substantial departures from symmetry, like the contrasted lengths of barchan horns. Usually erosional forms, like the nivation
hollow, have greater irregularity still, and several measurements might be
made to avoid ambiguities. Fig. I.4.E illustrates the description of a different
aspect of landforms, relating to their spacing and consistency of orientation,
rather than to their actual size. Although the geometrical regularity of drumlins makes these forms an obvious example, spacing and orientation studies of
more irregular forms and terrain units are feasible.

A type of description, coarser than measurement but easier to define, is the
ranking of features according to their size. The most widely used example of
ranking in geomorphology is stream orders, by the use of which R. E. Horton
hoped to improve runoff predictions but although some modifications have
been suggested, certain basic difficulties remain in applying the scheme in

geomorphology. Many field workers report that it is sometimes difficult to follow a definite channel in the field and to find its corresponding representation on a map where the restrictions of reduced scale limit the details of form which can be described. Besides being rather coarse, with difficulties at the outset in standardizing the concept of a first-order channel, the applicability of the stream-ordering scheme to dry valleys or to relict Pleistocene wadis indicates that it bears far less relation to contemporary hydrological characteristics than might be supposed.

When describing landform in section, the equivalent to the contour map which provides a plan view is the surveyed cross-profile, whether it be downslope, downstream, or a cross-section of smaller features. Attempts to convert the graphical description into numbers and indices are perhaps fewer in number and show less variety than descriptions of shapes in plan. Ratios can be used in a similar way, however, as between the height and length of a landslip area, but the same fundamental problem of complexity and irregularity in the form remains. Distinctive and regular landforms which are easily categorized in cross-section occupy only a small proportion of the landsurface area. Recently C. D. Ollier has suggested a nine-fold classification of relief cross-sections based on the recognition that both interfluves and valleys might be classified as rounded, flat, or angular. Fig. I.3.C shows that many diverse landscapes might be described, if not classified, by precise reference to these simple subdivisions, particularly if the scale and any asymmetry is also noted.

There are many difficulties in devising precise, objective, and unambiguous descriptions of the intricacies of landform, and despite the time involved in such attempts the results often appear to meet with limited success and may afford little insight. These factors may help to explain why there has been a reluctance to abandon completely the concise genetic descriptions in geomorphology. However, with growing realization of two fundamental limitations in genetic description, schemes like the classification of coasts developed by Davis, Gulliver, and Johnson, or the Cycle of Erosion itself are now less widely favoured. Firstly, the incorporation of new information and theories into existing interpretations may mean that the original terms are no longer the same in a descriptive sense. A classic example is the Irish eskers which gave their name to other long sinuous ridges of sand and gravel only to be denied the right, in the light of subsequent research, to be called eskers. Added to the difficulty of including new information in genetic schemes, is the growing awareness of the danger of circular argument. For example, it was formerly understood that with the attainment of 'old age' a subdued rolling topography results, and that conversely the presence of subdued rolling terrain implied that a state of old age had been reached in landscape evolution. Nonetheless the vigorous work in geomorphological mapping in Europe,

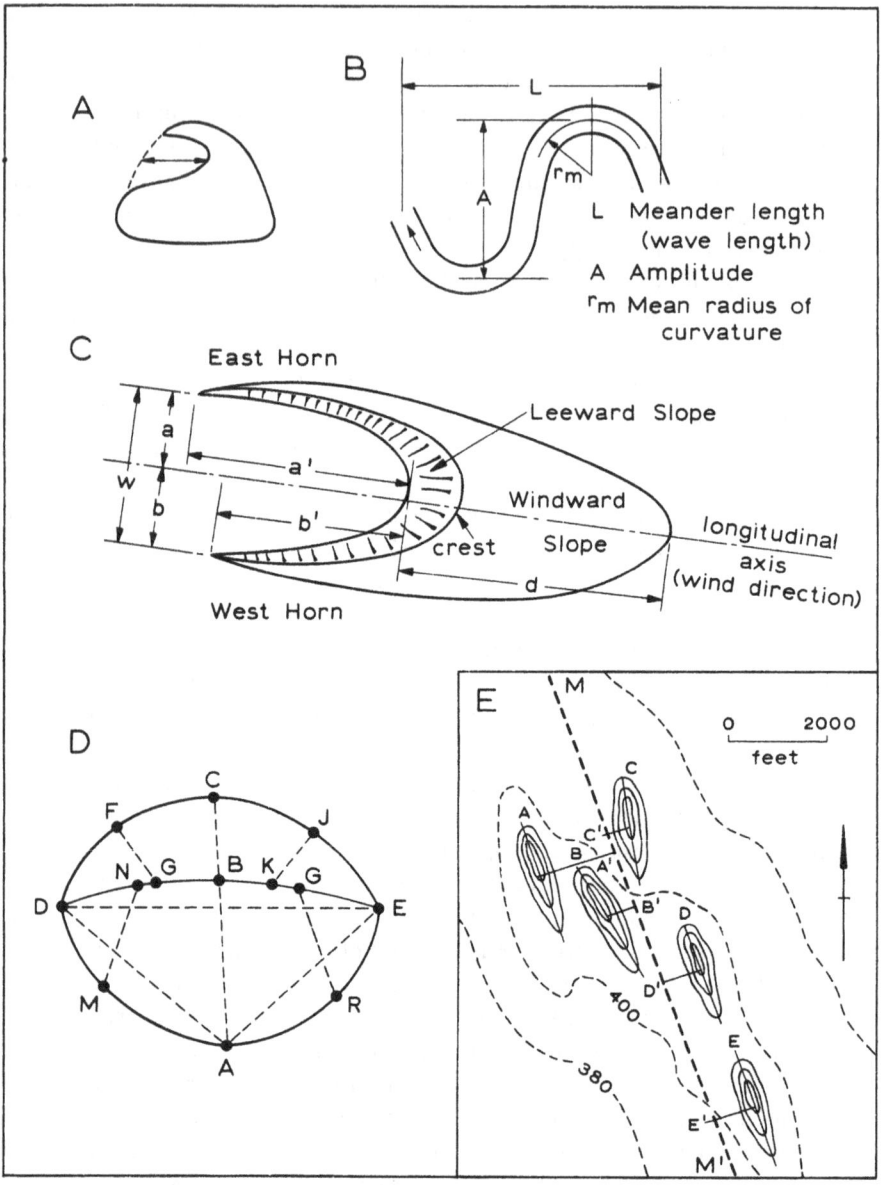

Figure I.4 Examples, from more symmetrical forms, of some measured dimensions in landform study.

A. Depth of cavernous weathering (*from Calkin and Cailleux, 1962*).

B. Meanders (*from Leopold* et al., *1964*).

C. Barchan dune (*after Finkel, 1959*).

distinct from the aims of purely descriptive morphological mapping advocated by R. A. G. Savigear, embodies decisions on the age, origin, and sometimes on the erosional agent of the landforms mapped.

Explanation in landform studies poses formidable difficulties which in part reflect the problems of achieving a precise basic description of even simple forms. A major problem in searching for explanations of land form is the intrinsic incompleteness of the geomorphological and geological evidence which does not usually form a sufficient basis for conclusions deduced. There is therefore no prospect of proof in the absolute sense as in mathematics. Interpretations tend inevitably to be qualified by such phrases as 'tends to' or 'possibly'. A further problem in the study of the inter-adjustments between land form and natural forces changing continually through time is that isolated causes are usually inseparable from distinct effects. All that can be recognized is a distinction between the antecedent conditions which affect subsequent conditions. In addition to the incompleteness of evidence, the sheer variety and intricacy of inter-relationships between biological, chemical, physical factors, and antecedent landforms mean that there are no invariable laws, as in physics, on which to base interpretations of present-day landforms.

Some geomorphologists who advocate the use of quantitative methods have solved this problem of explanation to their own satisfaction, by measuring their success in terms of the variability which they can account for in inter-correlations between parameters of land form. The explanation thus achieved is statistical rather than scientific, and in depending on the minimization of antecedent and subsequent conditions in order to correlate parameters simultaneously, eliminates the very aspect of landform study which requires a causal explanation. Sometimes these studies are clearly influenced by the technologists' responsibility to predict rather than the scientists' interest to understand, and any geomorphological explanations that emerge from these studies may be superficial and quite unsurprising.

In geomorphological literature uncertainty, disagreement, and debate about the origin of specific landforms abound compared with a dearth of measurements of both form and process. Possibly the influential views of W. M. Davis on the role of the imagination in geomorphological research may have encouraged his successors to admire too readily the stimulating spectacle of the highly assumptive reasoner juggling with scanty data. 'We may

---

D. Nivation hollow (*from F. A. Cook and V. G. Raiche, 1962*, Geog. Bull., *No. 18*). These dimensions were measured at Resolute, North West Territories, Canada. DCE is the top of the backwall, DBE the base of the backwall, and DAE the front of the hollow floor.

E. Drumlin spacing and orientation (*after Reed* et al., *1962*).

still give praise to those who apply themselves chiefly to gaining first-hand knowledge of observable facts,' he allowed, 'but we have learned to give greater praise to those who, on a good foundation of visible facts, employ a well-trained constructive imagination in building ingenious and successful

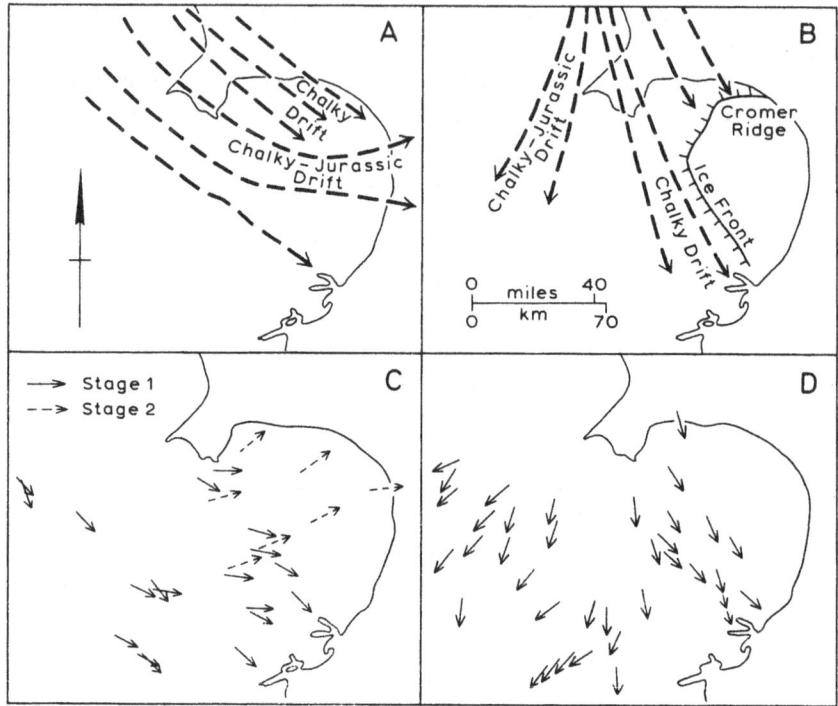

Figure I.5 An illustration of a principal mode of explanation in geomorphology – the independent use of different methods pointing towards the same conclusion. Direction of ice movement across East Anglia during the Lowestoft Advance (A) and during the Gipping Advance (B) deduced from changes in lithological composition of tills *(from D. F. W. Baden-Powell, 1948, Geol. Mag., Vol. 85).* Similarly, direction of ice movement during the Lowestoft Advance (C) and during the Gipping Advance (D), as indicated by preferred orientation in the tills *(from R. G. West and J. J. Donner, 1956, Quart. Jour. Geol. Soc., Vol. 112).*

theories which shall bring to sight the invisible facts.' However, Davis by assuming that the facts of earth science had, by the end of the nineteenth century, been 'abundantly acquired and so thoroughly systematized' was unwittingly advocating praise for speculating thinking. If geomorphology has taken Davis's disdain for the unimaginative to heart, perhaps his lesson should be revised so that the work of patient and painstaking description at any level in geomorphology receives more approbation and the unsubstantiated speculations of the lively mind a little less.

'Judge not according to the appearance.' St. John VII, 24

2. Process and form

Since 1945 there has been more detailed investigation of processes operative in landscape development. This trend has led to some disagreement about the degree to which process studies are an integral part of the study of land form. This problem of definition can be posed by contrasting two quotations. R. A. Bagnold (1941) did not think that he would see 'the geomorphologist rest content in his study for its own sake of the shape and movement of sand accumulations, until he knows *why* sand collects into dunes at all, instead of scattering evenly over the land as do fine grains of dust, and *how* dunes assume and maintain their own especial shapes'. In 1958 S. W. Wooldridge wrote in different vein, for he regarded it 'as quite fundamental that Geomorphology is primarily concerned with the interpretation of forms, not the study of processes. The latter can be left to Physical Geology.' In following this largely morphological approach it is necessary to infer the nature of the processes from the forms they are believed to have created. However, in many situations even the identity of the main process may be in doubt. Increasingly the argument appears that any generalization about past conditions and the possibility of suggesting evolutionary changes depends on the initial systematic measurement of processes in order to establish and to understand what is happening today. Together with an increasing emphasis on the importance of process studies come criticisms of the work of earlier geomorphologists, particularly that of W. M. Davis, for establishing an approach to landform study which oversimplified or neglected process studies. However, there is probably a limit to which W. M. Davis might be held responsible for any alleged inadequacies which also may underlie a possible weakness in S. W. Wooldridge's belief. This is the degree to which physical geology has not developed, in the interpretation of past processes and ancient results, comparative studies of modern processes which can be directly observed. As recently as 1966 W. H. Bradley in his address as the retiring president of the Geological Society of America suggested that there was a great deal more to be learnt about geology from observing present-day processes than had yet been explored. Geology is unique in offering for investigation, in most circumstances, only the results of natural processes, like plutonic igneous complexes. It is therefore not surprising that a strong tradition of inferring the nature of processes has grown up in many branches of geology, including geomorphology. However, geomorphologists are not alone in realizing that they can observe and measure the actual processes involved. For instance, in addition to the developments in the study of contemporary processes in sedimentology in the last two decades (J. R. L. Allen, 1968), there is currently an increased interest in studies of living foraminifera by palaeontologists.

In geomorphology the need for more information on processes has emerged in a variety of ways. Most significant is the growing appreciation of uncertainty if not total ignorance of the nature of most processes and the ways in which these processes and land form interact. For instance, little is known about sub-glacial mechanisms, and the complex problems of ice movement mechanics may still be regarded as incompletely solved, not to mention the mechanisms by which ice might erode bedrock. Several important aspects of fluvial dynamics, such as the nature of turbulence, are not yet well known, and nor are the relationships between fluviatile processes and the reworking of depositional landforms. Understanding and knowledge of the highly dynamic processes on coasts is grossly deficient, and the climate of the soil is still very poorly understood with the processes of rock decomposition and weathering in relation to the formation of soils needing much further study. In particular there is a lack of basic data concerning mineral solubilities and the factors affecting the stability of the dissolved constituents, particularly in tropical ground water and soil water.

In view of the rudimentary knowledge in studies of process, some workers consider it too optimistic to assume that the nature of processes can be inferred from the land form. There is also the risk of a circular argument using the evidence of forms to build an hypothesis of their genesis and then to explain the same forms by this hypothesis. There is also the neglected problem of 'homologies' or 'convergence' in landforms, where different processes may produce similar forms. For example, small circular weathering pits occur in varied environments and on a larger scale rounded ponds in formerly glaciated areas might be either kettle holes or ancient pingos. Rounded summits with tor-like eminences also occur in varied environments, so that those near Fairbanks in Alaska have to some observers appeared very reminiscent of tors on Dartmoor. In coastal areas there appears, for example, to be more than one genetic type of barrier island. In the absence of processes studies the investigation of homologous forms can lead to insoluble disagreement between a variety of hypotheses of origin and even in the interpretation of the field evidence on which these theories are built. Although many striking processes are usually significant not all fundamental processes are necessarily either conspicuous or even visible. Therefore, even the observation of processes, if made without measurements, is sometimes insufficient. For instance, the outward characteristics of a river channel are not usually diagnostic of whether a channel is in the process of aggrading or degrading.

Another reason for the increased interest in process studies stems from the need, in addition to understanding how a form was produced, to use this assumed knowledge as a step towards a broader interpretation or reconstruction. For instance, without understanding clearly the ways in which shore platforms develop at present, conclusions drawn from them about relative movements of land and sea are uncertain. Similarly, by studying the conditions under which a present-day sediment develops, such as frost action producing patterned features in frozen ground, the knowledge gained can then be used to interpret fossil structures of similar origin in reconstructing past environmental conditions.

There are innumerable difficulties in studying geomorphological processes and in attempting to integrate the information acquired into the interpretation of landform. The tackling of these difficulties would call for too great a diversion from the essential purpose of studying actual land form, in the view of some workers, whereas for others these difficulties merely re-emphasize the unlikelihood of making adequate inferences about the mode of landform development from a study of the forms alone. One difficulty is the slowness of operation of some processes in relation to the time-span which one investigator can expect to devote to their study. For example, one of the criteria in C. F. S. Sharpe's definition of soil creep is that it may be imperceptible except to measurements of long duration. Frequently the complexity of the forces involved is another major difficulty. It is the complexity, for instance, of beach movements that makes the detailed nature of its processes so little understood (C. A. M. King, 1959), for here there are four distinctly different hydrodynamic zones which are overlapping and compressed into a relatively narrow strip between the backshore and the nearshore. A third problem is that while most processes are theoretically observable, in practice it is difficult, if not dangerous, to do so. Moreover, the physical risk tends inevitably to increase at times when mechanical processes are at their most active and in areas where biochemical activity is most vigorous. Thus a simple and observable process like cut and fill in the zone of breaking waves and offshore, or in streams, is not well-documented because the points are difficult to reach even at low water.

There are several problems in evaluating the results of process studies. In many areas there is the difficulty, if interpretations of the past are to be made, of estimating the degree to which present-day erosion rates are due to artificially accelerated erosion since man has swung the scales of the natural balance between the forces previously involved. In all situations there is the problem of deciding which came first; whether the process essentially preceded the form and thereby determined its characteristics, the converse, or whether the two co-exist in a closely interdependent state of 'dynamic equilibrium'. Many workers consider that there have been too many changes particularly in the last 2 million years, for the last suggestion to be anything more than a theoretical, remote possibility. S. A. Schumm and R. W. Lichty consider that the problem of antecedent and subsequent conditions and the possibility of states of balance depend largely on the time-scale involved. They take the example of helicoidal flow measured at river bends and suggest that in the short term, the meander is an anterior condition and that the observed helicoidal flow exists because of the meander. However, the meander itself, if considered in a broader time context, may exist partly because of different antecedent conditions of flow. Even where a given form is seen to be the result of a known process, it may be difficult to ascertain how much the form owes to the present intensity of the process and how much to slightly different conditions in the past. An example of this problem is the size of the solutional nip on limestone coasts in the tropics where the size might in part be due to slight changes in past sea-levels. A similar problem is to decide whether the characteristic features observed along a river valley are in fact relics of infrequent floods and not the result of day-to-day conditions of normal flow. For slowly evolving micro-features in active periglacial areas it is difficult to apportion the contribution of the present climate compared with that of slightly different climates a century ago. On a larger scale, P. Birot has reservations about supposing that the inselberg is typical of dry-season tropical areas because it cannot be assumed that these forms were created under the same conditions as exist today.

In addition to the difficulty of being unable to infer processes from forms, there is the restriction that forms cannot be inferred from processes. In consequence the study of sediments has grown within the context of landform studies to assist in bridging the gap between process and form. The study of sediments involves not only depositional structures like gravel lenses but also the measurement of statistical parameters describing characteristics of groups of individual particles. The value of the study of sediment characteristics is that they are in part a product of the same processes which operate on the landforms, and these characteristics, in part reflecting the rock which makes up the landform, also may accelerate or retard the rate at which processes operate and may influence their mode of operation. For example, the dynamics of the swash zone on a beach, although difficult to observe in action, are revealed in part by grain-size distributions across the foreshore slope. In turn the size may influence the rate of infiltration of the swash into the sediment. Even in environments where mechanical action is much reduced, the fact that many sedimentary particles will then have spent considerable amounts of time in a weathering profile means that the importance of process, climate, and time might be inferred from chemical variations. Thus, just as the usefulness of a study of contemporary processes extends to interpretations of the past, so also can the study of modern sediments, by revealing characteristic properties developed in distinctive environments cast light on the nature of past environments discernible from the characteristics of fossil sediments. In Northamptonshire where sediments, recognized as fills in former ice-wedges, descend to 12 m together with other disturbances structures, it is possible to suggest that permafrost affected the ground to depths in the order of 30 m and to suspect that solifluction could have been a factor in the development of the present-day slope forms. Former terraces can be correlated by the statistical characteristics of sediments and by primary sedimentary structures in alluvial fills. In 1860 A. Geikie, from his studies of sedimentary deposits, made one of the first reconstructions of a former process, by advocating that glacial deposition of boulder clay seemed preferable to C. Lyell's concept of deposition from icebergs. In fact it was W. M. Davis who in 1890 was probably the first to recognize and describe backset bedding in his studies of fluvioglacial outwash deposits in New England, and one may wonder whether the definition of geomorphology by the pure morphologist may overlook the degree to which the study of sediments has always been an integral part of landform study.

The study of processes and sediments in geomorphology raises two further points. The first is the very real risk of not only venturing beyond what is strictly required for landform interpretation but also in consequence claiming for, or handing to, the geomorphologist studies that lie more appropriately in other hands. In the study of sediment transport in rivers, for instance, the hydrodynamicists and hydraulicists are often primarily concerned with energy relationships between particles and the transporting fluid, whereas the geomorphologist is more concerned simply with the amount of material moved and with the problem of locating from where in the catchment the material was eroded. Many geomorphologists will be happy to agree with a leading expert on the transportation dynamics of sedimentary particles who, while recognizing that the geomorphologist would inquire into how and why processes produce distinctive forms, held the personal view that 'the subject of sand movement lies far more in the realm of physics than of geomorphology' (R. A. Bagnold).

The limit is more easily drawn in the case of glaciology as the realization that glacier flow is one of the most useful illustrations of plastic flow of a crystalline material has led to very intensive study well beyond the bounds of the geomorphologist. The fact that the flow of ice is perhaps understood better than the flow of any metal or rock material and that geomorphologists still do not know how ice erodes is perhaps a significant demonstration of the degree to which the study of land form depends on the study of ice-flow physics.

A further point raised by the increased emphasis on process in landform studies is the nature of 'dynamic' geomorphology which in a general way relates to the interactions of form, process, and soil. However, there is a substantial contrast in emphasis between the American and the European view of 'dynamic' geomorphology. The dynamic basis of geomorphology, according to A. N. Strahler, depends on basic physical properties of both materials and processes and their varying responses to applied shear stress. The interest is in how the physics of a process operates and in speculations on how physics might be applied to geomorphology. In Europe the view of 'dynamic' geomorphology as is evident from the work of A. Cailleux, J. Tricart, and others, is more confined to inquiring *where* given processes operate and *how much* work they achieve at present and to what extent the areas affected and the work done were different in the past. A further characteristic of their approach is to stress the importance of the effects of chemical and biological as well as mechanical processes.

It can be concluded that while landform is the subject of primary interest in geomorphology and is the object of study unique to the subject, the interpretation of a given problem of regional geomorphology or the systematic study of a distinctive form involves the geomorphologist in the study of processes as well. Ideally the geomorphologist tries to infer processes from the forms and also to consider what forms the processes studied might create. Due to intrinsic limitations in the evidence available neither can be achieved satisfactorily and the danger of circular argument always threatens when the conclusions from one approach are confronted with those from the other. Lending confidence to the geomorphologist during his hypothetical leaps between form and process is the flimsy safety net provided by the study of sediments.

3. Artistic and scientific elements

Geomorphology might be described as a part-time science not just because nearly all its exponents have other commitments in science, education, or technology but also because artistic elements of geomorphology itself and in the temperament of the investigator have been a significant factor in its study. Fundamental to an understanding of geomorphology is an appreciation of the widely held opinion, which is at times a loudly pronounced belief, that familiarity with the countryside, coast, hills, and mountains initiates, nurtures, and develops an interest in land form. For many leading geomorphologists a deep sense of the beauty of the natural landscape has been the inspiration of a life-time's devotion to its study. Artistic sensitivity and ability is not, however, confined to the subjective investigator who strides exhilarated to the wind-swept summit to scan the horizon for evidence to fit his theory. Nor is there much evidence to suggest that recent advances in scientific aspects of the subject have been at the expense of these artistic elements. A. Rapp begins the presentation of his exhaustive records and measurements from northern Lapland: 'The valley of Kärkevagge is rich in interesting problems and in natural beauty. Nine summers have not been enough to make me feel tired of it.' An ardent concern for conservation was an expression of a love of nature felt by J. P. Miller, a pioneer in the use of quantitative methods in landform study, and A. N. Strahler shows in his textbook an outstanding artistic skill in the diagrammatic drawing of landforms.

A literary sense is particularly important in geomorphology with workers compiling a rapidly growing range of qualitative descriptions of unique features from contrasted environments. In developing new interpretations, skill is required in defining new terms with precision, clarity, and with due regard for previously proposed terms. Partly because of the fundamental importance of regional comparisons in landform study the linguistic ability of geomorphologists who have translated work from foreign languages is valuable. Judging from the frequency with which their work is cited in bibliographies, one of the most important contributions to geomorphology for many students in the 1950s was Czech and Boswell's translation into English of W. Penck's *Die Morphologische Analyse*. Despite the obvious need for literary skill in recording the facts of a qualitative science, and style in presenting its arguments, there may be some truth in criticisms that some geomorphologists have been carried away by a flow of words. K. Bryan (1940) wrote of 'mild intoxication on the limpid prose of Davis's remarkable essays' influencing the geomorphologist to consider the question of slopes in words, not knowledge, and with phrases, not by critical observation. Certainly Davis wrote his *Essays*, believed in the power of words and was fascinated by them. He contributed in 1926 to the journal *Science* an essay entitled 'The value of outrageous hypotheses in geology', writing at length on the interesting etymology of the word 'outrageous' in preference to using a less ambiguous synonym in his title. Commonly used in the writings of W. M. Davis and in those who similarly have strived for literary effect are two devices which, in appealing directly to the imagination, require scientific restraint in their interpretation. First, the appeal of brief, vivid, and perhaps poetic expression has led both popular and technical writers to animate the landscape by the skilful use of metaphor (LeGrand, 1960). The mental picture which these words portray may, however, be quite unrelated to the dynamics of the process. A more detailed inquiry into the mechanics of apparently self-explanatory headward erosion reveals little beyond the fact that it breaks, bites, or even gnaws back into a divide. Secondly, due to the intrinsic incompleteness of evidence for geomorphological interpretations the analogy is used so frequently that its well-known limitation is easily overlooked. Sometimes, however, ill-chosen analogies are unmistakable applicable in one or two senses only and the imagination may then be overloaded with innumerable impressions of how the geomorphological factor differs from the analogous phenomenon. Analogies extend beyond the written word into the domain of quantitative analyses, particularly where some reasoned basis for a mathematical model is sought. Again the analogies require very careful scrutiny. For instance, early estimates of the viscosity of lava 'streams' were based on the analogy between these flows and that of river water, and it was the mistaken assumption that hydraulic equations could be applied to lava.

The metaphor and the analogy have influenced geomorphology profoundly. The collection of drumlins into a basket of eggs may appear a harmless way of describing a distinctive form, like many of the other kitchen-sink images in geomorphology. Nonetheless R. J. Chorley (1958) was unable to resist the temptation to liken the process of drumlin formation to that of egg-laying, a suggestion incorporated into later serious studies.

Favouring the development of another artistic quality in geomorphology's character is the passage of time which makes the studies of those with an historian's expertise and objectivity increasingly useful. With the greater interest in processes has come the realization of difficulties in studying changes over a span of a few years, in artificially modified and scaled-down environments, or in theoretical speculations on possible time-sequence changes of form. There is therefore a growing interest in critical searches for records of events, like delta growth, datable over spans of decades or even centuries. For example, in the search for references to Spurn Head, a prominent spit at the mouth of the much-sailed Humber, G. de Boer (1964) has searched through records spanning more than a millennium. Historical analysis of the patterns of development and decline of ideas within the study of landforms also becomes increasingly valuable as the passage of time leaves a growing length of history of geomorphology stretching back from the present day. Detailed studies like that of R. J. Chorley, A. J. Dunn, and R. P. Beckinsale (1964), or of G. L. Davies (1968), provide helpful aids to an understanding of contemporary geomorphology and in contemplating a future growth which might avoid past mistakes.

It seems that recent advances in more scientific aspects of geomorphology have not diminished the need for some artistic flair among its exponents. Their usefulness has developed also, but has been urged less than the scientific element where the need for growth has been reiterated for more than two generations. One of the first in the post-1945 period was R. P. Sharp's suggestion that the practising geomorphologist, in addition to being a good geologist, should be well-versed in soil science and acquainted with principles of hydraulics, meteorology, oceanography, and botany. He also thought that the application of physics and geophysics to geomorphological research would offer the greatest promises for fundamental advance. It would seem that inevitably this philosophy would clash with the idiom of the Davisian word-picture approach. However, Davis himself had said at the beginning of the present century that the methods of mathematics and physics were 'so much needed in all parts of earth science' and I. C. Russell had added astronomy and chemistry to the list of desirable assets. Therefore if Davis and his followers appear to some contemporary critics to have had no more than 'a butterfly-catcher's sort of interest in scenery', the apparent absence of scientific rigour in the past and the sensation of its impact at the present are not easily explained without skilled and sympathetic historical analysis.

In understanding recent scientific progress in geomorphology it is helpful to recognize several distinct themes which equally can be described as scientific. First, scientific method, with its rigorous procedure from first-hand observation to interpretation has always been more important than some discussions, preoccupied with the Davisian approach, might either allow or illustrate. Secondly, the application to geomorphological problems of scientific techniques borrowed from neighbouring sciences, such as soil science and hydrology, increases markedly. This trend leads to a third aspect of science in geomorphology but one which is impossible to define in general terms, and is, to some extent, a matter of personal preference, with the definitions of some proving unacceptable to others. This is the degree to which the study of geomorphology by including the investigation of processes, thereby necessitates also the study of basic mechanical and chemical principles. The extreme view is that of some theorists who go so far as to believe that most geomorphological generalizations should in fact or in principle be deductively derivable from the laws of physics or of chemistry. Others would deny this on the grounds of the complexity, significance of changes with time and the importance of biological factors. It is only fair to note that the first view is that of only a few, and includes most of the theorists whose familiarity either with the work of those with broader definitions or with actual field experience is conspicuously limited.

With the increased adoption of scientific techniques and the attempts to apply the principles of certain sciences to the interpretation of land form, the distinction between science and technology has to be emphasized as a fourth aspect of scientific aspects in the study of landform. The geomorphologist has to make a deliberate effort to recognize this distinction in order to avoid claiming the engineer's role or adopting techniques useful in engineering but which have little relevance to the solution of his scientific problem. Compared with science, which proceeds from observation to understanding, technology proceeds quickly through observation and understanding and then on to meet the practical demands for application and control. The technologist cannot wait either for exhaustive measurements or for the development of a complete, substantiated theory. He has to economize on the first two stages which intrigue the scientist and make do with modifications to readily available data often already collected for other purposes, or derived from scale-model experiments, theoretical models, or trial-and-error methods. With certain simplifying assumptions, the engineer arrives at conclusions which, although in a scientific sense are significantly different from those existing in the actual problem, are well within margins set by safety and economic considerations, and allow work on the project to start on schedule.

Although there are many examples in recent geomorphological literature of the blurring of science with technology, the problem of defining the objectives of space research has recently provided a sharp and general reminder of the fundamental distinction between scientific and technological endeavour. Clearly there is a limit to what the geomorphologist might gain from the experience and methods of technologists concerned with engineering in the natural environment. For example, although for practical purposes in hydraulic engineering river discharge is the most important parameter, the hydraulic factors measured in conjunction with flow determination are not sufficiently inclusive for geomorphological or even for hydrological analysis of the river system (Leopold and Skibitzke, 1967). Conversely, where a geomorphologist contributes to an engineering project there is an equally clear limit on the degree to which he can claim that the technological achievement is relevant to geomorphology.

Geomorphologists also have to be cautious before using the term 'science' itself too liberally, because unlike the rigorous sciences of physics and chemistry landform study knows no laws, if the word law is used in its correct universal sense, as a concise expression of a group of inter-relationships that have been repeatedly observed to be consistent. G. K. Gilbert's 'law' of divides or R. E. Horton's 'laws' of drainage network arrangement, although giving the superficial impression of scientific simplicity are no more than useful generalizations. They admit many exceptions, and the areas to which they might be relevant may require careful selection and delimitation to exclude domains where immense complexity and infinite local variety prevails. Unlike the biological sciences there are no intrinsic features like genes with which to provide a known link for a series of forms through time or for a division between species, and unlike the social sciences there is no possibility of dialogue between investigator and investigated. Geomorphology certainly has an important and growing scientific content, but to claim for it the full status of a science would be not so much pretentious as premature and might even misrepresent its true unique nature.

4. Qualitative and quantitative aspects

The complexity and irregularity of forms and processes complicated further by changes with time, makes an essentially qualitative approach the only realistic way of approaching many geomorphological problems. However, despite the emphasis on the power of words laid by W. M. Davis and reinforced by his followers, and on the accumulation of a vast store of qualitative records by European geomorphologists, since 1945 there has been an increasing realization that in some instances forms and particularly processes are sufficiently regular for more exact approaches and, in other instances, that qualitative analysis has proceeded as far as it can go. The clarification brought to time-scale problems, whose solution was impossible to resolve on the basis of earlier vague controls, by the measuring of recent geological time, provides a considerable incentive for searchers for more exact methods.

Increasingly geomorphologists are defining their views on the desirability, scope, and hopes of this trend. Extremists include a few who are too hopeful in their claims, contrasting with disclaimers at the opposite extreme who occasionally lapse into hostility. Many, however, recognize that in attempts to retard or to promote quantification, the reactions provoked by extreme views tend to balance each other out at any instant in time and merely provide a scatter of points about an inexorable, upward trend. Most views therefore range from enthusiastic involvement to sceptical interest.

Mackin's words summarize the misgivings of many, particularly those of the older, more experienced workers:

> ... the very act of making measurements, in a fixed pattern, provides a solid sense of accomplishment. If the measurements are complicated, involving unusual techniques and apparatus and a special jargon, they give the investigator a good feeling of belonging to an élite group, and of pushing back the frontiers. Presentation of the results is simplified by use of mathematical shorthand, and even though nine out of ten interested geologists do not read that shorthand with ease, the author can be sure that seven out of the ten will at least be impressed. It is an advantage or disadvantage of mathematical shorthand, depending on the point of view, that things can be said in equations, impressively, even arrogantly, which are so nonsensical that they would embarrass even the author if spelled out in words.

An illustration of a more extreme, but still significant viewpoint comes from a note entitled 'Robot geology' in which Link (1954) confesses that 'Somehow I hate to see budding geologists feeding machines numbers. They should be out in the field learning to get the right numbers to put into the infernal machines.' From a geographical viewpoint, O. H. K. Spate (1960) feared that expertise in a fashionable technique might disguise an essential poverty of thought, and feared the possibilities of the dogged analysis of trivia, the smoke-screen of formulae and the elaborate discovery of the well known. The reader encountering the conclusion that 'large rivers have large bends and small rivers have small bends' may appreciate these forewarnings. More recently where computer package programs make intricate statistical analyses available to anyone with a few numbers, pointless complications increase to include the instant and enormous output of the unintelligible. In general there is a suspicion that taking to quantitative techniques dulls one's geomorphological common sense and perceptiveness. Clearly an understanding of contemporary geomorphology depends on some familiarity with the reasons which justify this unease as well as on illustration of the possibilities of greater precision which this approach affords. Before discussing quantification in geomorphology several basic facts should be stated. First, any antithesis between qualitative and quantitative aspects of geomorphology is largely false due to their essential interdependence. Quantitative geomorphology indicates an emphasis or even provides a slogan rather than a definition for a separate specialism within the subject. Secondly, the quantitative element in geomorphology differs substantially depending on whether form, process, or time is involved. In process studies long-established familiarity with quantitative techniques already exists in subjects like hydrology, hydraulics, agricultural engineering, soil chemistry, or sedimentology. In the measurement of time many disciplines like archaeology, botany, and geochemistry similarly provide a range of techniques and experience in their use. By contrast the quantification of land form is basically the responsibility of the geomorphologist. Consequently there is little experience available in related subjects and quantitative studies of form are more rudimentary and those of the functional inter-relationships between form and process very few. Thirdly, since many of the investigations made primarily to demonstrate that some geomorphological problems are susceptible to a quantitative approach tackle problems that are peripheral to the bulk of questions that the geomorphologist has traditionally asked, their results are often of limited interest and difficult to integrate into existing studies. Finally investigators may use the adjective 'quantitative' in widely differing senses. This is because quantification involves several distinct stages and the term could be applied to any of one of these. It has been used to describe work ranging from the careful measurements of an experimenter in the field to the speculations of armchair theorists.

1. MEASUREMENT

The value of qualitative reasoning behind the collection of data is one of the most important aspects of quantification, and the importance of design before rather than after measuring cannot be overstressed.

(i) *Definition of measurement.* S. A. Schumm (1960) suggested that one of the most difficult problems to resolve in studies of sediment types is the selection of a parameter representative of the physical property of the sediment. The difficulty of expressing the erodibility of soils by single numerals has also been stressed. In overcoming this initial difficulty, some arbitrary decision, such as deciding at which point a cirque floor is judged to pass upward into the steepened headwall, is necessary. Many geomorphologists unaccustomed to experimental procedure have remained entirely qualitative in their approach because they were unable to accept this arbitrary schematization, which would lead to the measurement of something that was neither 'natural' nor 'real'. Attempts to overcome this problem by defining a feature to be measured in some genetic sense, like the interpretation of erosion surfaces as relics of a once continuous erosional plain, can, due to the intrinsic uncertainty in geomorphological interpretations, usually be countered by arguments rejecting the genetic interpretation.

The fundamental problem of defining a parameter to measure is difficult enough for static forms, but if the phenomena move, as in process and sediment studies, some changes elude definition by being too slow or too rapid.

(ii) *Making a measurement.* In addition to the definition of a measurement, with its inevitable approximations, there is the problem of how to measure the parameter adequately, with hopes of accuracy limited by the intrinsic variability of the phenomena. Also, for some variables, like the cohesiveness of a sediment in relation to its resistance to erosion, no field method has been developed. Ceaseless motion, as on the shore, makes it difficult to measure sea state in any long-term consistent manner or to collect samples in the surf. Movements in some less dynamic environments, like the bottom of a glacier or river, are very difficult to measure. In fact, the most serious hindrance to a complete understanding of sedimentary processes in general is probably the lack of instrumentation by which short-term velocity and pressure variations can be accurately recorded. In this situation, as in many others where existing methods appear crude, some workers prefer to await the development of more refined methods. However, with more sophisticated apparatus the greater delicacy requires precautions and even restrictions in its use, may involve patient calibration or the risk of expensive breakages or that equipment and markers employed in continuous measurements may be accidentally disturbed. It may also be impossible to tackle some problems geographically with expensive apparatus as the number of fixed points at which records can be made is inevitably limited.

Another problem in making measurements is the huge scale on which measurements have to be collected in order to encompass the complexity of inter-relationships in the natural environment and the range of geographical variability. For example, experiments on the scale of those with which the engineer, R. E. Horton, was involved in 1939, relating to the effects of rain intensity, rain impact, and energy on infiltration capacity depending on 192 infiltrometer tests on 96 plots, could be very difficult to equal for geomorphologists who are mostly restricted to part-time research. Another example is the 1962 resolution of an international commission on snow and ice: 'Our primary aim is to encourage all countries which have glaciers to make very simple measurements on as many glaciers as possible. We wish to emphasize this point: measurements should not be confined to a few glaciers in each country, but should cover the maximum possible number.' Whether measurements are concentrated in a small area, or extend to include contrasted environments, longitudinal analysis through time is usually required as well, the 1000 years of recorded water-levels on the River Nile being one of the most sobering examples.

(iii) *Second-hand measurements.* One common way of making some progress where sufficient first-hand measurements are unobtainable for practical or financial reasons is to use measurements derived from information collected for another purpose. The derivation from maps of numbers to describe land form is a well-known example. Fluvial geomorphologists have used the data collected by water engineers in preference to actual measurements of the parameters which theoretically are most meaningful from a scientific point of view because only information of first, approximate kind is available in large volumes. Similarly nearly all geomorphology includes climatic data which being influenced by certain meteorological abstractions may give only an approximate indication of conditions in the soil where weathering processes operate.

(iv) *Criticisms of measurements.* Any set of geomorphological measurements may be open to the criticisms of vague definition, inaccuracy, triviality, or irrelevance. They may be too small in number, collected from too few places, and observed too infrequently. Critics will also correctly warn of the risk of unsuspected artificial disturbances to the feature or to the recording device. The collection of numbers may be biased unintentionally in one direction or may subconsciously tend to support a preconceived view, or the omission of concomitant measurements of possible related phenomena may reduce their value.

In 1896 W. M. Davis wrote, 'Verbal descriptions are so insufficient in geographical teaching that supplementary illustrations, in the forms of maps, views and models, must be employed as far as possible.' Significantly, he did not include measurements in this list nor in the publication of his research during the subsequent forty years. Perhaps he knew better.

2. CORRELATION

The measurement of the degree of association between two geomorphological variables involves many difficulties. A common problem is that within a range between the largest and smallest observations of a sample, the bulk of the measurements may bunch near one end of the scale. For instance, in a drumlin field the smallest drumlin may only be half the size of that of the majority, which the largest may exceed by several times. This raises statistical problems, starting with the simple fact that an average value is not really representative.

The relationship between two geomorphological phenomena, as is frequently the case in nature, is rarely that of a straight line. Behaviour may even change suddenly when a certain threshold value is exceeded or with the introduction of a new factor at a particular time or beyond a certain distance. For instance, clear quartz sand 0·08 mm in size was observed to slip when the surface is tipped at an angle very little different from that for larger grains, when dry, but when exposed to air of normal humidity it tends to stand in vertical walls.

The many curvilinear relationships where y is increasing with values in x further increases in x beyond a certain point may actually lead to decreases in y. Fig. I.6.A illustrates how, in streams incising an alluvial terrace, the grand mean for valley-side slopes increases with stream order up to the third order, then diminishes at higher orders; Frame B shows how progressively slower wind velocities are required to initiate the movement of smaller sand particles, while for sand particles less than 0·1 mm diameter, greater velocities are required. In fig. I.6.C soil loss increases with the length of time that rain falls until the inwashing of finer particles produces a protective crust which reduces the amount of erosion. Similarly, in fig. I.6.D, increasing annual amounts of effective precipitation in the United States produces greater erosion until values of 38 cms or more are sufficient to support a complete vegetation cover which arrests erosion.

Finally, even if a geomorphological factor and a given environmental variable are unlike these examples by being normally distributed and linearly related, the complexities of the natural environment will make it impossible to obtain an excellent correlation.

3. EQUATIONS

Proceeding from the demonstration of a relationship between measurements of two or more variables, the derivation of a mathematical equation may provide a useful, concise summary of the finding. However, the precision of invariable mathematical relationships rarely exists in geomorphology, and average statistical correlations as generalized descriptions are often as far as one can proceed without introducing simplifying assumptions that would reduce the relevance of the results to the natural environment. Mathematical relationships cannot allow realistically for the possibility of changes with time, like climatic changes in the Pleistocene or over much shorter intervals,

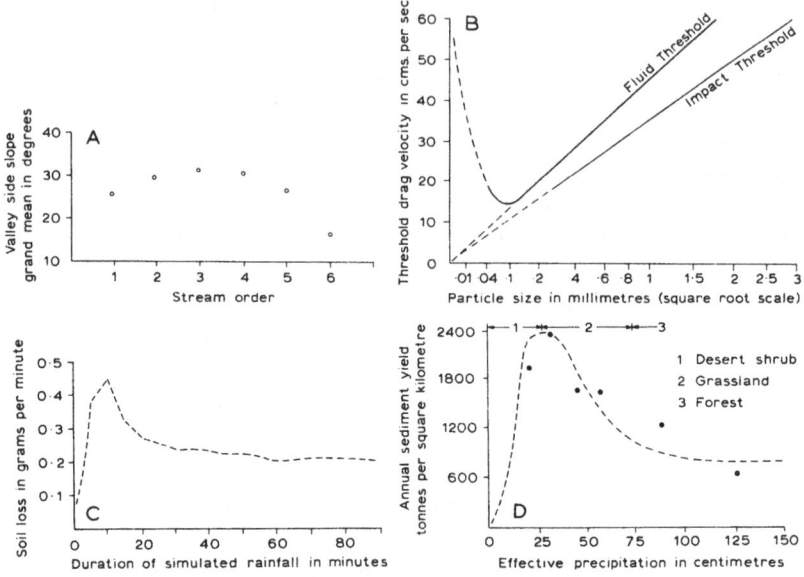

Figure 1.6 Some non-monotonic relationships in landform studies.

A. Variation of mean valley-side slope angles with stream order (*from Carter and Chorley, 1961*), for the Lighthouse Hollow drainage network, Poquonock, Connecticut.

B. Variation of the threshold drag velocity in air with particle size (*from Bagnold, 1941*).

C. Soil loss in runoff water during a 90-minute simulated rainfall application (*from E. Epstein and W. J. Grant, 1967*, Soil Sci. Soc. Amer. Proc., *Vol. 31*). The soil is a silt loam from Maine.

D. Variation in sediment yield with varying amounts of effective precipitation (*after Langbein and Schumm, 1958*). The data relates to all climatic regions of the United States.

as this means that certain parameters cannot be assumed to have remained constant, nor for the 'kaleidoscope of hydrologic conditions caused by heterogeneities of geology' (LeGrand, 1962). Even where an equation satisfactorily describes the situation in one case, such as the transportation of sediment in a river, variations in the stream-depth flow velocity, bed material, bedrock, and historical legacies make equations very different for each river. Rarely do predictions from one equation check with the data from another area.

Nevertheless, extensions of the work of R. E. Horton on the geometrical properties of drainage basins have led to demonstrations of an invariability in these properties which justifies mathematical treatment. For A. N. Strahler and his students Horton's work has been of particular and vital interest.

others it is an irrelevant appendix. Many are mystified by the invariability of the mathematical revelations when their own experience of landforms leads them to anticipate large statistical variations in nature. Since most of the mathematics applied to geomorphological data involves Hortonian studies, it is necessary to search for reasons to explain why some find invariable relations while others believe them to be the exception in nature.

First an air of over-optimism has surrounded the topic ever since Horton used the word 'law' to describe more statistical relationships and geomorphologists assumed that a technique designed primarily to improve runoff predictions would be suitable for landform study. Some workers, while observing that Horton's technique of classifying stream order is subjective and hard to replicate, still refer to the approach as a 'touchstone'. A great deal of the work has been spurred on by the initial assumption that mathematics can be applied to landform study and the deliberate adoption of Horton's ideas to demonstrate this.

Secondly, the over-optimism includes a lack of concern about the basic quality of the data. For the purist the orders of a ranking system would be too coarse to manipulate arithmetically. A more practical point is the readiness with which the same stream may differ by an order or more, depending on the scale of the map, or whether, on the same scale, the map is old compared with a recently revised planimetric version. Some channels observable in the field do not appear on maps.

Thirdly, the use of the term 'law' perhaps misleads students and makes them less appreciative of the full significance of what M. Morisawa (1964) termed 'aberrations' from Horton's laws. A stream may not, in fact, have its steepest gradient in the first-order segments 'as is necessary to comply with Horton's law'.

Fourthly, patterns develop clearly only on certain types of homogeneous rocks. Differing lithologies in a drainage basin introduce irregularities and within a homogeneous rock even weak structural control of stream networks may influence length ratios. Problems of lakes fed by multiple inlets and islands bounded by branching channels pose difficulties.

A fifth factor follows from the presence of geological and hydrological irregularities, in that most workers avoid these difficulties. However, not all workers are as specific as M. A. Melton (1958) in defining the criteria used to select study areas. In his study, basins with cliffs, well-developed floodplains for the main channels, obvious compound slopes, extensively gullied slopes or trenched channels were avoided. Many Hortonian studies relate to badland gullies in weakly consolidated sediments, where miniature patterns are organized quickly. Because Horton's scheme makes it necessary to avoid basins, lakes, and islands, an artifact of the technique automatically excludes 'ill-organized' drainage.

A sixth point is that some of the Hortonian relationships and constant ratios are unrelated to hydrological parameters. They apply not only to badlands in the United States but also to areas of tropical lowlands in S.E. Asia submerged by Pleistocene seas, to tidal channels, to chalk dry valleys, and to the relict pattern of Pleistocene wadis in the Sahara. The point to emphasize is perhaps not the wonder of this mathematical invariability which has confused some theorists about the fundamental nature of geomorphology but that techniques which cannot discriminate between such contrasted areas are of limited use to the geomorphologist.

In conclusion it seems that the mathematical regularity demonstrated by the Hortonian approach refers only to certain, limited areas and that some of the 'laws', like the 'Law of Stream numbers' is a statistical probability function, applicable to any regularly branching system, and is derived not from the landscape but simply follows as an automatic consequence of the definition of stream order.

The use of the analogy between a landform and a geometrically defined shape has a long tradition dating back to the delta and involves fewer confused issues. The degree to which a mathematically defined shape fits a form is obvious to anyone as is the fact that departures from that shape are real and not merely a statistically random scatter about that shape. Almost invariably only an analogy of shape between the curve and the landform is implied, like Galileo's suggestion that the long profile of streams approaches a quarter circle. Workers who have attempted to use empirically fitted curves to reconstruct extensions of former river long profiles soon realized that the expressions have no geomorphological significance. The inevitability of irregularities in most profiles due to resistant rock outcrops means that at best a profile is segmented into a series of curves that would each require separate equations. The same restriction applies to smooth coastlines between headlands or to most hillslopes. Even where a smoothed segment is close to mathematical simplicity a given set of constants usually apply only to one segment and so have little use for comparative studies.

An agricultural scientist, Troeh (1965) has considered briefly the application of three-dimensional formulae to landform and it is not surprising that he concludes that the complexity of a series of hills and valleys is so great that it would be impossible to describe them even approximately with a three-dimensional equation, that it would be difficult to give any meaning to the coefficients and that items of interest such as slope gradient would not be readily obtained from them.

If order in nature extends to the shape of landforms then, like the processes and the inter-relations between process and form, this approaches mathematical regularity too infrequently to be of more than academic interest. However, in recent years workers, assuming that the landscape and the processes operating on it should be definable in the terms of the laws of physics, have constructed theoretical mathematical models. These equations should be sharply distinguished from those which summarize large amounts of actual measurements because their values are not observations but assumptions and predictions from the laws of physics. This approach is mainly of interest to hydrologists and most geomorphologists will probably find the basic assumptions unacceptably oversimplified for their purposes. In addition to its inherent complexity, landscape evolution involves as much chemistry as physics and theoretical ideas about the nature of the landscape nurtured in laboratory-controlled sciences traditionally underestimate the biological factor. The huge cost of hydraulic engineering scale models and the trial-and-error procedures involved are a measure of the degree to which the inter-relationships between water and landform defy mathematization.

A third type of formula used mainly by technologists occasionally appears in geomorphological literature and may be based partly on measurements and partly on theoretical considerations. Its main use is for predicting approximate conditions where engineering problems are concerned, and, instead of summarizing measurements, is rather a substitute for them. Highly skilful work can reduce a comparison between prediction and observation to about a 100 per cent error in civil engineering which is invaluable to the engineer but of less use to the geomorphologist. However, when a geomorphologist is going to collect measurements it is sometimes very useful to have available some empirical formulae which might indicate the approximate range of values to be expected. One device used in deriving empirical formula for technological purposes should be guarded against in scientific interpretations. This is the use of the same measurements on both sides of an equation, which increases the degree of apparent association between two parameters. As an hypothetical example a graph may show the relationship between A and $B \times \sqrt{CD}$, when in fact A might have been defined earlier as $\dfrac{EF}{D}$. Stream width is one of many examples that might be cast in the role of D.

5. Laboratory and field work

The first geomorphologists were explorers or the field geologists preparing maps on reconnaissance surveys. Today the model geomorphologist is for many still cast in this mould, striding forth equipped with well-dubbined field boots, a sketch pad, and a well-trained roving 'eye for the country'. For larger-scale surveys the Land-Rover geomorphologist has become acceptable, driving vast distances across little-known terrain. With growing emphasis on measurement and scientific method the field geomorphologist wonders if laboratory experiments and remote sensing of the environment by satellite will confine him indoors, and the armchair theorist hopes for added justification for framing his view of landforms with four square walls.

The fundamental geomorphological experience is that there is an indefinable advantage of actually looking at the realities of land form. It is difficult to explain the logical basis for this belief beyond making the slightly arrogant but obvious assertion that the linking of the human eye to the human brain can provide the practised student of landform with an aid of potentially immense power. Because of the importance of actually observing landform, it follows that a wide range of experience is preferable both in the search for generalizations and also to heighten appreciation of the distinctive features in a given area. This essentially geographical approach was familiar to Herodotus whose remarkable generalizations on river activity included examples from modern Greece, Turkey, and Egypt, and innumerable fundamental geomorphological facts have been established in this way. For example, A. Cailleux, on visiting Alaska, observed the round lakes near Fairbanks, including one surrounded by half a dozen smaller round ponds 3–10 m in diameter, and could record that an identical pattern had been observed near Tartarin, in the vicinity of Paris.

The mere fact of being in the field increases the chances of the discovery of crucial evidence either as a distinctive assemblage of forms or an unusually complete stratigraphical section. As recently as 1960 one of the most complete Pleistocene successions in the western Mediterranean was discovered on an almost inaccessible section of the south-east coast of Spain (Butzer and Cuerda, 1962). There is also the chance that a field investigator might witness an infrequent event. W. M. Davis advised that it was hardly worth sitting down and waiting for a rare event like a sheetflood, but on the other hand, some of the observations that McGee (1897) was able to make, after being caught in one more than a metre deep, have been of lasting importance.

Since 1945 the emphasis in some aspects of landform study has tended to pass from exploration to more systematic research. A stream-dissected hillside, as well as being pleasant to view, becomes also a pile of unanswered questions riddled with problematic features. The problems are not so much those of interpretation but of how to approach interpretations. The eye for the country alone is no longer sufficient due to differing preconceptions in the minds of observers who are looking at the same feature. The decision to make measurements, however, introduces several difficulties which are scarcely soluble. To be as representative as possible and to allow him to use probability theory in stating the accuracy of his measurements and correlations, the geomorphologist knows that his sample must be random in either space or time. But this ideal approach is scarcely adopted when experience indicates the locality where measurements would be most significant, when it is possible to visit fifty localities near roads within a few hours and use bulky or sensitive apparatus instead of spending a day walking, with little equipment, to a remote randomly located point on a mountain top, and to have no comparable measurements for that day. On the other hand, experience both in the United States and in Russia is that nearly all hydrological instruments are sited conveniently in the lowlands, with the result that very few measurements are to hand from headwater areas that are hydrologically and hydrochemically of greatest importance. It also means that without sediment measurements there is no information for river reaches where gravel is the dominant bed material and incorporates unstable fragments or minerals. In turn this makes suggestions about graded streams speculative.

There is the added dilemma that in cooler environments it is far easier to make very intensive measurements in summer than to collect a single set in winter, and the wet season in the tropics is a similar limitation. In many cases the most significant geomorphological changes may occur at times when the study area is least accessible. For instance, the best time to inspect material from an eroding coastal cliff is in late winter after a severe storm. This, theoretically, is also a critical time for collecting samples in the zone of breaking waves. Even inconvenient hours in the day may be critical. Østrem, in excavating ice-cored moraines, found that pit-digging had to be done partly at night because with sun shining on the pit walls, melted masses slumped down. Time for surveying hot desert forms is limited by the rapid development of heat haze. The discomfort of extremes of temperature, rain, and wind, or sand in the eyes, can make accuracy in calibration and measurement a vexing task. Accidental disturbance of recording equipment is irksome to those working in populated areas. Several workers find that their enthusiasm for measurement of geomorphological phenomena is shared by members of the local fauna who invariably show keen interest and, where possible, inspect too closely any recording device.

The field geomorphologist sometimes has the additional problem of devising a method of measurement where no appropriate one exists or is too expensive to apply on a wide scale. Students of shore processes, with the ingenuity of castaways, are particularly adept at solving such problems, and without too much self-consciousness, have traced current movements with cement blocks, bricks, and seaweed pods, have observed wave motion with the aid of grapefruit and table-tennis balls and collected sediment samples with ice-cream cartons.

Traditionally, the geomorphologist, while happy to spend hours indoors poring over maps, has been averse to compressing the scale of his forms to create laboratory models. However, C. A. M. King (1959) considered that engineering scale models do simulate with reasonable accuracy the changes on natural beaches, and recently interest in the results obtained by engineers and geologists from scale models has grown, even if financial and technical limitations make feasible only the construction of very simple geomorphological scale models. One attraction of this approach involves a resort to the Baconian procedure of studying only one factor at a time, and keeping constant all others, which is impossible in field conditions where several factors vary simultaneously. Also attracted are those who are uncertain about how a landform changes with time and see in scale models a possible means of accelerating time. However, while the accuracy of scale models is adequate for engineering purposes, their aid in assisting a scientific understanding of the processes and inter-relationships involved is limited by too many artificialities to mention. Particle sizes cannot be scaled down because smaller silt-sized particles behave in a different way from sands. Clays are not used because flume distances are far too short for clay-sized particles to settle. The depth of a model has to be exaggerated sufficiently to make turbulent flow possible. Erosion in a flume or tank of randomly packed sieve fractions may not be very closely analogous to the natural erosion of an exactingly selected and packed natural sediment. On the other hand, not many studies used mixed sizes anyway, which is a serious handicap to an attempt to understand fully erosion, transportation, and deposition. Added to this is the fact that differently shaped particles react differently to water flow. Again the fundamental difference between the aims of the technologist to predict and the natural scientist to interpret must be stressed. Extension of interpretations of scale models to the context of natural conditions are difficult because of unanswerable doubts on how close the simulation of natural conditions might be. There is widespread agreement that laboratory studies will provide qualitative conclusions only, and that these are usually applicable within narrow limits only. The conclusion to be drawn from any scale model study is unequivocable; any interpretation depends on further measurements made in the field.

The observation of chemical, as opposed to physical changes in the laboratory has been standard practice in geology for many years. Scale problems are not involved, but one essential difference between laboratory and natural chemical reactions is the period of time over which the processes operate. Materials which show negligible solution under controlled laboratory conditions may, over long geological periods, be eliminated from a weathering profile. A major difference is the sterility under which most of these experiments are conducted, compared with the natural environment teeming with life and life processes, where chemical reactions can no longer operate in isolation. Experiments with physical properties not involving scale reductions, like soil shear strength, freeze–thaw cycles, or permeability have little quantitative significance if determined in the laboratory.

All laboratory experiments discussed so far involving the dynamic interaction of parameters should be distinguished clearly from laboratory analyses where certain properties, like the proportion of clay in a soil or the estimation of the amount of sediment in a river-water sample involve only sampling errors in the difference between laboratory and field conditions.

Space research, with promises of environmental observations from satellites, far from threatening to make the field geomorphologist redundant, reiterates his basic assumption by attaching some important to visual observations from manned spacecraft. With the increasing emphasis on the importance of field measurements and sample collection for laboratory analysis, the traditionally excellent geomorphological excuse for getting out of doors is still sound. Measurement and analysis, however, may be tedious and time-consuming, and sample collection may involve the restriction of systematic visits to the same point. The adventure and sporting challenge to physique and perseverance for those who prefer to spend nine-tenths of the day pitching and breaking camp is today perhaps all the more clearly defined by the growing awareness of the paucity of geomorphological field measurements and the possibilities of laboratory analysis of field samples retrieved adding clarity to explanations. As an engineer recently remarked, 'data collection in the mountains is unconventional, expensive, difficult and not without a certain amount of hazard to personnel and equipment' (Meier, 1967).

It seems arguable that scientific method in geomorphology does not involve the telescoping of time and the compression of size of problems in order to squeeze them inside the walls of a laboratory, but rather to take laboratory techniques and approach, with suitable modifications out into the field, and to return with measurements and samples. Increasingly, more time may be spent on analysis in the laboratory, but it seems that geomorphology remains fundamentally a field study, with an immense amount of field work yet to be started.

6. The role of geomorphology

One of the more frequent uses of geomorphological studies in the past has been as an aid to geological studies. S. W. Wooldridge, like W. M. Davis, considered that landforms were the best indicators of earth history in most parts of the earth's surface, and he believed that geomorphology should provide the answers to the last brief chapters of geological time. This was the aim of the reconstructions of many denudation chronologies in Britain in the 1950s. However, S. W. Wooldridge spoke only a couple of years after the realization of the potential usefulness of radiocarbon dating, and in the last two decades techniques of absolute age determination using radioactive decay rates have relieved the geomorphologist of much of the responsibility for chronologies and its perplexing task of attempting correlations of age, using as evidence only the form of the land or its altitude. In conjunction with information from other studies, facts about geomorphological features and their interpretations remain useful to geologists in a few specific situations. The degree of dissection of glacial features assists the Pleistocene geologist in his attempts to differentiate boulder clays and moraines of different ages. The same may apply to areas of recent volcanic activity. For instance, in the Pribilof Islands, Alaska, St George Island is nearly stripped of primary volcanic features, and is bordered by marine terraces and high sea cliffs. On another island, St Paul, the volcanic topography is still well-preserved. The youngest rock dated on St George has a potassium–argon age of 1·6 million years compared with an age of less than 100 000 years for the youngest dating on St Paul. Thus confirmed by radiochronometry, geomorphology shows clearly that volcanism terminated sooner on St George. Another example is the intention of W. Penck to use slope-form studies to elucidate tectonic movements. Although Penck's hopes are now largely regarded as unrealistic, W. Q. Kennedy (1962) still believes, with Africa in mind, that geomorphology might reflect the tectonic conditions that prevailed during the evolution of the surface relief. Also there are some specific instances where the interpretation of geomorphological features can provide evidence of tectonic movements. As early as 1886 Le Conte noted that in the Sierra Nevada volcanic rocks have preserved Tertiary river channels. These ancient alluviated channels have steep gradients of about 15–19 in/km compared with approximately a 6 m/km fall of the actively eroding contemporary rivers, a contrast strongly suggesting tilting. Also there is a differing ratio between the two gradients according to compass orientation which could indicate direction of tilt. To avoid over-emphasizing the broader significance of this striking example it should be added that, unlike most landforms, these channels contained gold. Interpretations of landforms are important in some geological mapping problems where failure to recognize a paleoform, such as the ancient landslides in the formations along the San Andreas fault, may cause stratigraphic confusion to puzzle more than one generation. In South Australia, the Eden fault was recently reinterpreted as a former sea-cliff. Intensive drilling for oil has shown that buried unconformities are not necessarily peneplane-like surfaces, but on tilted strata may retain the ridge and vale forms of a scarpland which provide oil traps. The detailed study of raised beach elevations in which there is a great deal of current interest in Britain (Walton, 1968) provides another example. From this data the reconstruction of postglacial isostatic recovery of formerly glaciated areas provides geophysists studying isostasy with invaluable information on the internal strength of the earth's crust.

In using interpretations of landforms the geologist, together with the geographer who queries all aspects of landscape, stand largely alone. Admittedly studies of soil formation, particularly those in the tropics, would benefit were the age of a land surface known approximately, as studies in Australia have shown; it might be speculated that the length of weathering time on a land surface could be related to hydrological studies of the permeability of an aquifer or of a dam site and a potential reservoir area. In general, however, much of the utility of geomorphological information to a large number of related studies depends on no more than the descriptive facts, often obtainable from a contour map. For example, the natural movement of water on the ground surface and into and out of the ground depends on landforms, as was demonstrated with measurements by water engineers in the 1930s, and today 'many of the observational features to be used in solving hydrological problems lie within the scope of geomorphology' (LeGrand, 1962). Within a given area of homogeneous rock-soil types tend to form a predictable sequence or 'catena' downslopes, as was first demonstrated by G. M. Milne in 1936. This relationship, and similarly where vegetation changes are involved, is often due largely or indirectly to slope form through its influence on moisture conditions in the soil. Some geographers, including several in the United States, would argue that the study of man–land relationships may require only a knowledge of the form of the land, although the importance of understanding processes in order to appreciate man's effect on these would be emphasized by other geographers.

As well as his description and interpretation of landforms the geomorphologist's study of processes may also be useful. However, interest in erosion, transport, and deposition is shared with other disciplines and often follows their approach and utilizes their techniques. In this context the work of a geomorphologist may be useful, but it might be confusing to claim that such work enhances any intrinsic usefulness of landform study.

Applied geomorphology describes the deliberate attempt to concentrate geomorphological expertise on the solution of practical problems. Although the term 'economic physiography' cropped up in this context more than sixty years ago, it is only since 1945, with the development of planned economies in East European countries, that geomorphologists have searched for practical problems to solve. Despite government policies, the incentive of possible financial support, the hope and will to do some practical good, or the fear that changing times are not going to be kind to the scientist who insists on seeking knowledge for its own sake, the possibilities of applying specifically geomorphological expertise to practical problems are limited. One reason is that problems like landform stability were already in the experienced hands of the civil engineer and engineering geologist before the word geomorphology was introduced. However, particularly by collecting field measurements to supplement theoretical predictions, some geomorphologists working alongside engineers have made important contributions to practical studies of landforms, like the erosion of coastal cliffs. The practical application of the detailed geomorphological map is one of the examples most frequently quoted as an illustration of applied geomorphology. East European workers describe how, from these maps, other maps can be compiled giving information about landforms favourable or unfavourable for farming, communication, or housing. However, with the growing familiarity with process studies and equipped with a distinctive array of basic inter-disciplinary techniques, the geomorphologist, while hesitating to define his work as applied geomorphology, finds that he can contribute to the study of many of the most pressing practical problems. One example is the disruptive effect of periglacial processes on man-made structures in high latitudes or at high altitudes. Another is the problem of sedimentation in rivers, reservoirs, and irrigation and water-power plant intakes. The supply of this sediment is frequently augmented by artificially accelerated erosion which constitutes a closely linked problem. In the little-known areas of both dense and sparse population, the geomorphologist can make useful contributions to the work of survey teams describing and evaluating land capacity.

By far the largest amount of effort and interest in geomorphology comes from teachers and students. This may justify the suggestion that by far the most practical role for geomorphology is in education. In its pre-measurement phase geomorphologists may have continued needless controversies about the relative merits of speculative theories on the grounds that lively discussions were stimulating. However, increases in knowledge within recent years, while by no means raising geomorphology to the standard of a discipline, have meant that landform study has begun to make distinctive demands on the mind as well as filling the lungs with fresh air. While some aspects are susceptible to the rigour of scientific method, imagination alone can provide some of the tentative links which might string fragmented evidence together. Amid the lawless irregularities of form and complexities of process, where the words cause and effect do not apply, where scales range from thousands of miles to a micron, where there is very limited prospect of mathematical regularity and where even measurement is often very difficult, one learns to assess possibilities and to be cautious of generalizations. One learns not only that the environment is dynamic, not static, but also that man can trigger off catastrophic changes very easily, and further, that the full significance of such triggered changes is usually only apparent some time later. To anyone who has studied the inexorable progress of erosion, this concept of change with time is elementary, whereas the challenge of day-to-day thinking in four dimensions is not a feature of many disciplines.

In addition to the chronic pressures of food and water shortage in many areas is the problem in others where general awareness of damage to the environment has only recently made demands for conservation and control arouse anxieties about the dangers of uncontrolled exploitation. As geomorphologists are increasingly interested in process studies and therefore need to adopt techniques which, although mostly borrowed from related specialisms, constitute by their distinctive range an invaluable aid for the study of practical problems involving soil and water in relation to slope, geology, and man, students of the subject, after their academic training, are beginning to make contributions of growing practical significance.

However, those who have pointed to the weaknesses in the work of leaders like Davis, Penck, and Wooldridge will recall that the first, in devising a 'Cycle of Erosion', wanted a geomorphology to use in teaching, for the second its use was interpreting geological structure and Wooldridge's aim was its use in completing geological history. It seems in retrospect that a deliberate attempt to make geomorphology useful may be mistaken. But by asserting that the fundamental role of geomorphology is simply the scientific one of description and interpretation of landforms, one not only keeps the unique core of the subject in full view, but also the growing use of interdisciplinary techniques associated with the increasing appreciation of the importance of the study of processes, equips students with some expertise that they could subsequently apply to the collection of data relating to some fundamental practical problems.

C. Basic Postulates

'The action of a single night of extraordinary rain has crumbled it away and made it bare soil.' Plato

'It is to be concluded that what may be natural for one part of the geologic column may not be for another.' M. M. Leighton, 1958

1. Catastrophism and uniformitarianism

The view of earth history proposed by the Catastrophists of the early nineteenth century was of a succession of abrupt upheavals culminating in a great Flood. These paroxysms were interpretated as the result of Divine intervention. In contrast, C. Lyell and J. Hutton favoured slow changes due to natural processes and considered that interpretations of earth history could be based on present-day evidence. Geology developed from their work, and A. Geikie's maxim, 'the present is the key to the past', is often quoted, perhaps partly because the phrase is little longer than the word Uniformitarianism.

However, work in geology and geomorphology is sometimes less consistent with Geikie's maxim, now essentially of historical interest, than its frequent quotation may suggest. For example, since endogenic processes are unobservable, for many geologists the present is not so much a key as one number in a combination lock; geomorphologists have reconstructed many ancient peneplains without the aid of an uncontroversial present-day example to serve as a model. There is growing appreciation of a non-uniform if not catastrophic element in the rates of both endogenetic and exogenetic changes. One reason for this is perhaps due to improved communications and with an extension of increasingly dense inhabited areas there is an accumulation of eye-witness accounts, added to every year, of some spots on the earth's crust **changing abruptly in a matter of seconds or minutes.** The second reason is that if the term abrupt replaces the subjective 'catastrophic' adjective, abrupt as opposed to uniform change is seen to be an essentially relative concept, depending entirely on the time span involved. For instance, within the span of geological time the Quaternary glaciations were abrupt. Compared with the 10 million years of Pliocene time, the huge Würm ice-sheet disappeared in an abrupt 10 000 years. If it is clear that time-spans much longer than a lifetime are implied, statements in the Catastrophic idiom become scientifically acceptable today. For instance, in a recent interpretation of the formation of **the Snake river basin in Oligocene times, D. I. Axelrod (1968) describes** 'the basin opening up as an ever widening rent'. Therefore reviewing the significance of abrupt events is made difficult by the inevitability of using terms which are relative, depending on a time-span which is often scarcely definable. Even a catastrophe, by implication a natural force capable of destroying life and property, is not necessarily moving more rapidly than an average human being can run. There are some very slow movements which, in the absence of self-regulating mechanisms, continue to move persistently in one direction. These may outstrip man's technical advance, and with cumulative effect may become a catastrophe. In this sense earth history as we see it, with a rise of 130 m in sea-level between 15 000 and 5000 B.P., did in fact terminate in a great 'flood', which for unimaginative, uninventive, and technologically backward islanders could have been catastrophic. To have any meaning, a description of the relative abruptness and intensity of a catastrophic event must also hint at its context in some broadly indicated period of time; similarly for the time over which an abrupt event recurs sufficiently frequently to be regarded as commonplace and part of a uniform rate. For periods of a few years or less, a recurrence interval might be stated accurately, especially in an annual context. At places where the same catastrophic movement has occurred more than once, it might be assumed that a recurrence interval can then be estimated on the assumption that these abrupt movements themselves recur with a certain uniformity in time. This has been observed to be the case for rockfalls on the Kiruna–Narvik railway between 1902 and 1960 and is a basic tenet in the prediction of floods. For longer periods, words appropriately lacking in precision include a lifetime, implying a number of decades, with a span of a few centuries implicit in the phrase 'historical time'. Occurrence within postglacial times implies a span of a few millennia, but events catastrophic in the context of a few 10 000s of years might have to be described as such. At the other extreme geological time implies upwards of a few million years and Quaternary time implies several 100 000s of years.

In the study of present-day surface forms and sub-aerial processes the geomorphologist follows more closely the notions included under the broad heading of 'Uniformitarianism' than the geologist concerned with reconstruction of the past. He has to consider how he may extrapolate from the present to the past and how abrupt as opposed to uniform rates of change could affect these extrapolations. Similarly he considers the degree to which the present, with its abrupt and uniform changes compounded, is typical of the geological past into which he has to delve. In another sense he may distinguish his approach as the reverse of that of the geologist by regarding the past record of sediments and paleoforms as a check on his interpretations of present forms and processes.

Extrapolation from the present to the past, in apposition to the reverse process of prediction, has been called postdiction. From observations of present processes, postdiction involves two distinct concepts. First, a calculation of present-day erosion is multiplied by some time-span, often 1000 years, to gain some idea of the time that an appreciable change in landform might take. Clearly the figures are notional, like the counts of modern pollen rain needed for fossil pollen studies. Their main value is in the order of erosion indicated and whether this indicates that associated forms are being produced at present or whether very feeble erosion rates suggest that they are relict forms. Studies of hurricanes off Florida provide an example of the use of arithmetic postdiction. In the last fifty years eight storms affecting south Florida have been of hurricane intensity. Postdictions from these observations suggest that about 160 hurricanes could occur in 1000 years, and that hurricanes were commonplace in the context of Pleistocene time when 160 000–320 000 might have occurred.

A second, distinct aspect of postdiction involves the application of the appropriate knowledge of contemporary process-form inter-relationships to the interpretation of past events when similar inter-relationships presumably existed, although perhaps with some difference due to unique local conditions and possibly on a different scale. For this reason the geomorphologist who studies landforms in formerly glaciated areas also makes first-hand observations on the margins of contemporary glaciers.

One difficulty in assuming present processes to be the same as those operating over a much longer time is in deciding on how much significance to attach to slow but uniform denudation compared with the intensity of an abrupt event. For instance, in the upper Rhine valley in Switzerland, the late-glacial rockfall near Flims moved material 1540 times farther horizontally and 140 times farther vertically than calculations of the annual movement achieved by present processes. In the 1953 storm surge 3-m-high cliffs receded 27 m along a 1·5-km sector of the Suffolk coast at Covehithe (Williams, 1956). In contrast to this abrupt change, field and laboratory studies suggest that the major portion of sand transport along some beaches at relatively slow rates throughout the year, rather than by occasional large storm waves which erode coastal cliffs. Clearly the ratio between the significance of uniform and abrupt mechanical processes varies from one environment to the next, depends on the time-span considered. Beyond indicating the problem for each individual case-study, general statements are not yet possible, although the uniformity of chemical processes is conspicuous.

In considering the nature of catastrophic changes at least five aspects are distinguishable. First, there is usually a protracted period during which a potentially unstable situation slowly develops. This would be the gradual accumulation of unconsolidated materials that would offer little resistance to a subsequent sudden onrush or downpour resulting from extreme atmospheric disturbances. In Louisiana the 1957 storm surge elevated and transported entire portions of mudflats, including one segment 2 km long, part-way across the adjoining marsh. Alternatively, where gravity is directly involved, a long period of weathering finally brings a potentially unstable geological structure to the brink of catastrophic disequilibrium, or snow accumulates to form avalanches in spring melts, or rivers and lakes may inch towards levels of abrupt overspill. The overflow of Pleistocene Lake Bonneville is a classic example, reconstructed first by G. K. Gilbert in 1890. Recent excavations suggest that downstream the river flood was over 120 m deep and of the order of 280 000 m³/sec and moved boulders of more than 6 m in diameter. The discharge compares with the present peak flood runoff in the vicinity recorded in 1910, 2800 m³/sec. A common type of abrupt movement of flows is where an initial small barrier, by intercepting more material, becomes self-accentuating so that when the barrier is finally overwhelmed, it is by a surge of comparatively large force. A carefully recorded example relates to a raft of logs in the Colorado river, Texas, first noted in 1690. By 1824 the raft was 4 km long and had reached 66 km in length by 1925. When released in 1929 a flood swept the raft into Matagorda Bay where a delta 2930 ha in extent was built within a dozen years. There is also the possibility of active structural movements leading imperceptibly to a gravity-driven catastrophe. The diversion of the Brahmaputra as partly the result of gradual tilting of the Madhupur block has been suggested. In short, some lengthy gradual process is usually a prerequisite for catastrophic movements.

A second aspect of abrupt events is the triggering mechanism which finally releases gravity-driven movements of material, snow, or water. Again either exogenetic or endogenetic processes could be responsible as either the result of unusual atmospheric disturbances or due to earthquakes in tectonically unstable areas. In situations of chronic disequilibrium the energy released by the triggering mechanism need only be small. In Hokkaido, Japan, falls of rain of only 50–100 mm are sufficient to release landslides. In many areas man unintentionally has become the main triggering geomorphological mechanism.

A clearly defined third aspect of catastrophic change is endogenetic, the volcanic eruption which combines the features of slow build-up to the point of eruption and some trigger to effect its release.

A fourth aspect of catastrophic events is the chain reaction which the results of an initial movement may initiate. This is often due to the constrictions of dissected mountainous relief where catastrophic movements are inevitably most common. For instance, in 1958, a landslide of 30 million m³ of material, triggered by an earthquake along the Fairweather fault, Alaska, fell into Lituya Bay. In turn this slide created a wave which swept up to an altitude of 530 m. on the opposite shore. The disturbance of ocean water too may follow earthquakes, landslides, or volcanic eruptions, and huge waves hundreds of kilometres in length may proceed at velocities up to 800 km/hour.

A fifth factor is that any ratio between the significance of catastrophic and uniform processes would vary from one environment to the next, even where these may be in close juxtaposition like the sea cliff and beach, the river cliff and bed, and would be very difficult to estimate.

Assuming that the present-day forms are adequately described and that an impression of processes includes a representative range of continuous uniform movements and discontinuous rapid ones, postdiction involves the further consideration of considering to what extent the present observations are typical of past conditions. This problem applies equally to investigations where the study area is believed to represent closely past conditions in another area as to studies interpreting the landforms of the study area itself. It is important to distinguish between the erosional and the tectonic aspects of this problem, as the latter will influence rates of erosion and deposition. The present discussion is confined to the problem of whether in a relatively stable tectonic area the present is typical of the past, and whether areas at present involving tectonic movement are typical of similar areas in the past.

Strakhov suggests that the average rate of sediment accumulation in basins of the geological past and the limits of variation of that rate generally fit very well within the norm of present-day sedimentation, but due to man's influence on land, present erosion rates may be on average twice that of rates prior to the advent of man (Douglas, 1967). In certain localities, like highlands in the humid tropics, man's interference was sufficient to produce chronic disequilibrium. Man's interference is not just restricted to the widespread ploughing of recent decades, and the difficulties of calculating erosion rates due to water abstraction, but extends back for centuries when man first started to destroy the natural vegetation. Farther back in time the Quaternary glaciations represent geologically abrupt events when phenomenal ice-sheets spread over nearly 20 million km^2 of the earth's surface, over-riding areas where formerly mild climates had prevailed. It is possible that the earth had never been so cold since the beginning of geological time. Scanning back into geological time, erosion on the unvegetated pre-Devonian relief must have been distinctive and although vegetation types expanded their range, grasses did not appear until the Miocene.

It seems that a geographical range of studies of the inter-relationships between present-day forms and the processes of erosion, transport, and deposition will still give a good idea of the way in which these mechanisms operated in the past. But the appearance of the past is not easy to imagine. It is difficult to conceive that where now there is the landslip scar and irregular toe of slumped material there was a smooth slope giving little indication that rain, molecule by molecule, was weakening its internal cohesion. It is difficult to imagine forests once cloaking grassy or rocky slopes and that in terms of geomorphological time a cover of ice disappeared overnight. It is hard to believe that only five hundred generations ago, sea-level was 130 m below its present stand and that it then started to rise at a rate of about 45 cm per generation. By the time that the rise halted more than one must have counted his livestock aboard two by two.

2. Stillstands and the mobility of earth structures

Erosive activity is usually limited by a base-level below which most mechanical processes cease to be effective. In upland areas a resistant rock outcrop frequently provides a local base-level which might retain a reasonably consistent relationship to land surfaces upstream. On a global scale, sea-level provides a regional base-level at which marine erosion may score a mark at the coast, and inland provides a datum at which, or at some distance above, mechanical erosion on the land surface ceases to work downwards. Rapid movements of the earth's crust tend to be somewhat periodic in time, separated by periods of relative stability. W. M. Davis therefore felt that a theoretical separation in time of uplift from protracted periods of stability would facilitate landform interpretation, and in particular the recognition of evolutionary stages of a progressive flattening of the land surface. From other quarters came the suggestion that when coastal movements did in fact occur they were upward on the continents and downward in the ocean basins, leading to a world-wide progressive relative elevation of the land surface in relation to the sea (Suess 1906; Chamberlin 1909). From a combination of these views it could be predicted that the continents would carry at increasingly higher altitudes, relics of progressively early phases of flattening or 'cycles of erosion'. Critics of Davis immediately asked how he could assume that long periods of comparative stability approached the stillstand of base-level on which his model depended. He was certain that stillstand was insufficiently long in certain regions, but for other areas pointed out that the existence of relics of levelled erosion surfaces were the very indication that stillstands had prevailed. The simplicity of this argument is more striking than its circularity, and an immense amount of geomorphological effort has gone into testing this prediction. However, when Davis and his critics discussed the problem of stillstand neither had available the measurements which can now reveal the exact age of some land surfaces nor the knowledge that changes in relative sea-level are rapid enough to be observable within a lifetime in some areas. Fishermen in localities around Bonavista Bay, Newfoundland, can point to rocks that are above water at low tide today which they remember as being always below water half a century ago. In the Buena Vista oilfield, on the west side of the Grand Valley of California, a thrust fault has shortened the distance between two power-line pylons by 60 cm in thirty years.

Although the progress in separating the crustal movements from eustatic changes will be discussed, the scale of total movements is illustrated first. There are two types of measurement which express crustal movement. The first is the result of actual measurements of elevations separated by a known time, which can therefore be expressed as actual rates, like the levellings along the Trans-Caucasian railway, where a known increase of 730 cms/1000 years at one point contrasted with a depression of 640 cms/1000 years in the Kura depression. Secondly, there are measurable displacements but for which the time involved can be inferred only approximately. It seems that measured rates of present-day epeirogenic movements, for both platform and folded zones are of the order of tens to hundreds of centimetres in a thousand years. For example the Atlantic coast of France rises at 90–280 cm/1000 years, parts of Britain at 120–130 cm/1000 years, and for Africa rates of elevation are anything from nil to 420 cm/1000 years. Uplift in the Puerto Rico–Virgin Islands area is about 500 cm/1000 years. In the Los Angeles basin terraces are rising at about 400 cm/1000 years. Occasionally, more extreme values like elevations of 1630 cm/1000 years along parts of the Caspian Sea, or depressions of about 1000 cm in a few hundred years in the Sylhet Basin are recorded in the most active tectonic regions of the world.

With the aid of radiocarbon dating it is even possible to trace variation in the rates of movement in late-glacial and recent times. For example, uplift at the entrance of Chesapeake Bay seems to have been continuous, averaging 850 cm/1000 years between 15 000 and 8000 years B.P. By 6000 years B.P. it was possibly about 6 m above the present mean low water but then crustal subsidence at 230 cm/1000 years followed, allowing extensive flooding of

Chesapeake, until crustal uplift apparently resumed about 2000 years B.P., and probably continues to the present day.

Stratigraphical displacements indicate some areas where movements active today are part of huge disturbances starting in late-Tertiary and within Quaternary times. A classic example of this type of area is the Pacific states of the United States. In the Raton Basin, southern California, several episodes of later Tertiary vertical uplift, largely along faults, but with some warping, produced vertical displacements of at least 5 km. In the Carson Range, east of Lake Tahoe, post-late-Pliocene displacements of lower Pliocene andesites amount to 1200–1500 m. In the Andes, work in the Cordillera Oriental, Colombia, suggests that there the main uplift occurred as recently as late Miocene times.

More tentative arguments, restricted to areas known to be tectonically active, include references to early Pleistocene or even Pliocene deposits not consistent with the existence of the present mountains at that time. If significant mountain uplift in Turkey took place in very recent times, it might be possible to attribute the apparent absence of early Pleistocene glaciations to the mountains being substantially lower at that time. D. A. Axelrod has recently argued that fossil floras in Pliocene and Pleistocene deposits in the Sierra Nevada are anomalously high and that a reconstruction of their optimum climate might indicate subsequent elevations of some thousands of metres.

Rates of accumulation in subsiding basins provide an indication of rates of crustal movement. For instance in the Chilean longitudinal valley, great thicknesses of clastic fill suggest a net downward movement of about 2000 m during the Pleistocene.

A separate consideration from rates of uplift is the question of whether changes in these rates within a relatively short distance cause warping. There are several indications of axes of warping both approximately parallel and at right angles to coasts. Indications of the first case, that of marginal down-flexures of continents, is widespread, the down-warping of the coastal plain and continental shelf of the eastern United States being one of the areas documented in greatest detail. This area, particularly in the margin of the Mississippi delta includes the complication of a hinge line seaward of which sediments down-warp, whereas inland they have been uplifted. Isostatic adjustment to sedimentary loading appears to be an important factor. In the second case features identified with high confidence as former shorelines may show changes of level along a coast, due to warping in that direction. On a small scale and within recent time, a marine bench near Night Cliff, in the vicinity of Darwin, drops from 2·4 m to 60 cm within 180 m but within 2 km rises to 4·5 m and some benches rise as much as 6 m in 1·5 km. On a larger scale, spanning Pleistocene time and mantled with deposits rich in a prolific molluscan fauna, are three major erosion surfaces in the coastal region of north-west Peru. These are the Mancora, Talera and Lobitos tablazos, which are approximately 270, 150, and 20 m above sea-level at Cabo Blanco. However, when traced along the coast substantial changes of level occur. The Mancora tablazo, for example, falls 180 m in a distance of 72 km. In California there are Pleistocene sediments which are not only folded but may even stand vertically.

Lateral displacements of the crust are of some indirect significance in the measurement of vertical movements partly because they reinforce the evidence of a dynamic crust and partly because they inevitably include some vertical displacement. In the north-east Pacific Ocean, two transcurrent faults involve a lateral displacement of 1425 km, and along the San Andreas fault there has been a displacement of 320 km since early Miocene times, with present-day movements measured at 50 m/1000 years. In New Zealand, the East Ruahine fault has moved at approximately 7 m/1000 years since the last glaciation.

On stable platforms rates of movement are slow and on a small scale, and rapidly subsiding trenches like the present-day African grabens are uncommon. However, tectonic history can still be sufficiently varied to give recorded altitudes on land surfaces or coasts little meaning in absolute terms, as their regularity of recorded former sea-level heights along the south coast of Australia suggests. Moreover, the position of shorelines on low-lying platforms is very unstable because even small vertical fluctuations of sea or land will cause substantial changes in their configuration. The development of platforms of moderate elevation appears to lead to the chemical weathering producing crusts and planation surface but at altitudes unrelated to sea-levels. One of the best examples of this situation occurs in central and southern Africa, unfolded since mid-Mesozoic times, where L. C. King and F. Dixey (1938) were among the first to map extensive surfaces of planation.

Superimposed on the structural movements are elevations of the land in areas relieved of the load of Pleistocene ice, which amounts to 5 per cent of the earth's surface. Isostatic rebound is appropriately named for such rates exceed in many areas those of tectonic movements. They may be largely complete before a region is entirely ice-free. These rates are difficult to compare because rate of recovery diminishes rapidly with time, any given rate being approximately halved in a span of about 800 years. Also the possibility exists that rebound is not necessarily a continuous phenomena and that maximum rate of uplift in areas more centrally located with respect to the ice cover may occur later than in the more peripheral areas. This last possibility may explain why over the past 3000 years, uplift on the west coast of Baffin Island averaged 5 m/1000 years compared with 3 m/1000 years on the east coast.

Some examples of rates of emergence, using the middle third of the emergence curve to obtain an average value, include 29 m/1000 years for Spitzbergen, 24 m/1000 years for eastern Greenland and 40 m/1000 years for the west coast of Baffin Island. Examples of maximum rates suggested include 70 m/1000 years for Boston, which is similar to that for north-east Canada. For the north of Scotland a rate of 8·5 m/1000 years may have occurred 12 000 years ago slackening to 0·7 m/1000 years in the last four millennia. The most spectacular present-day movement is in the Gulf of Bothnia which continues to rise at 10 m/1000 years, with still one-third of the expected total postglacial uplift to come. Uplift so far is between 240 and 550 m. In Hudson Bay, a recovery of 270 m may have occurred and a maximum depression of 600–900 m for the central areas is possible. Amounts of rebounds vary, being largest where the former ice thickness was deepest. Thus in Norway uplift was greatest at the heads of deep fiords and least on the open coast. The highest strand line, marking the marine limit is 220 m in the Oslo and Trondheim fiord areas. At places near the open coast, an uplift of apparently only 10 m has occurred.

In the Quaternary the incredible spreading of ice over 20 million km^2 of the land surface, inducing on melting catastrophic rates of elevation of areas depressed during glaciations, should not divert attention from the extraordinary glacio-eustatic fluctuations of sea-level, with amplitudes of at least 146 m. Although present levels are as high as might be expected in an interglacial, only 16 000 years B.P. the sea was probably about 140 m below its present level and possibly still 30m below this about 8000 B.P. Complicating the picture of a progressive rise, even on structurally stable coasts, unaffected by isostatic rebound, is the possibility that the weight of water added to continental margins has been sufficient to deform isostatically the coastal areas in proportion to the average depth of water in the vicinity. The study of the postglacial submergence history of 5 points on the eastern coast of the United States supports this hypothesis (Bloom, 1967).

Despite the intricate problem of sea-level changes and, inevitably, considerable differences of opinion in detail, two facts on the wane of the last major glaciation meet with widespread assent. The first is that since about 15 000 years ago sea-level has risen from depths at least 75 m and probably 135 m below the present and that a few millennia ago, all ranges between two and six thousand having been quoted, the present level was obtained. From evidence in the stable southern Florida area it appears that sea-level has not risen appreciably above its present position in the last 4000–5000 years. Considerable disagreement exists for the period between 35 000 years and 18 000 years ago, during which time the sea-level may have risen continuously from its low stand of approximately −140 m. Alternatively, using evidence from many sources, including former lake levels in the Great Basin,

Broecker concludes that the rising sea-level fell between 25 000 and 18 000 years ago.

Because of glacio-eustatic changes in sea-level and the varied tectonic complications, sea-level positions before late-Pleistocene can only be reconstructed where a wealth of evidence exists and then may be of significance only to areas in the immediate vicinity. Even generally agreed indications of sea-level positions in earlier Pleistocene times are now realized to be much more difficult to give than was widely believed a decade ago. It appears that the glacio-eustatic changes in sea-level corresponding to the major glaciations were approximately the same order as the last, that the changes have occurred at decreasing time intervals, but that there is no reason to suppose that sea-levels would necessarily maintain a constant level during an interglacial. However, evidence grows to suggest that a few localities may have been demonstrably stable before and during Pleistocene times, and can give some indication of relative sea-levels in pre-Pleistocene times before glacio-eustatic complications began. At such a site in Manatee County, in Florida, estuarine deposits contain a varied fauna of Hemphillian (Middle Pliocene, 10–4 million years B.P.) age. Compared with their Pliocene estuarine environment these deposits now stand between 1·8 m and 3 m above mean sea-level, suggesting that sea-level then stood only a few metres above its present elevation (Webb and Tessman, 1968). This conclusion contrasts with that of previous workers who, on the basis of less precise evidence and sometimes retaining the mistaken notion that the capacity of ocean basins was increasing progressively, considered that sea-level was in general higher at this time. Impressions of sea-level farther back in time are few, although the interpretation of bevelled landforms as indicating former base-levels has inspired a wide range of guesses. P. Kuenen (1950), from evidence not relating to altitude of land surfaces, considers that sea-level may have fallen about 250 m since late Cretaceous; R. W. Fairbridge (1961) estimates a possible 200 m.

If it appears an unrealistic and misleading simplification to separate uplift from erosion, some other concept of the inter-relationship of erosional processes and changes of crustal stability of sea-level position may appear necessary. For huge flows of water to balance uplift with the necessary downcutting is possible, as the down-cutting of a rising rock barrier between two of the Great Lakes shows (fig. I. 7), but in general it appears that equivalence between rates of erosion and of uplift are rarely achieved.

Although it may be difficult to justify the separation of uplift from contemporaneous erosion, the figures cited show that where there are orogenic movements of the earth's crust these substantially exceed rates of erosion as the latter, when calculated for the land surface as a whole, rarely exceed a few centimetres per thousand years. In contrast, epeirogenic movements involving vertical movements of stable platforms may be small, but since subsidence of only 20 m would involve the flooding of huge areas, and since uplifts of the same amount lead to extensive marine regression and the drying out of equally large areas, the enormous lateral changes produced make erosion rates look negligible. Incidentally, these continental transgressions and regressions are unrelated to the capacity of ocean basins, as was thought at the beginning of the twentieth century. Similarly, sedimentation rates are commonly also of an order of a few centimetres per thousand years, so that for areas like the post-Quaternary Baltic, already 200 m deep, in places, there is scarcely a veneer of sediment on its floor.

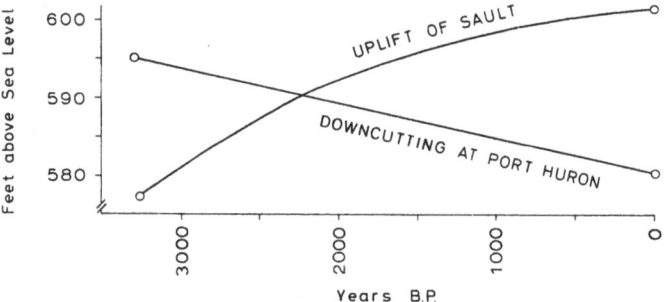

Figure I.7 Illustration of how downcutting has matched the rate and amount of emergence of the Sault rock barrier in St Mary's River, Ontario (*Farrand, 1962*). The rock barrier emerged, due to isostatic rebound, since the Algorna water-level stage (3200 BP).

In some places, therefore, a separation of tectonic movement or of changes in sea-level or shoreline position, from erosion or deposition, has been a virtual reality. In other places a sequence of events would be unrealistic. In either case the major difficulty appears to be rather that of postulating, at some arbitrary point in time, some 'initial' form which resembles more the appearance of the present day instead of the unrealistic flatness of an idealized, hypothetical plain. Apart from suggesting that it is the nature of a land surface before movement which is the greater problem than deciding on how erosion and crustal and sea-level changes interact or whether one essentially precedes or follows the other in its subsequent history, only one generalization can be offered. The nature of all the factors involved in these movements must be considered and evaluated for each particular case.

3. The Cycle of Erosion

W. M. Davis, at the end of the nineteenth century, proposed a Cycle of Erosion in which initially steep-sided incisions into an abruptly uplifted surface gradually broadened and flattened until, if continued stillstand of base-level were assumed, an erosional plain of faint relief resulted. The reason why he often referred to this sequence as the Geographical cycle may not be immediately clear. However, W. M. Davis believed that the races of mankind were obviously determined by the larger features of the lands and even that 'religious ideas themselves ... are seen as if in a mirror held to nature'. It seems that he termed the Cycle a geographical one because he believed that it would explain more than the landforms themselves, but it is not clear to what extent Davis's knowledge of human history, combined with his primitive deterministic view of its progress, influenced his concept of landform evolution. However, critics immediately argued that Davis's theoretical model did less than explain landforms, and controversy continues to the present day. Four points might be made in an attempt to put this debate in some perspective. First, the general erosional reduction of rapidly elevated structural features has occurred repeatedly in geological time. Hercynian folding, for instance, was approximately levelled within some 20 million years of comparative stability. A few hundred metres of relief on such structures is trivial compared with the geometrical reconstruction of the hypothetical structural relief. Although not approaching the levelling of the hypothetical peneplain, the concept of an erosional reduction of orogenic structures to a relief of about 300 m, with local instances of 1000–1200 m, provides an acceptable datum for the study of many problems in stratigraphy. On the scale of the African continent, L. C. King's highest 'pre-African' or 'Gondwana' surface, attributed to cyclic pedimentation, is usually above 1200 m and is thought to be traceable in the Abyssinian Highlands at 2400–2700 m. Rough levelling down is undoubted; it is with attempts to identify and to correlate fragments of erosion surfaces within very narrow altitudinal ranges that disagreement has arisen.

A second point is that Davis's Cycle was never adopted on the scale that the amount of parochial controversy in some quarters might suggest. Despite the translation of his work into German in 1912, and his extended visit to Germany, his ideas left little impression there, and in the Soviet Union and much of eastern Europe geomorphology has progressed largely unaided by the Cycle. In 1924 S. W. Wooldridge observed that British workers had been slow in following the lead given them by W. M. Davis. In the United States many geomorphologists lost interest even in debating about peneplains during the 1930s and took little notice of H. Baulig's work on the Cycle which aroused interest in France and Britain. By comparison, a wide range of studies of landforms and processes were pursued with little or without any reference to the Cycle concept.

A third point is that there has always been debate on the reality of the peneplain. For instance, W. M. Davis saw 'a complete lack of sympathy between structure and form' of a former peneplain in Connecticut, whereas

R. S. Tarr supported the observations of J. G. Percival, who, in 1842, had seen that this upland surface corresponds 'very exactly with that of the geological formations'. It seems that arguments about the peneplain are largely insoluble because the complexity of landform makes proof or disproof unobtainable. With varying heights in a dissected surface it is very difficult to demonstrate effectively the existence of old denudational surfaces, but equally difficult to deny that a peneplain, however utterly remote might be the chance, is theoretically possible.

A fourth factor is an increasing awareness that regardless of the possibilities or disadvantages of the Cycle, sentimental attachment, or dislike, a growing bulk of geomorphological work proceeds without reference to the debate. It is possible, without much reference to the tedious seventy-year-old catalogue of reasons for attacking or dismissing the peneplain on general grounds, to suggest some reasons why the Cycle appears to be in relative decline.

One reason is the frustration experienced by exponents of the Cycle when they find that cyclic interpretations within one area do not agree. P. Birot (1966) takes the example, where agreement is rare, of an escarpment in a crystalline massif separating two plane surfaces of different altitudes, and the difficulty of deciding whether the breaks of slope separate two successive cycles of erosion or represent a comparatively recent fault dislocating one former surface. The Appalachian Plateau has become a classic area for disagreement on the number and correlation of erosion cycles. General opinion varied between three and five surfaces, although some workers have recognized a dozen within a small area. There has never been any substantial measure of agreement as is shown by hundreds of controversial and speculative pages. In parts of Europe the cyclic but non-Davisian ideas of stepped erosion surfaces postulated by W. Penck, termed Rumpffläche or Piedmonttreppe and attributed to the control of interfluve elevation by an accordance of valley thalweg altitudes, have led to no greater measure of agreement. In Africa, from the sweeping success of the back-wearing scarp and pediment one gains the superficial impression of greater unanimity as 'the dating of cyclic landscapes in southern Africa is now established beyond challenge' (L. C. King, 1956).

Another problem is the increasing difficulty of recognizing remnants of a peneplain now that the eye for the country of some past old masters can no longer be turned to the problem. The major difficulty is the spanning of wide open valleys where there is no continuity of erosion surfaces and increasing doubts as to whether erosional horizons can be treated stratigraphically by using the analogy with remnants of a definable but dissected lithological horizon.

A third difficulty is the growing realization that with the effectiveness of erosion that would be necessary to produce a later cycle, remnants of progresively earlier stages will be increasingly few. J. L. Rich (1938) thought that the presentation of two or more cycles would be rare and N. M. Fenneman (1936) that correlations based on altitude alone were liable to error in proportion to the antiquity of the surfaces concerned. In identifying two peneplanes in the Ozarks, J. H. Bretz (1962) observed that the evidence for the earlier one had been largely destroyed. In southern Tanganyika, H. Louis (1967) observed that surface erosion by continuing on all present surfaces meant that peneplanation was persistently recent. As a final example, which is more specific since it relates to an uncontroversial feature, solifluction debris obliterates older shoreline sections of the ancient Lake Bonneville at Franklin, Idaho.

A fourth problem concerns deposits which could provide independent evidence of the age and origin of erosion surfaces. This evidence is being examined in more detail but in some instances it only casts doubt on the interpretation of the erosion surface as a remnant of a formerly more extensive plain. In a classic area, on the dip slope of the South Downs, field evidence indicates that the basal surface of the Clay-with-flints represents the approximate northward extension of the Eocene basal plane (fig. I.8). This plane, soil scientists claim, coincides in its northward rise with some of the surfaces of supposed sub-aerial or marine origin described by B. W. Sparks. They conclude that 'careful field investigation of every such surface listed by Sparks showed no trace of marine or other deposits except for Clay-with-flints, and most of them are ridge crests or summits occurring either at the intersection of the secondary escarpment slope and the sub-Eocene surface or between adjacent dip-slope dry valleys' (Hodgson *et al.*, 1967).

Fifthly, there is the awesome example of Pleistocene chronology. With increasing evidence of the complexity and problems in deciphering, correlating and dating even late-Pleistocene features, the unlikelihood of precise interpretation of earlier forms is more apparent.

A final factor is the advent of techniques of absolute age determination which have removed one of the main purposes for attempting to establish the age of land surfaces from their appearance and altitude alone.

If the Cycle is in decline there are, in addition to the factors contributing to this, other factors operating positively in other directions. There is an increasingly wide array of alternative explanations in non-cyclic terms for land surfaces bevelled across geological structures, although some relate only to restricted localities or to restricted instances. There are several ways in which geological structures themselves can give an impression of erosional levelling. On the largest scale, and contained within the stratigraphic column, are unconformities now believed to be due only partly to erosion and with tectonic gravity sliding as the major cause. At the surface and on a smaller

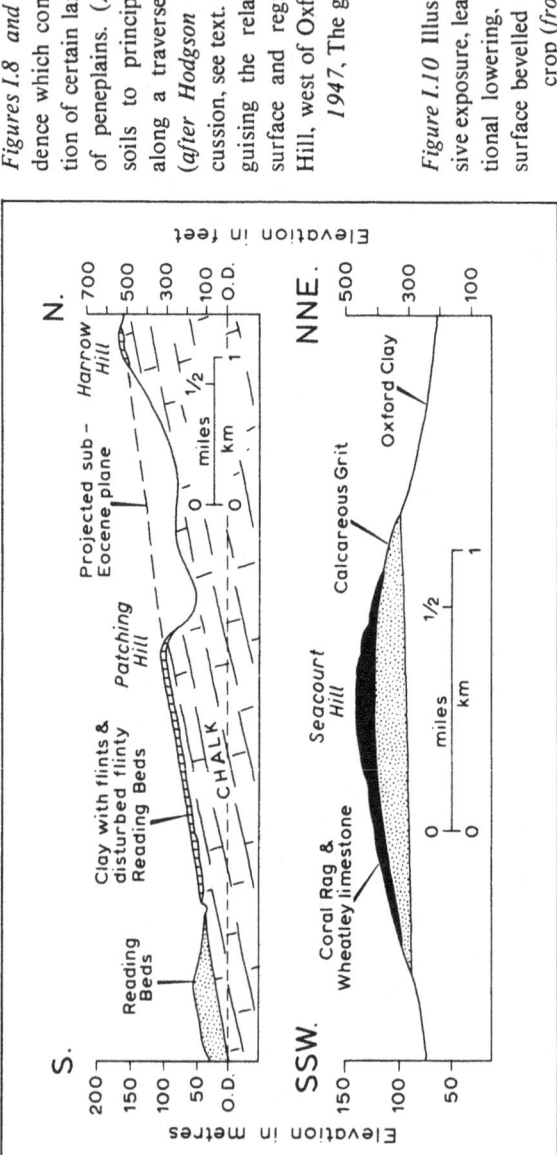

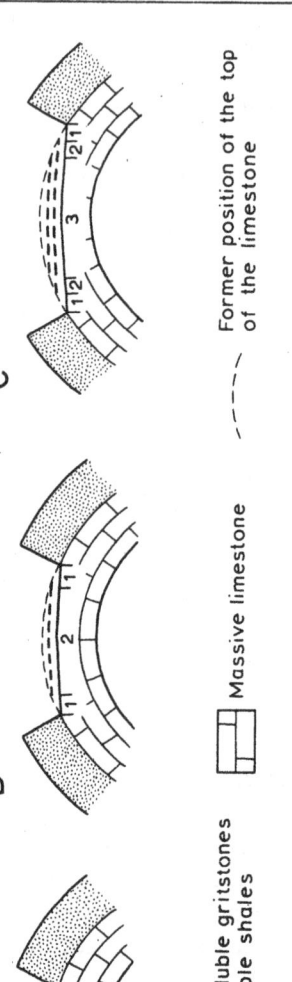

Figures 1.8 and 1.9 Examples of evidence which complicates the interpretation of certain land surfaces as remnants of peneplains. (*Above*) Relationship of soils to principal geological features along a traverse of the South Downs (*after Hodgson et al., 1967*); for discussion, see text. (*Below*) Cambering disguising the relationship between land surface and regional dip on Seacourt Hill, west of Oxford (*after W.J. Arkell, 1947,* The geology of Oxford).

Figure 1.10 Illustration of how progressive exposure, leading to differential solutional lowering, could produce a land surface bevelled across a limestone outcrop (*from Pitty, 1968*).

scale structural instability of rocks like shales underlying more competent caprock in an escarpment can induce sliding and sagging in the overlying strata (fig. I.9). For instance, in limestone–shale scarplands near the Kurkur Oases in Egypt, local bedrock along the scarpland edges dip east at 1–3 degrees, giving the impression of discordance with the overall westerly dip. Also extensive flat land surfaces, often developed on flat-lying resistant beds are described as stripped structural surfaces. D. Easterbrook (1969) suggests that such surfaces are perhaps the non-cyclic flat areas most commonly confused with cyclic erosion surfaces. Geological considerations are important in an indirect sense too, where resistant strata create a local base-level. N. M. Fenneman's view (1936) was that the influence of resistant strata had probably been under-estimated and the number of regional base-levels over-estimated, particularly in dissected plateaux and specifically in the Appalachians where the altitudes might owe more to two Carboniferous sandstones than to any succession of cycles. More recently Meisler (1962) describes the example of the Little Swatana Creek where a resistant andesite lava at the outlet of a basin, underlain by shales, resulted in mean and summit elevations being set off sharply 12 m above a lower shale basin. Where a lake upstream is lowered progressively, the importance of local base-level and its stepwise lowering is often clearly seen. In the Lake Victoria basin, strand-lines, associated with downcutting of the outlet, stand at 18 m, 12–13 m, and 3–4 m. A radiocarbon date of 3720 years for the lowest may suggest about 20 000 years as the span for the three stillstands.

For erosion surfaces not necessarily related to a local base-level, there is a range of explanations for specific situations. Pediplanation is the main example and surrounds itself in its own system of controversy. For instance, any suggestion that planed land surfaces are preserved at average elevations, successively higher with increasing age has, according to E. S. Hills (1961), no relation to reality in Australia. Altiplanation, a concept formulated by Eakin in 1916 in Alaska, is sometimes used in an attempt to explain level surfaces in areas of active or former periglacial areas. There is the suggestion that so-called Tertiary peneplain remnants in the southern Ardennes are altiplanation surfaces. It is not usually certain whether these benches are bedrock features or include a layer of debris. On the specifically erosional cryoplanation terrace there may be wash process removing material from the bench and nivation attacking the riser at its rear. According to J. Tricart (1967) altiplanation is a theoretical notion seldom realised in nature. An alternative explanation to peneplanation suggested for escarpments and plateaux of limestones formerly overlain by an impermeable cover is that the concentration of solutional activity at the surface could lead to differential lowering, the result of progressive exposure accompanying the gradual removal of the impermeable cover (fig. I.10). It seems possible that the ground surface of any rock type more susceptible to any type of weathering than an overlying cover rock which is being weathered backwards, might similarly show differential lowering due to progressive exposure.

A further factor contributing to a relative decline in cyclic studies is the realization of the increasingly wide range of aspects to landform study in which measurements and observations can be made without the necessity to interpret the feature beforehand. The expansion of interest in aspects neglected in the Cycle provides the more obvious examples of interests commanding increasingly large followings. These include Quaternary geology, slope studies, and climatic and general process studies. An associated factor is that the mixed blessing of modern facilities offers opportunities for work with sophisticated instruments which may tend to invite some researchers away from topics, like the Cycle concept, to which their application is not readily apparent.

4. Morphoclimatic zones

Geomorphologists who consider processes then infer the probable effect of these on landforms, may come to one of two main conclusions. On the one hand, the physical principles involved in weathering and in resistance to weathering, although they may vary in intensity from region to region, are immutable. The resultant forms therefore tend to be similar, and latitudinal contrasts in climate are consequently of little significance. This physical approach is that of some leading American geomorphologists, like A. N. Strahler, who makes relatively few references to climate, and L. B. Leopold, M. G. Wolman, and J. P. Miller (1964) who conclude that 'evidence indicates that the forms of hillslopes may be identical in all climatic regions'. This approach might be traced back to that of the engineer, R. E. Horton (1945) who said, 'The geomorphic processes we observe are, after all, basically the various forms of shear, or failure of materials which may be classified as fluid, plastic or elastic substances, responding to stresses which are mostly gravitational but may also be molecular ... the type of failure ... determines the geomorphic process and form.' In addition to the American workers, L. C. King saw that this physical approach could explain his direct observation that '... all forms of hillslope occur in all geographic and climatic environments'.

In contrast to the physical principles approach, the second conclusion is that reached by an environmental or climatic approach to geomorphology whose interest in processes recognizes differences in physical *factors* and extends to include the *unobservable* chemical and biological processes. The hypothesis is that the well-known zonal contrasts in climate, with their distinctive effect on vegetation, mean that physico-chemical processes combine in varying ways and operate at different rates, with inter-related vegetational influences emphasizing the distinctiveness of morphogenetic processes in contrasted regions. The immutability of physical laws is less important than the environmental contrasts for landform development, and it is assumed that the latter should leave some discernible distinctive traces on the land surface. The emphasis is therefore on studying process and landform in various environments, with accent on establishing the changes in the relative effectiveness of biochemical to physical weathering. This approach has, by describing and comparing the distinctiveness of regions, a strong geographical emphasis.

A minimization of the importance of latitudinal contrasts in climate is perhaps an inevitable conclusion which follows from the initial assumption that physical considerations provide the basis for landform and process study. The neglect of biological and chemical considerations means that the very factors which do change with latitude are omitted. So far the physical and the climatic approaches, apart from J. Tricart's (1958) and J. C. Frye's (1959) rejection of L. C. King's (1957) notions on the uniform nature of hillslopes, have co-existed with the fundamental differences being scarcely apparent. This is probably in part due to the fact that the languages of climatic geomorphology are German and French, and those of the physical approach, English, and the occasional equation.

Even if eventually landforms were demonstrated to have many predominantly azonal characteristics, the study of the inter-relationships between process and form in a range of environments much wider than the 'normal' mid-latitudes would remain a fundamental task in geomorphology. One has only to think of coral islands to realize that some element of zonality exists in landforms. Therefore some knowledge of the difficulties faced by the climatic geomorphologist, as he probes further, is an important aspect of geomorphology.

The distinctiveness of climatic regions and its possible implications for landform study were first considered by A. Penck in 1883, and he attempted to classify correlations between climatic areas and surface relief in 1910. In that year, E. de Martonne was probably the first to introduce the term 'climatic geomorphology'; and awareness of the distinctive impression of different climates on landform had already led W. M. Davis to Turkestan and to the formulation of an arid cycle of erosion. The progress of climatic geomorphology owes much to German workers after A. Penck, like C. Troll and J. Büdel, and more recently to J. Tricart and A. Cailleux in France. Today the phrase climatic geomorphology covers a very broad range of approaches. The work of A. Cailleux and J. Tricart, searching for the essential characteristics of an area, rather than speculating about hypothetical and indefinable boundaries, is very much in the tradition of French regional geography. Their attempt is to recognize distinctive morphogenetic processes and to link these to landforms. Others are sometimes less specific and merely produce a physical geography catalogue of climate, vegetation, soils, and landforms for a recognized climatic zone, with little attempt to link these into a morphodynamic system. There is the tacit assumption that from the distinctiveness of the environment it follows that the forms must be distinctive. In this way some morphoclimatic types have been defined on climatic data alone without initial reference to the forms which are then fitted, or assumed to fit, into the scheme.

The irregular distribution of moisture where precipitation provides an inadequate water supply for the weathering and transport of rocks in either hot or cold climates makes any clear zonation of weathering processes with latitude unobtainable. A major problem for morphoclimatic geomorphology is that the fundamental proposition, climate, is not a single definable factor, but consists of two variables which are partially independent but largely inter-related in most complicated patterns. The net denudational efficiency of any one pattern is difficult to predict from climatic data alone. A further complication affects runoff of water as denudation involves evacuation as well as the weathering of material. At higher temperatures the water supply may be largely returned to the atmosphere by evaporation and evapotranspiration, the intensity of this loss being a function of the rise in temperature. Therefore, although higher temperature may speed up reaction rates, the rate of evacuation of material from a weathering zone may be relatively slow as the higher temperatures also reduce runoff. A further complication is that higher temperatures speed up reprecipitation processes by as much as they accelerate weathering, so for this reason temperatures above a certain level become a self-arresting element in denudational loss processes. In cold areas the influence of snow and ice cover is similarly difficult to evaluate. Although temperature may appear to be the critical factor, it is in fact whether precipitation falls in winter or summer that determines if the ground will be exposed to periglacial weathering.

In addition to the unpredictable influence of temperature changes with latitude, distinctive geological characteristics are often sufficiently dominant to make any zonal generalisation difficult. The ineffectiveness of climate alone to cause a uniform course of rock weathering was shown by the differences observed in weathering rinds of fourteen igneous rocks, six sedimentary rocks and two metamorphic rocks weathering under the climate of the north-east of the United States. In structurally active areas tectonic movements change erosional processes in a given latitude, and makes a climatic interpretation difficult. This is the case in the Amazon basin where large-scale uplifts took place in the Andean headwaters in late-Tertiary time. Even the age of rocks can make zonal generalizations difficult. In Kenya, Pliocene–Pleistocene basalts are not weathered to the same extent as basalts in Scotland.

Altitude introduces manifold changes at a given latitude. In central Africa, Kilimanjaro, 5879 m in height, Mt Ruwenzori and Mt Kenya all carry present-day glaciers, and well-developed cirques, with floors at about 4750 m, surround the most easterly of Kilimanjaro's volcanoes. In the equatorial Andes the snowline is about 5000 m compared with 4500 m on Ruwenzori.

Along a given latitude climate changes substantially The disposition of mountains on the western side of continents and the direction of prevailing winds combine to create arid areas as discrete patches rather than in belts. In some areas like Indochina, southern China, and Atlantic Brazil the forest cover is continuous from equatorial areas to the humid mid-latitude areas, whereas in Africa, north of the equator, a broad arid area intervenes. In higher latitudes periglacial activity is practically without zonality, partly because in the southern hemisphere there is virtually no land in the appropriate latitudes, and in the northern hemisphere there are contrasts like those between southward jutting Greenland and the coasts of north-west Europe influenced by the north-east extension of the Gulf Stream.

Long rivers can introduce substantial changes in flowing through areas which contrast with their headwater area. In the northern hemisphere there are fundamental contrasts between north-flowing and south-flowing major rivers. In tropical areas several smaller streams create conditions like the Nile on a local scale. In Equatorial Somalia, the vegetation is generally dry savanna, through which the river valley is a 2–3-km-wide densely forested strip simulating the conditions and processes of the humid tropics.

In many ways relief and drainage combine to create weathering conditions which may vary dramatically in geographically adjacent localities. The valleys and mountains of the Burma range provide an example, with variations in precipitation ranging from 500 to 5500 mm annually, and all types of tropical soils occur within this narrow latitudinal span.

The concept of morphoclimatic zones includes the assumption that present-day forms at least approach an equilibrium with present-day processes. Two factors make it necessary to treat this assumption with caution. First, although it is difficult to establish clearly how climate varied during the glacial periods, the instability of climate rivals that of regional base-levels. In tropical areas like West Africa abrupt climatic changes produced coarse gravel terraces, thus simulating the effect of the periglacial episodes of the temperate and Mediterranean zones. In highlands, like the East African Mount Elgon, Aberdare Mountains, and the Abyssinian Highlands, glaciers were present. Weathering crusts, in active formation only in tropical regions with a pronounced dry season, occur as relict features in forests like those of the Amazon basin. Conditions at the margins of these latitudinal shifts of climatic zones are not easy to reconstruct. In central North America there was a southward shift of isotherms by at least 1200 kms in the ice-covered areas, but palaeontological evidence suggests that temperate zone flora, as in central Russia, existed virtually to the ice front. Farther back in time there were pines and firs in Greenland in the Oligocene, and evidence of tropical Tertiary flora has been found on the Canadian Shield.

Recently an added problem has become increasingly apparent, as it now seems possible that marked latitudinal zonality could be a specific characteristic of Pleistocene and Recent times. Badgley (1965) suggests that the North Pole migrated into the Arctic in Pleistocene times, isolating the Arctic from other oceans. At the same time the South Pole migrated from the free circulation of southern oceans to the Antarctic continent. Both polar regions became sources of cold polar air to provide an unusually marked contrast with warm air from equatorial regions.

Several processes are multizonal. Outside permanently frozen areas, running water is a fundamental process in all climates. Even in arid areas of today, relict drainage patterns like the former wadi patterns in the Sahara, reflect this fact. Distinctive regimes too can be azonal. Intermittent flow characterizes semi-arid areas due to seasonal drought or periglacial areas, and the heavy sediment load associated with great fluctuations in discharge favours aggradation and the development of braided stream courses. On slopes with permafrost producing the same effect as hardened clay or solid rock in arid-region slopes, and with neither type protected by a permanent vegetation, sporadic or seasonal supplies of surface water produce sheetwash. Wind transportation is active throughout polar deserts, particularly where no desert pavement has developed to protect sandy material, and bedrock may become grooved and polished by wind action. Various processes combine to produce changes in volume. Cracks are functional in the development of some soil patterns in both cold and warm climates, and expansion leads to bumps as widely separated as the Icelandic thufurs and the Australian gilgai mounds. Even chemical weathering processes can have a certain uniformity. In the absence of water and strong evaporation salt incrustations and salt lakes occur in north Greenland as well as in hot arid deserts. Some vegetation characteristics which can have an important influence on landforming processes are multizonal. For instance, the predominance of roots in the tundra communities (70–80 per cent of the biomass) are like arid communities in this respect because in both environments conditions are so harsh at the surface that living matter is best able to maintain its activity in the soil.

As well as process effects, many distinctive landforms are reported from a broad span of latitude. In some cases it is uncertain whether these forms are the result of the same azonal processes or whether two contrasted sets of distinctive zonal processes can produce the same form; or whether they are essentially paleoforms, relics from when a given set of processes different from that of the present-day once prevailed. To avoid havoc in the interpretation of landforms, the possibility of different processes producing essentially similar forms cannot be overemphasized. L. C. King (1953) wrote of homologies in landform, H. Wilhelmy (1958) of convergence. The latter word appears often in the writings of J. Tricart and A. Cailleux. Some examples of homologies include microforms produced by rock weathering in all types of climate, as described by Wilhelmy. In unconsolidated material, also, similar forms appear, like the close similarity between polygons produced by desiccation in clay and those in frozen ground. On a larger scale there are the semi-arid and periglacial ramp-like lower hillslopes, and P. Birot (1966), in describing the dissection of the Serra do Mar escarpment in Brazil, notes that the valleys have some similarity in appearance to the U-shaped glacial valley.

From coastal studies some clarification of the climatic approach is possible, for it is clear that the general physics of the beach and shore with constant supply of water and energy has substantial uniformity in operation as well as in principle. Swales, cusps, spits, tombolos, and similar features are found in many latitudes. However, beaches in cold climates have additional features. Pitted beaches occur where buried ice melts, frost cracks some hundreds of metres long are common and stone circles and polygons may occur. Ice-pushed ridges also exist. Mechanical erosion by waves may be restricted by the lengthy periods when coastal fringes remain frozen as spray from surf, and swash freezes to form a crust of ice on the shore and lower parts of the cliffs. However, periglacial weathering of the cliffs may produce solifluction lobes which encroach on to the beaches. Boulder-strewn coastal flats are restricted to cool-temperate regions, like the St Lawrence estuary, where drift ice is an important agent of transport and deposition. Any area that has experienced frost-weathering or glaciation will tend to have relatively large quantities of coarse material on its shores. By contrast, hot arid lands, where runoff is limited and allogenic streams absent, have little inorganic material on the shores amid sand derived from coral fragments. Coral growth and beachrock development and similar biochemical phenomena are restricted to lower latitudes. As early explorers soon found, intertropical coasts are fringed with huge cordons of sand with swamps inland, due to the abundance of fine sediment in the rivers. Headlands are infrequent and coarse particles rare, due to the deep chemical decomposition of many rocks.

Although contrasts similar to those observed on coasts have yet to be established for erosional landforms inland, the precedent seems set that different latitudes do have some distinctive features. It is important to stress, however, that these may be relatively minor touches, and that an attempt to construct homogeneous belts of distinctive processes, let alone the assumed resultant differences in forms, would be unrealistic and misleading. However, bearing in mind that it is only by implication that some landform characteristics might, to a greater or lesser degree, be related to sets of processes implied under certain broad climatic headings, it is probably justifiable to sketch briefly the tendencies shown in the areal distribution of certain distinctive present-day processes.

If one neglects the thickness of ice and its implications, and disregards the distinction between ice that is or is not frozen on to the underlying bedrock, and the probability of contrasted reliefs beneath the ice, areas at present covered by ice are distinctive. Where areas in high latitudes are very dry, with perhaps 5–7 cms of precipitation or less, polar deserts may occur. With greater moisture available, the ice-free areas are dominated by periglacial environments which include very diverse combinations of processes, although larger scales in patterned ground are probably restricted to the more severe climates where permafrost exists. Large areas in Russia can be described as steppe where there is a sharply expressed moisture deficit, absence of forest vegetation, and chemical processes accumulate weathering products. For the humid temperate climates, W. M. Davis's definition is amply precise. He described the 'normal' climate as not too dry as to lead to intermittent drainage and warm enough for winter snows to disappear. In tropical regions with temperatures usually above 20°C, there is a fundamental distinction between seasonally humid areas and those with a savanna climate characterized by a marked dry season. The specification of a savanna or Mediterranean-climate environment is made more difficult by man's disturbance of the natural processes. The distinctiveness of hot arid areas is well known.

The formulation of more precise statements on morphoclimatic geomorphology, which are not inferences from climatic data, awaits the measurement of forms. While the existence of certain subtle differences in landform with latitude suggest that more traits may be discovered, it is equally well-established that some earlier claims for morphoclimatic subdivisions in geomorphology were over-optimistic, disintegrated on the irregularities of local conditions and configurations, and became confused because of convergences in both forms and processes. The realistic morphoclimatic approach, on the basis of evidence to hand, is almost synonymous with the phrase 'process geomorphology', but also denotes that it is process geomorphology studied in an environmental context.

5. Structure, process and stage

No phrase has been repeated more often by students of geomorphology than W. M. Davis's statement that landforms are a function of structure, process, and stage. Yet on how to interpret this dictum and which factor to emphasize, there is such a wide range of opinion that views are almost diametrically opposed, with in between a spectrum of personal or synthetically grouped opinions. Difficulties arise when one or two elements of the trilogy are systematically placed in a subordinate role. By emphasizing the importance of one element interpretations easily become self-strengthening as the investigator tends as a result to collect evidence relating to the element he has chosen to consider. It is possible to approach this problem in a detached way by manipulating Davis's simple formula. If the dominating influence of structure is assumed, discovered by investigation, or if a structurally controlled land surface is deliberately selected for study, one assumes, discovers, or selects a situation where landform is essentially a function of structure. Inevitably one therefore assumes, discovers, or selects situations which are essentially independent of process and stage. W. M. Davis's Cycle of Erosion has frequently been criticized on the grounds that it included little mention of structural complications or that his work included too many assumptions about processes. To some extent this inevitably follows from Davis's interest in stage and in his belief that this was the most important and useful of the three considerations. Similarly, the current emphasis on the study of erosional processes and their effects introduces the possibility of neglecting the importance of structure and stage.

There are two further considerations that might be added to structure process and stage. The first, scale, was not apparent in Davis's qualitative accounts, yet the relative importance of structure, process, or stage depends very much on whether the study area is a portion of a single slope or a subcontinent. The second is the antecedent conditions which in appearance may be far more diverse than the idealized flat erosional surface produced as the hypothetical end-point of an assumed previous cycle or the smooth veneer of sediments of the abruptly uplifted coastal plain. However, evidence is too scanty to restore easily past environments and land surfaces, like a prefaulting surface, and the advantages of oversimplified initial assumptions very tempting.

For a given scale and with some unsatisfactory assumptions made about antecedent conditions, discussion on striking a balance between structure, process, and stage is made easier for two reasons. The first, a purely technical point, is that the study of process has already been discussed. Secondly, whereas adherents to an historical approach advocate stage, and others emphasize process, promoters of the importance of structure are comparatively few. There are several possible reasons why neither the dynamic tectonic aspects of structure nor the static lithological ones are emphasized. Perhaps one reason is that the importance of structure is obvious for all to see. Tectonic alignments or lineaments and other tectonically and lithologically controlled elements of the earth's crust can be found in almost every square mile of the land surface. Yet here and there land form does not accord with structure, and it is these features, the departures from structural control, that intrigue the geomorphologist. Compared with the fascination of dynamic erosional processes or the challenge to abstract thinking posed by the reconstruction of stages, the interpretation of a structurally controlled landform might seem dull, self-evident, and dismissible in a couple of words. Areas like the Canadian Shield, where the surface has been lowered very little and where the pattern and appearance of bedrock landforms have remained unchanged at least since Ordovician time, do not attract much geomorphological interest. Perhaps the geomorphologist's awareness of the significance of structure in landform studies is diminished in proportion to his readiness to define areas like the Canadian Shield as outside the limits of his studies. Another possible reason for underestimating structure is that ever since C. Lyell (1833) showed that rivers could erode valleys, any suggestion of the importance of structures in valley development appears to be a reversion to outdated ideas. E. S. Hills (1961) on observing the control of stream courses in Australia by faults and recent warps and folds, realized that, in suggesting that these rivers were determined by crustal disturbances, he might appear to be taking a retrograde step. A third factor is that many of the geographers who comprise the bulk of the followers of landform study, have neither the specialized expertise nor the facilities to examine the strength or weakness of structures on a macro-scale nor to examine microscopically the detailed constitution of rocks. Instead there is a tendency to infer these characteristics from the form of the land, although there are some exceptions, like the work in P. Birot's laboratory in Paris.

When the difference between two opposed views appears difficult to reconcile, it is wise to suspect the importance of more than two factors. Perhaps some of the differences of opinion engendered by the promotion of process or of stage studies arise because both approaches regard structure as a secondary or even temporary influence to be finally erased by process or by time.

While the idea of two coherent schools of thought would be quite misleading, there are two approaches in considering the inter-relationships between landforms, processes, and time.

Some theorists believe that landforms develop in stages and they may follow an historical approach in the study of the sequence of these events. The classic example is Davis's Cycle of Erosion. By considering an end-form that could take such lengths as those of the Cretaceous or Tertiary eras to achieve, and by beginning with an idealized flat surface, this view inevitably exaggerates the significance of changes in landform through time. Also the reconstruction of former stages of peneplanation involves a theoretical ambiguity in believing both in the efficiency of levelling processes and the inter-dependence of all parts of the advancing Cycle, yet also in the survival of relict erosion surfaces unadjusted to the new Cycle. The assumption of abrupt change in base-level is the only way to avoid this ambiguity. The approach by emphasizing stages in erosional levelling has led to the neglect of vital structural considerations like the significance of local base-levels where hard rock outcrops across a valley floor, or to the construction of chronologies with limestones making up the hills on which key levels have been identified. A conservative estimate of the lowering by surface solution of such hills would be 30 m in a million years. L. C. King's pediplanation hypothesis is also essentially cyclic but is distinctive in the assumed semi-independence of the component forms of the landscape. It is easier to comprehend his suggestion of the retreat of scarps towards a high central area where the oldest surfaces remain theoretically unintegrated with the advancing lower surface.

Workers who study processes as a major interest or those who deduce that processes will be of primary importance, although by approach less concerned with the possibility of changes in form with time or convinced by the great difficulties in doing so, usually allow for such changes in their interpretations. Differences of opinion exist among process students on whether changes in form with time will necessarily lead to greater interdependence of forms in a landscape. Some leading American workers who have measured form and processes, like S. A. Schumm and M. A. Melton, believe that this may be likely. Some European climatic geomorphologists, however, regard the fundamental inter-relationship to be between climate and form, with the main changes in form reflecting adjustments to the last major climatic change. Both J. Tricart and P. Birot have observed situations where hillslope and channels are evolving with little reference one to the other. P. Birot observes that the humid tropics are distinguished by decomposition of rocks at rates more rapid than transportation on slopes, and that transportation on slopes is more rapid than linear erosion.

In addition to students of stage or of process there is a small school of thought which has attempted to develop an analogy between landscapes and physical systems. Although this group is very small and the terminology and concepts unfamiliar to most geomorphologists, their main conclusion attempts to force itself to the attention of all concerned with landform interpretation. For present purposes it is difficult to decide not to agree with many workers who regard this conclusion as sensational rather than sensible, and dismiss it as irrelevant, but instead to risk continuing needless controversy by examining the statement in detail possibly disproportionate to its significance. The conclusion is that landforms do not show appreciable changes during time, are stage-less or time-independent. In part this suggestion attempts deliberately to expose weaknesses in aspects of the Davisian approach which emphasize a sequential change in landform during time. Two separate lines of thought arrive at the conclusion of time-independent landforms. The first is the revival of the concept of dynamic equilibrium introduced from physics to landform study by G. K. Gilbert in the late nineteenth century. Although the recent revival owes much to J. T. Hack (1960) and a desire to seek for an alternative to the Cycle of Erosion as a fundamental postulate in landform study, it is confusing to find the concept of dynamic equilibrium, which supposes that there is an approximate balance between work done and the imposed load, reflected in the concept of grade in the Davisian Cycle itself. Also the inflows and outflows mentioned by Davis are not measured by the contemporary theorists who assume their approximate balance from the appearance of the forms. In fact where attempts to measure the balance have been made, the results of workers like A. Cailleux (1948), who studied rillwash in the Dourdan area, show lack of balance. A. Cailleux found present-day processes too ineffective to account for the present relief form, which he concluded must therefore be a relict feature. However, in the ridge and ravine terrain which may appear to suggest quasi-equilibrium between stream lowering and the erosion of slope and divide, there is no room for relict features such as occur on the broad interfluves of areas like the Ozarks evolving semi-independently of stream channel activity. A further point which is difficult to follow is that while the assumptions of base-level stillstand in the Cycle of Erosion may now appear unrealistic, the assumption that certain landform properties remain essentially constant over long periods of down-wearing appears to ignore the problem of base-level altogether. In fact measurements of sediment accumulated offshore from the Appalachian ridge and ravine area show that recent rates are about one-eighth that of post-Triassic times (Menard, 1961), demonstrating a substantial change in the balance between process and form. These measurements which fail to support the analogy between ridge and ravine topography and physical dynamic equilibrium could, in fact, be explained by the opposite effect, that of reduction in relief. The suggestion that landform properties and processes have been relatively unchanging may, however, be valid. In the 1930s both G. H. Ashley and N. M. Fenneman described 'non-cyclic' erosion causing the straight, horizontal Appalachian crests to be lowered hundreds of metres yet looking the same afterwards as before. Their interpretations were based on the structural control of landform development due to the parallel belts of rocks with differing resistances to erosion, but were unpopular at a time when most investigators were searching for constant altitudes as evidence for former stages of peneplanation. If landform is essentially a function of structure, it follows from Davis's equation that landforms will be essentially independent of process and stage.

The visual impression of balance of unchanging process and time-independence that one gains in the Appalachians indeed suggest an analogy with dynamic equilibrium. This does not alter the fact that the *explanation* of these phenomena is probably structural control. It is perhaps significant that

J. T. Hack, in studying areas of domed topography in crystalline rocks in the southern Appalachians, attributes these distinctive forms to expansion by sheeting, a structural control.

A second line of investigation has led to the conclusion that landforms may be time-independent. This is the study of drainage basin networks in homogeneous rocks which has revealed very close correlations between parameters like stream numbers, lengths, and the frequency of confluences between tributaries. These same correlations have been observed in many areas and have strengthened the belief that Horton's 'laws' do describe unvarying properties of drainage basins, the development of which is therefore independent of changes with time, including climatic changes. By visual inspection alone of the distinctive even texture and delicate pattern of drainage networks on homogeneous rocks it is possible to appreciate intuitively the invariable geometric pattern which the statistical analysis reveals. However, the invariability of pattern, although a striking and important geomorphological feature, does not demonstrate that landform evolution is time-independent. Although the homogeneous nature of the rock, which in most Hortonian studies is non-cohesive or weakly resistant, may appear to make the pattern independent of structure too, in fact the very lack of structural or lithological guidance to drainage in a homogeneous rock introduces a distinctive type of structural influence. This influence is that which permits surface runoff to drain downslope in directions determined purely by chance. The stream parameters and their inter-relationships are therefore determined by statistical laws of averages, chance and probability. It is these which provide the regularity described by Horton's 'laws', and it is their strength which shows through, despite the poor quality of the basic data, not some intrinsic and general property of landform development. The pattern persists perhaps because the random lines along which drainage first became established when a rock is first exposed provided the antecedent condition of reduced resistance to the transport of material in these directions in subsequent phases. Thus the regularity of drainage networks, although apparently demonstrating a lack of structural 'control', is none the less a function of structure, due to homogeneous materials permitting runoff patterns to develop according to laws of chance. From their being a function of structure it follows from Davis's equation that in such circumstances, process and stage have an unchanging influence.

It seems that provided the importance of scale and the problem of reconstructing antecedent conditions, are appreciated, the dictum on structure, process, and stage provides the fundamental basis of geomorphology. It also seems that assumptions about the dominance of any one or even none of the three have perhaps been made too hastily in the past and that more emphasis on the careful study of all three elements in contrasted environments, without preconceived notions about the dominance of any one, might form a useful diversion from the continuing tradition of speculative discussion.

'The height of folly is to indulge in wishful thinking and fail to face reality.' J. B. Bossuet

6. The necessity for simplification of geomorphological complexity

With the many difficult and complex problems of landform study beyond the confines of artificial experiments, a degree of simplification is a necessary step towards some appreciation of the full realities of landforms. The necessity for simplification raises three problems. First, it means that no generalization can be dismissed outright, on the grounds that it is oversimplified, without a careful re-analysis of, and probably also with additions to the evidence. Because the necessary evidence might be drawn from widespread areas, a feature of geomorphology is the long persistence of innumerable simplifications whose sole support is subtlety of phrase or plausibility. For instance, W. H. Hobb's 1904 model of four cirques encroaching on an upland area from four points of the compass is still a textbook favourite, despite the fact, established in half a dozen scattered areas, that cirques tend to concentrate on one side of an upland. Secondly, distinction between a useful simplification and abstract speculation is partly subjective in a scientific inquiry, and the simplification which, for practical purposes, allows the engineer to build a technically adequate and financially viable structure is likely to be far too generalized for purposes of scientific investigations.

Thirdly, there is conflict of aims in a community of scholars between interpretation and explanation, with simplifications to aid inquiry and the extension of knowledge on the one hand, and simplifications to make this knowledge comprehensible to students on the other. The dual role clearly does not belong to the simplification, yet geomorphology contains many examples of confusion on this point, particularly the widespread use of the Cycle of Erosion concept designed explicitly for students, as a research tool. There are many ways in which geographical knowledge can lead to a simplification of landform studies. One is to use information from the classic area where a given phenomenon is particularly well displayed. For instance information of global significance on the inter-relationships between tectonic movement and relief, and their influence on erosion and deposition might be derived from detailed studies in California, where contemporary movements are active. Iceland, apart from being one of the few subaerial parts of a Mid-oceanic ridge, is also the classical land of outwash plains and also for waterfalls of very varied origins. Possibly the world's greatest concentration of pingos lies in the Mackenzie delta area and Spitzbergen may be the classical land of the soil polygon, but it is in Poland that Quaternary periglacial features are perhaps best known. This last illustration exemplifies the difference between the classic area and the type area, the latter taking on some of its significance from the intensity of existing detailed investigations in an area often where certain features were first recognized. With changing interpretations, however, the features of some type areas may become of historic interest only. In 1947 Western Australia was thought to be a natural eustatic gauge on account of its structural stability. It is no longer useful as a type area in this respect because there is now evidence to suggest late Tertiary epeirogenic warping and major folding movements in the Pleistocene.

In some situations there is a natural control which eliminates a variable. The best example of this is a comparison between lake and coastal shores. The absence of tides in the former, and the smaller size of waves means that terraces form better and deltas are simpler. The shore zonation has not yet the complexity of tidal areas. In the Nile, unlike most streams, there are no complications to downstream progressions because little water and sediment enters the river during its long traverse of the desert area. Most caves afford examples of stream environments unaffected by frost.

Complementary to natural controls like these is evidence accumulating from several areas where the same factor or feature is present or absent. For instance, meandering currents in the Gulf Stream and on ice are very similar in appearance to river meanders, the significant factor being that ice or ocean currents exhibit meanders in the absence of sediment load. The importance of an abrasive in the development of stream meanders might therefore be minimized. As the example of the Gulf Stream meanders show, there is the likelihood of comparative evidence not just from other areas but also from other contact surfaces. It was suggested a long time ago that the ocean floor would contain forms produced entirely by constructional and tectonic agencies, free from the complications of sub-aerial erosion. For example, beneath the Arctic ice, the Lomonosov Ridge may be a unique great fold in the earth's crust preserved in almost pristine condition. Space scientists would be quick to point out that forms on the lunar surface have developed in the absence of water and atmosphere, and under the influence of a small gravitational pull, and also of their intention of obtaining information from contact surfaces farther afield. Certain rocks introduce important simplifications into the study of landforms. Limestones, well known for subterranean forms some of which are known in other situations like permafrost, peat, or glacier ice, and described as pseudo-karst, afford an ideal subject for studies of solutional processes due to their soluble, monomineralic composition. In tropical areas, limestones offer the best opportunities for studying the recession of cliffed coasts. A conglomerate offers an unusual opportunity for the study of rock-weathering as all conditions are constant other than the parent material provided by the individual fragments. Areas of unconsolidated sediments are particularly sought after by investigators who wish to apply statistical methods. Without the necessity for a preliminary weathering phase, water as a transporting agent has sufficient energy to remove unconsolidated material and the statistical regularity of integrated erosional topography develops comparatively quickly.

As climatic geomorphologists would point out, studies within some climatic types introduce a certain control. For instance, arid areas in high and low latitude might be devoid of vegetation. This certainly was an advantage of barren dunes in the Sahara desert plains compared with coastal dunes for R. A. Bagnold's (1941) study of wind action on sand. The comparing of characteristics in different areas is difficult due to the inescapable problem that changes other than those in the object being studied are likely to occur as well. However, while geographical comparisons may be very difficult to interpret, the examples quoted above give ample illustration of why the development of geomorphology has been closely linked with geography, as it is by geographical search that the natural situation which affords the only realistic controls on a landform problem will be discovered.

Another problem of simplification of particular concern to regional studies is that of the antithesis between the uniqueness of the individual case and the global generalization. The increased use of statistical averages of a large number of observations of small individual items leads away from the first extreme. There is also a trend away from the other extreme of over-optimistic hopes for universal generalizations. As long ago as 1930 W. M. Davis said, 'On shifting residence from one side of the continent to the other, a geologist must learn his alphabet over again in an order appropriate to his new surroundings.' Although Davis may have remained a determinist to the last, it is increasingly appreciated today that no single theory can explain all features and that each area is better surveyed and studied individually and the evidence evaluated, at least initially, for that area only.

Amid the accumulation of information is the need to achieve some simplicity by structuring diverse facts into a coherent system. Classifications are seldom realistic owing to the importance of local conditions. Instead theoretical systems or conceptual models have been favoured in landform study as repositories for facts which offer at the same time an interpretation of the facts. W. M. Davis's was a major achievement in structuring much of the thinking and fact-finding of the nineteenth century into his normal Cycle of Erosion. However, he quickly realized that the scheme lacked universal generality and devised specific cycles for areas with some 'non-normal' individuality, like arid lands. In recent years, while the awareness of the impossibility of making universal generalizations has grown, the acceleration of the accumulation of facts makes the need for systematizing all the more keenly felt.

This problem applies to knowledge in its entirety, and workers in special embryo study groups devote increasing attention to the study of information in a search for general principles about the patterns of knowledge. A. N. Strahler was the first to apply some of the initial ideas of systems theory to geomorphology, and interest has grown. For instance, R. J. Chorley has done much to familiarize geomorphologists in Britain with the approach of Strahler and other workers in the United States (Chorley and Haggett, 1967). However, occasionally the use of contemporary systems theory jargon, instead of its concepts being inconspicuously incorporated into methods of

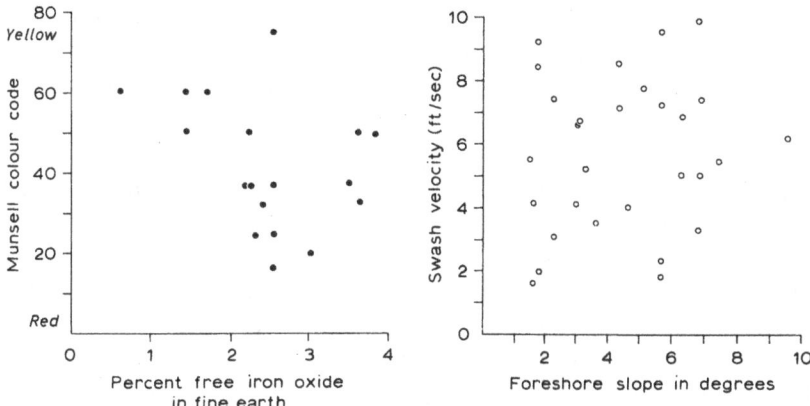

Figures I.11 and I.12 Examples of relationships for which *a priori* reasoning suggests a correlation but which measurements do not substantiate. (*Left*) Relationship of free iron oxide to Munsell colour code (YR hue × value) in *B* horizon of six soil profiles (*after J. M. Soileau and R. J. McCracken, 1967*, Soil Sci. Soc. Am. Proc., *Vol. 31*). (*Right*) Apparent independence of swash velocity and foreshore slope (*Ingle, 1966*). The wide scatter is in part the result of variation in wave heights and periods.

approach and interpretation, have complicated rather than simplified some issues. There are other ways in which attempts to simplify geomorphology by considering systems analysis approaches complicate its study considerably, because frequently in geomorphology one does not know all the factors involved nor are the regional boundaries which limit their extents easy to define. Also, as figs. I.11 and 12 show, correlations which seem reasonable to assume may not in fact exist or be demonstrable, if actually measured. Thus with incomplete knowledge and with ill-defined boundaries it is difficult to construct a meaningful model by theoretical reasoning. In addition the models of systems developed from biological and social sciences data differ fundamentally from technical models in engineering systems. It is difficult to avoid confusion in geomorphology where the former appeal to those who still retain the analogy between landforms and organisms, while the latter appeal to those whose approach is based on the principles of physics. However, a specialized interest termed General Systems Theory represents an attempt to search for features common to all kinds of systems whether organic or mechanical and faces the challenge of devising a robot with an electronic mind of its own. R. J. Chorley (1962) has explored analogies between geomorphology and General Systems Theory, but since the latter has so far failed to further its unifying mission, more detailed investigations of the analogy are held up.

Although many may consider that complication, not simplification, has accompanied systems analysis in its introduction into a few restricted aspects of landform study, and that it is of interest to only a small group of theorists, some discussion of this trend must be attempted.

Within a so-called 'open system', equilibrium can be achieved if arrival of one material equals the rate of escape of another. For instance, as described by W. M. Davis, the outflow of waste material from a hollow on a slope may balance that of inflow. Where this state of balance or equilibrium can be demonstrated a 'steady state' could be recognized. Over a hillslope as a whole the hypothetical situation is much more complicated. Rates of departure of material from a free face or underlying bedrock, or over a slope or from the base of the slope into a stream may vary and these changes in turn influence the way in which the other interactions occur. A situation of continually shifting or 'dynamic equilibrium' is then recognized. For many workers these simplified situations are too hypothetical in view of the known changes in erosional processes, climate, and base-level which were particularly frequent in Quaternary times, with human activities adding further profound complications in recent times. With these frequent changes where sets of forces are opposed in geomorphology one set is almost invariably stronger than the other, so that the result is evolution of forms, at least for a time, not equilibrium. Even in the highly dynamic beach environment where easily reworked unconsolidated materials are involved, inherent instability in wave and sediment parameters makes only a modified form of steady state conditions possible. On some beaches, even in the short span of recent years, the decrease of sand supply in relation to amounts removed down submarine canyons is appreciable. In fact where budgets have actually been measured in landform study, similar disequilibria are often found. It seems that without actual study of the dynamics of a process it is impossible to infer the nature of a system by reasoning alone. Laboratory measurements of soil-water dynamics tend to support this conclusion, where a series of moisture contents were established for soil samples for a range of pressures with a valve either open to admit an inflow to balance outflow, or closed. For either system the results were the same at the same pressures (fig. I.13). It would be impossible therefore to comprehend from properties alone, in this case soil moisture values, the nature of the system of inter-relationships involved in the production of these properties.

Among some workers who may regard equilibrium concepts as unrealistic, there is still recognition of some concepts defined in systems analysis. These are the so-called 'feedback' mechanisms which describe modes of change and include situations where processes are, up to a point, self-enhancing (positive feedback) or self-regulating (negative feedback). One of the first examples with which British geomorphologists became familiar was P. Pinchemel's concept of 'auto-assechement' in chalk dry valleys. An example of a self-enhancing process is the drying which produces contraction fissures in clays

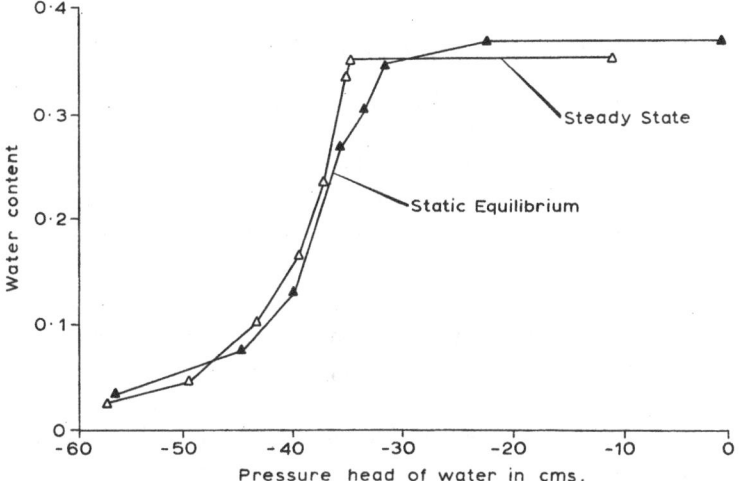

Figure I.13 Water content–pressure head drying data (*after G. C. Topp* et al., *1967*, Soil Sci. Soc. Amer. Proc., *Vol. 31*), showing the close agreement between the static equilibrium and steady state cases in the amounts of water retained.

and soils. Once produced a fissure promotes the drying and aeration of deeper horizons and the fissure deepens. Eventually a complicated set of self-arresting processes limit the depth of the fissure, and although one might then expect some quasi-equilibrium to develop between these two sets of processes, the dry spell might well have ended before this condition was approached. A good example of a self-arresting process is that which limits the heights of sand-dunes, because as these become higher the greater the contrast becomes between wind speeds on the dunes and over the troughs. Therefore the higher the dune, the greater the tendency for sand to be carried over the dune crest and deposited in the troughs.

The spans of time taken for erosion of consolidated materials and the frequency of changes in processes are infinitely greater than in unconsolidated materials. It may be more realistic to consider how feedback mechanisms are involved in adjustments to disequilibrium conditions introduced by the latest changes rather than to generalize about approaches to hypothetical equilibrium conditions. The concepts of systems analysis may prove of some use in simplifying the study of landforms, but so far difficulties have arisen due to the application of systems to landforms rather than by the investigation and measurement of the nature of landforms and process, to establish first which systems might be of greatest intrinsic importance. In consequence, the geographical approach of scouring the land surface for naturally controlled situations is likely to retain its strong appeal in landform study as the most realistic means of simplifying its infinite complexity.

D. Some Conclusions about Geomorphology

After considering the nature and basic postulates of geomorphology certain conclusions may be drawn which might to some extent reflect the aptitudes, training, and preferences of the individual. It is hoped, however, that those suggested below if not universal truths are a reasonably fair assessment, as they to some extent form the basis of the subsequent sections of this book.

The first conclusion is that the geomorphology, as the description and interpretation of landforms is an important study because it focuses attention on a factor which has an influence on everyday affairs. Forms like those of the multiplying branches of delta distributaries or of cols and gaps through ridges and mountains have had an indefinable but real influence in the courses that human history has taken.

Secondly, a more immediate and specific point about present-day geomorphology is that the closing up of border zones between academic subdivisions of knowledge has been much to its advantage, and the increasing awareness of the complexity of landform study has necessarily involved the use of information from distant fields. Team projects combining the expertise of more than one training have appeared, like the work which demonstrated that the Norfolk Broads were man-dug medieval peat cuttings (Jennings, 1960). Another example is the attempt to summarize and evaluate the recent volcanic history of Mount Rainier, which included the searching of records of libraries and local history societies, the stratigraphical study of ash falls, and the tree-ring dating of forests buried beneath flood debris.

Although in an inter-disciplinary sense geomorphology has made considerable advances, in another sense a third feature is the contraction in both time and space of the problems studied. The size of study areas diminishes so that measurements, often requiring a great deal of time, can be concentrated in detail and made frequently if necessary. Also the increased recognition of difficulties in correlation and of the difficulty of identifying homogeneous regions or systems leads to a contraction of the area to which conclusions are applied. More conspicuous has been the contraction in the time-spans which interest the geomorphologist. At a geological extreme workers like L. C. King remain with their centre of interest focused on past geological eras and for whom Quaternary landscapes are 'a multitude of microforms and too detailed and recent for lengthy discussions where the aim of the work is to demonstrate the surface history of the globe in geological terms'. The decreasing influence with time of past actions of the environment upon the land surface means that the remoter the past, the less may be inferred about past conditions and external variables. It is natural that knowledge of the last glaciation should be most detailed. Many British geomorphologists who, in the 1950s, were searching for remnants of Tertiary erosion surfaces, now specialize in the study of late Quaternary phenomena. The contraction in time-span is partly a function of the smaller size of areas under review, whereas the broad lineaments dating back into geological time are evident only in the review of larger areas. Also the increase in the study of contemporary processes and the mapping of forms is the description of the present day and a mode of thinking usually distinct from the geological view of a chronological sequence of successive landscapes.

A fourth conclusion is that while there has been a shift of emphasis in both space and time in the objects and problems studied in geomorphology, the subject probably has changed less than some advocates for change would claim. One often reads of how nearly all the old masters of landform study could, in advanced years, walk farther and faster than their students. Yet, with a possible subtle reflection of the degree of change, it was said that J. P. Miller, a leading pioneer in the use of quantitative techniques and measurements, '. . . could lift the largest rock, and dig a hole with a shovel faster and deeper than anyone in the party'. For many teachers an aesthetic appreciation of landform and an enjoyment of outdoor exercise have provided an incentive for inquiry into geomorphology, and for their students these factors are a ready source of inspiration. Recent trends in geomorphology have done less to change this aspect of the study than some commentators imagine. In fact, the growing significance attached to the systematic measurements in the field heightens the importance of contributions that can be made by those who would enjoy walking long distances to make measurements in inaccessible areas. In the past decade clubs have been founded in Sweden by interested laymen to assume responsibility for certain measurements on glaciers. For those to whom such pilgrimages would offer little pleasure, there are sound arguments for justifying a laboratory or chair-bound existence. Some of the urgent tasks in the last category also involve an artistic ability. Geomorphology in its scientific advance remains dependent on many qualities like intuition and imagination, on literary and linguistic skills or on a sense of time, which might be superficially described as non-scientific.

A fifth conclusion relates to those geomorphologists who utilize mathematics and advanced scientific techniques to solve problems of landform interpretation, rather than as an aid in a mission to infiltrate geomorphology, with mathematics, statistics, and scientific techniques, and are free to take a realistic view of the subject. They point to the limitations to an advance in the use of quantitative and scientific techniques and therefore from another source comes the impression that the nature of landform study has changed less than some believe or are led to believe. For instance, a leading exponent of quantitative techniques, W. C. Krumbein, asserts that although quantification is advancing steadily in some sub-fields of geology, the subject is basically a qualitative science and that geologists are most concerned with observational data. A. Jopling who has specialized in sophisticated studies of sediment transport in flumes describes as the greatest barrier of all, the fragmentary and incomplete record available to the field geologist. M. A. Melton (1958) who was one of the first to attempt the correlation of numerous geomorphic properties of soil, slope, and hydrological parameters in drainage basins, and to discuss positive and negative feedback mechanisms in this context, suggests that to argue that all the variability in any natural environment could ever be entirely explained is absurd. He also stresses the difference ignored or forgotten by some, between geology and physics or other exact sciences in the kind of variability encountered as well as the amount. The fifth conclusion, therefore, is that statements exaggerating the scientific and quantitative elements in geomorphology are sometimes traceable to lack of first-hand experience with the actual nature of problems in landform study.

A sixth point relates to W. M. Davis and how the position he retains is a matter of controversy which absorbs a great deal of introspective thinking in geomorphology. The energy put into this debate is perhaps misplaced. In one sense W. M. Davis's hypothesis of the Cycle of Erosion came at a comparatively early stage in the development of geomorphology. At that time there was very little accurate knowledge available about stream discharge data. Petroleum lay undiscovered and the piston engine had only just been invented, so that aerial views were unknown. Studies of the interiors of the great deserts awaited the motor transport developments in the 1930s. It was only in the 1930s, following the development of innovations like X-ray diffraction techniques that the general character of clay minerals was worked out. Despite the progress which geologists had made in the study of the earth's history by the 1890s it still remained a history without years, apart from the quite misleading predictions based on Lord Kelvin's physical models, until 1912, when de Geer's study of varves proved to be a first step towards establishing a scale of actual times. It was less than a generation after the idea of sub-aerial denudation had become widely accepted that Davis formulated his cycle of erosion hypothesis. More than two generations later we can see that the apparent simplicity of his scheme is more likely a reflection of limited knowledge at that stage than that Davis's brilliance could conceive of a scheme which would anticipate the facts still to be discovered. It seems, on the basis of an increasingly large body of evidence, no disservice to W. M. Davis to say that his simplified teaching model is out-of-date, but also too easy to criticize his work simply because of the time at which he was writing.

A seventh point relates to perhaps one of the most confusing aspects of present-day geomorphology. This is the parallel that exists between the weaknesses of Davis's approach to geomorphology and that of workers who have committed themselves to 'quantifying' the study, thereby to oust 'Davisian' geomorphology. The first parallel is that of oversimplification. If it is, as J. Dylik (1953) suggested, that one of Davis's greatest errors was excessive generalization, it is equally true that any attempt to reduce landform problems to equations demands extravagant simplifications. A second parallel is in postulating unchanging conditions. Davis postulated a protracted period with little change in base-level; J. T. Hack's concept of 'dynamic equilibrium' makes no reference to base-levels. It is said that Davis paid little attention to changes in climate or to changes in weathering processes in different climatic zones; these are scarcely mentioned in many studies committed to a quantitative approach. A third and close parallel is in the element of abstract reasoning in both approaches. Although W. M. Davis studiously avoided mathematics in the presentation of his ideas, his speculative way of thinking, more drawn to abstract reasoning than to the establishment of the facts was, according to J. Tricart, essentially that of the mathematical reasoner. S. A. Schumm also notes how the quantitative theoretical models of slope development are open to the same criticism as those of both Davis and Penck. Without data upon which the assumptions of a theoretical approach should be based, theoretical models are little more than speculations, whether elaborated by mathematics or by sonorous prose. Tricart advises that mathematical reasoning must not become a convenient means of camouflaging our ignorances as this would immediately lead to restarting on the laborious road of Davisian mistakes. A fourth parallel is widespread; it relates to Davis's view of youth, maturity, and old age which can be criticized because the analogy between organisms and landforms, understandable in the immediately post-Darwin age, is seen from greater distance to be misleading. The stages he defined may now be largely discarded yet the analogy between organisms and landforms lives on. L. B. Leopold, M. G. Wolman and J. P. Miller (1964) suggest: 'A river basin is like an organic form, the product of a continuous evolutionary line through time.' Fundamental to Cailleux and

Tricart's climatic geomorphology is the concept that geomorphology can be presented as an ecology of forms. The analogy is also to the fore in the 'open system' approach, as this concept from the biological systems point of view depends on the belief that living systems have to be analysed as systems that are 'open' to matter-energy exchanges with the environment. Davis saw that the systematic interdependence of land and water forms was so perfect 'that one must grudge the monopoly of the term "organism" for plants and animals, to the exclusion of well-organized forms of land and water'. It seems that several present-day workers still experience similar feelings, particularly when confronted with the well-organized electronic circuits simulating organisms, but that geomorphologists might as well accept, perhaps with a lump in their throats, that landforms are not alive. A fifth parallel is between the equilibrium that some present-day workers postulate and the near-perfect systematic interdependence of land and water forms which Davis observed. A final point of similarity is that the two approaches are similar in the degree to which their opposed views on time in geomorphology are unrealistic. For many reasons committed quantifiers have to assume the relative unimportance of change with time. Historical events are usually highly distinctive and therefore cannot embody laws, defined as recurrent, repeatable relationships. A steady state condition is much easier to deal with quantitatively than the shifting inter-relationships between antecedent and subsequent conditions and the changes and sequences of forms produced. Thus the committed quantifiers nose towards the noose of circular argument.

The seventh conclusion is therefore that there are far more similarities between W. M. Davis's approach and that of present-day geomorphologists who are committed to the systematic application of a quantitative approach than the apparent difference created by the absence of mathematical symbols in Davis's writings. It seems that the term 'Davisian' can have no precise meaning in geomorphology since in style some of Davis's most persistent critics are the closest followers of his idiom.

An eighth conclusion relates to the unease voiced by some that geomorphology's development has been somewhat slow. If this is true, one of the main reasons for retardation in the past and quickening pace at present is the stress of economic demands. In the second half of the eighteenth century the study of geology was founded, as coal became the principal source of energy, under the leadership of William Smith, a civil engineer. The great leap forward for American geology began with the mining boom in the West. Geomorphological observations were an important aspect of pioneer geologists' work because the landforms often provided important clues in delineating geological boundaries on reconnaissance mapping exercises. A study of geomorphological processes was also important because explanations for unconformities in coal-bearing strata were needed to assist the mining engineer.

In Russia, geomorphology emerged with the founding of Soviet soil science, and this close interdependence has remained. However, as the hunt for the treasures of the earth gathered pace, with the discovery of oil in Oklahoma in 1913 as a major turning-point, more sophisticated techniques of field observation and laboratory analyses became increasingly important. The search for oil, which has absorbed a major portion of geologists' efforts since that time, has depended little on geomorphology and for a generation, with diminished manpower and financial support invested in directions from which geomorphological facts might flow, the growth of information depended largely on those concerned with education. Since 1945 applied work has expanded to specialisms in which landforms and land-forming processes are more directly involved. The stimulus of the lead of the engineer R. E. Horton is one example. Another is that Swedish geomorphologists have found that in recent years the development of hydro-electric power stations has been accompanied by greater prospects, from the financial point of view among others, of studying deltas. Chronic pressure on land and water resources has led to the intensification of land and water utilization studies in marginal areas. Directly from such studies, or indirectly from the communication systems which support them, geomorphologists acquire increasingly large stores of data on water flow, erosion, and deposition as priorities for tech-nological endeavour shift towards problems of conservation rather than of exploitation in the natural environment. Therefore, while it is necessary to avoid a confusion of emphasis by distinguishing the technologists aims from that of the scientist, it is reasonable to suggest that as an eighth conclusion the rate of progress in geomorphology is closely linked with economic demands.

As a final observation, it seems that many discussions about the nature of geomorphology centre around contrasted views where differences may be exaggerated, and the existence of more than two facets to a problem obscured in dualistic debates. Conversely, differences in technique or in initial assump-tions may overshadow a similarity in method of approach. The divergence between the views of W. Penck and W. M. Davis on the evolution of slopes, conceived as the subject for a special debate by American geographers in 1940, diminishes in importance with the growing realization that the fact that neither measured the actual forms or processes involved is a more fundamen-tal issue. Few landforms, if any, are understood with certainty. The lively discussion which therefore results between various theories has traditionally been highly valued, particularly for the educational purposes of training open minds. Particularly influential has been the stress which Davis laid on the importance of imagination in geomorphological investigation. There are over eighty published theories on the movement of ice from which to choose. On a smaller scale, practically every idea concerning beach cusps advanced by one author is directly contradicted by that of another. Opinions are greatly divided on the existence of a y general lower limit for blockfields. Examples like the ideas themselves are easy to multiply endlessly. In recent years, partly due to the assistance of workers in related disciplines, the value of such controversies has been questioned and formerly esteemed 'imaginative interpretations' pale with the cold light of factual evidence to 'sheer speculation', as the number of converts to the discipline of experimental method grows. For example, in a study in the Southern Uplands of Scotland soil scientists Ragg and Bibby (1966) note that it is possible to speculate whether erosion now dominant after an apparently long stable period, was the result of a modification in climate or of overgrazing and burning. They conclude that 'only observations of the future course of development and the acquisition of quantitative data on rates or erosion and deposition can confirm or refute such speculations'.

A final conclusion therefore is that while ideas are easy to come by in landform study, it is much more difficult to conceive of ideas that can be put to practical tests and to gain the conviction that the idea and tests are sufficiently sound to sustain the investigator through the arduous tasks of acquiring the relevant data. A. Rapp (1960) urged that 'we must go out to the slopes and make our measurements on them if we wish to evaluate their actual development in quantitative terms'. This sort of exhortation is commonplace in geomorphology; the remarkable aspect of Rapp's statement is that he actually went to carry out this task. Work like this reflects one of the positive trends in landform studies at the present time, as the drift away from speculation gathers pace. Speculations are not confined to the nature and development of landforms. It is also easy to speculate about the nature of geomorphology itself and on the course of its development past and future. It is hoped that the discussion in this first chapter may assist the reader in his evaluation of further reading in geomorphology, including the remaining chapters of this book. These are largely devoted to illustrating some of the actual work that has been done in the past decade to further our understanding of land forms.

Landforms and structure

A. Geophysical Considerations

1. Earth movements and mountain building

Most of the major units of the earth's surface owe a great deal to uplift of land over very extensive areas, achieved in many different ways during Tertiary times. The configuration of the broader relief features, like mountain ranges and plateaux, reflects the differences in folding and faulting, and the spatial relationships of broad areas of rocks of differing resistances produced by these recent geological events. Although geotectonics, the study of the broad structural lineaments of the earth's surface, falls outside the geomorphologist's scale of operations, and to some extent that of many geologists too, some brief familiarity with the work of specialists like seismologists and geodesists is important in the investigation of the smaller land-surface sizes which the geomorphologist studies. Apart from understanding rock dispositions which underlie landforms in structurally complicated areas, or of appreciating that rates of structural movement can exceed by many times those of erosion, one's approach to landform study can be influenced fundamentally by assumptions about structural behaviour.

Three main points concerning earth structure seem reasonably clear. The first relates to information about the earth's mantle, which stretches 2900 km from beneath the thin skin-like crust of the earth, half-way to its centre. The core itself is dense molten iron with nickel and silicon, but the mantle is now known to be very complex, as yet unobserved directly, and in which many of the factors influencing thermal convection currents are unknown. Due to the present inadequacies of knowledge about the mantle even experts' attempts to explain mountain building movements in terms of subcrustal processes are inevitably speculative. Therefore any detailed attempt to summarize these diverse views within the context of geomorphology would be if not irrelevant, certainly inappropriate.

A second point is that there is now general agreement on the permanency of ocean basins which differ in petrology, sedimentation, and structure from the continents, at the margins of which the granitic layer thins abruptly to nothing.

Thirdly, crustal thickness of continents is closely related to land-surface elevation. As has been appreciated for more than a century, mountains are not loads supported on a rigid substratum. Instead, the height of a mountain block tends to be compensated by a root, made up of the same lighter material which projects by as much as 70 km down into the underlying denser mantle

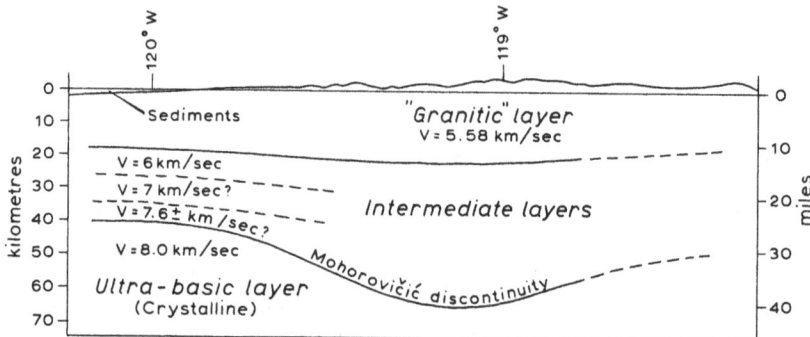

Figure II.1 Structural cross-section, southern California (*after Gutenberg and Richter, 1954*). The horizontal and vertical scales are equal.

(fig. II.1). On average this lighter crustal layer is 35 km thick, but shrinks to as little as 5 km beneath ocean deeps, sharply defined at its base by the M discontinuity, named after A. Mohorovičić who first identified its existence during studies in the Balkans in 1909. With changes in the distribution of weight of this sialic raft, the solid rock material of the mantle may yield to large stresses somewhat like a viscous fluid, as below approximately 80 km its strength is only one-hundredth of that near the surface. Under stable conditions the large portions of the crust come to project downward by an amount in proportion to their surface elevation. The term isostasy describes this ideal state of flotational equilibrium. The strength of the supporting mantle is regarded by some as due to elastic rather than viscous properties, but an important characteristic of these so-called rheological properties is that they vary in mode of response according to rates of change of strain as much as to the total strain which might eventually build up. Recent work shows that the properties of the upper mantle vary from place to place, and as its rheological properties are very sensitive to changes of temperature or in composition, isostatic balance is not wholly related to crustal thickness. The ideal condition is also probably only rarely attained because adjustments in weight may involve millions of years. Recovery by buoyancy may not operate if forces within the crust are sufficient to retain a block below its equilibrium position. Conversely, depression by sediment accumulation may be delayed until loads exceed the strength of crustal rocks. Appreciation of the nature of isostatic balance is fundamental to an understanding of landform evolution in areas on continental margins, particularly those in higher latitudes repeatedly depressed and re-elevated during Quaternary times. There is also the theoretical possibility of the erosion of highlands causing isostatic rise. The buoyant force is such that if erosion spread over millions of years removed 1000 m by downwearing the net lowering might be only 200 m due to the replacement from below by heavy rock as the denuded block was buoyed up. As yet there is no information to indicate whether this theoretical possibility is realized or not.

In the middle of the nineteenth century the conception of the geosyncline emerged, subsequently to be used in many different senses. In general terms, a geosyncline is a huge linear or elongate portion of the crust which after sinking below sea-level and becoming filled with sediments may become an orogenic belt. Usually a geosyncline is fringed by a foreland on the outer side and a mountainous continental hinterland from which the sediments are supplied. The mechanism involved in subsidence is not clear, but any suggestions as to its nature must take into account the crustal thinning or increase in density at the base of the crust that would be necessary to compensate for the lower density of the accumulating sediments. Suggestions about the motive power that leads subsequently to folding in the geosyncline are several and encompass a wide divergence of opinion. Since the discovery of radioactivity made the concept of a contracting mantle no longer tenable, nineteenth-century ideas about compressive forces generated in this way are possibly applicable on a local scale only. Increasing doubt is also cast on the classic tectogene view that Tertiary uplifts are due simply to isostatic recovery of deep sialic roots that had developed in geosynclines. Instead the energy for lateral compression may come from drifting continents piling up and picking up folded mountains, like the Andes, on their leading edge. Where continents pile up against one another and mount to great elevations, as in the north of India, compression proceeds at a rate of a few centimetres per year. The viscous drag of hypothetical convection currents might achieve a similar effect, a suggestion supported by indications from belts like the African rift valleys and the Red Sea that areas of thick crust may be split apart and that new crustal material may be created in the intervening gaps (fig. II.2). A hypothesis which combines these suggestions and other information acquired in recent decades is that of Heezen's (1960) sea-floor spreading. The suggestion is that new oceanic crust keeps appearing above a hot upward convection current in the mantle, and owes much to the discovery of remarkable parallel zones of remanent magnetism reflecting past reversals in the earth's magnetic field, with progressively earlier zones at increasing distances from the mid-oceanic rise. Their width suggests that the ocean floors might be replaced during a span of 200–300 million years. From this, and from many other lines of evidence, there is increasing support for a continental drift–convection current hypothesis of mountain building. In some situations it seems possible that upward movements in a geosyncline might be due to other causes. It is possible in some cases that granitization of the root zone leading to reduced specific gravity and increased volume causes the whole orogenic belt to rise as the batholith forms. However, there are many instances of

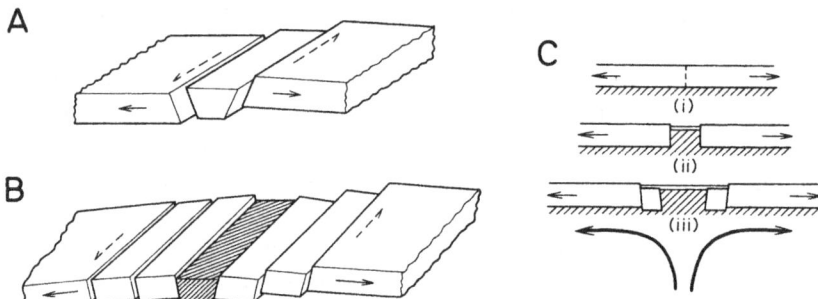

Figure II.2 The explanation of a rift system as an integral part of the theory of continental drift (*from R. W. Girdler, 1965*, Phil. Trans. Roy. Soc., Ser. A, Vol. 258).

A. The formation of a simple rift valley due to a relatively small amount of crustal extension; there may be considerable transcurrent movement.

B. The formation of the Red Sea structure, due to relatively large crustal extension.

C. A possible mechanism for creating the Red Sea structure and the separation of the crust. Rising mantle convection may induce tensional stress.

orogenic movements without a batholith at its core and others where in the association of orogeny and batholith, the orogeny may be the earlier event. By whatever mechanism an orogenic belt rises it is becoming increasingly clear that as it does, gravity movements limit the height to which the central lobe can rise as a mechanism, termed gravitational spreading, creates a flattening effect. The evolution of areas like the Alps are now examined in terms of such glides. Folds or nappes may develop where these gravity glide structures press against forelands. In this way the buttressing effect of the Vosges and the Black Forest massifs may have been important in the folding of the Jura mountains, and in turn this tangential pressure may have contributed to the elevation of these massifs. Gravity gliding may also be a factor in the tectonic 'denudation' of some areas and may also be involved in the initial sinking phase of the geosyncline.

Epeirogenic movements of vertical uplift which do not involve folding of structures, have produced basins and plateaux in many parts of the world with the results of more recent movements still clearly defined as structural relief. The rigid shield areas have experienced epeirogenic movements since pre-Cambrian time, and there is increasing doubt as to whether there is a clear separation in time of orogenic and epeirogenic movements. It seems possible that both types of movement are differing expressions of similar sub-crustal forces, differentiated according to the stability of the local tectonic framework. For example, in the Basin and Range province it appears that compression can take place in one area while tensional forces are dominant only a few miles away, while in other areas the pattern of late-Tertiary epeirogenic movements appears independent of the younger orogenic belts. Furthermore, it is now realized that motion along strike-slip faults driven by non-parallel sub-continental and sub-oceanic mantle flows, may cause repeated subsidence and elevation and could produce displaced blocks like the horst and graben of the Basin and Range province. There, in fact, strike-slip faulting is of late Tertiary and Recent age, and in earlier geological times, the Great Basin moved south-east in relation to the Sierra Nevada by 130–190 km.

From the point of view of landform development the contrast between the two variations of the effects of mountain building is quite striking. Compared with the clear vestiges if not prominent features of essentially vertical movements, any structurally identifiable units in areas of intense orogeny are unlikely, partly because of sub-aerial and tectonic denudation during mountain building and partly because many complicated folded structures may be formed at an early stage beneath the sea when the sediments were still being deposited.

Earthquakes associated both with mountain building and with vulcanicity are probably the result of deformation in the lithosphere which finally ruptures abruptly, releasing stored elastic energy. The earthquake epicentre is that point on the earth's surface vertically above the point, or earthquake focus, where energy of a given magnitude is released. Approximately 80 per cent of earthquake energy develops in the uppermost 60 km of the earth. When rocks are strained to breaking point and rupture, earthquakes may be triggered off by some exogenic force such as flood or high tide. Conversely, earthquakes trigger off catastrophic events in exogenic processes and are for this reason important in landform studies. One example is the 1870 rockslides triggered off on Mount Tacoma, when an area of about 32 ha fell away. A less disastrous result of the 1964 Alaskan earthquake was a persistent dust-cloud 600–1400 m high which developed and hung about a reactivated fault to the south of Lake Eklunta. A number of spectacular glacier surges occurred in Alaska after the 1899 earthquake. Tsunamis, or seismic sea waves, arrive on coasts, particularly in the Pacific, with great force. The effect can also be important in unconsolidated materials. There is the possibility of earthquake trigger mechanisms releasing tensile desiccation stress which produces fissure patterns on playa floors. Quick clays lose their strength by shocks and vibrations, assume a liquid state and begin to flow. Although there may be an order of a million earthquakes a year, and the total energy released a negligible fraction compared with incoming solar radiation, the main release of seismic energy is highly concentrated in a few great earthquakes which is clearly sufficient to trigger off profoundly significant changes in land-forming processes. It is difficult to say whether the artificially generated shocks of underground explosions which may equal major earthquakes in the amount of energy released will accelerate the importance of triggered exogenic movements as the possible effect of these added releases of energy may take some indefinable time before becoming apparent.

2. Continental drift

Continental drift has swung back into favour supported by palaeomagnetic and other evidence which within a decade has made the theory widely acceptable, and for some an 'increasing certainty'. The fit of continental shelves on either side of the Atlantic, remarkably exact at depths of 900 m (fig. II.3), and the symmetry of the mid-Atlantic ridge in relation to the two sides have also added to the stratigraphic evidence used in the hotly disputed arguments of F. B. Taylor, A. Wegener (1922) and A. du Toit (1937).

Geological evidence suggests that if large continental movements took place they have occurred comparatively recently, possibly with initial separation about 170 million years ago. In the Iceland area current present-day rates of drift might average about 2 cm/year, involve 1200 km of new sea floor since about the end of Cretaceous time and possibly 300 km within the last 4 million years. The Atlantic might have opened up in 50 million years. Strike slip faults tend to characterize lateral edges of drifting continents, that of the San Andreas fault moving 250 km in as many million years. The rates of movement observed are so great compared with those of other sub-crustal processes that, as has been seen, mountain building cannot be discussed without reference to continental drift. The mechanism involved may be divergent flow of convection currents made more effective by the downward projection of the continents. The rates involved are also orders in excess of landform change by sub-aerial processes. L. C. King has always maintained that land surfaces still in existence may antedate the present continents, and saw the initiation of the African cycle not as a result of uplift but due to parts of the dismembering Gondwana super-continent (fig. II.4) drifting away. He considered that the Gondwana landscape in Brazil sloped west and north-west

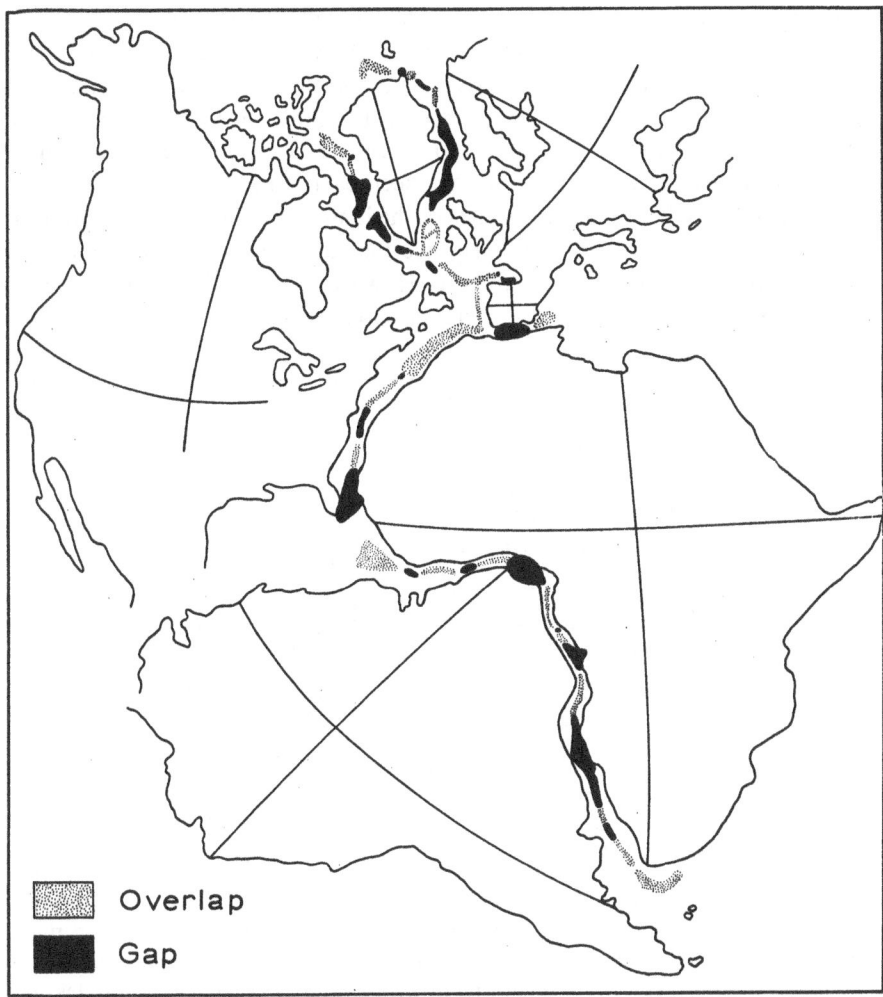

Overlap

Gap

Figure II.3 The fit of the continents bordering the Atlantic at the 900 m contour (*from E. Bullard* et al., Phil. Trans. Roy. Soc., Ser. A, *Vol. 258*). This computer-calculated best-fit reconstruction, drawn on a transverse Mercator projection, represents the possible situation prior to continental drift.

to the Proto-Pacific, and that the Amazon basin below 200 m widens west-ward right to the foothills of the Andes. In 1927 A. du Toit had pointed out that when Africa was joined to South America, the rivers must have flowed towards the Pacific, but the rise of the Andes across the outlets, due to continental drift, must have compelled the basins on the South American side to drain back towards the Atlantic. R. L. Sherlock (1933) was struck by the Amazon basin being for the most part only slightly above sea-level, and that while the present mouth had no real delta, the great extent of river deposits inland expanded westward, delta-like in outline. Clearly the geomorphologist studying larger features like extensive plains or major river valleys may now have to search for part of his evidence on the other side of the ocean. One looks at fig. II.3, and may muse about where the St Lawrence flowed, and if the Central plain of Ireland might have been part of its path.

The similarity between the outlines of the land on either side of the Atlantic is so striking that it was at an early age that one formerly learnt that their drifting apart was considered unlikely. However, for all the efforts to avoid the obvious conclusion and the heights of controversy that surrounded it,

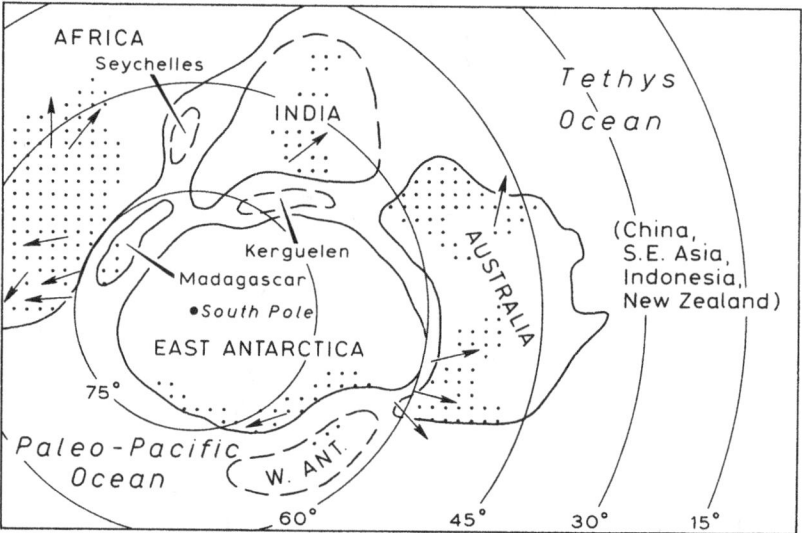

Figure II.4 Reconstruction of Gondwanaland at the beginning of Permian time (*from W. Hamilton and D. Krinsley*, Bull. geol. Soc. Am., *Vol. 78*). Tillite (stippled), ice-flow directions on glacial pavements (arrows), and paleolatitudes are shown.

continental drift is now a topic of sufficient stature to command many pages of the 1965 *Philosophical Transactions of the Royal Society*. As salutary as the recent findings are informative and incisive, are the implications of the chequered history of the theory itself.

3. Vulcanicity

Because the study of volcanic processes is a highly specialized aspect of geophysics, the geomorphologist tends to take the form of the erupted volcanic material as his starting-point or even to avoid the complication introduced by sub-crustal processes by studying non-volcanic areas. However, even if the study of sub-crustal processes properly belongs to other specialists, volcanic landscapes offer many unique opportunities for tackling general problems in landform study because the rate of ejection may be millions of times greater than increments from other depositional processes or rates of removal by erosional processes. In consequence processes of erosion may be presented with an initial approximately ideally shaped landform on which stages in the history of dissection may become well displayed. C. K. Wentworth (1927) took advantage of this type of situation in his studies of the amounts of erosion in Hawaii. Subsequently, the development of potassium-argon dating methods, being largely restricted to igneous rocks of geologically more recent ages, has made Tertiary and later extrusions invaluable horizons on an absolute time scale. Over shorter periods of time dates of more recent eruptions and extrusions are often adequately documented in historical records. The rapidity of weathering of some volcanic materials, although atypical in rate, is for this reason valuable in a study where the general slowness of most weathering rates is the source of many difficult problems. The value of volcanic materials as stratigraphic horizons is not just restricted to datable lava flows but includes volcanic ash deposits which may be carried hundreds of kilometres, like the material from the 1963 eruptions of the Bali volcanoes which travelled to east and central Java. In other areas volcanic dust may add scarcely noticed contributions to terrestrial sedimentation but even relatively minor amounts may influence the chemical composition of accumulating terrestrial clays. It is not yet clear whether such dust in the atmosphere could have existed in sufficient concentrations to induce climatic changes. Finally some brief familiarity with volcanic forms and processes is required because in some comparatively restricted parts of the earth's surface vulcanism dominates the present or the recent past. For instance, Mount Ararat (5170 m) is one of many huge Pleistocene volcanoes in north-east Turkey. In the Cascade Range extending from northern California to the Canadian border several volcanoes rose above the plateau surface during Pleistocene and Recent time, including eleven above 3000 m. In stable areas, apart from along and close to rift zones, volcanic landscapes are few. In the Soviet Union there is only the Kamchatka area which is volcanically active today. Vulcanism is also rare in fold mountains of the Alpine–Himalayan type, where superficial layers of sial are greatly thickened. In contrast, andesitic volcanoes, which tend to be explosive, are closely associated with island arcs and mountains made of recently folded geosynclinal sediments; most of the effusive basaltic material flows out from points on mid-Oceanic ridges like the Hawaiian and Icelandic areas. However, further generalizations are not helpful because some volcanoes may change their types of activity and contrasts between eruptions may be observed in adjacent volcanic areas. More than most landforms, the study of volcanoes tends to be that of the unique form and the unparalleled history.

The outpouring of lava probably originates from molten regions in the upper mantle which are quite small in comparison with the total volume of the mantle (fig. II.5.A). As volcanoes are associated with earthquake belts, it is possibly the heat released in the shearing of rocks which causes local pockets of melting. The proximal or immediate reservoirs of most central volcanoes,

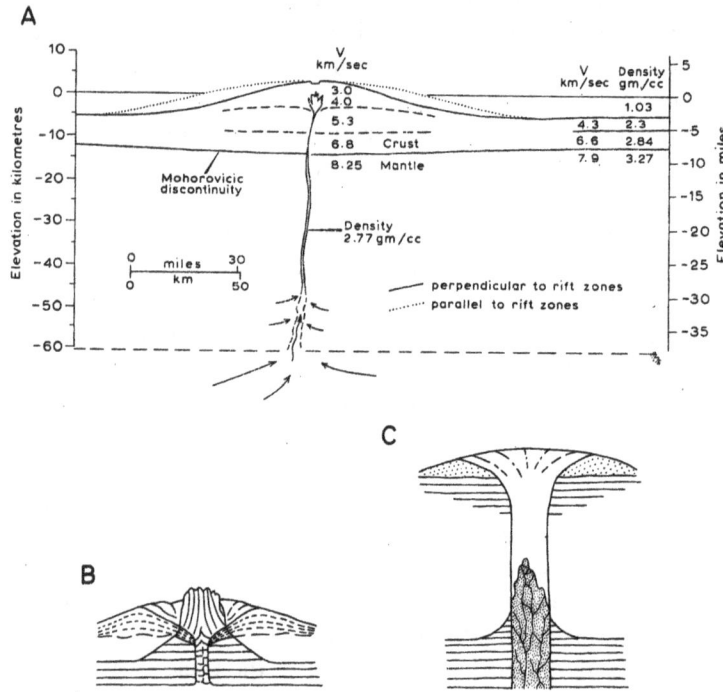

Figure II.5 Indications of the range of volcanoes as landforms.

A. An active Hawaiian volcano (*from Eaton and Murata, 1960*). Magma from a source some 60 km deep moves up to a shallow reservoir beneath the caldera. Occasional discharge of lava from the shallow reservoir through dikes that split to the surface constitute eruptions. Note the elongation of the volcano along rift zones and the slight depression of the M discontinuity beneath the volcano.

B. Common type of Hopi neck (*from H. Williams, 1936*, Bull. geol. Am., *Vol. 47*) with diverging columns of lava resting on inward-dipping tuffs. Possible original form of the now-eroded crater is suggested.

C. Common type of Navajo neck (*from Williams, 1936*). Dikes branch through the tuff-breccia shaft. Possible original explosion pit form at the surface, now eroded, is suggested.

replenished from depth may be relatively close to the surface. That of Vesuvius may be about 6·5 km below the surface. Of the forces driving magma upwards, hydrostatic pressure may be the most important because magma columns are lighter than solid rock.

If the ejected material consists entirely of fragments, known as pyroclastic material or tephra, coarser sizes will develop around a central vent to form a cinder cone, characterised by steep 30–35 degree slopes according to the angle of rest of the fragments. These cones may be up to 150 m in height. Composite cones, or strato volcanoes, result from alternating phases of explosive activity and outpouring of lava. The strata are partly tephra and partly igneous rock which may include intrusions as well as extrusive lavas.

Distinct from the size of normal craters which are usually less than 600 m in diameter, are the huge depressions of calderas, due to the collapse and subsidence of a volcanic cone. In Iceland a stratovolcano, about the size of the largest active volcano on the island, disappeared in late Pleistocene times to leave the Askja caldera with walls nearly 400 m above its floor. It is estimated that 6500 years ago about 70 km³ of material sank back into the Crater Lake caldera, Oregon. The walls encircling the present 8-km-broad caldera are 600–1200 m high.

In the absence of explosive activity, basalt flows may build up huge gently sloping cones, termed volcanic shields. Those in the Hawaiian Islands slope at 2–4 degrees at their submarine base, whereas the summits, which may project more than 4500 m above the sea, slope at 5 degrees or more (fig. II.5.A). Although several of these relatively low cones rise more than 9000 m above the ocean floor, all rocks are probably less than a million years old. The outpouring of basaltic lavas on land have formed huge plateau areas, like the Deccan trap area which covers 650 000 km² in India. In north-west of the United States an area only slightly smaller was covered with 150 000–180 000 km² of lava in Miocene times. More recently fissure eruptions at Laki, in Iceland, in 1783 poured out 12 km³ of material. In 1961 lava poured out of the fissures in the Askja caldera at an estimated maximum rate of 1000 m³/sec., with the lava front advancing at about 1 km/hour. In general, fluid lava flows very rarely advance at more than 10 km/hour.

In contrast to the fluid basaltic lava, silicic lavas move more slowly, and extruded in a nearly rigid condition, pile up in domes or solidify in the neck of the volcano to form an irregular spine on weathering (fig. II.5.B and C).

4. Hydrothermal activity and meteorite craters

One or two areas are world famous sites of present-day hydrothermal activity, resulting mainly from ground water coming into contact with a source of heat at depth (fig. II.6). This process must not be forgotten in landform studies of any area where this process might have operated in the geological past because the same basic physical and chemical principles apply to hydrothermal reactions as to weathering, and in the first case may occur without accompanying mineralization. The results may therefore be difficult to tell apart. However, W. D. Keller (1964) considers as convincing evidence of

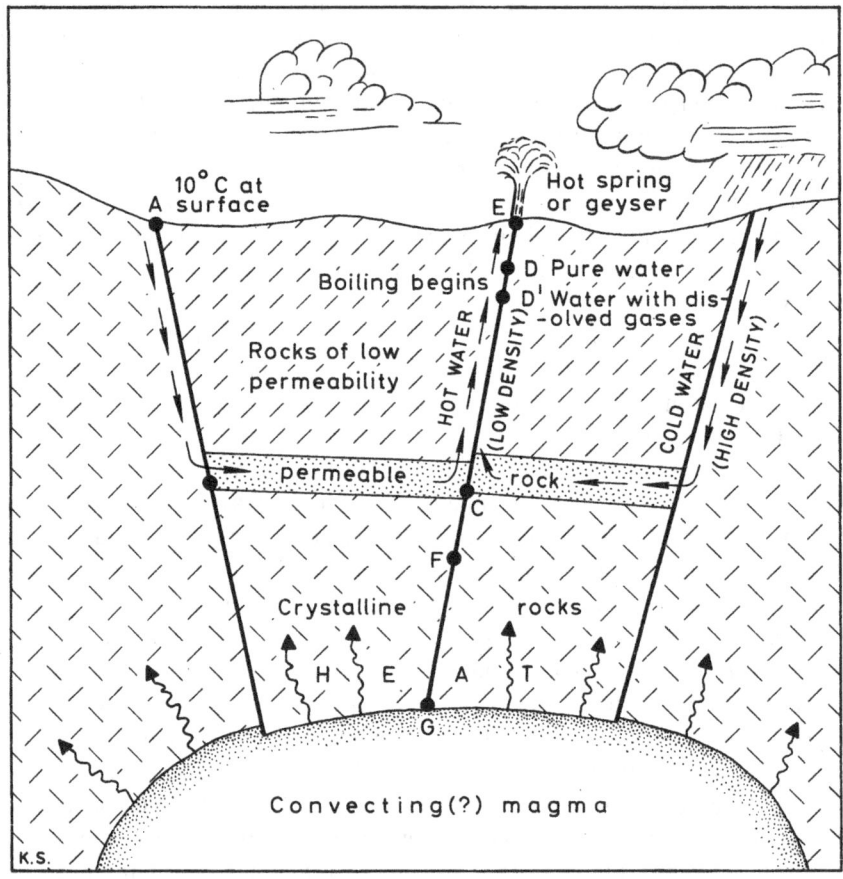

Figure II.6 Model of a high-temperature spring system with deep convective circulation of meteoric water (*after White, 1967*). Thermal conduction is the main source of heat rather than magmatic fluids.

hydrothermal origin, the occurrence beneath relatively unaltered sills, of the kaolin in Cornwall. In Yellowstone Park, much of the silica deposited around the hot springs is due to the decomposition of rhyolites as the hot water rises through them. The initiation of such openings in solid rock prior to sub-aerial weathering attack cannot be entirely overlooked. A more important consideration is that a fundamental process in all volcanic eruptions might be the conflict between ground water and magma. Even a cubic foot of water heated to 900°C. at a constant volume yields a pressure of about 7000 kg/cm², and there is the possibility of the seismic opening of fissures allowing super-charged ground water to explode into steam. In 1886 a 14-km fissure was opened in a couple of hours near Tarawera Mountain by this process.

Unrelated to hydrothermal activity but too infrequent to justify a separate heading are the some two dozen known 'authentic' meteorite craters. One of the more recent falls was in 1908 in the Tunguska tundra of the Soviet Union. This felled trees over a 30-km radius. The most clearly defined crater is the Barrington or Meteor crater in Arizona. It is about 1·2 km in diameter, 170 m deep and believed to be about 50 000 years old.

B. Geological Considerations

1. Resistance to erosion

Viewed from all angles, landforms almost invariably reflect in some degree the differing resistances to erosion of the underlying rocks. In cross-section a relatively resistant rock may occupy low ground simply because it is low in a horizontal or gently dipping stratigraphical sequence, but there are often reasons for inferring that higher areas owe their altitude to greater resistance of the rocks to weathering, particularly in areas of older or unstratified rocks. For instance, in pre-Triassic rocks of Connecticut quartz-rich rocks stand relatively high, whereas carbonate rocks, foliated rocks rich in micas and poor in quartz, and plagioclase-rich rocks generally underlie lower areas (fig. II.7). Gaps in the volcanic hills in Central Scotland, near Neilston and between Johnstone and Dalry, correspond with downfaulted blocks of sedimentary rocks. Along a valley bedrock of varying resistance may account for much of the differences in form of both valley and channel. In addition to creating steps in the longitudinal profile, resistant rocks may be related to valley constrictions with the stream constricted in gorges or in incised meanders, compared with broader valley reaches in the less resistant rocks where lateral erosion is unimpeded, like the Brazos river in the High Plains of central United States. On coasts, viewed in plan, practically any cliffed section will reveal more resistant features contrasting with weaker parts cut back by marine erosion into coves and embayments. North-east of Saigon the coast is rugged with granite and other igneous rocks forming resistant points between short sections of coastal plain. In cross-section wave-cut benches in igneous and crystalline metamorphic rocks are narrow compared with broader benches in less resistant sedimentary rocks.

Although resistance of a rock to weathering is obviously a fundamental issue in the study of landforms, the relative resistance of a rock is difficult to specify because it comprises several complicated features, each one of which may respond to either mechanical or to chemical weathering in a different way. Also, while solutional and mechanical weathering could exploit and enlarge a line of weakness, it is not always obvious how the initial weakness originated.

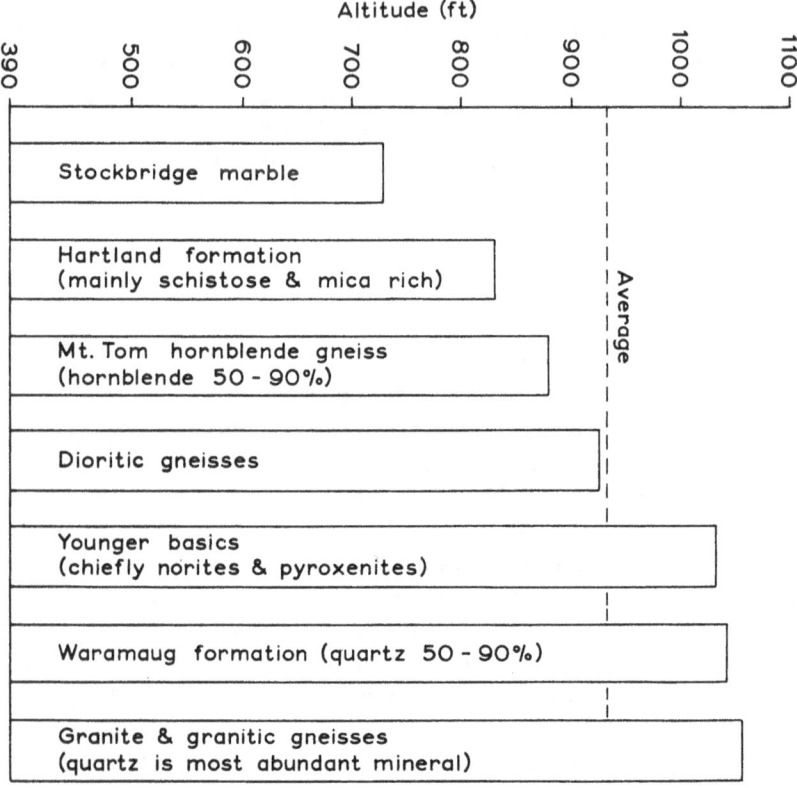

Figure 11.7 Differential resistance of rocks to erosion (*after Flint, 1963*). The bar graph shows the mean altitude of seven mapped rock units in the New Preston Quadrangle, Connecticut.

Theoretically, mineralogical attributes are important in resistance to chemical weathering. The coherence of crystalline rock is due to the bonds of the chemical reactions, inter-molecular forces, crystal growth, and mutual inclusion of crystals. Within the feldspars resistance varies; in orthoclase atomic disorder results in weak spots where water can attack, whereas the plagioclase group and microcline have more ordered structures and have weak spots only in cracks. Acidic rocks are often erosionally very resistant, whereas chemical bonds in more basic crystalline rocks with higher calcium or ferrous iron content are easier to dissociate. Grain size, however, can make a crystalline rock resistant to chemical weathering, regardless of its mineralogical composition. For instance, on basic igneous rocks in south-east Scotland, fine-grained basalts give rise to a soil depth of 23–30·5 cm, whereas rocks with medium-grained equiangular crystals like dolerite lie beneath 50–60 cm of soil. In coarser sedimentary rocks, with the bulk of the constituent grains usually of insoluble quartz, cohesion often relates to the nature of the cement, sandstones disintegrating where their cement contains calcite, iron oxides, or weathers to clayey material. Porous or poorly cemented rocks are very susceptible to frost-weathering. In sandstones and shales porosities may be about 20 per cent compared with 1 per cent or less for igneous rocks. In finer-grained sediments the resistance, even in unconsolidated material, is a function of the mass of individual particles which adhere due to mass attraction, intermolecular and electrostatic forces. However, in some unconsolidated materials the forces, such as the electrolytic effect of salts, which tend to hold the mass solid, may be more than counter-balanced by the influence of a peptizer which may render the mass fluid. A variety of organic and inorganic compounds can liquify flocculated clay. Resistance is equally an inverse function of the size of pore spaces through which water can move only by laminar flow. Within the very small interstices, frictional and buoyancy forces are negligible, nor at the surface is there the frictional drag which makes loose coarser sand-sized particles susceptible to removal by water or by wind. Pronounced fissility or schistosity, more than mineralogical attributes, may be a significant source of weakness in metamorphic rocks. These characteristics cause a rock to split into thin layers lying parallel to a particular direction, and may be due to preferred orientation of clay minerals and mica flakes. In shales films of organic material provide fissile planes (fig. II.8.A). In consequence rocks like slate are susceptible to frost-splitting. For instance, north of Oslo the lower limit of blockfields in areas of fissile rocks at 900 m is 100–300 m lower than in adjacent areas of igneous rock.

Both for mechanical- and for chemical-weathering the crucial factor is the ease with which water penetrates into a rock, whether this be essentially along crystal edges in crystalline rocks, through the pores of sedimentary rocks, or along the fissile planes of a metamorphic rock. But in this respect all rocks, regardless both of chemical composition and of origin, may be weakened in a similar way by jointing and fissuring (fig. II.8.B–E).

The fundamental fracture pattern of a rock mass, it seems, is established by extension early in its history, as soon as it is susceptible to brittle fracture, because joints are frequent even in recent sedimentary strata. In igneous rocks joints can form during or shortly after cooling. However, tension is impossible below a few hundred metres. The fracturing process may be related to fluid pressures, with systematic joints developing perpendicular to the axis of least compression. Although no theory on the origin of joint initiation is yet entirely satisfactory several descriptive facts are established. In many areas systematic joint patterns may be persistent and extend to appreciable depth, but for reasons unknown they may be related to the disposition of the more rigid rocks in the area. These patterns often consist of two directions which intersect at angles between 10 and 30 degrees. Systematic joints are usually planar, continuous, and perpendicular to the upper and lower surfaces of rock

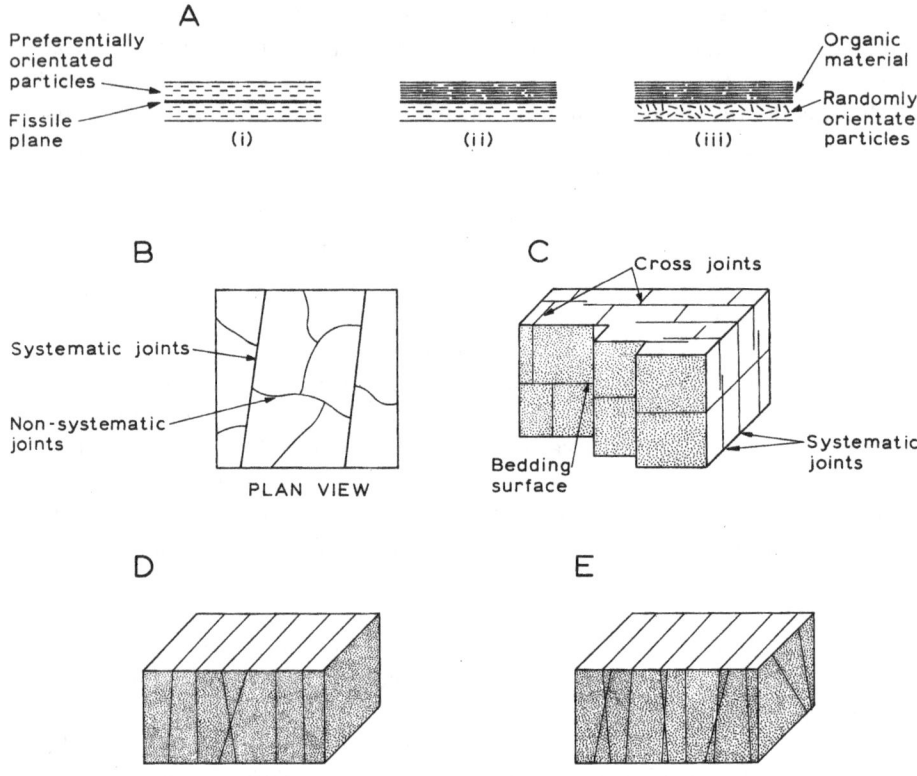

Figure II.8 Lines of weakness intrinsic in many rocks.

A. Fissility in shale (*from M. Gipson, 1965,* Jour. Sed. Petrol., *Vol. 35*): (i) fissile plane occurring in a zone of preferentially orientated particles, (ii) fissile plane occurring at the interface between zones of preferentially orientated particles and organic material, (iii) fissile plane occurring at the interface between zones of randomly orientated particles and organic material.

B–E. Joint patterns (*from Hodgson, 1961*).

B. Plan view showing a typical pattern of non-systematic joints and their characteristic termination against systematic joints.

C. Block diagram showing relations between cross-joints and prominent bedding surfaces.

D. Mode in which joints of the same set intersect in section.

E. Mode in which joints of different sets intersect in section.

units. By contrast, non-systematic joints are generally curved fractures which do not cross other joints and which frequently terminate at bedding surfaces. Micro-fissures, although ordinarily invisible on a fresh surface may occur as hairline fractures a millimetre or less apart, and are best developed on more massive and more quartz-rich rocks. They increase considerably the area of rock with which water can come into contact. For a capillary of 0·02 mm radius there is practically no penetration while one of 0·25 mm permits a penetration of over a metre. Joints may also develop essentially on the surface due to tension, as in the cooling of igneous extrusions or in the desiccation of sediments, and occasionally may approach the ideal hexagonal shape.

Joint directions may have a profound influence in guiding the course of erosional processes which is reflected in alignments in landforms. In the schist and granite area around Alice Springs several hills show combinations of straight, joint-determined salients and deep parallel-sided re-entrants formed on cross-joints. N. Caine (1967) has described Ben Lomond, Tasmania, where the polygonal joint pattern of the bedrock has been a major influence on the form of tors. Above many steep slopes the headwall of scars left by rockfalls are often the plane surface of a major joint, and avalanche chutes generally coincide with zones of structural weakness which facilitate disintegration. On valley floors streams follow joint directions as do many valleys themselves, particularly in jointed igneous rocks and especially in flat-lying sedimentary rocks where joints are often most markedly developed, and some linear or aligned lakes may focus attention on this fact. Observations of this relationship include those of Daubrée who, in 1879, noted the striking influence of joints upon the drainage lines in the upper Mississippi. In southwest Utah part of a system of deep canyons is strongly controlled by joints and faults of small displacement. In Yorkshire the Millstone grits, Magnesian Limestone, and Corallian have joint planes in five well-defined classes with means at 24, 55, 84, 115, and 163 degrees. Harrison and Thackeray (1940) considered that these directions occur far too frequently in segments of rivers like the Wharfe, Ure, Derwent, Low Lindrick, and Newtondale to be merely accidental, and noting that 25, 115, and 163 degrees were especially predominant, concluded that these valleys are controlled by joints. In tropical areas differential erosion is often less obvious, but in the Serro do Mar escarpment in Brazil, the dissecting valleys are always orientated by the boundary fractures and include abrupt turns. Also very close adjustment to structure may develop within the weathering zone, as the numerous artificial exposures in Hong Kong have revealed. More unusual in origin is Meteor Crater, Arizona, which is squarish in outline, with the diagonals of the square corresponding to the direction of the two principal joint systems.

Several lines of evidence indicate that joint frequency apparently decreases rapidly with depth and that many joints form near the surface. Fig. II.9 illustrates the common experience of water engineers who find that most open fissures in jointed rocks have been encountered by the time they reach depths of 30 m. It seems that joints must persist downwards but only open up when unloading occurs. In view of the importance of joints in landscape develop-

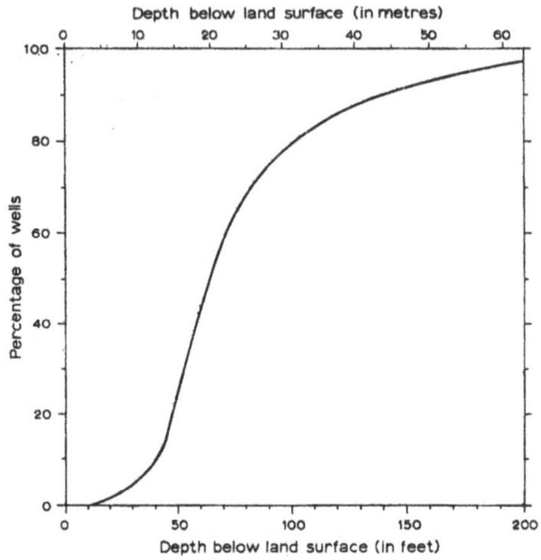

Figure II.9 Cumulative curve of depth to water-bearing crevices in 146 Kentucky wells (*after E. H. Walker, 1956,* Bull. geol. Soc. Am., *Vol. 67*), showing that 75 per cent of the wells meet water-bearing crevices before penetrating more than 90 feet below the land surface.

ment and the dependence in turn of joints on unloading to provide lines of weakness of exploitable size, the phenomena of unloading emerges as one of the most crucial considerations in geomorphology. Unloading can be of two main types, due either to linear erosion allowing lateral expansion or to downwearing removing the superimposed load. To some extent these two types of unloading may operate in conjunction with each other, but there is a tendency for large-scale exfoliation, or sheeting, to characterize areas of homogeneous crystalline rocks in which primary joints are not apparent. In this case with the initial relief of load in a vertical sense, the horizontal opening of joints follows at a distinct subsequent stage. By contrast there are no distinct phases where linear erosion and relief of load, essentially in a lateral sense, characterize horizontally disposed sedimentary strata, because both operate in the same direction and some of the most pronounced joint patterns result. Sheeting, which is like large-scale exfoliation in appearance, is often independent of all primary structures in a crystalline rock and may cross rock boundaries. Sheeting affects massive rocks like granite and also quartzites, and affects highly foliated rocks like gneiss and schists as well. In granite quarries in New England freshly exposed sheets on quarry floors have been known to split suddenly into two or more sheets, and at Barre, Vermont, expansion of several inches have occurred during quarrying. As G. K. Gilbert observed in 1904 these partings in a natural setting were always adjusted to the general shape of the land surface. Sheeting joints trending roughly parallel to the outline of the slope have since been observed in many parts of the world and been mistaken for a primary feature in others. Below the surface the configuration of the sheets is increasingly flat but still conforms roughly to the broad outlines of the landform above (fig. II.10.A). In bedded rocks lateral expansion tends to be more significant (fig. II.10.B) and joints become more abundant and open as one approaches cliff faces. In the massive limestone area of north-west Yorkshire one of the localities where open joints are most pronounced is above Malham Cove, the highest sheer face in the area. On the shoulders of cliffs in granitic rocks, the sheeting joints may still tend to curve round parallel to the topography.

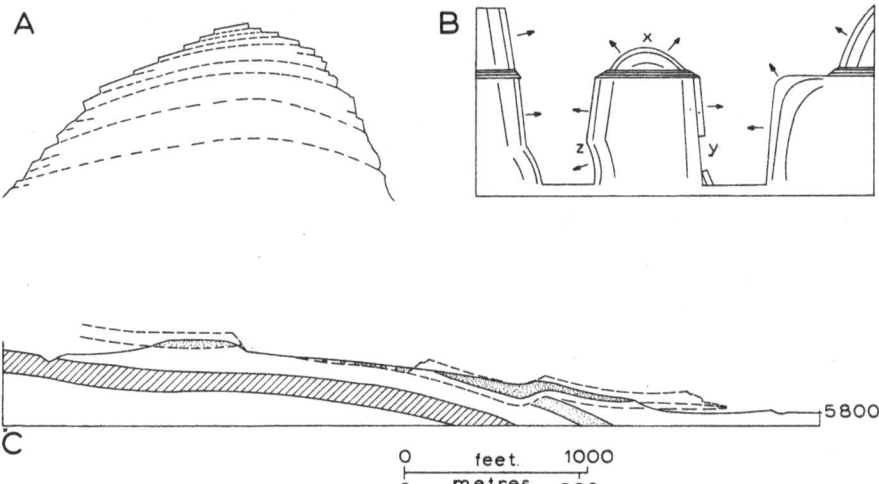

Figure II.10 Inter-relationships between landform and the surface expression of geological structures.

A. Sheeting arched nearly parallel to the land surface slope (*C. A. Chapman and R. L. Rioux, 1958*, Am. Jour. Sci., *Vol. 256*). The rock is a hornblende granite, observed in Acadia National Park, Maine.

B. Large-scale sheet structures and joints in massive Colorado Plateau sandstones produced by expansion (*Bradley, 1963*). Arrows indicate inferred directions of expansion; X is an exfoliation dome, Y an exfoliation cave, and Z is an overhanging exfoliation plate in a meander scar.

C. A block-glide landslide, Front Range, Colorado (*Braddock and Eicher, 1962*). An upper sandstone (stippled), exposed on hogback dipslopes, has slipped along bedding planes.

Two additional types of joint enlargement, of less general significance, might be added. On higher latitude coastal areas there is the possibility of ice-unloading favouring fissure development, and in eastern Canada excavations on the Quinze river, Quebec, and at other localities, have revealed extensive near-horizontal fissures a few centimetres wide 6–20 m below the surface for which glacial unloading is thought to be the primary cause. Secondly, joint widening by river-water solution on a valley floor of soluble rocks might be as significant as the lateral expansion of unconfined situations of cliffs on the valley side. This applies not just to massive limestone areas. H. E. LeGrand (1962) notes that many joints are enlarged by the action of solution especially in gneisses and schists containing silicates of calcium, and that many of these enlarged joints are associated with linear sags in the surface relief.

It is not easy to state the precise significance of joints in facilitating erosion because it seems that their development to exploitable dimensions by unloading is itself a product of erosion, while in turn, as in the east-central Sudan, the depth of unloading limits the effective depth of water penetration. However, because the effectiveness of erosion at a given period is influenced by the effectiveness of past erosional conditions in producing unloading, and the pattern created, perhaps one of the fundamental characteristics to recognize where joints develop, is that over a period erosion would seem to be self-perpetuating in pattern, but may work along joint-controlled directions. Also, because expansion takes place in the direction of least resistance, with sheeting joints related to existing landform, there is probably a tendency for distinctive forms like domes, canyons, and cliffs to persist.

2. Rock disposition

Faults are fractures in rocks along which opposite sides have moved past each other. If this movement is relatively minor, it may be hard to distinguish a fault from a joint, and to some extent the influence of either in landform development may not be easy to detect. Quite often, however, not one single fracture is involved, as faults tend to occur in zones where the dislocated strata may form a belt, perhaps 300 m wide, as on parts of the San Andreas fault. Such situations facilitate linear erosion, as in Trinity Bay in Newfoundland, an unusually deep elongated basin which is attributed to structural weakness rather than to glaciation, as the ice moved at right-angles to the orientation of the bay. The fiords of the west coast of Norway, according to some investigators, largely coincide with crush zones, joints, and similar tectonic or geological contacts, indicating that fluvial erosion and subsequent glacial erosion were largely tectonically controlled (fig. II.28.A and B). In other regions, faults stand out boldly, or truncate a structural grain sharply, but not always as clearly as on the bare rocks of the Canadian shield. In plan the outline of many inselbergs in several parts of Australia and Africa are coincident with fractures. Vertical movements on faults may exert a control on the orientation of drainage, like the fault bounding the north-east flank of the Barind, in India, where the Karataja river and its tributaries flow south west into the Jamuna river, probably due to south-west tilting, and examples of lateral offsetting of streams occur in the areas of active transcurrent faulting.

Faults may be an important influence on landform development, but since it is possible to describe faûlts according to a variety of geometric, geological, and genetic criteria, an attempt to present a simple classification would be misleading. A fault displacing strata essentially in a vertical sense may be rotational. On a pivotal fault the vertical slip and the height of an associated escarpment gradually decrease along its strike, then gradually start to increase beyond a given point to produce a scarp facing in the opposite direction. In a hinge fault there is also a rotational movement, but one which produces increasing displacement along a fault-line in one direction only. All faults have a certain rotational movement, but where this is insignificant a translational fault is recognized. The fault may be 'normal' where the dominant movement is downward on one side, with the dip of the fault plane usually exceeding 45 degrees, or the dominant movement may be the thrusting upward of one side, often along a fault-plane inclined at less than 45 degrees, to produce a reverse fault. By contrast in strike-slip or transcurrent faults the principal stress and resultant movement is essentially horizontal, although vertical displacements, both non-rotational and rotational may be produced. In a number of particular tectonic settings, possibly associated with rising convection currents in the mantle, faults in approximate alignment may produce not just steps in the landscape and diversions or alignments in the rivers but the distinctive fault-bounded blocks and depressions of horst and graben relief. However, the vertical movements which leave horsts upstanding either by actual upward movement or due more to the sinking of grabens, may be an integral part of dominantly large-scale strike-slip movement, as already discussed.

Faults are sometimes described by the trace of the fault-line in plan, seen in relation to the disposition of the bedrock. In sedimentary strata these are strike, dip, oblique, or even bedding faults. Peripheral and radial faults occur on domes like those of some igneous intrusions. In relation to regional structure as a whole large faults may either be longitudinal striking parallel to the long axis of the major structure, or transverse, if striking across it.

Three points about the relation of relief to faulting might be added. First, there are many references in the literature to the distinction between fault scarps where any difference in altitude was due to vertical displacement, and a fault-line scarp which is essentially an erosional feature due to the removal of less resistant material from one side of a fault. A second point, related to the fault-line scarp, is that altitudes on either side of a fault may not correspond to the direction of initial vertical movement if less resistant material is on the side of relative uplift. Thirdly, a large number of faults exist without trace in the land form.

In areas of more gently folded rocks the structural geologist differentiates between symmetrical and asymmetrical folds, the limbs of the latter differing in dip. Where folding has been more severe three degrees of increasing asymmetry are recognized, the limbs in an isoclinal fold being essentially parallel, in an overturned fold the steeper limb has passed through the vertical and may be essentially horizontal in a recumbent fold. In an area of essentially low dips, a monocline is a fold which produces a local steepening of dip. In addition to these hypothetical, idealized cross-sections of folds, in plan and more recognizable in less disturbed areas are anticlines and synclines, characterized by the width of the fold being narrow in relation to its axial length. Where numerous smaller folds are compounded into broad structural arches or basins, an anticlinorium or synclinorium might be recognized. In areas where folds are less systematically developed domal structures, which are broad in cross-section in comparison with their short axial length, may occur. Structural salients and embayments on the margins of folded areas may complicate a systematic pattern of folding, major transverse faults may disrupt it, and strike-slip faulting tends to drag along a series of folds *en echelon*. Moreover, with the exception of one or two classic examples like the half-cylinder of the Zagros Mountains in south-west Iran, erosional processes appear to be sufficiently effective during slower rates of folding to restrict topographic elevation to less than its hypothetical structural elevation and to reduce it comparatively rapidly to a fraction of this height after folding ceases. In addition there is the complication of an inversion between structural relief and land surface form, with the axial lines of some anticlines deeply trenched whereas synclinal folds in thick resistant strata resist erosion and within a short period of geological time after folding commences, become the highest points in the area. Because of the complications in pattern of folded rocks and the significance of erosion during folding, attention to rock disposition in landform study of these areas often focuses on the scarpland relief of individual limbs of folds.

The inter-relations of three aspects of the disposition of strata in the limbs of folded structures, the inclination, lithological sequence, and relative thickness, are particularly significant in landform study. These distinctive aspects are also apparent in structures that are essentially tilted blocks. Although none can be considered realistically in isolation one important effect of inclination, or dip, of the strata is that the relative erodibility of a rock is expressed most accurately in land elevation when dips are vertical and least accurately where dips are low, where if a resistant horizon is exhumed, it shields the underlying strata. Structural benches then become significant landforms, like those in the Old Red Sandstone of Caithness in northern Scotland, or more extensive structural plateaux may develop. Moderate dips accentuate the importance of gravity in erosional processes, particularly where the lower end of a tilted block or limb of a fold is eroded away. This may induce landslides along the bedding planes in deeply dissected relief (fig. II.23.C), in quarries or along coasts where the dip is seaward, like the slips of Chalk and Upper Greensand, overlying the impermeable Gault Clay, seen along the south coast of England. In addition to the susceptibility of mass movement at the unsupported lower end of an inclined block, there is the accentuation of erosional attack at its upper end, concentrated laterally by gravity against the exposed edges of less resistant strata. This may apply to ice movement, tongued and grooved into the strike ridge and vale relief as much as to stream erosion, as strike valleys in glaciated areas in the Seal Lake area of eastern Canada or the Cairnsmore of Fleet area in southern Scotland show.

The sequence of varying lithologies of rock is a dominant factor of land-forms developed in regions of folding. A stratum can play one of three roles. First, a more resistant stratum may produce a cap-rock effect, protecting less resistant material buried beneath from erosion and may also provide resistant fragments of debris which move down the scarp slope from its outcrop, which reduces erosion of the less resistant material forming the lower part of the down-dip valley side. It follows that the presence of an inclined resistant stratum between two less resistant lithologies tends both to protect the underlying stratum from erosion and to accentuate the erosion of the overlying one. A third effect is that of an unstable horizon. This may be as a slip-surface which appears to be present as a distinctive horizon in all cases where mass movement takes place in the direction of dip, or which by its local absence, may make certain points along a valley-side stable. A useful example is the much studied slide, dislodged by the 1964 Alaskan earthquake on to a tributary of the Sherman glacier. Generation of the slide of massive coarse sediments was favoured by the presence of weak fractured beds of phyllites and slates dipping steeply and almost parallel to the original surface of the slope. A period of chemical decomposition of the material at the slip-surface may be necessary before a basal plane becomes a potential slip-surface. The schist and gneiss with foliation dipping steeply downslope which on failure led to the 1959 Madison canyon rock-slide avalanche in south-west Montana, was highly sheared and deeply weathered. Broader outcrops of incompetent material by weathering or by shearing under the weight of the cap-rock, cause the latter to camber over, with the consequent opening of joint, favouring reduction in the prominence of the cap-rock edge.

In folded rocks along coastal margins the attitude in plan of the strike of the structures in relation to the direction marine erosion is analogous to that, seen in cross-section, of the inclination of strata to sub-aerial and sub-glacial erosion. Like the elevations and depressions seen in land in cross-section and related to near vertical dips, structures striking perpendicularly to the coast

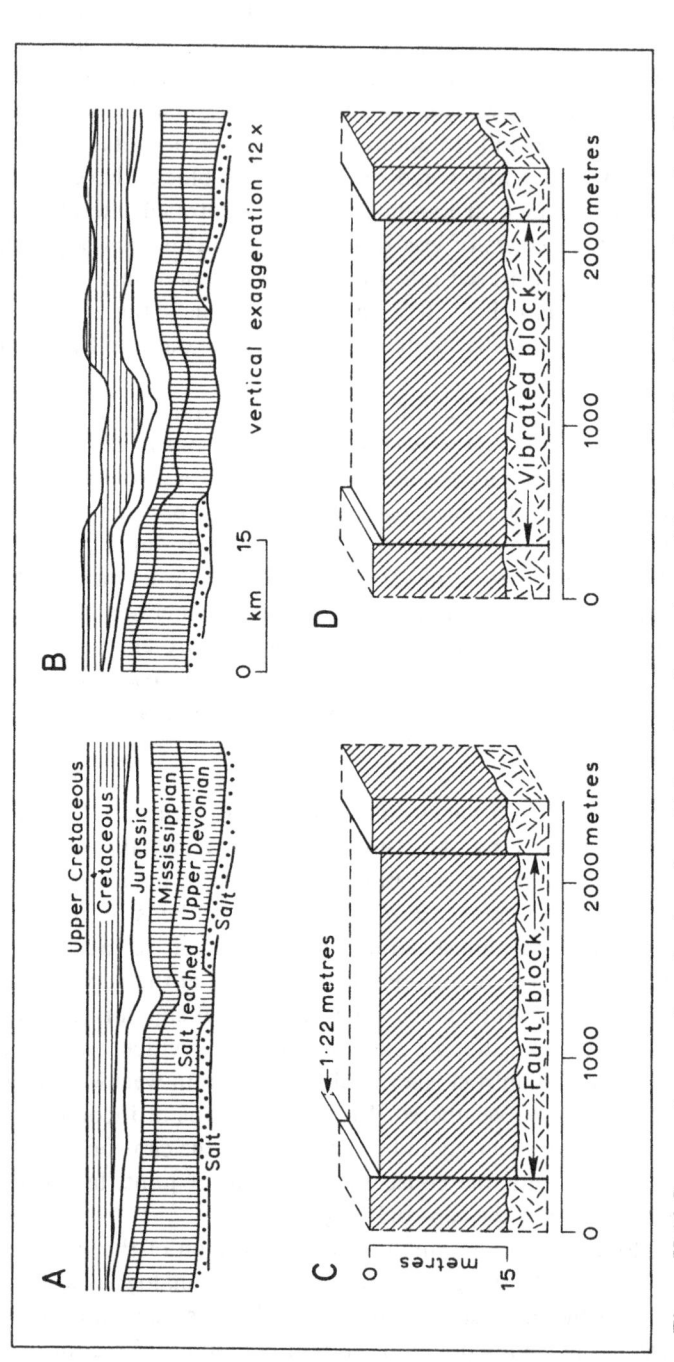

Figure II.11 Some underground mechanisms by which surface depressions might form. (*Above*) Collapse depression due to underground solution of Prairie evaporites in the Regina—Hummingbird trough. (A) Late Cretaceous time. (B) Present-day, with much broader area of salt removed, approximately equalled by the amplitude of the ground surface depression (*from De Mille et al., 1964*). (*Below*) Two possible mechanisms by which surface depressions might form in unconsolidated deposits (stippled) (*from Plafker, 1964*). (C) Block faulting. (D) Compaction, as suggested by the study of superficial, orientated features in the Beni basin in Bolivia.

produce in plan promontories and coves in close relation to the erodibility of the rocks. Like the horizontal strata of a plateau, resistant ridges striking parallel to a coast will protect less resistant rocks inland or provide resistant fragments to add to the resistance in coves where the sea may have broken through to less resistant rock.

A third factor is the ratio of the stratigraphic thickness between strata. For a given dip, it seems theoretically sound to infer that the thicker a cap-rock is in relation to the underlying less resistant material the slower will be the erosion of the cliff or scarp, although there are not the measurements available to demonstrate this relationship. With changes in the angle of dip, however, the outcrop width of a given stratum as opposed to its stratigraphical thickness will vary. Thus a prominent scarp former like the Chalk forms broad relatively unbroken escarpments where the dip is low, as in eastern England, but where tilted into steep hog's-back ridges, as on the Isle of Wight and adjacent areas in southern England, the narrow Chalk outcrop is broken through at several points by valleys which run transverse to the strike.

In addition to landform inter-relationships with structural movements at the surface, there are some limited areas, particularly in unconsolidated sediments where surface forms are related to the upward propagation of structures that exist or move at depth. For instance, the alluvial plain of the Beni basin in north-east Bolivia is an area with pronounced surface lineaments and orientated lakes. The latter have neither inlets nor outlets, but have distinctive flat floors and relatively abrupt sides and may be up to 19·8 km in length. Some lineaments can be traced from the Brazilian shield outcrop into the basin. G. Plafker's suggestion is therefore that the orientated surface features in the Beni basin form through slight movements in the basement blocks although faulting may be a more likely explanation for some orientated features in other areas (fig. II.11.C and D). Increasingly there has been recognition of the possibilities of intrusion of plastic salt into a restricted opening causing deformations in overlying strata and forming some form of surface disturbance or perhaps some surface definition as a dome or lineament. Pressures of only $150-200$ kg/cm^2, applied at room temperatures, are sufficient to induce flowage deformation of salt rocks, and this is an important consideration in the landforms on Triassic and other strata including evaporites throughout the world. There is also the possibility that the shape of some static structure at depth will influence landform development on the surface. Fig. II.13 illustrates an instance where a valley running transverse to the general alignment of surface ridges follows the pattern of sub-surface contours of an underlying rock.

Surface features due to the presence or movement of sub-surface structures may have been neglected in landform studies. Rather more widespread but perhaps also somewhat neglected are features related to the action of water on

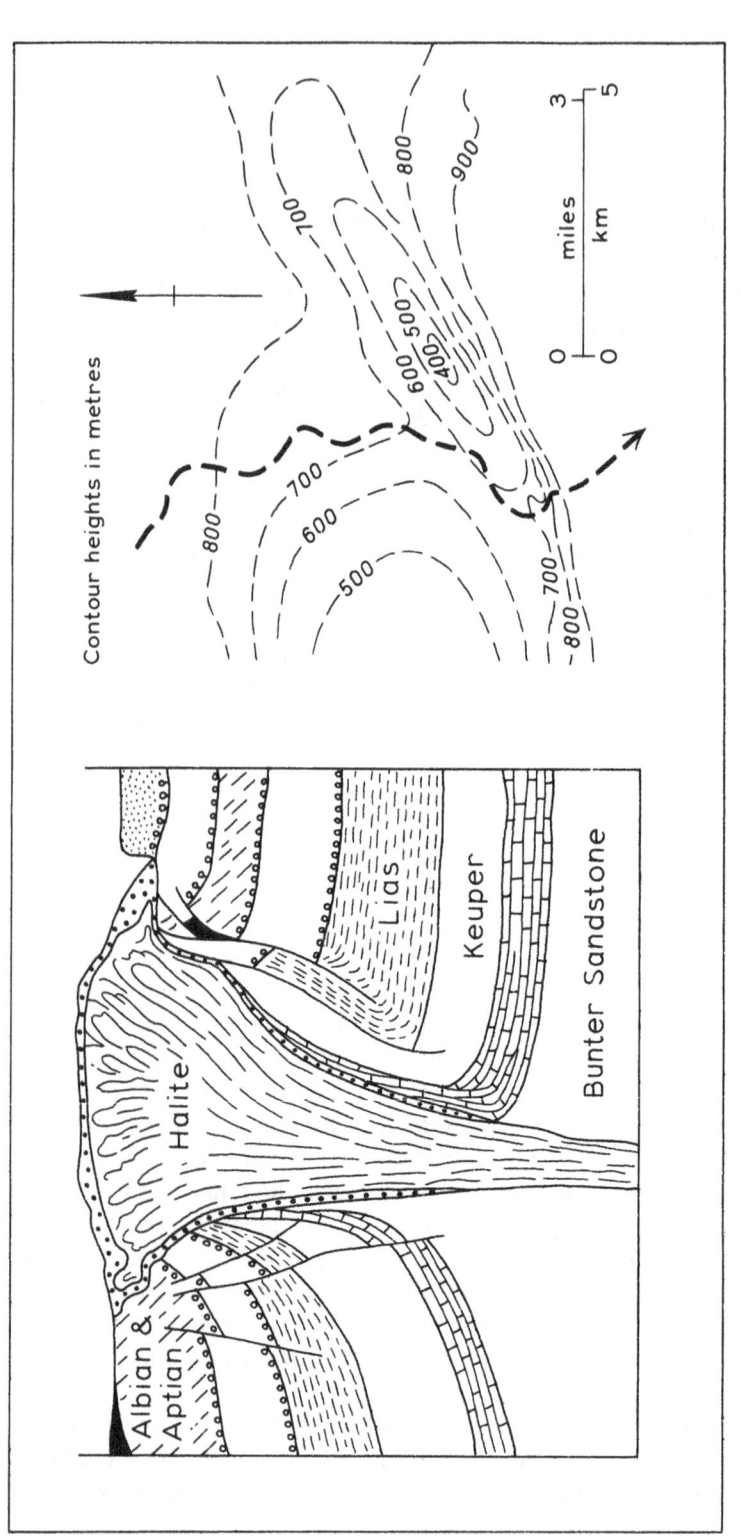

Figures II.12 and II.13 Further examples of sub-surface structures influencing land surface form. (Left) The upward expansion of a salt dome (H. Bochert and R. O. Muir, 1964, Salt deposits). The example of the Wienhausen–Eicklingen salt dome in Germany shows, according to A. Bentz, tangential faults almost parallel to the margin of the intrusive salt body and the overturning of enclosing sediments. (Right) A surface valley following subsurface contours which define a structural col in Cretaceous limestones (Coque, 1962). This map by P. F. Burollet shows the Wadi Tseldja, the only large transverse valley to cross a major anticline in the southern Atlas, following the sub-surface contours at the base of the Campanian strata.

strata at depth, termed interstratal solution, because it affects carbonates as well as evaporites. In central Canada evaporites reach a thickness of over 240 m and may be covered by 300–1500 m of younger rocks. Over a belt of 1900 km structural lows are depressed by as much as 115 m and may be up to 160 km long, approximately equal to the amount of salt removed by water circulating downward through fractures (fig. II.11.A and B). Southward in Kansas and Oklahoma, along the eastern edge of the High Plains, a similar process has produced more obvious surface basins, formerly attributed by G. K. Gilbert (1895) and others to deflation. Beneath the North Crop of the South Wales coalfield, extensive interstratal solution of Carboniferous Limestone produce pronounced collapse features on the ground surface of the overlying Millstone Grit, the enclosed depressions being up to 45 m in

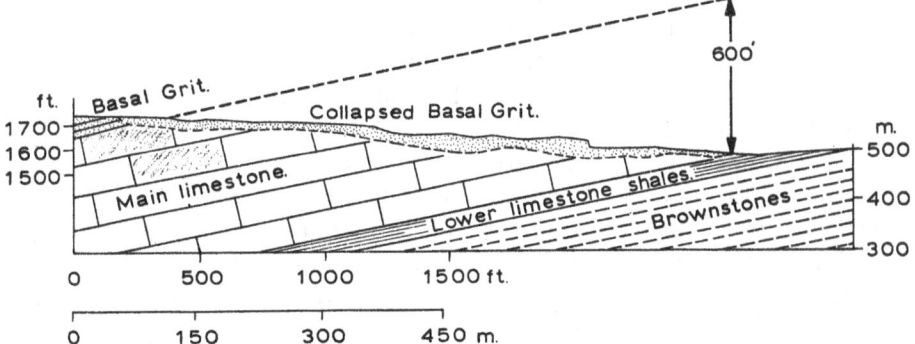

Figure II.14 The collapse of Basal Grit on the North crop of the South Wales coalfield due to interstratal solution of limestone (*T. M. Thomas, 1963*, Trans. Inst. Brit. Geogrs, *No. 33*). The tongue of collapsed grit lies immediately to the north of Carreg Cadno.

diameter and as frequent as 8–12 per ha. Similar features occur on the ground surface of many similar so-called 'covered karsts', and may be the product of interstratal solution over millions of years, and it is possible that some features in present-day bare-limestone karsts might have developed beneath a now-vanished cover. In fact, in existing landform studies there are probably far more references to the influence on present-day relief of former covering structures that have now disappeared, than on invisible sub-surface structures and lithologies, and there are many examples of the surface characteristics of a given stratum superimposing some imprint of their pattern on the underlying strata and the structures.

The importance of rock disposition at the earth's surface in influencing, or even in essentially controlling, landform is at times too obvious to offer much challenge in interpretation. On the other hand, sub-surface features, because they cannot be directly observed in most geomorphological investigations, or now-vanished former covers, because of uncertainties in their reconstruction, illustrate how geomorphological evidence is intrinsically incomplete.

3. Distinctive lithologies

It is well known that the pioneer of British geology, William Smith, recognized unhesitatingly the Chalk outcrop in east Yorkshire by the appearance of the Yorkshire Wolds when he viewed these for the first time from York Minster tower, two dozen miles away. From a review of geological considerations in landform study it becomes clear that not only do the relatively few rock types possess distinctive lithological characteristics and a certain indefinable distinctiveness in their landform expression, but that they add further distinctiveness to landscapes by tending to occur in certain areas only, where distinctive structural settings are conducive to their formation. Therefore, although the importance of rock types has been assumed rather than studied in geomorphology, their more clearly established characteristics should be summarized briefly.

In sandstones the particles have a reduced chemical range due to their detrital origin and, if well-cemented, are relatively insoluble. Their dominant characteristic is their well-defined joint surfaces and bedding planes which makes them susceptible to wedging by tree roots and clay fills. Their susceptibility to frost wedging is less clear due to free drainage through their porous texture and down the joints themselves.

Shales in all climates are frost-susceptible and changes in water content alone are sufficient to exploit their fissility. Weathered shale transformed to clay moves readily downslope. Clays, by their impermeability are susceptible to surface erosion and due to their ability to absorb large quantities of water are prone to mass-movement. Earthflows tend to be closely related to clay minerals which expand on wetting, thereby decreasing the strength of the layer. Kaolinites have two properties which do not encourage slipping. One is the small coefficient of shrinkage which means little fissuring on drying and little expansion stress on wetting; the second is its high liquid limit, which means that high water contents are necessary to induce flow. In contrast, illite clays have much lower liquid limits and montmorillonite clays have a higher coefficient of shrinkage.

Chalk, being essentially calcium carbonate in composition, is readily soluble in acidulated soil water, and if saturated its high porosity makes it so susceptible to frost shattering that a layer of pulverized debris some centimetres thick can accumulate in a few years on an initially bare surface. The well-known characteristic of massive limestones is also solubility. However, these rocks are often less micro-fissured than crystalline rocks, porosity is absent and joints are often widely spaced, all three factors reducing susceptibility to frost action both on small and large scales. One' well-marked characteristic, due to the near-perfect cleavage of their principal mineral, calcite, is low resistance to impact. As limestone valleys are typically stream-less or have flows regularized by underground flow, the apparent neglect of this characteristic may be justified. However, in examining the origin of gorges which typify extensive limestone terrains in mid-latitudes, the low resistance of massive limestones to impacts from gravel carried by spring-melt floods in cooler phases of the Pleistocene, might be important. Dolomite, although again soluble, is made up of crystals not firmly bound together by an interstitial cement, and if saturated, will undergo intense granular disintegration like chalk.

Granite stocks or batholiths simply due to their mode of emplacement tend to produce domed forms. On smaller scales sheeting tends to produce domal forms too, with joints developing after sheeting to outline sharp angular relationships between plane surfaces. Rarely are cliffs produced which have the sheerness of vertical joint-controlled sandstone, limestone, or basalt cliffs. In consequence of the distinctive properties of granite, R. J. Russell (1967) can cite examples of granite coasts south-east of Gäule in Sweden, near Rio de Janeiro in Brazil, or near Albany on the southern coast of western Australia which show striking similarities. On the other hand, with more intense micro-fissuring, granites may form gentler boulder-covered or even sand-covered slopes, and in porphyritic granites the large felspathic crystals are vulnerable to chemical weathering and they crack easily under impact due to their excellent cleavage. Other acidic crystalline rocks like diorite and porphyry share some of the distinctive properties of granite but due to smaller crystal size tend to be more resistant.

The main characteristic of extrusive basalts is the highly distinctive horizontal surface of recent flows. Where interbedded, like intruded dolerites also, they are often distinctive as well-jointed cap-rocks. With soft mineral constituents like chlorite and olivine, chemical decomposition is often rapid.

Of the metamorphic rocks, gneiss and schist tend to split along mica-rich planes and weather into readily transported fragments. In consequence, in dissected relief, these metamorphic rocks and fissile slates tend to produce long relatively even slopes of moderate steepness close to the angle of repose. Quartzite, cemented by crystalline quartz, can approach indestructibility and is the rock to emerge most clearly from the deep weathering mantle of the humid tropics which effectively reduces the influence of most distinctive rock properties in landform development. The highly metamorphosed rocks of the ancient shields are very resistant and preserve a surface form that may date back to pre-Cambrian times.

4. Variations within an individual rock

An inevitable feature of rocks, of fundamental significance in the study of landforms, is that whatever may be their properties at a point these will change in degree if not in kind laterally. Inevitably, at some other point a stratum or rock body when traced laterally will disappear. This characteristic is perhaps most readily appreciated with reference to igneous rocks fed from a central pipe; a phacolith is by definition a lense-shaped intrusion at a fold axis, a laccolith a dome-shaped intrusion. The more elongate igneous bodies, like the vertically intruded dike or the horizontally intruded sill or the extrusive lava flow also inevitably die out laterally. The effect on associated landform is often clear. Near the northern boundary of the Central Lowlands of Scotland, J. B. Sissons (1967) describes how the volcanic rocks reach their greatest altitude in Ben Cleuch in the Ochils (718 m), where the lava beds are at their thickest. These beds then fall in altitude north-eastwards towards Montrose as the lavas become thinner and increasingly interbedded with sedimentary rocks.

Reflecting changes in channel positions in their depositional environment, many sedimentary rocks show marked lateral variations in grain size, compaction, and cementation, as well as changes in lense or stratum thickness. Within the Lower Greensand series in the Weald area of south-east England, the outcrop of the Hythe Beds, extending east and west across the Dorking and Reigate district, is seldom over half a mile wide; in the Leith Hill area it widens to 5·5 km as chert seams present in the sandy formation in this area provide a cap-rock effect. Compared with the prominent salient of Leith Hill, its summit at 294 m being the highest point in the area, the land around Dorking, where the Hythe Beds consist almost entirely of sand, is low. Even in plateau areas, lateral changes are significant in influencing the details of landform. In the Arizona plateau where many sedimentary rocks are persistent units across the area, others are lenticular on a broad scale, or show marked lithological changes laterally, indicating the effects of local basins and swells in the sedimentational history. Along the eastern escarpment of the Chuska Mountains in north-west New Mexico the front is cuspate in plan at several places. The concavities are relatively thin units and the points of the cusps are developed where the units are locally thickened. Even in unconsolidated material like morainic debris local concentrations of boulders and blocks will form small promontories on a shore like the north coast of Galway Bay, close to the town, or along parts of the margin of Lake Vättern in southern Sweden. Fig. II.15.C illustrates a situation on a Cotswold dip-slope where incision of a stream shifting downdip on insoluble clay appears to deepen into underlying limestone where the clay dies out. While sediments are deeply buried, lateral changes may be induced by folding, particularly of incompetent strata, producing thinning on limbs and thickening in axial areas by flowage. Jointing is of fundamental importance in weathering and since systematic joints tend to be more widely spaced in thicker more massive units, the significance of lateral changes in joint frequency is redoubled. In the Appalachians there is a general decrease in the size of joint surfaces in shales and coal from east to west, the range of spacings from centimetres to several metres occurring on a local scale also. In the extensive strip mines of the Houtzdale–Snowshoe area, narrow zones of closely spaced joints are separated from one another by broad areas of few or no joints (Nickelsen and Hough, 1967). Observations like these are particularly valuable because variations in joint spacing are frequently inferred in the interpretation of erosional features without being able to demonstrate that the apparently weaker zone was in fact crossed by closely spaced joints before it was eroded away. Similar zoning of joint frequency may occur in igneous rocks. In the southern Beartooth mountains of Wyoming, erosion appears to have cut elongate ridges parallel to zones of maximum microjoint development. D. U. Wise (1964) suggests that the microjoint stresses coincided with the closing phases of block uplift of the Laramide Rocky Mountains. When lifted free of the confines of adjacent basins, the local fault pattern permitted expansions, but in taking place in different directions, local dead spots in the expansion pattern left some granite masses without microjoints. Therefore, although difficult to demonstrate due to the removal of the required evidence, there is a logical basis for suspecting that some inselbergs could be the more resistant portions of a crystalline mass that possessed few if any joints. There are many effects of joint zones other than guidance of linear erosion. In the Skjomen area, northern Norway, postglacial weathering has been insignificant on the less-jointed solid rock and there may be no talus at the foot of cliffs. At other probably well-fissured points talus may accumulate to heights of 20 m. As well as changes along a cliff or slope, differing densities of jointing are an important consideration downslope too, particularly in crystalline masses (fig. II.15.D). In north-western Eyre Peninsula, Australia, some granite slopes are determined by huge curvilinear sheets of massive granite, while those on closely jointed granite are boulder-strewn. Differential weathering where joint frequencies change over short distances produces sharp breaks of slope. In the Vosges, irregularities appear to be related to joint density. The more densely jointed areas are susceptible to frost-weathering while more resistant prominences gradually emerge from the surrounding more fissile material. Fig. II.15 illustrates some situations in which landforms appear to be related to lateral variations in resistance of rocks.

In the study of smaller features vertical changes in properties of rock have to be considered. These are most pronounced in sedimentary rocks where the term basal conglomerate itself is perhaps sufficient illustration of vertical

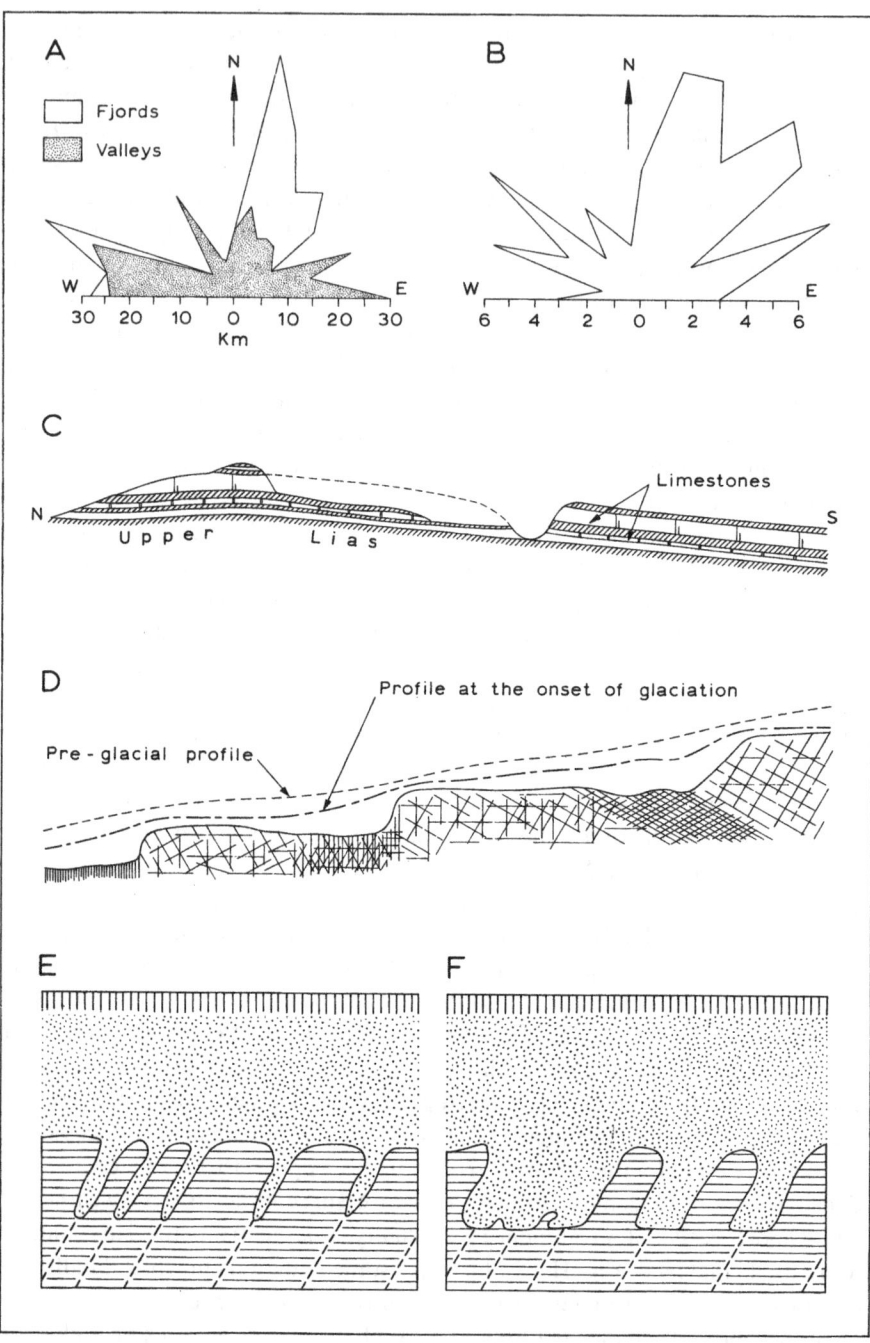

changes. In unconsolidated sediments the pressure needed to shear one horizon might be twice that required to shear another a few metres above or below it. In consolidated sedimentary rocks unevenness in cementation leads to corresponding irregularities of form including curious pedastal rocks formerly attributed to wind action, like those near Montpellier-le-Vieux in Languedoc.

Figure II.15 Lateral variations in rock resistance and associated erosional and weathering forms.

A and B. Vector diagram of valleys and fjords (*left*) and strike frequency diagram of fractures (*right*) in South Lyngen, near Troms in northern Norway (*from Randall, 1961*).

C. Vertical incision of a strike stream shifting down dip related to the lateral disappearance southward of a clay beneath the lower limestone (*after L. Richardson* et al., *1946*, Geology of the country around Witney); the river is the Windrush, south of Leafield, and the disappearing stratum, the Jurassic Estuarine Clay, underlying the Taynton Stone.

D. Steps in the longitudinal profile of a glaciated valley with over-deepening in well-jointed parts (*after F. E. Matthes, 1930*, U.S. Geol. Surv. Prof. Paper 160) as suggested by studies in the Yosemite valley.

E and F. Density of jointing controlling the distribution of penitent rocks at the sub-surface weathering stage (*E. Ackermann*, Zeit. f. Geomorph., *Vol. 6*).

III

Physical, chemical and biological basis of geomorphological processes

A. Regional Climate

1. Temperature

Solar radiation received by the earth is the transfer of thermal energy by means of electro-magnetic waves, no material medium playing an essential role in its transmission. This is the energy which drives the chemical, biological and mechanical forces of sub-aerial weathering processes. This amounts to 13×10^{23} calories of radiant energy each year, only 47 per cent of which reaches the earth's surface due to absorption, scattering, and reflection of the incoming waves by the earth's atmosphere. Furthermore, snow and ice may reflect nearly half of this fraction, grassland nearly a quarter, bare ground 7–20 per cent, and woodland may reflect 10 per cent or less. As all surfaces reflect more and absorb less energy when the angle of incidence of the light waves is low, the amount of energy received equatorward of 38° latitude leads to a net heating, compared with net cooling towards the poles. Oceanic and atmospheric circulation counteract this lack of balance and prevent temperatures at the poles and equator from becoming progressively more extreme. Part of the total input of energy thus becomes an azonal influence on geomorphological processes because the water evaporated is distributed the world over, maintaining ice-sheets, glaciers, streamflow, and winds are created. The other portion of the incoming energy, involved in the actual heating of the ground surface shows a pronounced latitudinal trend. The annual amount of energy eventually filtering through to the ground surface is 60 000–70 000 calories/cm² in the humid tropics compared with a mere 2000–5000 calories/cm² in arctic deserts and tundra. The ground surface of mid-latitude forest, grassland, and steppe receive between 10 000 and 40 000 calories/cm². In cold areas, although the energy input is minimal, there is in less extreme climates enough heat to melt ice. This makes refreezing possible.

This process involves the reorganization of water molecules into a rigid crystalline network with an associated 9 per cent increase in volume. Refreezing thus generates stresses in confined spaces in soil and rock. With higher temperatures although water molecules are increasingly in the random positions of a liquid and have a greater energy than the molecules of the solid, ice, turbulent flow dissipates much of this energy as frictional heat. However, at higher temperatures chemical, biochemical, and biological weathering becomes increasingly important, rates doubling for every 10° C rise in temperature. It is impossible to predict how effective the heat reaching the ground surface is in land-forming processes. With the equatorward increase in clay decomposition products and the increasing relative importance of solutional weathering, thermodynamics have little to do with the transport as well as the erosion of the bulk of the weathered material. Poleward, snow is a very poor conductor of heat, particularly if in a dry state. The thermal conductivity of snow may be almost one-tenth that of the mineral part of the soil, markedly limiting frost penetration in winter. In Siberia the greatest differences between air and soil temperatures occur everywhere in January as soon as the snow cover begins to persist. Snow cover may even keep the ground-surface temperature at or near melting point if there is heat transfer from the sub-soil. As a result, in areas like the greater part of the high mountains in north Sweden, permafrost is unlikely to be widespread although it is not possible to establish a clear relationship of soil freezing with air temperature and depth of snow cover. Even lakes may be deep enough to keep bottom deposits unfrozen and the ground beneath may remain unfrozen at all depths creating punctures through frozen ground even in areas of extensive permafrost. In the Taylor valley, South Victoria Lands in Antarctica, where average annual temperatures are −20° C some of the deeper ice-covered lakes like Lake Bonney remain liquid at depth. In summer in cold regions much of the small amount of solar energy is absorbed merely in melting snow and ice, and increases in soil temperature to promote biological and chemical activity are minimal. In European U.S.S.R., the total inflow of heat between April and August is 1000 calories/cm² in the north-west, rising to 3000 calories/cm² in the south-east; the expenditure of this energy in thawing accounts for all that supplied in the north-west and half of that supplied in the south-east. In humid and cool temperate latitudes extensive peat deposits exist because summer temperatures do not provide adequate supplies of energy to evaporate the colossal volumes of water held in these highly porous organic veneers. In warmer humid latitudes photosynthesis and respiration of plants route an appreciable proportion of the solar radiation input into the biological cycle. In the humid tropics the task of pumping out water from dozens of metres of deeply weathered mantle absorbs much energy. In addition there is not just an acceleration of weathering reactions but equally pronounced accelerations in processes involving the reprecipitation of dissolved solids, sometimes referred to as back-weathering, which also makes it very difficult to assess without measurements how much energy is converted into the work of net removal of material from the land surface.

Several speculations about the influence of temperature on land-forming processes are possible. The greater solubility of carbon dioxide at lower temperatures has been used as an argument for expecting greater solution in higher latitudes, whereas the increased reaction rate at higher temperatures appears in qualitative terms as an apparently sound reason for expecting the reverse. A decrease in temperature, by lessening the settling velocity of suspended sediments, might suggest that sediment transport is more efficient in higher latitudes. On the other hand, with viscosity decreasing with higher temperatures, so that of tropical rivers is about half that of temperate rivers, the greater turbulence could suspend more easily the smaller-sized sedimentary particles of those regions, thus offsetting the more rapid settling velocity. Clearly the mere statement of these temperature-controlled chemical and physical relationships can, without measurements from comparable natural situations, lead to contradictory statements. A very large number of measurements will have to be made before it will be possible to understand how the immense energy of solar radiation is converted into the work of sculpturing landforms.

B. Amounts and Motions of Water and Ice

1. Hydrological considerations

Water, the only compound that occurs naturally at the earth's surface in gaseous, liquid, and solid states, is indispensable for weathering, erosion, and for the removal of weathering products. In many situations, whatever may be the detailed mechanisms involved in rock breakdown, it is ultimately the bulk of water in runoff which is a major factor in influencing rates of denudation (fig. III.1). Climate determines the availability of this agent and its volume and period of flow in streams, through soils and down slopes. Three hydrological characteristics of water available for geomorphological work are more significant than the total precipitation.

First, the effective precipitation may, after the demands of evapotranspiration have been met, be an increasingly small fraction of the total precipitation where the supply of energy to vaporize water increases. At higher temperatures, air can hold a very large quantity of water without its relative humidity increasing. Evapotranspiration in the tropics is therefore potentially very intense, and a month with only 20 mm is a dry month. The coefficient of flow describes that percentage of precipitation which enters river channels, the volume of running water increasing at a slightly greater rate than that at which the temperature decreases. Examples include the forested River Njong basin in the Cameroons where only 16 per cent of the 1800 mm precipitation enters the river and on the Ivory coast coefficients are similar. About 25 per cent of the 1500 mm of precipitation falling in the Congo basin enters the river. In low flat areas like the Yucatan platform, coefficients may be 10 per cent or less. In the areas of more accidented relief the coefficients may exceed 50 per cent, and in the Amazon basin, due to the altitude of its Andean

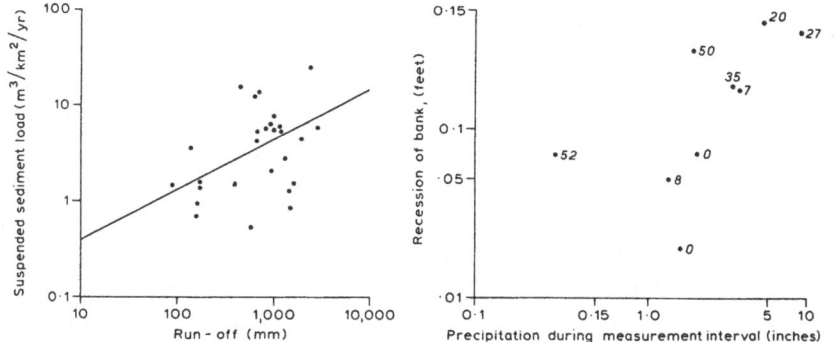

Figure III.1 Some hydrological and temperature factors leading to greater removal by fluvial transportation. (*Left*) Relationship of suspended sediment load to runoff for catchments in eastern Australia (*from Douglas, 1967*). (*Right*) Relation of stream bank recession to precipitation and frost (*from M. G. Wolman, 1959*, Am. Jour. Sci., *Vol. 257*). The numbers are percentages of the observation interval during which mean daily temperature was below freezing. The observations were made on the Watts Branch, Maryland.

headstreams, 40 per cent of the 2300 mm precipitation flows off. In general, however, evapotranspiration in equatorial regions could account for about 1200 mm of precipitation, and runoff varies between 250 and 1000 mm according to precipitation and slope. Great actual losses in semi-arid or arid areas are experienced along wooded margins of allogenic streams or even from the margins of ephemeral streams. For instance, in the Tombstone area of Arizona losses from such situations are the equivalent of 300 mm of precipitation per month. In the United States 70 per cent of the average precipitation of 750 mm returns to the atmosphere due to evapotranspiration. Comparable with this deficit is that of French river basins, which range from 475 to 520 mm. In the steppe zone of Russia precipitation decreases from 400 to 200 mm in southward and eastward directions, exceeded by potential evapotranspiration figures of 700–800 mm. As an instance of mid-latitude conditions in the Far East, the Han river basin in Korea has a deficit of 540 mm compared with a total precipitation of 1200 mm. In cooler environments, like northern Sweden where precipitation may be of an order between

600 and 2000 mm, evapotranspiration is about 150 mm/year. In the tundra of European U.S.S.R., precipitation of 250–450 mm/year compares with an evapotranspiration loss of about 100–150 mm/year. The Lena leads off about 175–190 mm of the 370–385 mm of precipitation and the loss in the Yenessei basin is similar. To the south in the taiga part of the forest zone precipitation of 400–600 mm/year similarly exceeds evapotranspiration of 150–300 mm. In the milder mixed broad-leaved forest zone, precipitation increases to 500–700 mm, but because evapotranspiration increases also, the moistening of the ground is no greater.

Intensity of precipitation is a second important factor in landform study because not only is the importance of mechanical impact increased on bare soil but also a high percentage of a downpour will run off with little lost to evapotranspiration if it exceeds the infiltration capacity of the soil. Tropical areas are prone to repeated downpours. In the high Guinea area of the Congo, ten downpours of more than 50 mm falling at a rate of 5 mm/min have been recorded. At Ibadan near the Nigerian coast, more than 70 per cent of the annual total of 1200 mm falls in downpours of over 12·5 mm. At Bulawayo, over a period of 13 years, falls sustained for at least 15 minutes included one with a rate of 200 mm/hour, eight at 100 mm/hour, and fifty-two at 50 mm/hour. Even within a year downpours may be frequent. Over much of West Africa falls in the 150–200 mm/hour range may be sustained for very short spells. Monsoon Asia is another area experiencing huge downpours. In southern China precipitation in excess of 400 mm may fall within 24 hours. Occasionally extremely heavy downpours occur in other environments. For instance, in 1966 heavy rain in Italy included falls of more than 500 mm in 2 days in the Alpine parts of the Tagliamento basin. During the Exmoor storm in August 1952, which produced intensive erosion in the Lyn valley, precipitation in excess of 185 mm was recorded. In the Crimea midsummer torrential rains of between 50 and 100 mm are sufficient to release mudflows. While intense erosion is a feature of the occasional once-a-generation downpour in higher and mid-latitudes, in equatorial and monsoon areas intense downpours are experienced several times a year or even annually.

A third factor adds to erosional capacity of surface water in cool environments, the melting of snow and ice. For example, the channels leading to the Kvikkjokk delta in Swedish Lapland are generally frozen from the end of October to the second week in May. Annual total discharge is 870 million m³ to 1080 million m³, with daily discharge during ice-free periods of 25–40 m³/sec rising to 200 m³/sec in floods. As melting reaches its peak, rivers may increase in volume by 300–400 per cent in 24 hours. On the River Colville in Alaska, 42 per cent of the total annual runoff was discharged during 4 weeks of ice break-up. During spring flooding on the Mackenzie discharges in excess of 500 000 cusecs (14 000 m³/sec) have been recorded, more than twice the maximum observed for the Colville. These figures compare with the Yenissei where the annual flow of 17 400 m³/sec is the flow of only part of the year. Despite the huge width of these valleys the Mackenzie may rise more than 6 m during the spring melt and the Yenissei by 10 m. Ice jams may cause overtopping of adjoining levels.

Although the chances of abrupt and substantial downpours or snow-melt surges of flow are not high in temperate latitudes, floods are still an important consideration in landform studies. Interest centres round the recurrence interval. This is the average interval of time within which a flood of a given magnitude will be equalled or exceeded once. It also is a prediction of the chances of a flood of a given magnitude occurring in any one year. For instance, a flood with a recurrence interval of 50 years has a 2 per cent chance of occurring in any one year. The ratio of discharge for some recurrence interval to the mean annual flood gives a useful indication of possible geomorphological effects of the extreme event in relation to average conditions. In the United States, bankfull stage occurs once in two years out of three. Similarly a recurrence interval of 1·5–2 years for the bankfull stage has been observed in north Sweden, and likewise the inundation of the Ob floodplain does not take place every year. In England and Wales discharges greater than bankfull may occur on two days each year on average.

2. Hydraulic considerations

The quantity of water leaching through weathering profiles or draining down slopes is the most important single factor in controlling the amount of breakdown of parent materials and in their removal. Large volumes of water, on average, even if dilute in sediment and solutes, remove by their sheer bulk far more material than small flows with a dense sediment load and high solute concentrations. However, in the study of certain distinctive localities, usually those where unconsolidated sediments are being reworked, the lack of cohesion in the alluvium makes some understanding of the motion of the water important in an understanding of the features developed. Also the geomorphologist who might be interested in the hydraulicist's studies of alluvial channels requires some familiarity with the terms describing the ways in which water moves.

In a thin boundary layer just above the static floor of a stream channel water particles move in a parallel way with no cross-currents. This motion is termed laminar flow. This is usually a mere boundary film with stream or slopewash velocities nearly always rapid enough for the cross-currents of turbulent flow to develop. Due to the rough surface of channels in mountainous areas there may be almost as much backward as forward motion in the cross-currents and eddies. Turbulent motion consists of two types, the division depending on the Froude number

$$F = \frac{v}{\sqrt{gR}}$$

where v is the mean velocity, g the acceleration due to gravity, and R is the hydraulic radius of the channel. The last term is equal to the depth of water for wide shallow channels. Streaming turbulent motion occurs when the velocity is relatively low and is less than the square root of the hydraulic radius times the acceleration due to gravity. Velocities with a Froude number greater than 1 are those of shooting turbulent flow. In deeper channels with smooth perimeters in cross-section and sinuosities in plan, cross-currents are arranged in a circulatory motion because in a bend the faster-moving water particles nearer the surface are subject to greater centrifugal force than cross-currents near the bed. These motions combined with the overall downstream flow cause a spiral motion termed helicoidal flow. This motion is observed in meanders and at channel forks but is easily obscured by localized turbulence caused by channel irregularities. Where a confined stream of water enters a broad area of tranquil water it is useful to consider the analogy with submerged jet flow, an hypothesis developed by hydraulic engineers, in considering the geomorphological implications of the sharply defined discontinuity where the incoming flow is lined with intensely turbulent eddies and flanked by zones of complementary reverse flow. However, the way in which a sediment-laden stream decelerates and diffuses laterally is different from that of a piped supply of sediment-free water discharging below a water surface.

Changes in stream velocity are a difficult characteristic to discuss because of the empirical constraints in the calculation of mean velocity from a series of points in the cross-section. One constant which varies over a considerable range is von Kármán's constant of turbulent exchange which compensates for the damping effect of turbulence by dense concentrations of suspended sediment. In fact, formulae for velocity distribution in an open channel are not valid at high concentrations of suspended sediment. Another variable factor is channel depth. For a given volume of water velocities will be less in a broad shallow channel than in a relatively deep narrow channel where channel resistance is much less because the wetted perimeter is shorter. Increasingly large volumes of water tend to move more quickly because, until bankfull stage is reached, the relative proportion of flow involved in overcoming frictional resistance of the channel becomes less. For instance, water flowing into the Laitaure delta in Sweden with a volume of 253 m³/sec, nearing bankfull discharge, had a mean velocity of 1 m/sec whereas a summer low flow of 39 m³/sec. flowed at 0·4 m/sec. With flood discharges on the Ob, velocities are between 1·5 and 2·0 m/sec. The average velocity of the Colville river, Alaska, is 1·5 m/sec; for several streams in the United States mean velocities vary between 0·9–1·2 m/sec. Increases in the depths of channels downstream improve the efficiency of the channels to transmit faster flows and larger flows moving over diminishing gradients in lowland areas. The study of these inter-relationships is of crucial importance to hydraulic engineers concerned with flood control.

3. Waves and offshore currents

The description of waves advancing on a shore involves two basic measurements of scale. The wave length, L, is defined as the horizontal distance between corresponding points on two successive waves. The wave height, H, is the vertical distance between the wave crest and the preceding trough. Waves consist of orbital movements of water which are not quite complete, with water particles moving forward slightly as each wave passes producing a slight displacement of water. It is the transmission of energy through water waves that is the important consideration rather than the relatively slight displacement of the actual water particles. The diameter of the orbital motion diminishes rapidly downward in a geometric progression related to wave length. The orbital diameter is halved for each increase in depth of $\frac{1}{9}$L. Wave velocity in deep water is primarily dependent on wave length, but high waves of a given length run somewhat faster than low waves. On coasts the wave height gives a good indication of the total energy which is usually proportional to the square of the mean wave height in most theoretical equations and is largely a result of wind velocity, the fetch or the distance of open water over which the wind blows, and the water surface gradient created by the tidal range. Thus waves greater than 2 m in height are not frequent off the surf zone in the Mediterranean where the mean tidal range is only about 25 cm. Tidal range in the Great Lakes is as little as 7·5 cm. In contrast the classic case is the Bay of Fundy funnelling tides of 3 m at its entrance up to a range of 15 m at its head.

Waves where the depth to the bottom is less than half the wave length, L/2, are defined as shallow-water waves. L/2, referred to as the wave base, is often about 9 m below the surface. In entering shallow water the horizontal particle velocity at the wave crest becomes increasingly greater than that retarded near the bottom. At a depth equal to 2H the wave profile becomes very peaked and asymmetrical. Breaking of the wave finally occurs when the depth is about 1·3H. If the bottom gradient is gentle the crest of the wave spills over the advancing front of the wave without completely destroying the wave form. In contrast to the spilling breaker, a plunging wave forms above steeper bottom gradients when heavy swells pitch the wave crest into the preceding trough. After a wave of oscillation breaks, a wave of translation bores across the surf zone until it meets the increased slope of the foreshore or swash zone, where it exhausts its energy either by running up the foreshore slope as swash, or by collision with the returning backwash which tends to form a layer on the bottom in the swash zone. Where the bottom contours in shallow water are not parallel, waves slow down first where the plan of bottom contours projects seawards and the crests of waves swing round and tend to approach an alignment tending to parallel the bottom contours, thus concentrating energy on the headlands.

Currents differ from waves in that there is a progressive movement in one direction. Due to the hydraulic head of water piled against a shore by waves, compensating seaward currents may form narrow lanes which exist right to the water surface and cut through the breakers. These rip currents, although first described only in 1941, are now recognized as a major element of nearshore and surf-zone dynamics. They may maintain velocities of over 1 m/sec for periods of minutes. Waves approaching a shore obliquely cause a unidirectional movement alongshore. The zig-zag pattern of these longshore currents may, however, be interrupted by seaward-flowing rip currents. The greatest longshore velocities, which may be a few dozen metres per hour, occur midway between the two high-energy zones, the breaker and swash zones, which cause drag at the seaward and shoreward edges.

Tidal currents, produced by the ebb and flow of tides rarely exceed 3·2 km/hour, but where the flow is channelled through some constriction in coastline form, currents may at times be as much as 16 km/hour. Currents seaward of the breaker zone do not usually exceed 30 cm/sec, although seas around the southern half of the British Isles are notable for the strengths of their currents which are 2–3·5 km/hr over large areas, and 4 km/hr in smaller areas. Off the New South Wales coast, a variable southerly current moves at an average of 5–7 km/hr 8–15 km offshore with a more indeterminate and variable current running north closer to the coast. Ebb tidal currents move through Hell Gate in the East River at New York at 9 km/hr. Near-bottom currents in the Bering Straits average 5·4 km/hour. Other similar currents are not restricted to the surface. In the Indonesian Strait, strong current movement exists to a depth of 3000 m and a deeper flow through the Straits of Gibraltar is in a direction opposite to that at the surface, as dense saline water, a hundred times greater in volume than the Mississippi's discharge. The Gulf Stream moves at rates of 5–12 km/hour, with the eastward-flowing equatorial counter-currents moving at three times these speeds. Smaller-scale circulations due largely to the strength and direction of the wind may move a few kilometres a day like those moving anticlockwise in the North Sea. In the Mediterranean where tides are insignificant there is nonetheless a south to north current in the Levantine basin which is about 0·9 km/hour, probably influenced by the Nile's discharge. Although these currents may all play some part in redistributing river-borne sediment entering the sea, they do not exert sufficient drag on the bottom, below 10 m down, sufficient to resuspend deposited material.

4. Ice and snow

Glaciers cover about 11 per cent of the land area of the earth, 30–50 per cent of the land area is covered with snow and about 23 per cent of the ocean area is covered by sea ice. The Arctic ice, although it has a profound influence on heat exchange in the oceans is only 3 m thick. The Greenland ice-sheet covers 1 700 000 km^2 with a depth of 3 km at one point, and the Antarctic covers 12 950 000 km^2, and in thickness exceeds 4 km at places. These two masses account for about 99 per cent of the volume of the ice on the globe and their marginal tongues of ice may be larger than valley glaciers, like the Beardmore glacier in Antarctica, which is 200 km long and about 40 km wide. There are also some very much smaller ice-caps on uplands in higher latitudes. At 70° N the Barnes ice-cap in Baffin Island covers 5900 km^2 and is about 165 km long and 22–62 km wide. In the Icefield Ranges the most extensive ice-cap in the North American continent is 240–670 m thick. In contrast to the dome shape of summit ice-caps, glaciers occupy valleys and a few continue beyond the confines of the valley to broaden out as piedmont glaciers. The Malaspina glacier near Yakutat emerging from several narrow valleys in the St Elias Mountains spreads out in a broad lobe about 50 km across and about 40 km long. Valley glaciers may reach lengths of 30–50 km and 3–4 km in width, whereas in other areas the ice-covered area consists of a large number of small glaciers. In Transcaucasia there are 487 glaciers with a total area of only 635 km^2.

The process of the conversion of snow to ice starts with the recrystallization and partial consolidation of porous fresh snow, transforming its low density of about 0·1 gm/cm^3 to the densities of 0·4–0·8 gm/cm^3 of granular snow termed firn or névé, and finally to ice with a density of about 0·91. Transformation of névé to ice takes place at depths of about 30–40 m but may be nearer 100 m for ice-sheets. In temperate zones higher temperatures speed up the transformation process which takes about 50–100 years, whereas in polar ice-caps recrystallization, taking place without passage through a liquid stage, takes 200–300 years.

An important concept in discussing the distribution of snow and ice is the 'snowline', recognized in 1887 by Brückner and today commonly defined as the lower limit of perennial snow. The definition of the climatic snowline disregards local influences like wind, insolation, and aspect which make any close correlation between temperature–precipitation averages and the snowline unrealizable, whereas the orographic snowline links the points at which the preceding winter's snow will just disappear. In Spitzbergen, at 78° N, the snowline is approximately 500 m. In Iceland at 65° N the snowline rises towards the south-east from 600 to 1000 m, is about 1600 m in southern Norway at 60° N, about 2700 m in the Alps, and at 4250 m on Mount Kenya. A very important fact is that east–west changes in snowline are usually more abrupt than the general equatorward rise. The snowline on the dry eastern slope of the St Elias Mountains, Alaska is 1500 m higher than on the wet west slope.

Close to the observable firn line is the theoretical equilibrium line, dividing the zone of ice accumulation near a glacier source from that of net loss in the zone of ablation in its down-valley part. Studies of glacier regimen, considering the balancing processes of accumulation and ablation, are simplified in comparison with water-balance studies by the absence of significant evaporation losses. Such changes, together with the condensation of ice from water vapour known as sublimation, are less than 1 per cent. This is a result of the very large supplies of energy needed for vaporization, being 540 calories per gram, and because absorption of moisture by cold air is very limited, as amounts of water vapour range from 4 per cent by volume in hot tropical climates to only 0·01 per cent in glacial climates. Near the snouts of many Scandinavian, Icelandic, and Alaskan glaciers the ablation loss of ice may amount to 10 m or more in one season. In a more southerly location, in British Columbia, the Salmon glacier at 56° N, with an accumulation zone at 1650 m, loses about 5 cm/day. One of the more striking characteristics of glacial budgets, and one which emphasizes how sensitive some interactions are to changes, is the degree to which ice-caps generate the local climatic conditions to maintain a supply of moisture and snow. The Barnes ice-cap, for instance, is just a low dome on an even upland surface, generating its own nourishment in the form of orographic precipitation. Later workers have enlarged on Leverett's (1916) view that the North American ice-sheets grew westward because of storms coming from the south-west. Warm moist Tropical–Gulf maritime air masses would have been cooled both by coming into contact with the ice-front and by being forced upward over the glacier and its associated wedge of cold air, creating a major zone of precipitation along the ice-front. Similarly, self-enhancing conditions in the development of the Greenland ice-cap were suggested by Wegmann in 1941. He surmised that as the climate cooled at the end of the Pliocene, the relief of Greenland was that of an embayment, surrounded by bordering chains on which glaciers developed which spread into the interior lowlands as piedmont lobes. It seems possible that if such sheets coalesced and the ice surface rose to the permanent snowline névé supply could have then added directly to the surface.

Like the effect of channel floor on modes of stream flow, the irregularities

of the sub-glacial floor make the movement of ice a very complicated phenomenon, with the variables influencing the mass budget and névé formation
changing with altitude, aspect, latitude and in a downvalley direction. From
early attempts to describe ice-flow, using laws of viscosity, impressions were
gained of a spreading body unable to conserve its form indefinitely. Recent
measurements exclude the possibility of viscosity existing in glacier ice.
Instead J. F. Nye (1952) suggests that ice yields to compression stress and
strain rate imposed by an external force like a plastic solid which retains the
form it took under the effect of deformation. It seems that due to pressure
within a glacier increasing with depth, melting and refreezing facilitate intergranular movement, crystal growth, and transfer of material. On a smaller
scale grains tend to be drawn out in response to the forces acting on them,
while on a larger scale, intergranular movement takes place. Melting at grain
boundaries reduces friction between the grains and facilitates movement of
the grains relative to each other. In some temperate glaciers ice aggregates
may be surrounded by a weakly saline membrane which, by tending to remain
liquid when refreezing occurs, may facilitate inter-granular movements. This
mode of mechanical movement is termed plastic deformation and depends on
a certain depth of ice, about 20 m for temperate glaciers and on some downvalley gradient. Under a laboratory stress of about 1 kg/cm^2 ice flow becomes
appreciable, but in field conditions with loads imposed for much longer
periods of time, loads of less than 100 gm/cm^2 are sufficient. Probably much
of the movement of thick temperate–latitude glaciers on gentle slopes occurs
by plastic deformation, although the rate of movement due to plastic deformation decreases rapidly near the base of a glacier. However, a second process,
that of sliding on the bed, is an important, if spasmodic, addition to the total
movement of the glacier. It is accompanied by tensions which, if shearing
occurs, create large fissures at the surface. The French geophysicist,
L. Lliboutry, and J. Weertman, have done a great deal to clarify concepts of
basal sliding. This mechanism may account for as much as 90 per cent of the
total movement, particularly where there is a steep gradient, a high absolute
velocity and if the ice is relatively thin. Clearly, temperatures at the base of
the glacier must be near melting point for sliding to occur and in very cold
areas all but the most thin glaciers are frozen solid to the rock at their base.
B. Kamb and E. La Chapelle (1964) made the first direct field observations of
mechanisms involved in the sliding of a glacier over its bed. The observations
of N. A. Ostenso (1965) have added to evidence which suggests that melt-
water lubrication is an important factor. As water has a greater density than
ice, water-filled spaces in and below the ice cannot be compressed without
being previously emptied. At the base of the Casement Glacier, the winter slip
rate in 1966–7 was 2·3 cm/day compared with summer rates of 2·9 cm/day
in 1966 and 2·6 cm/day in 1967. Over a 24-hour period there is a direct

relation between slip rate and water available at the base for lubrication. The breaking up of the glacier into blocks makes it possible for the ice to flow by still another mechanism, moving like a powder on a giant scale, particularly if water fluidizes the ice. Meltwater circulating in the ice in tunnels, running under pressure in siphons and enlarging fissures provides nearly all the water at the base of a glacier. Even in Iceland where the geothermal heat supply is two to three times the normal supply, this minimal source of heat accounts for only 5 per cent of the runoff in mid-winter. However, in the case of some glaciers, ground water may be an important part of the discharge at the winter minimum, according to observations in Switzerland and northern Sweden, and may make significant contributions to the lubricating effect.

In high latitude glaciers and ice caps where the ice is welded on to the bedrock, the slow movements take place along a zone of shearing in the lowest 100 m, and in formerly ice-covered areas of deep narrow valleys some experts find it conceivable that ice sheared across the subjacent ice filling these valleys. Near the termini of glaciers shear planes similar to thrust faults are often observable, but this process appears to be of only very local significance. It is important to distinguish ice-carapaces from glaciers as the area on which the first may occur, as seen in Alaska at the present day, could easily lead to an overestimation of the thickness of a former glacier. For instance in central parts of the Trinity Alps, California, there are extensive glaciated rock surfaces on sloping valley walls far above the level of the former ice streams. However, here striae, scoured and plucked rock surfaces and schrund lines show that the valley walls in these areas had a mantle or carapace of ice possibly no more than 60–90 m thick, that flowed towards the valley axis.

The velocity of displacement of ice tends to be proportional to the cube or fourth power of stress. Velocities usually range from a few millimetres to a few metres per year. The very cold glaciers may move only a metre or two in a year despite their immense thickness, although the Ferrar glacier near the Ross Sea moves at 5 cm/day and the Beardmore glacier at 0·8 m/day. On the west coast of Greenland the velocity of several glaciers entering Disko and Umanak bays is about 1·5 km/year. In the Alps speeds of ice surface flow are of the order of 40 m/year, small glaciers in China move 10–30 m/year, in Patagonia the Grey glacier moves at 450 m/year, and extreme cases in Alaska move at over 60 m/day. In addition to the factors like ice thickness, slope, and temperature, mean velocities may change for a variety of circumstances. Velocities, for instance, are not constant down-valley; the Tasman glacier in New Zealand flows at a rate of 0·5 m/day 19 km from the snout and at 0·35 m/day 10 km from the snout. Mean velocities may change from one decade to the next. In Swedish Lapland, the Mikka glacier which now terminates just below 1000 m altitude had a mean velocity of 6·2 cm/day in 1895–

7; in 1899–1901 it was 7·3 cm/day but was half this figure in 1958–9 and only 2·9 cm/day in 1961–2. In fact, during the last century most glaciers have been shrinking. Distinct from flow velocities are rates of change in the position of the ice-front. Studies of varved clays suggest that the ice-cap front of the last glaciation retreated at 100–150 m/year from Scania, the southern-most province in Sweden, while in vacating areas beyond Stockholm, rates increased to 200–300 m/year. Retreats of ice-fronts, however, are essentially due to glacial thinning rather than to processes operating at the ice-front itself.

Movements of frozen snow and ice, in addition to glacier flow, merit some consideration. Snow-creep is a slow continuous glacier-like movement of a snowpack. Stress is too small to produce shear failure as in avalanches but appreciable slip may take place along the interface between the ground sur-face and base of the snowpack. Investigations at 1100 m on Mount Seymour in British Columbia found movements of 60–105 cm within a winter with maximum rates of movement of 12·5 cm/week. The localities where ava-lanches occur are usually those where precipitation, increasing with altitude, leads to great thicknesses of snow cover which shear under their own weight at altitudes where atmospheric circulation multiplies temperature oscillations through freezing point. In consequence mountains in temperate oceanic areas are the domain of avalanches and may occur there within an altitudinal range of less than 500 m. Wet snow may slide on slopes as low as 15 degrees.

C. Mechanical and Frictional Forces

The mechanical and frictional forces involved in the physics of denudation are simply those of the shear stress exerted by the moving agent and of the resistance of the land-surface material to that stress. Most of the essential features of flow in fluids and deforming solids can be represented as in fig. III.2 by a sheet of infinite width and length flowing down an inclined plane. If

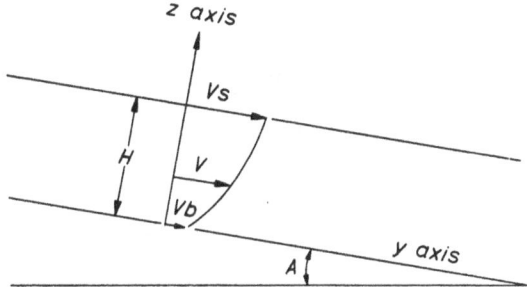

Figure III.2 Stresses and strains for laminar flow on an inclined plane. H is the thickness of the flow, Z the height above the inclined, A is the slope angle, V the average velocity, Vs the velocity at the surface and Vb the velocity at the base.

the flow is assumed to be essentially laminar the shear stress acting on a plane perpendicular to the z axis is equal to

$$\text{Density of flow} \times \text{Acceleration due to gravity} \times \left(\text{Thickness of flow, H} - \text{Height above inclined plane, } z\right) \times \text{Sin of slope angle, A}$$

The maximum shear stress occurs at the base of the flow, where z equals 0, although offshore in depths greater than 9 m this case occurs above the basal plane and no shear stress is exerted on the sea-floor. Shear stresses within actual flows are usually too complex, even in theory, for most geomorphologists to consider that the study of the greater complications within actual flows in the natural environment fall within their sphere of competence.

The force of resistance to movement of land-surface material involves two quite contrasted aspects. The first, the breakdown of solid rock by weathering processes may take millions of years and is considered in some detail in Chapter IV.A. The second concerns the initial dislodgement of a weathered particle or the reworking of unconsolidated deposits. Here much of the resistance is simply internal friction. As the shear stress exerted by some flow on its basal plane increases, there comes an instant when a few of the loose fragments on this plane are entrained. Fig. III.3 illustrates how the force of a flow turns an individual particle about its leeward points of contact with subjacent particles. The force of the grain's weight, or its immersed weight in a fluid, acts through the centre of gravity of the particle to pull it backwards and downwards. When the drag force of the flow just exceeds this, movement takes place. On a scree, steeply piled sand, or similar accumulation, the value of the angle A is very close to that of the angle of repose.

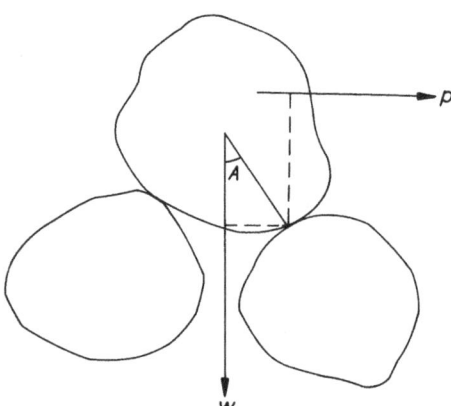

Figure III.3 Physical forces acting on a particle at the threshold of movement (*after Bagnold, 1941*). *W* is the grain's immersed weight, *p* the drag force due to the fluid acting parallel to the general surface, and *A* is the angle at which the grain is poised on adjacent grains, close to the angle of repose.

D. Sediments and Mechanical Characteristics of Soils

'On beaches . . . with the motion of the waves, oval pebbles are driven to the same place as oval and round to round . . .' Democritus

There are four reasons why the geomorphologist may find some knowledge of the physical properties of soils and sediments useful. First, sediments may, as already discussed, develop properties characterizing their depositional environment, and it is often a simpler, if not the only approach, to measure sedimentary parameters rather than the depositional processes themselves. Secondly, there is a close interdependence between soil formation and landform development, and, since Barton in 1916 described accelerated weathering below old soil lines in Egypt, several ways have been recognized in which soils and sediments can either accelerate or retard further weathering of bedrock. These inter-relationships will be discussed in Chapter IV.D.9. Thirdly, fossil soils are a record of phases of erosion and stability over thousands of years and of environmental conditions during periods of stability when soil formation took place. Fourthly, because the spans of time during which soils change tend to be shorter than times involved in landform change, notions on rates of the latter might be grasped from studies of soil adjustments to changing conditions, which could guide thinking on the degree to which landforms might adjust from one change in environmental conditions to the next.

Soils and sediments become an integral part of landform study, not simply because they may afford useful aids but mainly because any advantage to be gained depends on a full awareness of their complexities or of limitations on interpretations which might expose over-optimistic approaches or superficial generalizations.

1. Particle sizes

Since the publication in 1914 of J. A. Udden's attempt to relate particle-size modes to different kinds of particle movement, the significance of particle size has been much studied. Sizes determined by sieving and settling velocities indicate approximately the diameter of a sphere of equal volume to that of the particles. Particles smaller than 2 mm, usually taken as the upper limit for sand-sized particles, are often described in microns. A micron is one thousandth of a millimetre, often symbolized by the Greek letter mu, μ. References to particle sizes may appear in three contexts. First, there are boundaries between particle-size classes which are arbitrary but necessary in establishing convenient subdivisions. There are several schemes of subdivision, but most use a reversed log scale with each equal subdivision spanning a progressively narrower size range, so that a small number of large particles does not

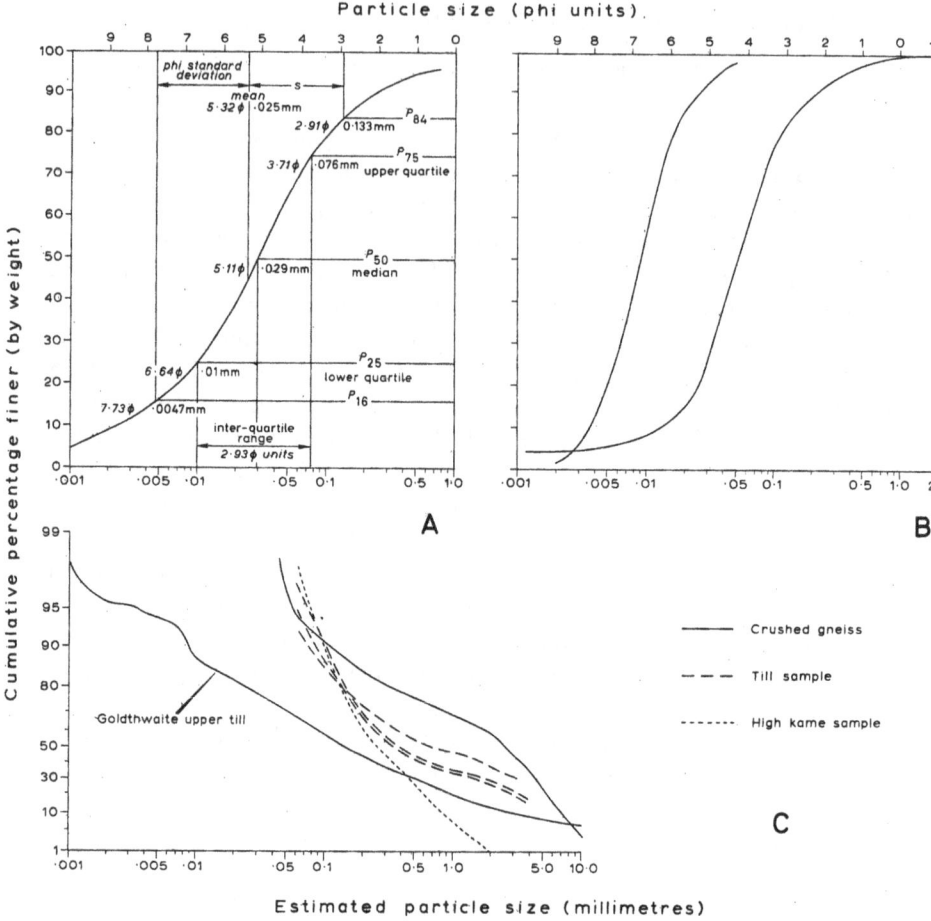

Figure III.4 Cumulative frequency curves showing particle-size distributions of sediments.

A. Percentage points used in the calculation of indices to describe particle size distribution.

B. Illustrations of the distinctive sorting achieved in wind-transported sediments; the fine silt is Saharan dust, wind-blown to Britain (*A. F. Pitty, 1968*, Nature, *Vol. 220*) and the coarse silt, from the Derbyshire limestone plateau is attributed to wind-transport during former periglacial conditions by C. D. Pigott (*Pitty, 1966*).

C. Particle-size distributions drawn on Rosin's Law of crushing paper (*from J. T. Andrews, 1963*, Geog. Bull. No. 20), an alternative to the popular semi-logarithmic paper, illustrating the similarity between crushed gneiss and tills from cross-valley moraines.

dominate the distribution. Secondly, there are specific diameters to describe a cumulative frequency distribution of a sediment sample, of which the usefulness of the P_{50}, the median indicating the central-size value of the transformed distribution (fig. III.4), is most readily apparent. This value may indicate the average velocity conditions of the depositing agent. Thirdly, there are critical sizes that appear to have some functional significance. For instance, frost-shattering appears to be incapable of splitting grains finer than 10 microns; the higher fraction of particles smaller than the 60 microns range in river sands compared with beach sands appears to be the main distinction between these two sediments. However, in general it is inadvisable to make direct and simple inferences from diameter measurements alone because so many variables are involved. Equally diagnostic are sorting indices which describe the range of sizes involved in the rearrangement or adjustment of particles to specific dynamic parameters, like current velocity. For the purposes of sedimentary petrology, the Greek letter *phi* is often used to describe the logarithmic transformation in terms of units of equal arithmetic width,

$$\phi = -\log_2 d \text{ or } d = (\tfrac{1}{2})$$

where d is the particle size expressed in millimetres, as suggested by W. C. Krumbein in 1934. However, partly because tables of negative logarithms to the base 2 are not readily available, a certain amount of difficulty in the use of the *phi* transformation has arisen. Fig. III.5 is an attempt to provide sufficient detail on the micron and *phi* scales to permit interpolations of accuracy adequate for general purposes. For single expressions of central tendency, a description in millimetres or microns is probably more explicit. For sorting indices values of *phi* are obtained from cumulative curves for certain cumulative percentages (fig. III.4). The most commonly used sorting index is Krumbein's (1938) *phi* deviation, $\dfrac{\phi_{75} - \phi_{25}}{2}$, whereas A. Cailleux subtracted ϕ_{50} from ϕ_{75} and ϕ_{25} from ϕ_{50}, and used the smaller value, representing the steeper part of the curve as an index, *Hé*. In more recent work more efficient indices are based on a wider spread of percentiles, R. L. Folk and W. C. Ward suggesting a modification to Inman's (1952) statistic—

$$s = \frac{\phi_{84} - \phi_{16}}{4} + \frac{\phi_{95} - \phi_{5}}{6 \cdot 6}$$

Fig. III.4 illustrates the steepness of curves reflecting the efficiency of wind transportation as a sorting process. A further measure is skewness, an expression of whether material is predominantly in either the coarser or the finer side of the median diameter or symmetrically distributed on either side. In fig. III.4 the median is coarser than the mean and the particle-size distribution is therefore positively skewed. This skewness index can have some value where the finer tail of a distribution might be removed, like the fraction smaller than

62 microns winnowed from beach sands or gravel lags which, due to the truncation of sizes, tend to have negative or near zero skewness values. However, negatively skewed beach sands become positively skewed when either the coarse tail of a symmetrical log-normal curve has been removed or

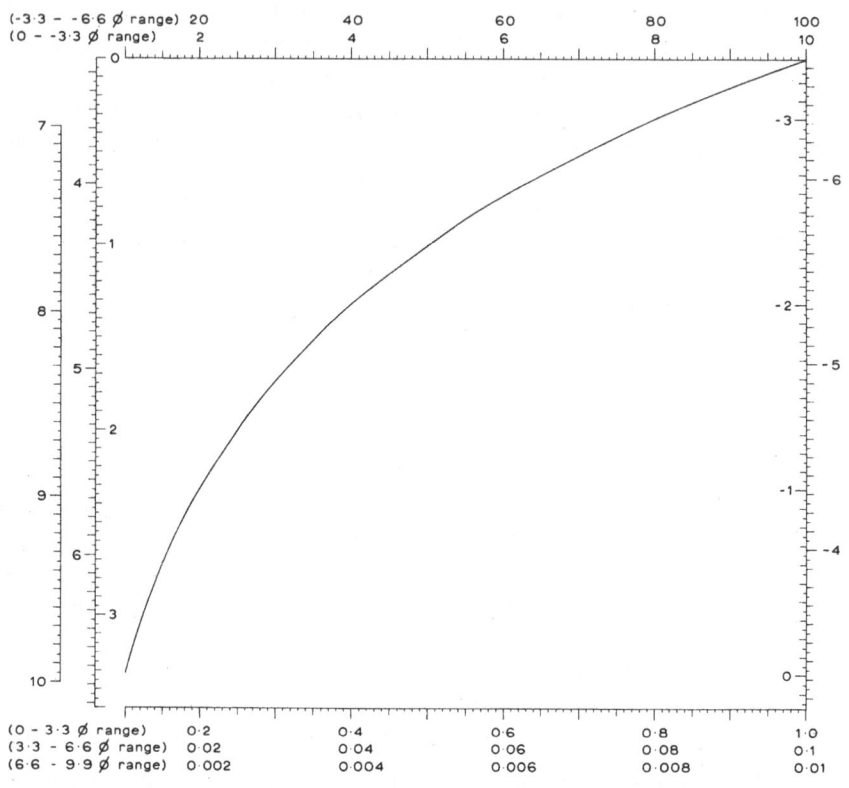

Figure III.5 A metric–*phi* unit conversion chart. The rows are metric units, preceded by the *phi* range (in brackets) to which they correspond; the latter are read off the appropriate column on the ordinates.

a fine tail added by finer material filling up a porous frame of coarse material. In contrast to beach materials, transport by unidirectional flow tends to produce positive values of skewness as is often the case with river and dune deposits. Folk and Ward suggest a *phi* skewness index

$$Sk = \tfrac{1}{2}\left(\frac{\phi_{84} + \phi_{16} - 2\phi_{50}}{\phi_{84} - \phi_{16}} + \frac{\phi_{95} + \phi_{5} - 2\phi_{50}}{\phi_{95} - \phi_{5}} \right)$$

In addition to a reversed log scale sediment size distributions are occasionally plotted on Rosin's 'law of crushing' paper, as both mechanical and chemical disintegration of rocks has a distribution following this 'law' for artificially crushed material which apparently results from random breakages of material (fig. III.4).

Great care and caution are necessary in interpreting textural parameters. In addition to reflecting environmental conditions they are influenced by the availability of materials of a given size. In some environments several processes may operate simultaneously or occur within a short interval. Moreover, even at one point more than one population may be present. It may include grains deposited from the bedload, material settling later from the suspended load and contributions from an underlying veneer deposited under different conditions. There is also the possibility of post-depositional resorting by frost or by organisms. These problems are magnified by the difficulty of recognizing polygenetic soils and sediments in the field. In consequence there is no guarantee that textural parameters will necessarily establish clear differences between dune and beach, glacial and proglacial, blockfields and boulder clay, or between alluvial and colluvial deposits.

2. Particle shape

There have been many attempts to use particle shape in reconstructions of erosional and depositional environments. Indices are basically of two types. First, in coarser materials, indices are based on measurements of the three axes length, breadth, and depth, which might be abbreviated to L, B, and D respectively. A. Cailleux introduced a flatness index $\dfrac{L + B}{2D}$ which has some discriminatory power, as fig. III.6 shows. E. D. Sneed and R. F. Folk (1958) suggest a detailed classification of a sample into several shape categories. The second type of index concerns the smoothness of particle outline. Powers approach depends on the use of a set of reference images with which each particle is compared and classified. For larger particles part of the circumference of pebbles is measured. Where corners are rounded the radius, r, of an inscribed circle is estimated by superimposing the corner on a set of concentric circles. A. Cailleux and J. Tricart have made extensive measurements of an index of wear, $\dfrac{2r}{L}$.

Although criteria distinguishing the effects of distinctive environments on shaping pebbles are necessarily somewhat tenuous, one or two tendencies have been observed, or significant distinctions established (fig. III.7). It seems that most well-rounded boulders of glacial transport may retain traces of original concave surfaces to a far greater extent than is generally observed on stream or beach cobbles. The most important factor governing pebble shape is usually, like the initial triangular form of some glacial boulders, lithological composition. Also a doubling of the mean size of particles may

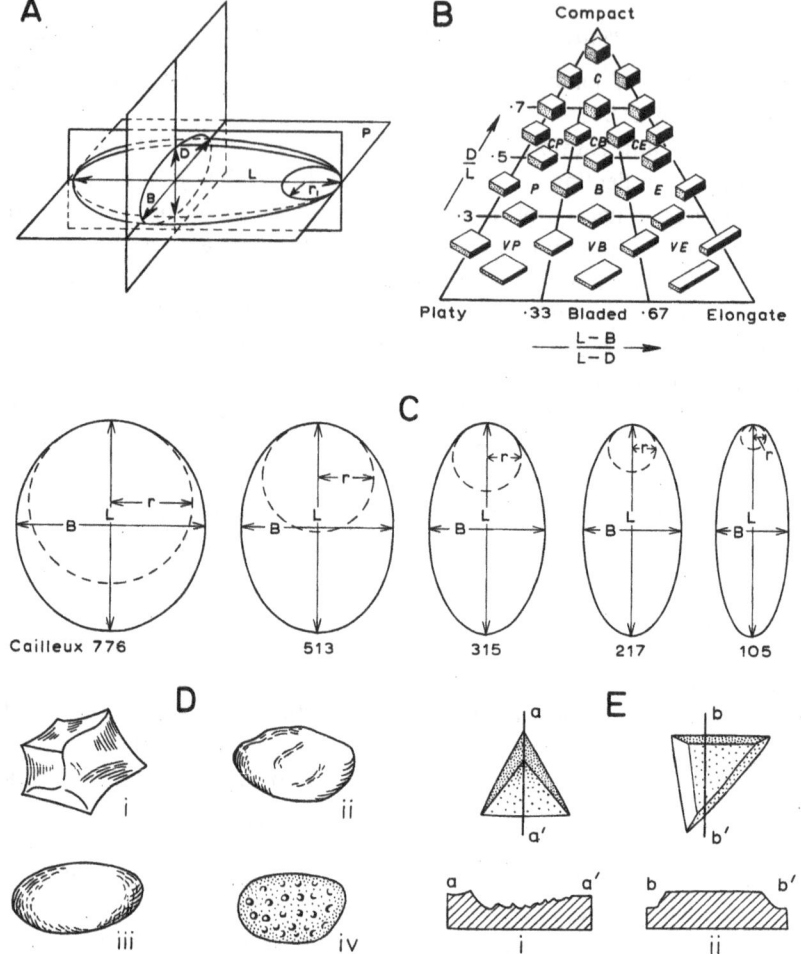

Figure III.6 Some morphometric properties of sedimentary particles.

A. The main axes of a pebble. These are L = length $\rangle$ B = breadth $\rangle$ D = depth. B and D need not intersect at the same point in the horizontal plane, P; also shown is the radius of the circle inscribed in the least rounded corner in the horizontal plane, P.

B. Triangular graph of pebble shape (*from Sneed and Folk, 1958*). Shapes of pebbles falling at various points on the triangle are illustrated by a series of blocks with axes of the correct ratio. All blocks have the same volume.

C. Various pebble shapes and their Cailleux indices of wear, $\dfrac{2r}{L} \times 1000$ (*after Kuenen, 1956, Jour. Geol., Vol. 64*).

D. Examples of the appearance of sand grains under the microscope (*after F. Ottman, 1965, Introduction à la geologie marine et littorale*); i = angular and unworn, ii = sub-rounded, iii = rounded and polished, iv = pitted.

E. Examples of the appearance of sand grain surfaces on election micrographs (*from Krinsley and Donahue, 1968*); (i) mechanical V, showing central depressed area (*above*) with rough surface in cross-section (*below*) and (ii) chemically etched V (*above*), showing a flat raised centre in cross-section (*below*).

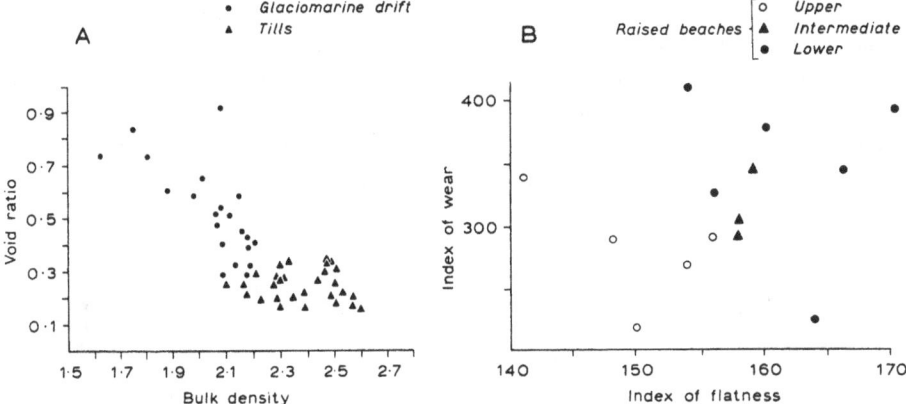

Figure III.7 Differentiation of depositional environments by sedimentary properties.
A. Contrasts in compaction between sediments (*after Easterbrook, 1964*), the tills
being the Vashon and Sumas tills in the Puget Lowland, Washington.
B. Contrasts in the Cailleux indices of wear and flatness for different levels of
raised beach (*from Godard, 1965*). The measurements were made in the vicinity of
Helmsdale, Sutherland, where the lower raised beach is about 1–5 m above sea-
level and the upper raised beach averages 40 m.

affect form and sphericity of pebbles to a greater extent than several hundred
kilometres of river transport in extra-tropical areas. It seems that in general
larger cobbles are less susceptible to breakage and therefore may become
more rounded. It is often very difficult to estimate whether progressive
change of shape might take place with transport, whether a balance of rounding and fragmentation is established, or whether there is a continuing supply
of fresh material, perhaps frost-shattered or solifluction gravel, downvalley or
along the shore. A major factor appears to be the presence of sand because,
like ventifacts polished and shaped by wind-blown sand to the most distinctive of pebble shapes, it seems that gravels in water acquire their shapes when
stationary by wet sand-blasting. Global generalizations about pebble shape are
probably unobtainable partly due to changes in parent material, varying combinations of suites of rocks in a sample and latitudinal changes in weathering conditions. These may range from periglacial areas where the supply
of coarse angular material on slopes or entering stream channels and beaches
is abundant to tropical areas where rapid chemical weathering makes coarse
material rare on coasts and inland recementation makes it impossible to
separate much of the highly weathered material into discrete particles.

3. Surface appearance of particles

The surface appearance of sand grains may show degrees of dullness or
shininess due to distinctive modes of etching of the surface (Cailleux and

Tricart, 1959). Apart from classifications by surface appearance, these surface textures were first studied in detail with an electron microscope by R. L. Folk and C. E. Weaver (1952). Recently a range of distinctive patterns has been established, like small V-shaped indentations, characteristic of beach environments and possibly due to grain collisions which have a maximum density of about two notches per square micron. Combinations of four or

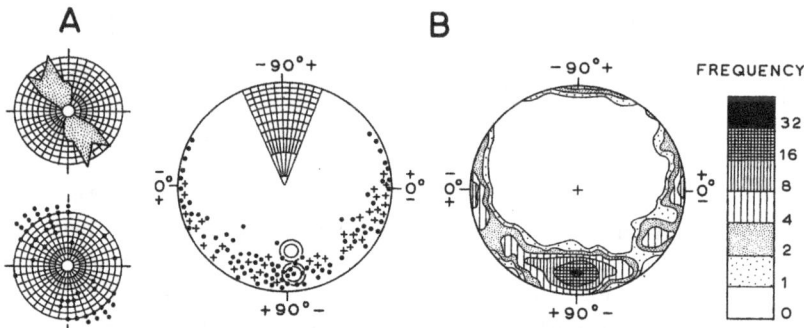

Figure III.8 Particle disposition and orientation.

A. Rose and dip diagrams showing pebble orientation in an East Anglian till *(from R. G. West and J. J. Donner, 1956, Quart. Jour. Geol. Soc., Vol. 112).* In the rose diagram *(above)* each circle represents half a stone in a given 10-degree class. Half of each frequency is placed on opposite sides of the diagram. In the dip diagram *(below)* the radii indicate long axis direction and the concentric circles the angle of dip, from 0° at the circumference to 80° on the innermost circle.

B. The Schmidt net (lower hemisphere), a three-dimensional fabric diagram illustrating data from a laboratory delta *(from Johansson, 1963).* *(Left)* A point diagram for the *L*-axes of a foreset bed. (The crosses indicate pebbles 0–10 cm from the delta lip, dots those settled farther away.) A segment of co-ordinate counting network is included and two pairs of circular counters (1/100 and 1/180) are also illustrated. *(Centre)* Data plotted on the point diagram frequency contoured with a 1 per cent circular counter, centred at square grid intersections on a co-ordinate counting network. *(Right)* Frequency, as points per unit counter area.

more distinctive markings serve to separate glacial, beach, and dune environments. It has also proved possible to distinguish between high, medium, and low energy beaches. These microfeatures are identifiable in ancient sediments (Waugh, 1965), are reproducible experimentally, and in Quaternary studies in East Anglia have provided interpretations generally in close agreement with stratigraphical fossil evidence.

4. Disposition of particles

Textural parameters alone may not provide the information required for certain reconstructions, and the geomorphologist may at times have to venture even farther into sedimentary petrology and soil studies and consider bedding structures and soil fabrics with his textural studies. Disposition of large fragments is increasingly studied since K. Richter (1932) made the first detailed studies of stone orientation in boulder clays. It is now established that the orientation of stones in tills is parallel to the direction of transport, although the mechanisms involved are little understood. It is also appreciated, however, that the orientation may change within a few metres vertically or over a few dozen metres horizontally, and that a later ice advance may complicate a pattern by inducing some reorientation in an underlying till deposited during an earlier glacial phase. The selection of a plane in the till for orientation analysis poses a further problem. In solifluction deposits the preferred orientation of the long axes also tends to parallel the direction of movement. In streams the preferred orientation tends to be on average perpendicular to the direction of flow. However, in some flood deposits orientation may be markedly longitudinal. It seems that two forces are involved and that where gravity is important, orientation tends to be longitudinal, while the dominant influence of hydraulic tractive forces tends to generate perpendicular orientations. For smaller particle sizes in the sand range, however, parallel orientation is widely observed, whether the current be backwash, stream, or air flow. A rose diagram is useful for representing orientation data (fig. III.8). Classes 10 degrees in width are commonly used and the number of stones in a given orientation class are counted. Half of these are placed on the diametrically opposite class on the diagram because, unlike a wind rose, there is no movement in either direction along the axis of orientation. The departure of each peak from the centre of the rose diagram indicates half the total frequency of pebbles lying along a given orientation.

Any process which produces some preferred orientation in a horizontal plane in sediments must also affect its disposition or inclination in a vertical sense. Near Wadena in Minnesota, an up-glacier inclination of the long axes of stones was observable in some parts of drumlins, whereas in other parts where the orientation was still well-marked there was no preferred inclination. In East Anglian tills, the latter situation prevails with no preferred inclination in the orientated stones. A systematic upstream inclination is widely recognized in river gravels, whereas on beaches changes occur over short distances, due to the abrupt and fundamental changes in the characteristics of water movement. This is an important and widely recognized feature of sedimentary deposits. S. B. McCann (1962) was able to suggest that two large gravel deposits at Corran near the entrance to Loch Etive in west Scotland were unlikely to be marine gravels, but fluvioglacial fans because the pebbles dipped systematically in the opposite direction to the bedding sets and had apparently escaped the reworking which marine action would have caused. In screes, debris, if sufficiently elongate, tends to pack with an inclination less than that of the angle of repose, but instances where the converse holds may indicate important contrasts in the mode of scree material movement. In the arrangement of rock fragments in superficial soil layers on slopes it might be assumed that preferred orientation or inclination would not be well-marked, but few measurements have been made.

Diagrams to show both the inclination and the orientation of a pebble sample may be based on a series of concentric circles. Radiating lines, as in a rose diagram, represent the orientation, with the concentric circles representing the angle of dip, horizontal at the outermost circle and diminishing to 90 degrees at the centre point (fig. III.8). A dot on this diagram represents each stone. Polar co-ordinate paper makes a similar but more sophisticated display of pebble fabric data possible. Here the radius of the concentric circles varies so that each sector bounded by these and the radiating lines has equal area. This is an equal-area azimuthal projection, more simply described as a hemisphere. The radii of each n circles r_n, is on a square root scale so that $r = R\sqrt{\frac{n}{10}}$, where R is the radius of the whole projection and n the selected angle of dip. In drawing contours round the frequencies, the number of points counted in the four cells adjacent to any intersection is used.

5. Soil fabrics

Recently fabric studies have been extended to features indicative of movement, reorganization and concentration in soil materials. The primary units of study, analogous to the coarse fraction in a pebble fabric, is the ped, a cluster of soil aggregates separated from neighbouring peds by recognizable surfaces of weakness or voids. Soil fabric analysis, therefore, deals with the size, shape, and arrangement of soil peds and voids, and as this approach, initiated by R. W. Brewer (1964) begins to specify more accurately the processes involved, it may become of increasing relevance to landform studies.

6. Soil hydrology

The mode and rate of movement of soil water is of fundamental importance in influencing the proportion of surface runoff in an area. It is well known that heavy downpours are particularly efficient geomorphic agents when antecedent precipitation or the thawing of ice, concentrated in the surface soil by freezing, has already saturated the soil layer. The amount of water in a soil, expressed as a percentage of dried soil in a sample, is closely controlled by the porosity of the soil. Porosity in turn is influenced by size distribution, but varies with different degrees of compaction and will vary with the moisture content itself if colloidal particles are present. Also, there appears to be a distinction of size orders of pores between inter-aggregate porosity, as occurs between peds, and intra-aggregate porosity. This distinction was observed in studies of chernozem soils with 42 per cent of their particles in the 50–10-micron range which revealed a marked deficiency of pores in the 60–5-micron range. Within the pores, water is held by both the mutual attraction between soil and water, termed adhesion, by cohesion, due to the mutual attraction that exists in water molecules, and by absorption on to the electronegative surfaces of colloids. Capillary water can exist in pores as fine as 0·03 microns.

Permeability in turn tends to be closely related to porosity, but rates change during the course of one shower, as fines are washed into open pores, and rates at the surface are in part influenced by the nature of the sub-soil which might be quite different. In cool temperate conditions there is often an abrupt reduction in permeability on descending into the illuvial horizon. Fissuring due to desiccation or to root pressure or burrowing by animals accelerates permeability. In many free-draining situations water may move down through a metre of sandy soil in about 3 minutes. If there is an admixture of silt and clay in a sandy soil permeability rates may be up to a dozen hours, while soil water may take longer than 4 months to move through heavy clays. Permeability rates for intermediate textures fall between the two extremes. In the loess of Voronezh Province, U.S.S.R., permeability is 0·15–1·30 m/day. Similar field measurements in the forest steppe of the Trans-Volga region indicated rates of water penetration into a ploughed silty clay of 0·1–0·9 m/day. Here thawing of the sub-soil caused lateral flow of soil water in spring at a rate of 71 m/day and other measurements suggest that 3 m/hr is a reasonable indication of rates of lateral flow in soil water when vertical filtration is arrested in the sub-soil.

Although there are many reasons why particle size alone is an insufficient indication of the mode of water movement, it is a fundamental factor in all aspects of soil water movement and the importance of particle size analysis in geomorphology is seen to have redoubled value.

7. Mechanical properties of clays

Clays can absorb huge quantities of water partly because the total surface area of tiny clay-sized particles is vast and water content is the main consideration in many dynamic physical properties of soils. This is partly because clays increase in volume as they absorb water. Montmorillonite can take up to ten to fifteen times its initial dry weight and increase in volume by more than a third. This swelling introduces abruptly increased pressures and swelling clay minerals appear to be an inherent factor in landslips. Engineers studying soil mechanics make frequent reference to the Atterberg limits, which describe water contents at which soils tend to change their state. A soil is said to be in a plastic state when the water content is sufficient to permit changes in shape without the production of surface cracks, the latter characterizing volume changes in the semi-solid state. At moisture contents less than the shrinkage limit no further reductions in volume accompany diminishing water content. This limit is usually between 10 and 12 per cent for a wide range of soils. When the difference between the plastic limit and the liquid limit, the Plasticity index, is small, soils are highly susceptible to erosion by rainwash and running water. For clayey soils in a semi-solid or solid state the shear strength of the material is the most important mechanical property influencing its susceptibility to erosion. In unconsolidated material the shear strength is directly dependent on the soils' compaction achieved mainly by wetting and drying processes acting on the clay portion to produce cohesion and on a range of particle sizes to increase internal friction. Some examples of values of compression shear strength include $7800 \, kg/m^2$ for postglacial clays susceptible to earthflow near Ottawa; $1700 \, kg/m^2$ for laminated silty clays at Rutherglen, Glasgow, with extremes for such materials ranging between $1000-4000 \, kg/m^2$. Values for weaker consolidated rocks are much higher. At Mam Tor in Derbyshire, the site of one of the larger landslips in Britain, the compression shear strength of shales, with illite and kaolinite the main clay minerals, was $48-67 \, kg/cm^2$ and in the Ashop valley to the north where postglacial landslipping is again spectacular, values are $67-89 \, kg/cm^2$. Another soils mechanics concept, defined by K. Terzaghi (1965), is that of sensitivity which is the ratio of the shear strength of a soil in an undisturbed condition to that of the same material remoulded at the same water content. Sensitive clays have values between 4 and 8 and in the extreme case of quick clays, values are over 16. The regain of strength during remoulding is termed thixotropy and is favoured by the very small particle size and high water-absorbing capacity of montmorillonite clays.

So far attempts to apply methods of soil mechanics in landform study have not been noticeably successful. However, it would be a mistake to assume that the physical analyses by civil engineers for technical purposes and measured by empirical tests, are involved in the production of particular landforms. Most of the measurements are synthetic, refer to a situation where a soil is burdened with an artificial load, and are not realized in nature. Even liquid limit values determined by laboratory tests may define volumes of water greater than the porosity available in the field. Therefore, while landform study may make increasing reference to the mechanical properties of soils and weak rocks, these will be made in full awareness that in the natural situation the crucial factors for variations in the properties of a given material are water content and gradient and the other factors like permeability which influence it, as water content and gravity are the fundamental influences on most of the dynamic mechanical properties in soils in natural settings.

8. Frozen soils

One group of dynamic mechanisms in soils which have been part of geomorphological investigations for at least as long as they have required the engineer's attention is that concerning frozen soils. Soil materials have a specific heat ranging between 0·3 and 0·8, thus gaining or losing heat on average twice as quickly as the same volume of water. For this reason temperature effects in a soil depend on the amount of soil water rather than on differences in the soil material itself. Depths below 0·75 m in soil profiles on Reading Beds are often structureless, impermeable, and unaffected by former frost-heaving activity. The absence of water-filled fissures, unlike the chalk areas, is possible the main factor limiting this depth. In most soils, for temperature above freezing, a wave of temperature change advances 2–3 cm/hr, whereas the rate of frost penetration is approximately 5 cm/day. This rate may be nearer to 7 cm/day in more porous, drier sandy material or slowed down to 3 cm/day in damper constricted pores of a humus loam.

Segregations of ice form where there are large supplies of water. In samples taken from frozen ground in the Mackenzie delta area, the average ice content may easily reach 500–1000 per cent in relation to the dry soil material. Water may be held in a frozen layer that would normally have drained. Freezing of the soil may have a desiccating effect on adjacent unfrozen layers due to the emptying of the capillary-sized pore spaces (fig. III.9). Osmotic phenomena are important in the freezing process, ice attracting water because of its higher electrical potential. In Victoria Land, Antarctica, rates of growth of wedges range from 0·3 mm to more than 5 mm per year and average 2 mm. Segregations grow particularly rapidly in silts. This is probably because water moves appreciable distances in a short time through the pores of silt-sized materials, which, particularly in the 20–50-micron range, provide the most efficient passageways for capillary movement. Capillarity, the phenomena of forced ascension of water in fine tubes, operates equally along lateral moisture gradients in soils. Sands are too coarse for capillary-sized channels to be extensively developed and in clays the pores are too small for rapid movement, whereas water might move 2–2·5 m/month by capillarity in silts. Sands and coarse materials do not freeze as a block because ice occupies only part of the pore space and without capillary circulation, no ice segregations develop. Where they do develop the sandwiching of the seasonally thawed layer, or mollisol, between closing layers or lenses of frozen ground, leads to the development of pressure structures such as involutions and injections. Their development requires a minimum seasonal frost penetration of perhaps a metre, which appears to be related to nearness of mean annual temperature to 0°C rather than to the intensity of winter cold. A

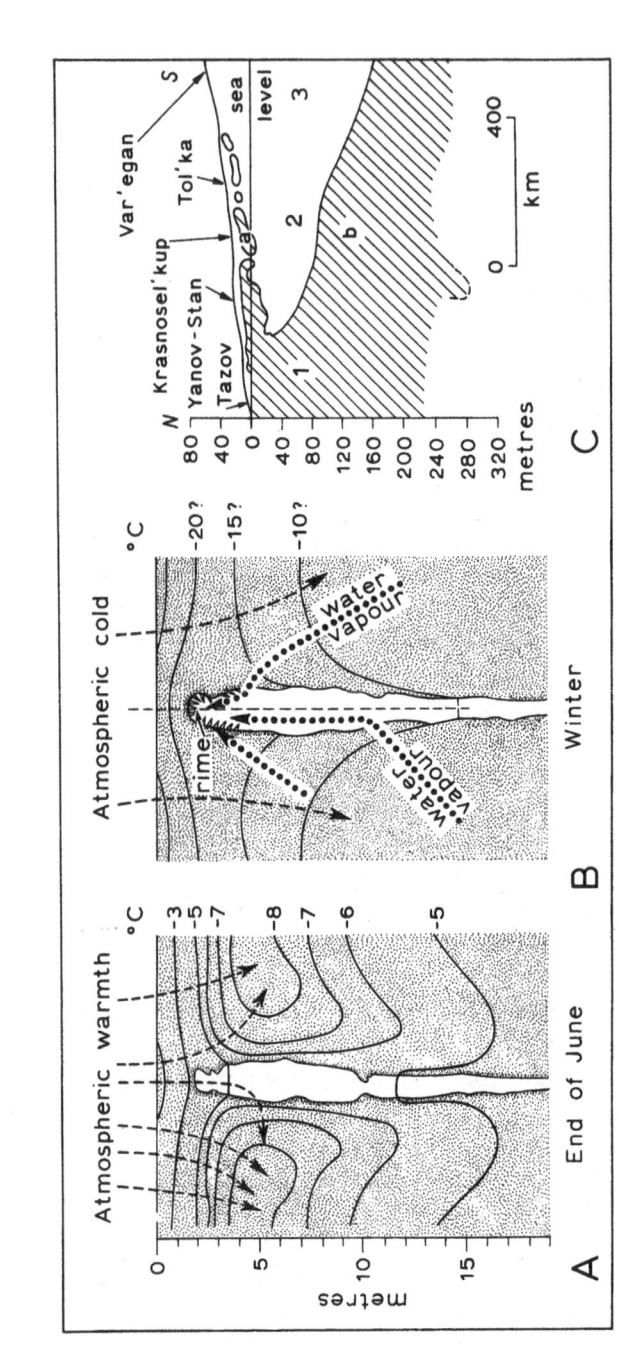

Figure III.9 Ice in the soil.

A and B. Temperatures around an ice wedge (*from R. M. Bone, 1962*, Geog. Bull. No. 17).

C. Profile of permafrost in Western Siberia (*after Lliboutry, 1964*). In this profile, based on work by A. A. Zemtsov, *a* indicates an upper body of permafrost, *b* an extensive lower layer of frozen ground, and the numbers 1, 2, and 3 indicate continuous, discontinuous, and relict permafrost respectively.

distinctive feature is the persistence of frozen soil as a relic from a past colder phase.

E. Geochemical Considerations

When rocks, being formed at higher pressures and temperatures than exist at the surface, are initially exposed to the sub-aerial environment, reactions occur to produce new compounds of greater volume and lower density. This mobilization and redistribution of elements during weathering is a complex and variable process open to few generalizations. Although the geologists and soil scientists who study geochemical processes involved in chemical weathering report on their complexity, it is an advantage to the geomorphologist following their studies that the number of rock types, mineral groups, and chemical elements involved is relatively few. There are only half a dozen main rock types, and eight main elements, oxygen (47 per cent by weight), silicon (28 per cent), aluminium, iron, magnesium, calcium, sodium, and potassium. Moreover, no other liquid can compare with water, either in abundance or in the number of substances it can dissolve, or in the amounts that it can hold in solution.

1. Physicochemical factors

One important factor influencing the pattern of selective loss of minerals is ionic potential which is in part dependent on the dipole nature of water molecules, with cations attracting the negative ends of water molecules and anions the positive end. The bigger the ion and the less intense the charge on its surface, the greater the number of water molecules that gather around it. The factor Z/r, where Z is the charge and r the ion radius, is known as the ionic potential. Where the concentration of positive charge on the surface of a cation is sufficiently high to repel the protons in water molecules, precipitation of an insoluble hydroxide may result. In contrast, elements like sodium, calcium, and magnesium, with low ionic potentials of $1\cdot0$, $2\cdot0$, and $3\cdot0$ respectively, remain in solution during processes of weathering and transportation. Elements with intermediate Z/r values are precipitated by hydrolysis, although elements with still higher ionic potentials form anions containing oxygen which are usually again soluble in this form. Ionic potentials of other elements commonly involved in rock weathering include aluminium, $5\cdot9$, and silica, $9\cdot5$. In general, although monovalent ions are leached out more quickly than divalent ones, and the latter more quickly than trivalent ones, the behaviour of some elements in weathering and transportation processes is very complex and often at variance with work on theoretical solubilities in distilled water. A comparatively well-understood example is iron, which in the ferrous state is stable in solution (Z/r for $Fe^{++} = 2 \cdot 7$) so that oxidation to the ferric state (Z/r for $Fe^{+++} = 4 \cdot 7$) must precede precipitation of iron.

Some of the most significant physiochemical processes in geomorphology are due to materials, clays in particular, and also humified organic matter, which exist in the fine subdivision of the colloidal state. Colloidal systems consist of distributed particles approaching molecular dimensions and a solvent, the terms being the 'disperse phase' for the former and the 'continuous phase' for the solvent or dispersion medium. A colloidal solid–liquid system incorporating a small amount of colloidal matter in a much larger quantity of water is termed a sol. Like liquids a sol is not rigid and flows readily. Where a relatively small quantity of water is enmeshed in the interstices and crevices between and within colloidal particles, the viscous mass shows some rigidity and is called a gel. Pastes are systems in which the concentration of discrete particles is such that they form the bulk of a plastic mass. Coagulation of colloids is governed by the presence of free ions. If the number of free ions is high they coagulate the colloidal particles into 'clots', crumbs of greater volume which reduces the contact forces responsible for cohesion of the colloidal substance which thus loses its consistency when flocculated. The distinctive effects of sodium are seen in the precipitation of river-borne clays on entering the ocean and in the instability of recently uplifted marine clays. Colloidal systems in the weathering profile are normally in the gel condition because of the limited amount of water usually present, although clay minerals are usually suspended in a colloidal solution when they first leave a weathered mineral. On the other hand changes of state which accompany relatively short-lived additions of abundant water are some of the most significant events in the evolution of many land surfaces. There are also two highly significant properties of a solid–liquid colloidal system in addition to changes of state. First, a characteristic of the surface of contact of a colloidal particle is its interface actions, due to its electro-negative charge. As a result colloids can fix and retain by adsorption gaseous, solid, or liquid particles of very small size. Secondly, as the average figure for the size of colloidal particles is less than 1 micron, their numbers are in hundreds of billions per gram, with a surface area of thousands of hectares in a cubic metre of clayey material on which interface actions, like the adsorption of water, may take place.

2. Weathering reactions

Base-exchange involves the mutual transfer of cations like Ca^{++}, Mg^{++}, Na^+, and K^+ between a thin film of water rich in one cation and a mineral rich in another. Exchange reactions are reversible, and different ions may replace one another. The rate of exchange depends on the acidity, organic matter, temperature, and other properties of the solution as well as on the abundance of the various cations and their chemical activity. Generally the rate of exchange is rapid, requiring only a few minutes. Increases in cation-exchange capacity (CEC) is one of the more dramatic results of clay formation, and there is a strong direct correlation between changes of concentration of certain elements and variations in the proportion of the clay fraction.

Many clay minerals are hydrated, involving the addition of the entire water molecule to the mineral structure. The disruptive effect of minerals expanding due to water incorporated by hydration is again essentially mechanical and one factor causing granite decomposition. In a 10-year experiment of actual weathering, a 9 per cent decrease in the amount of material in granite fragments larger than 200 microns was attributed to hydration. An example of hydration is the conversion of hematite to limonite:

$$2Fe_2O_3 + 3H_2O - 2Fe_2O_3 . 3H_2O$$

but such reactions are rather easily reversed by heating and there then has been no fundamental chemical change.

In hydrolysis, by contrast, the hydrogen ion, H^+, becomes part of the atomic structure of a clay mineral. This is the most important weathering reaction for silicate minerals, even in deserts. The physical effect is an exchange of H^+ ion from the water for a cation of the mineral, leading to expansion and decomposition of the silicate structure. A chemical effect is to increase the pH of the water. The reorganization of the silicate structure makes it possible to accommodate even more absorbed water in the crystal lattice. The way in which hydrogen ions and water decompose potassium felspar illustrates the effectiveness of the H^+ ions:

$$2KAlSi_3O_8 + 2H^+ + H_2O \rightarrow 2K^+ + Al_2Si_2O_5(OH)_4 + 4SiO_2$$

Orthoclase + Hydrogen + Water → Potassium + Kaolinite + Silica
felspar   ions   ions

Here the H^+ ions force their way into the potassium felspar structure and displace potassium ions which then leave the crystal lattice. Water thus not just dissolves and alters minerals but also acts as a source of H^+ ions, although carbonic acid is a much better supplier of H^+ ions for hydrolysis than pure water would be. Carbonic acid ionises to form hydrogen ions and bicarbonate ions:

$$H_2O + CO_2 \rightarrow H_2CO_3 \rightarrow H^+ + (HCO_3)^-$$

Water + Carbon → Carbonic → Hydrogen + Bicarbonate
dioxide   acid   ion   ion

Commonly in the weathering of felspar, hydrolysis and carbonation operate together:

$$2KAlSi_3O_8 + 2H_2O + CO_2 \rightarrow Al_2Si_2O_5(OH)_4 + K_2CO_3 + 4SiO_2$$

Orthoclase + Water + Carbon → Kaolin + Potassium + Silica
felspar   dioxide   carbonate

Here hydrolysis produces the clay residue, anhydrous aluminum silicate while potassium carbonate, the result of carbonation, is carried away in solution.

Carbonation is the reaction between carbonic acid and minerals, water acquiring its acidity largely from the carbon dioxide generated by humification processes in the soil. Fresh rainwater is also slightly acid because it dissolves a trace of carbon dioxide from the atmosphere. The effects of carbonation are best known in the solution of calcareous rocks

$$CaCO_3 + H_2CO_3 \rightarrow Ca^{++} + 2(HCO_3)^-$$
Calcium   Carbonic   Calcium and bicarbonate
carbonate   acid   ions

The solubility of iron is depressed in carbonated water but that of silica enhanced. These responses favour a theory of iron oxide accumulation by water rich in carbon dioxide dissolving silica and concentrating iron oxide.

Oxidation is common in freely draining parts of the weathering zone, as it is the oxygen dissolved in infiltrating rainwater which oxidizes metallic iron and changes ferrous iron to the more oxidized ferric state. Although oxidation commonly involves combinations with oxygen as part of the weathering process, this is not always the case. By definition a substance is oxidized when it loses electrons and reduced when it takes on electrons. The combination of iron or other elements with oxygen weakens the original mineral structure, freeing the remaining minerals for participation in other chemical reactions, although hydrolysis usually precedes oxidation. Iron is a very common chemical element and its distinctive rusty colour on weathering is widely observed. For instance, in weathering at Marble Point, Antarctica, the only measurable chemical change is the progressive loss by oxidation of ferrous oxide in iron-bearing minerals. During chemical weathering ferrous silicates, such as the rock-forming minerals pyroxene, amphibole, olivine, and biotite form hematite, and hydrous iron oxides like limonite and goethite, by oxidation.

3. Physicochemical characteristics of weathering environments

Some important physicochemical factors relate to environment rather than to the elements themselves. The degree of acidity, for instance, is of great significance in chemical reactions in the natural environment, reflected in a weathering profile by the hydrogen ion concentration of its aqueous suspensions. The concentration of dissociated hydrogen ions in this aqueous suspension is in equilibrium with those actually adsorbed on the weathering products. Thus the hydrogen ion concentration of an aqueous suspension from a weathering profile is an index of the intensity of acidity in the weathering environment. Its expression is in terms of pH units which is the negative logarithm of the hydrogen ion concentration. Thus if a suspension has a pH of 6, this value indicates that its hydrogen ion concentration is 10^{-6} or $\frac{1}{1000\ 000}$ of a gram per litre. In a neutral solution like pure water pH is 7; for an acid solution the number is less than 7, reflecting the greater hydrogen ion concentration. The pH of an aqueous suspension is particularly significant in controlling the precipitation of hydroxides from solution, the solubility of iron being 10^5 times greater at pH 6 than at pH 8·5. The solubility of ferric iron is so low, 0·01 ppm in pH range 5–8, that significant solution and transport of iron probably involves reduction to the ferrous state. At pH of less than 4 alumina is readily soluble, which may lead to the preferential removal of alumina in extremely acid environments, like those of podzolization, whereas from pH 5 to pH 9 the solubility of silica increases considerably but alumina is practically insoluble. However, due to the very high binding cation-exchange energy that clay has for alumina, the mobility of alumina due to solution in an acid environment may be cancelled out by immobility due to cation exchange. An increase in pH aids precipitation of carbonate.

The oxidation-reduction potential varies with varying concentrations of the reacting substances. This is in part dependent on pH as most reactions involve hydrogen or hydroxyl ions, with oxidation generally proceeding more readily the more alkaline the solution. In reducing environments, compounds including hydrogen ions are common and organic matter tends to accumulate. For instance, air trapped in waterlogged interstices of an anaerobic environment may develop substantial concentrations of methane, CH_4, compared with the combination of carbon with oxygen to form carbon dioxide under aerobic conditions.

4. Mineral solubilities

From the discussion of the three preceding geochemical aspects of weathering, it will be clear that it is not easy to generalize about the solubility of various minerals. Even though it is useful to distinguish between the several different reactions involved, in practice, and as an unavoidable consequence of the composition and properties of water, they operate simultaneously. On the other hand a particular geochemical environment may dictate that a distinctive pattern of weathering reactions be followed. Thus, although perfect mineral grains fall into a sequence according to their susceptibility to decomposition, regardless of the rock type in which they occur, an agreed absolute scale is still wanting, and there is probably a need for different scales for different environments. In general monovalent ions are leached out more quickly than divalent ones, and the latter more quickly than trivalent ones. The lattice site of an element is in many cases an important factor in determining its relative mobility during weathering. Calcium generally exhibits much greater mobility, especially during the early stages of weathering, than any of the other elements, like magnesium, being taken up into solution at rates tens to hundreds of times faster than for other elements. It has been observed in the weathering of volcanic ash on St Vincent that it was the pyroxenes, olivines, and plagioclase of more calcic composition that altered more quickly. As the first phase of felspar decomposition is rapid disintegration by hydration, no loss of material is involved initially. In fact in 1947 P. Birot (1947) found that a 5–7-month treatment of samples of acid crystalline rocks under alternating 12-hour spells of wet then dry conditions at 70° C resulted in a 0·1 gm increase in weight of a 20-gm sample due to hydration. Ordinary room temperatures did not produce this effect. Deep chemical weathering is a phase subsequent to hydration. The first sign of plagioclase weathering is surface pitting and widening of fractures by solution. Alteration to clay minerals results in an expansion of the plagioclase lattices, particularly with alternate wetting and drying and rock-shattering results. Despite the presence of highly mobile cations, the alkali felspars are relatively stable because the lattice restricts access of water. The bases, followed by silica, are leached out of a surface zone of weathering, leaving sesquioxides of iron, aluminium, and manganese as a residual accumulation. A residuum of iron sequioxides depends on high temperatures and does not take place under continuously dry conditions. Mica weathering is essentially a process of potassium depletion. There is hydrolytic interplay between water and the interface K^+ of the mica lattice. In an acid environment this cation is displaced and the bonding effect between mica sheets loosens with the dissolution of magnesium and the oxidation of iron, producing further loosening and expansion of the crystal lattice. The weathering to clays of biotite and, to a much lesser extent, plagioclase, is a vital factor in the breakdown of granite. Like plagioclase, pyroxene and olivine minerals may show etching, indicating that dissolution is part of the weathering process. The solubility of silica depends greatly on whether it exists in the form of quartz or as amorphous silica, the latter including hydrated and dehydrated silica gels and the skeletal remains of silica-secreting organisms. At 5° C the theoretical solubility of quartz is about 6 ppm compared with 60 ppm for amorphous silica. The undersaturation of terrestrial waters may be attributed, in addition to the slowness with which silica or silicates dissolve, to the activity of organisms that use silica.

Specific examples of observed differences in mineral resistance to weathering include studies by W. W. Smith (1962) who found a mineral sequence in the weathering of basic igneous rocks in Scotland, in order of increasing stability, of olivine, labradorite, augite, magnetite, ilmenite, and hematite. Studies in Maryland provide an example of a scale of resistance to weathering of heavy minerals. The scale, related to a garnet value of 1, includes zircon at

100, tourmaline about 80, sillimanite and monazite at 40, kyanite 7, hornblende 5, staurolite 3, and hypersthene less than 1.

An important geochemical concept in the study of weathering is that of weathering ratios, calculated either between the proportion or amount of a given element in a soil compared with similar quantities in the parent material, or between concentrations of different elements in a soil or clay. Examples of the latter include the molecular ratio between silica and the sesquioxides:

$$\frac{SiO_2}{Al_2O_3 + Fe_2O_3}$$

A related expression is the $SiO_2 : Fe_2O_3$ ratio. Indicative of the general concentration of silica and alumina in Cuban weathering crusts, for example, are values of this ratio at 3–5 and frequently higher. For latosols in the humid tropics, values of the ratio $SiO_2 : Al_2O_3$ are generally between 1 and 2.

F. Biological Activity

Ever since 1882 when Charles Darwin's measurements showed that earthworms could move 10 tons/acre/year, attention has been drawn, from time to time, to the tendency for studies of sub-aerial processes to emphasize physical and chemical factors, perhaps at the expense of biological considerations. But in many investigations bacteria and other forms of life, an irksome contaminant in controlled laboratory experiments and often a source of variation irreducible to the confines of an equation, remain excluded. Today awareness of the risks in neglecting the biological factor are keen, as leading authorities in a range of disciplines independently come to emphasize its importance. For instance, it is organisms that keep the oceans so low in dissolved silica; experts like W. D. Keller regard bacteria as one of the most important and essential adjuncts to the processes which result in argillation; the lichen *Caloplaca* was found at 85 degrees south; it was found impossible to weather biotite in laboratory conditions until tree seedlings were incorporated into a recent experiment; in Florida Bay, south of the Everglades, sediment size and physiochemical parameters were not strongly related, and even the physiochemical parameters were not strongly inter-related, but that there was a close correlation between the distribution of sediment size and turtle-grass (*Thalassia restudinum König*).

Biological factors provide an aid to landform studies, and, as an expression of climatic controls and local environmental factors, are an important intermediary agent in land-forming processes. Plants and, to a lesser extent, animals are able to contribute substantially to rock weathering because the sun transmits to them the necessary energy, whereas inorganic reactions must be driven by built-in chemical energy with fewer direct contributions from solar radiation. These processes require very careful consideration and observation because it is difficult to predict whether the intermediary agency of plants will accelerate or retard erosion. Mechanically plant roots can bind or break and, simultaneously might enhance solution or alternatively induce precipitation and storage of chemical elements. In addition, it is necessary to consider the organic factor from three separate viewpoints, that of its bulk or the biomass at the surface, the expression of different environments seen in different ecological types of plant, and thirdly, the variations in physiological processes in contrasted environments.

1. Acceleration of mechanical processes

It is not certain how much mechanical work the expansion of growing tree roots might perform. P. Birot (1966) states that a living root 10 cm broad and 1 m in length is capable of moving a block weighing 40 tons. Where soils are stony and in the upper horizons of bedrock where root space is reduced, roots are forced to develop around and between boulders and to penetrate into the openings in bedrock. In valley-side cliffs in Magnesian Limestone a dozen miles south-east of Sheffield, the rock-disrupting roots of yew trees (*Taxus baccata*), after seeding some distance back from the cliff may grow to be 15 cm in diameter at depths of 6 m (fig. III.10). Penetration depths of more than 6 m have been observed in sandstones elsewhere, and in fractured granite in the Colorado Rocky Mountains, Ponderosa pine roots penetrate to depths of 10–12 m. Extreme depths of 30 m have been suggested, but in spruce and pine woods near Moscow, 75 per cent or more of the tree roots lie within 40 cm of the surface. In the taiga spruce forest roots are confined to the upper 30 cm, and in deciduous forest still only to 50 cm. A significant fact may be that in deserts where vegetation is apparently scant there is an exceptional preponderance of the biomass in the root fraction, on average 80 per cent, which may not spread in the upper horizon which dries out rapidly. In the shrub and arctic tundra where conditions are also severe at the ground surface, up to 90 per cent of the biomass may be below ground. In all zones of deciduous trees and coniferous forests roots are 15–24 per cent of the biomass.

Apart from the prying action of growing roots and the movement of superficial material when trees are under the stress of high winds, the main contribution of trees to mechanical weathering is probably due to the tremendous leverage exerted by the trunks of falling trees. H. J. Lutz (1960) studied wind-thrown trees, 30–60 cm in diameter in New England forests and found several examples where rock masses more than a cubic metre and weighing over 0·25 tons were moved vertically or horizontally 0·5–3·5 m often with

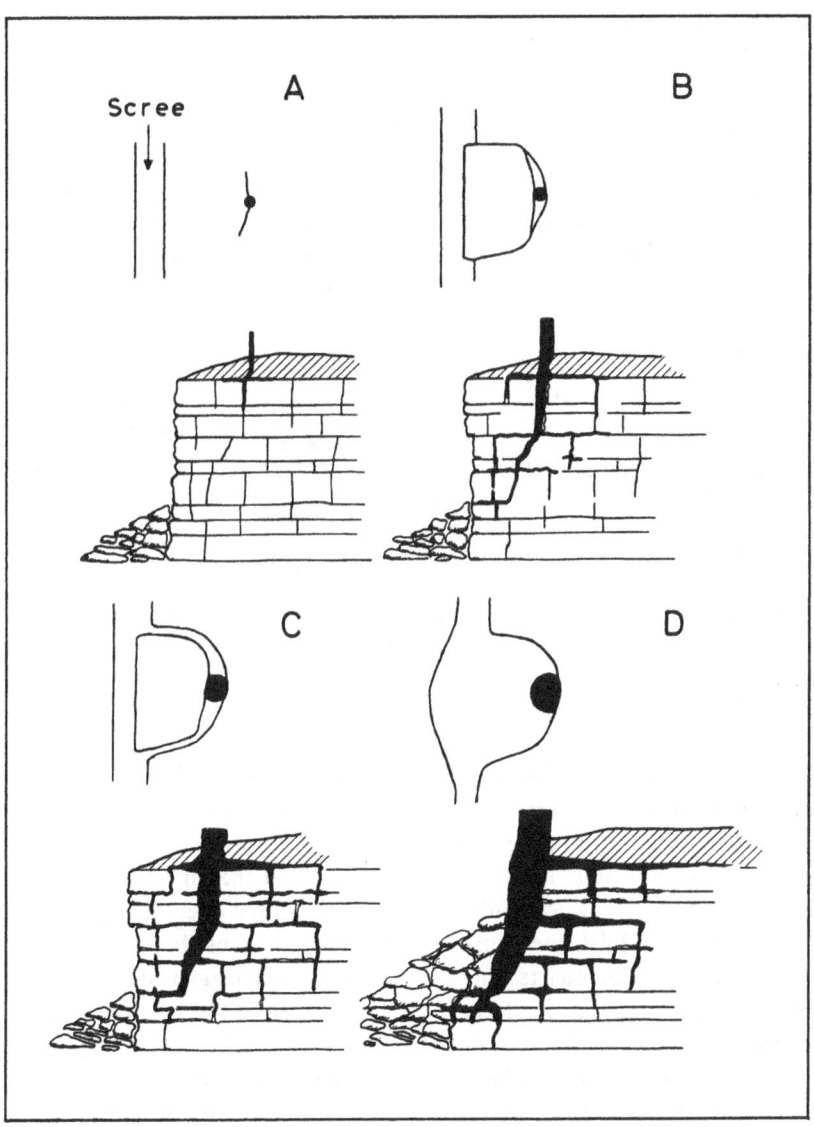

Figure III.10 Inter-relationship between tree roots and cliff breakdown (*from G. Jackson and J. Sheldon, 1949*, Jour. Ecol., *Vol. 37*), as suggested by a study of yew trees (*Taxus baccata*) on Magnesian Limestone cliffs at Markland Grips, south-east of Sheffield.

fresh fracture surfaces in the hollows. Mounds produced by wind-throw are common in the deciduous forests throughout the eastern United States. In Guyane, A. Caillieux counted ten wind-thrown trees per hectare. The Peruvian Andes is another locality where wind-throw is reportedly common, facilitated by shallow rooting, 0·2–0·3 m in depth, due to the supply of nutrients being limited to litter fall on the ground surface. Similarly, but usually because permafrost creates a hard impervious substratum, trees in arctic conditions are easily overturned. The possibility of some latitudinal effects due to root action has been suggested (fig. III.11).

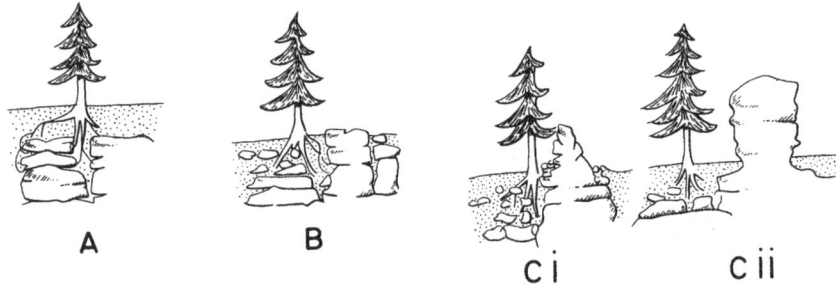

Figure III.11 The role of trees in the sub-surface development of tors (*from I. Gams, 1966*, Geografski Vestnik, *Vol. 38*). A, initial stage. B, disintegration of more closely bedded and jointed blocks by tree root activity and lowering of the soil surface, C. Further disintegration and lowering of the soil surface: (i) cool climate, (ii) warm climate.

Three important facts about the distribution of wind-throws are the tendency for them to occur on exposed sites like ridges and escarpments; their importance particularly in the tropics, due to tree colonization of bare rocks where nutrients are most readily available; and that the ultimate fate of most trees if not broken by wind would seem to be to topple over. It might be most illuminating to establish the degree to which the contrast between the slopes which typify tropical coasts and the sea-cliffs and jagged stacks of extra-tropical latitudes reflects the descent of dense forest to sea-level in the tropics compared with the treeless coastal strip on spray-bathed extra-tropical sea-cliffs. In fact the degree to which the landform in any forested environment differs from that of a treeless domain might be worth considering in terms of root action and wind-throw.

The amount of energy expended and the mechanical reworking achieved by soil fauna in the soil is substantial. Their action in moving material from the upper layer of soils on to the surface is probably the most important method of truncating soil horizons, especially in the tropics. In eight experimental plots at Rothamsted total consumption of soil by earthworms ranged from 10 to 90 m. tons/ha/year, involving up to 8·7 per cent of the total weight of the top 10 cm of soil. Thus all the material in this depth would pass through the alimentary tracts of the earthworm population in $11\frac{1}{2}$–80 years. Although the temperature for optimum earthworm activity is as low as $10°$ C, in tropical latitudes the termite takes over a comparable role, consuming similar volumes of earth if its preferred diet of vegetable debris is deficient. In the savanna, termite mounds may be 2–4 m high with a volume of 2–4 m^3 (fig. III.12.B). Sizes in the cerrados of Brazil are similar, with a spacing measured north-east of Andarai, averaging 15 m. Near Elizabethville the mounds cover 7·8 per cent of the ground surface. One estimate of the net work by termites is a complete reworking of the top metre of the soil every 1000 years.

Ant mounds may be 1–15 cm in height and from 7 cm to 4·0 m in diameter. The number of mounds per hectare ranges from 62 000 for small sizes in humid temperate forests to 25–50 for large ones in steppe areas. Commonly they occupy 1–4 per cent of the total area and the ants may move the top 6 cm to the surface once every 100–300 years. In Silwood Park in Berkshire, mounds are up to 40 cm in diameter and 15 cm high and may occupy up to 10–11 per cent of the ground with a density of 5300/ha in parts. In the Wisconsin prairies ant mounds are 30 cm high, 60–90 cm in diameter and number 100–125/ha. Therefore each point on the surface could have been occupied by a mound at least once in 600 years. The combined activities of earthworms, rodents, and ants could turn over the upper 60 cm of grassland once a century.

Mole hills studied near Moscow were about 10 cm high and about 30 cm in diameter, ejected at a rate of 11.1 m^3/ha/year which represents 0·3 per cent of the weight of the soil in the top 10–40 cm. In the recycling of unconsolidated material even the contribution of elephants, where appropriate, might be taken seriously. In the Wankie National Park, the volume of solids removed by an elephant at each wallow amounts to a cubic metre.

The large scale of reworking of coastal sediments by benthic animals has only been appreciated for a couple of decades. However, it has now been established that *Arenicola* will, for each square metre of sediment surface, rework 150 000 cm^3 of sediment in a year, or rework the entire thickness to depths of 20–30 cm in 20 months. Another worm *Clymenella* can rework to a depth of 20 cm in 2·4 years. In subaquatic environments like the Hudson estuary, man-made debris occurs to a depth of at least 2 m into the bottom sediment, believed to represent the scale of overturning of debris by aquatic worms.

A significant implication of these immense expenditures of energy is the amount of incoming solar radiation diverted into the reworking of unconsolidated material in the soil layer rather than directed to the breakdown of rocks. However, on slopes the heaping up of loose material would inevitably lead to some downslope displacement by gravity alone. On the other hand a pronounced effect of many of the reworking activities is to increase porosity

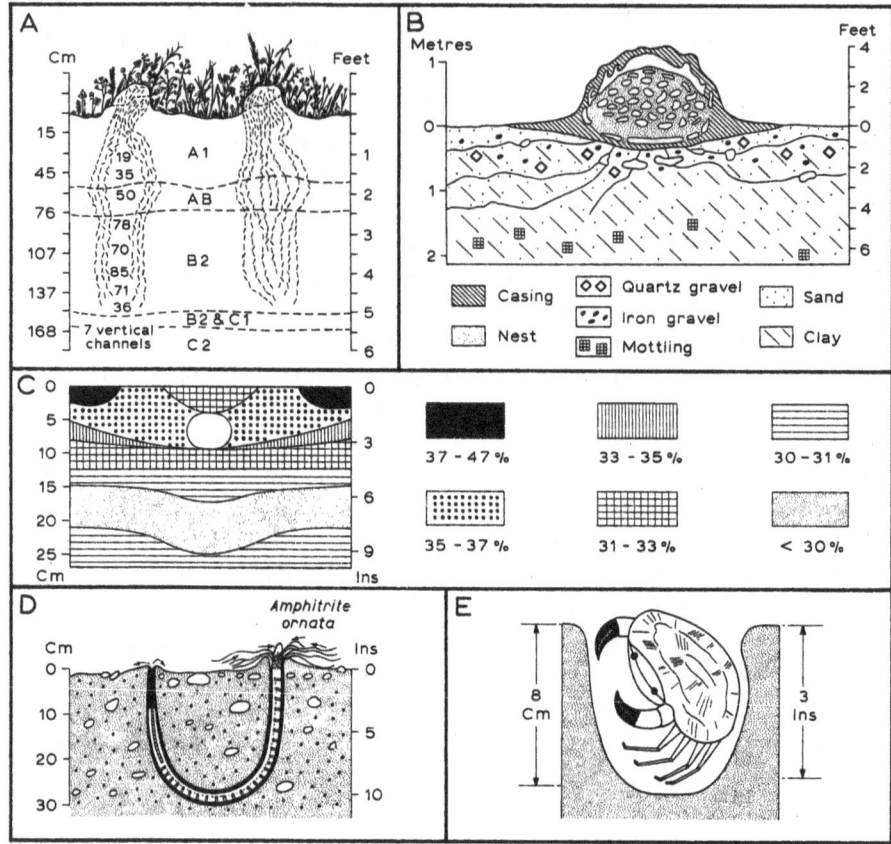

Figure III.12 Reworking of superficial material by organisms.

A. Ant mounds in the Wisconsin prairie (*after F. P. Baxter and F. D. Hole, 1967,* Soil Sci. Soc. Am. Proc., *Vol. 31*). Figures under the mound on the left are the numbers of ant channels observed on the floor of the pit at the depths indicated.

B. Termite mound in Nigeria (*from P. H. Nye, 1955,* Jour. Soil Sci., *Vol. 6*).

C. Effect of mole tunnels on soil moisture content (*from A. B. D. Abaturov and L. O. Karpachevskiy, 1966,* Soviet Soil Science) observed in the Krasno-pakhorskiy Forest near Moscow.

D. Sediment sorting by *Amphritrite ornata,* Barnstable Harbour, Massachusetts (*from D. C. Rhoads, 1967,* Jour. Geol., *Vol. 75*). Grooved ciliated tentacles of *A. ornata* collect and transport sediment 1 mm and less in size to the mouth. Non-ingested manipulated remains as a conical deposit around the anterior opening of the U-shaped mud tube and egested sediment surrounds the anal end.

E. Reworking of intertidal sediments by the burrowing of a crab (*from Evans, 1965, after W. Schaefer*).

which therefore reduces surface runoff. The activity of moles can reduce the bulk density of a surface soil from 0·8 to 0·5 g/cm³ and increase porosity by about 10 per cent. Measurements made on an experimental farm near Moscow revealed a dense network of passageways 5–7 cm in diameter, averaging 255 cm/m² in the top 10 cm of soil. This represented 15·3 per cent of the area and 7·2 per cent of the volume. The effect on soil-water conditions is evident from fig. III.12.C. The sum of the volume of earthworm burrows in a sandy soil in eastern England was the equivalent of a 4·4-cm pipe beneath each square metre. On the shore, burrowing activities may lead to a 10 per cent increase in porosity, a figure closely comparable with the results of moles' activity in the soil.

2. Retardation of mechanical processes

Increases in soil porosity are just one indication of a remarkable variety of ways in which organic influences work to retard some erosional mechanisms. The involvement of vegetation in retarding erosional processes starts with the interception of precipitation by foliage. Proportions vary, but a threshold value of precipitation of at least 2 mm before any moisture penetrates to the ground beneath Canadian hardwoods illustrates one aspect of interception. On the Ivory Coast 50–95 per cent of the total precipitation may be intercepted. Even vegetation as lowly as mosses, being vigorously hygroscopic, absorb moisture not just intercepted from precipitation but from atmospheric vapour too. In addition to interception vegetation indirectly influences the infiltration rate into a soil because of water losses due to evapotranspiration. If a plant uses up large proportions of soil moisture before and during a rainy season, there is more space available for the intake of the subsequent rains. Most trees draw up water from depths of 6 m or more. Studies in the Carmel mountains, Israel, showed that the maquis shrub can utilize moisture from depths down to 7·5 m. The average annual moisture withdrawal was 450 mm from ground covered by maquis, 410 mm by pines, and 330 mm from beneath pasture grasses. These figures compare with an annual precipitation of 700 mm and a loss of 265 mm from bare ground. Moisture depletion in the upper metre was similar under the various cover types.

The many experiments comparing controlled plots on agricultural research stations give the clearest impression of the scale on which differences in vegetation cover, by influencing surface runoff, regulate the erosional potential of precipitation. Table III.1 illustrates a typical example. These figures suggest that in drier climatic areas where grassland prevails most of the flow may be on the surface with little soil-water flow compared with small volumes of surface flow under forests where infiltration capacity is very high. Under some oak forests surface runoff has never been recorded even in years of

Table III.1 Surface runoff coefficients for spring precipitation in the mixed forest zone of European U.S.S.R. (*M. I. L'Vovich*, Soviet Soil Science, *1966*).

	GRASSLAND	PLOUGHED LAND	FOREST
Sand	0·20	0·10	0·01
Sandy loam	0·33	0·23	0·03
Clay loam	0·53	0·39	0·19

maximum runoff. Where the surface is bare, peak flows may be up to twenty times the volume of peak flows from comparable forested areas. Associated increases in erosion rates are usually substantial if not catastrophic. A typical example was the seven-fold increase of soil slippage on grass-covered areas in the San Gabriel Mountains, California, in areas formerly covered with a natural chaparral vegetation.

As effective in reducing the erosive potential of precipitation as the living foliage above the ground is the organic matter decaying at and below the ground surface.

It is important to differentiate between thoroughly decomposed residues, termed humus, and organic matter in less complete stages of decomposition, the humus having colloidal properties. Humus in the soil, because of its electro-negative properties, not only retains large proportions of absorbed water within the soil after precipitation but is a major bonding element in the delicate cohesiveness of soil aggregates or crumbs, the porosity of which increases infiltration while the cohesion of the crumbs themselves makes them resistant to erosion. Table III.2 illustrates the decline in water-stable aggregates and the diminution in pore space which accompanies the removal of vegetation.

Table III.2 Measurements of porosity and water-stable aggregates in the 0–7·5 cm soil horizon in three types of tropical forest clearings, Ghana (*from* Cunningham, *1963*).

	POROSITY			WATER-STABLE AGGREGATES PER CENT AIR-DRY SOIL 3 MM	
	Capillary pore space	Non-capillary pore space	Total pore space	Total structure	True crumb structure
Shade	37·0	14·7	51·7	55·3	39·8
Half exposure	35·0	16·4	51·4	50·2	28·9
Full exposure	32·6	10·1	42·7	48·7	29·0

Perhaps more significant than humus itself is its rate of production. Humification of organic material, an oxidizing process, is more rapid at higher temperatures, partly because it is a temperature-controlled reaction and partly because high temperatures minimize the quantity of water in the soil that would impede air circulation. In consequence there is little accumulation of organic residues in the humid tropics, despite the very high biological productivity, and thick accumulations in cool temperate climates despite low productivity. The result is a definite tendency for the thickness of the organic-residue blanket at and beneath the ground surface to increase towards cooler latitudes, one of the few geomorphological parameters to show some semblance of a systematic latitudinal trend.

Due to rapid rates of humification the surface of soils in the humid tropics are significantly low in litter accumulation despite the prolific leaf fall. Amounts range between 1000 and 5000 kg/ha but reserves as low as 400 have been measured in south-east China. On average the amounts are half or even a third those in deciduous forests of the temperate zone and the chances of erosion of the forest floor correspondingly increased. In savanna areas the accumulation of organic matter is smaller, amounting to 50–150 kg/ha in many areas. This is less than in steppe areas where organic material is conspicuous in soils because rainfall is too low to remove it from the soil horizons. For instance, beneath the relics of a climax vegetation isolated on two steep-sided plateaux in North Dakota, where precipitation is about 375 mm/year, organic material in the soil amounts to 17 000 kg/ha. In the Russian steppe decomposition rates similarly lag behind supply of plant residues and a characteristic steppe matting accumulates to about 4000–5000 kg/ha in meadow steppe and 1500 kg/ha in arid steppe. The role of an organic mat in retarding erosion in steppe conditions is conspicuously delicate in balance, for two reasons. First a very large proportion of the total biomass cycles annually, being 50–55 per cent in meadow steppe and down to 40 per cent in arid steppe. Secondly, there may be a sharp decline in the ground mass in dry years, which in extreme conditions might be only one-tenth that of very wet years. In arid areas the amount of organic matter at the surface is many times less than in steppe areas. However, it is not yet possible to suggest the degree to which landforms in desert areas are distinctive and the degree to which any distinctive traits are related to the absence of an effective organic blanket between the bedrock and the sky.

In humid temperate areas temperatures are such that the organic residues are effective in reducing potential erosional activity. On the one hand oxidization rates are rapid enough to produce ample quantities of colloidal humus which increases water-holding capacity and water retentiveness within the soil and creates water-stable aggregates. On the other hand more resistant constituents of forest litter remain undecomposed for several years and, by filtering out silty particles from runoff waters, assist in keeping pore spaces and passageways open in the underlying soil, thus favouring high infiltration rates.

In cool temperate environments distinctive conditions prevail as the accumulation of organic matter begins to exceed decomposition rates. This is the result of the low temperatures being ineffective not merely in oxidizing organic material rapidly enough to balance supply but in evaporating sufficient quantities of ground-surface moisture to create an oxidizing environment. However, poleward of the damp cool-temperate areas temperatures soon become too low for biological productivity to make any appreciable contributions to the litter fall. In eastern Canada, peat thicknesses diminish in thickness from about 9 m near the St Lawrence valley to about 2 m northwards at 55° N. Similarly the net primary productivity of tundra willow and dwarf birch thickets is only one-seventh to one-fifteenth that of birch woods in the central taiga. However, the organic residues remain little altered for so long that a mass reminiscent of peat layers still accumulates. Organic mats are often widespread covering, for instance, 10–15 per cent of Labrador–Ungava. They retain water, three times its own weight in the case of peat, due to absorption by decomposition products, the swelling of plant residues and the filling of the porous space between the macrostructure. One geomorphological consequence is that this prevents substantial proportions of precipitation from coming into contact with the underlying mineral soil and bedrock, and also reduces runoff on gentler slopes. On the northern edge of the taiga peat covers all slopes less than 10 degrees, but may disappear with more rapid water flow on steeper slopes. It is perhaps only a coincidence that slope-angles of 11 degrees are highly characteristic of the lower parts of many rectilinear slopes in Britain, but it does seem possible that the importance of the inter-relationships between a peaty or organic-residue blanket and water draining off slopes has been underestimated. Perhaps there is wider significance in the fact that in the Alaskan tundra it is the protection of shores by drifting mats of vegetation debris that appears to determine lake basin morphology.

In cold climates where saturated soils and runoff are the rule due to reduced evapotranspiration losses and to frozen sub-soils, perhaps the major effect of diminished vegetation cover, which becomes discontinuous when annual means fall below 6° C, concerns temperatures. An organic cover prevents thawing temperatures, which would produce solifluction movements on slopes, from reaching underlying frozen soils, and it has been observed that the removal of the insulation of a moss cover does greatly hasten the thawing of frozen ground. A water-saturated organic cover layer conducts freezing temperatures downward when frozen, but all the energy of temperatures above freezing are absorbed in evaporating a portion of the melting water, warmer temperatures do not penetrate to the underlying soil and the cold reserve in the ground thus persists. However, in some permafrost areas where the organic mat is thick enough, frost-heaving can be confined to this superficial peaty layer. In slightly milder conditions of cool temperate environments the insulating role of a humus or peaty layer is reversed, as sub-zero temperatures, in freezing only the topmost layer of water saturated organic material, do not penetrate to depth. In South Wales at an altitude of 600 m peat was observed to be frozen to no more than 5 cm depth. Living plants and trees also affect temperature conditions in the soil. By decreasing air current velocities they impede heat radiation from the soil to the cold air. Their cover checks ground cooling by nocturnal radiation particularly when it retains a snow cover. J. Tricart (1967) quotes measurements in Germany where frost reached a depth of 47 cm in a field, 45 cm beneath Norwegian Pine, 38 cm under beech, and 34 cm under Scots Pine. In Paris basin depths of frost penetration recorded were 27 cm in well-drained sand, 18 cm in poorly drained sand, 13·1 cm beneath grass mat, and 11 cm under forest.

In addition to the variety of blanket effects of roots, litter, and humus layers, vegetation has a highly significant braking effect on the velocity of water and air movement by dissipating their energy. This may induce deposition of material in transport from areas where higher velocities prevail, and may have a filtering influence on the lateral distribution of sediment sizes. One of the reasons why this effect requires careful consideration is because equations for calculating stream-flow velocity, being inapplicable to conditions where vegetation grows in natural channels, have necessarily led to the systematic exclusion of information from vegetated channels in measurements made by water engineers.

In the deeper part of the outlet section in Lake Laitaure, sparse vegetation grows to 0·3 m above the bottom at the time of high-water discharges in summer, and as fig. III.13 shows, the braking effect is such that the flow velocity may be proportional to the log of the water height above the vegetation but not to the log of the water height above the channel bottom. In southwest United States, the gradients on some valley terraces are steeper than modern arroyo gradients, particularly in head-stream areas. This contrast exists despite the existence of large cobbles and boulders on the arroyo floors, which is the reverse of expectations based on hydrophysical principles which indicate that a steeper gradient is an adjustment to a coarser load. M. A. Melton considers that this apparent contradiction is due to the dense grass 0·6–0·9 m high, which occupied the former river valley floors, unlike the canal-like cleanness and straightness of the modern arroyos.

Vegetation in channels is limited by basin altitude, gradient, stream velocity, bottom material, and stream depth. Although the broader geomorphological implications of its role have not been investigated, examples of a few critical measurements are suggestive. In European U.S.S.R. stream channels are only overgrown at elevations up to 300–400 m above sea-level. Critical longitudinal gradients are 0·1–0·2 per cent, but occasionally are as steep as 1 per cent. Velocities which prevent vegetative growth are 0·56 m/sec in canals in India and 0·61–0·76 m/sec in canals in Spain. However, in sandy channels, the abrasive action added to the velocity of flow makes critical velocities of the latter 0·45–0·65 m/sec in the plains rivers of European U.S.S.R., whereas in bouldery or gravelly channels vegetation grows if velocities remain less than 0·65–0·95 m/sec. Yet again, in this context of the effect of sand in streams, one catches a glimpse of a more subtle aspect of the self-

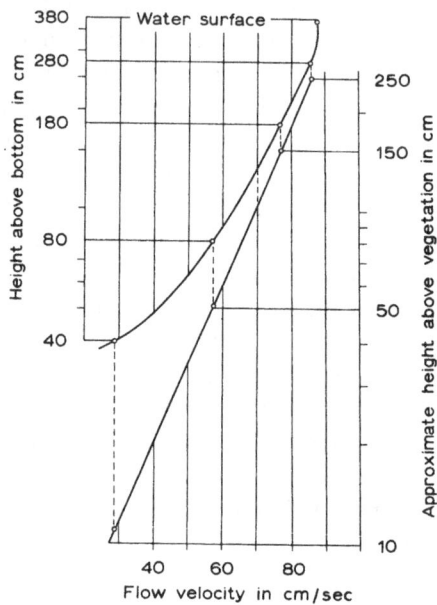

Figure III.13 The damming effect of the summer growth of vegetation on water-flow through the outlet of Lake Laitaure, northern Sweden (*from Axelsson, 1967*). The observations, made on 26/8/1957, show that the velocity was proportional to the logarithm of the approximate height above the 0·3-m-high vegetation and not to that of the height above the channel bottom.

accentuating nature of accelerated erosion unleashed by man's disturbance of the environment. Vegetation growth in downstream reaches depends on low water not exceeding 1·0–1·5 m and is in consequence usually absent when catchments are greater than 8000–10 000 km². In general it seems that channel vegetation will tend to occupy middle reaches of many streams, limited in headstream areas by the low temperatures at the higher altitudes and by steeper gradients, and by water depth in the lower reaches of larger streams. For ephemeral streams, limiting conditions will be those which affect the amount of growth during the intervals between floods. The occasional flood in the Gebel Akdhar hills near the North African coast may encounter trees several decades old in its path.

In coastal areas plants resistant to high concentrations of salt may have a braking effect on water movement. If Fucus become established on wave-cut platforms, their presence reduces the force of the waves. In estuaries, plants like Salicornia and Spartina reduce water movement to the extent that accretion may take place at a rate of 10 cm/year. Sand trapped by eel grass (*Zostera sp*) forms irregular mounds, as on the Wash flats, and carpets of algae like *Enteromorpha sp* can trap sand also.

On hillslopes completely covered with tough grass the braking effect may be sufficient to keep sheetflow below erosive velocities, and even-contoured slopes remain undissected. The streams of such an area in the Ngong Hills, Kenya, are sediment free, and on forested slopes a solid bed of dead leaves is a substantial obstacle to rillwash.

In sub-aerial situations, as is well-known, vegetation decreases substantially the velocities of winds close to the ground surface. The planting of an open stand of dune grass may increase the zone of virtually calm air close to the sand surface more than thirty-fold. On a larger scale it seems that many seif dunes originate where vegetation forms sand-traps rising above the general surface of a sand sea, as illustrated more clearly on a smaller scale by the nebka (fig. III.14).

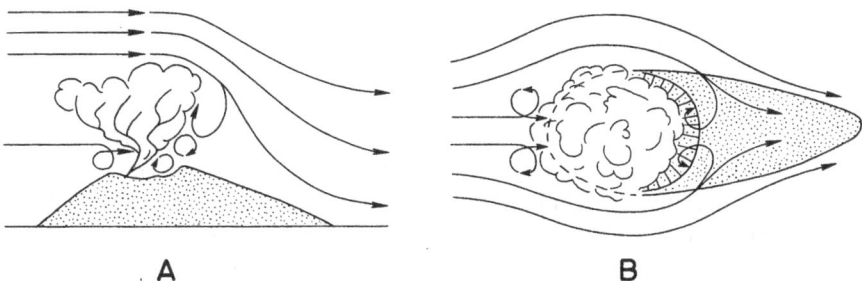

A   **B**

Figure III.14 Deflection of air currents by a nebka (*from Coque, 1962*).

The braking effect of vegetation on rates of water and air movement is clearly substantial and produces distinctive geomorphological effects, although its importance, particularly in slowing down water flow has perhaps been underestimated. One reason why the braking effect of vegetation requires careful consideration is because below a certain velocity it is a self-enhancing process, since overgrowing of a channel bottom, by increasing channel roughness leads to slower velocities and hence creates more favourable conditions for further growth.

Like the biomass above the ground surface slowing down movement of water and air, roots may hinder soil movement. The roots which may suddenly wrench up a ton of material on wind-throw may throughout the tree's life act as a binding and anchoring agent in environments ranging from deep tropical soils to the mollisols of periglacial areas. Their significance is particularly marked in tundra conditions where up to 90 per cent of the biomass may be below the ground surface. Vegetation may make scree angles somewhat steeper. In northern Sweden a scree 65 m high, 37 degrees in the upper part and 38 degrees lower down has scattered *Oxyria digyna* rooted in finer stable material 20–30 cm beneath the scree surface. Roots resist lateral erosion as well as downslope movement. On the shores of Lake Vättern in southern Sweden the Rosenlund cliffs retreated at 0·21 m/year in the east and 0·10 m/year in the west between 1795 and 1908, whereas there was no retreat in the forested section. On tropical coasts the mangrove, adapted to very saline conditions and to oxygen deficiency in its root zones, is a pioneer community, particularly in many Australian and African estuaries and adjacent open coasts. With roots amounting to half the biomass, the mangrove plays an important part in fixing banks of unconsolidated sediments. The effect of colonizing pioneer communities in stabilizing dunes is particularly well marked on tropical coasts where vegetative growth may be rapid and may extend close to the shoreline. It is unknown whether the mechanical resistance afforded by roots, which account for up to 60 per cent of the biomass in savanna grasslands and 80 per cent in desert areas is geomorphologically significant.

3. Biochemical weathering

The involvement of organisms is a fundamental characteristic of many aspects of chemical weathering. In as much as the scale of chemical denudation is essentially a function of the volume of river water discharged into the sea, plants are involved indirectly to the degree in which they restrict the coefficient of flow by returning moisture to the atmosphere by evapotranspiration. In this sense they are involved in a fundamental mechanism which limits the amount of dissolved materials as well as suspended sediment that is removed from the landmass and is of increasing significance towards lower latitudes. It is difficult to generalize about the more direct influences of organisms on chemical processes. Although every 10° C increase in temperatures introduces an approximate two-fold increase in rates of biological and biochemical reactions, it is only in certain situations that these reactions contribute to uptake of solutes. In many situations, particularly in lower latitudes, there is an equal acceleration of biochemical processes leading to the precipitation of dissolved material.

4. Acceleration of chemical weathering

The role of organic acids in weathering is the subject of occasional speculation but scarcely is investigated systematically. It seems that some more complex organic compounds are powerful chelators which are capable of forming soluble co-ordinate covalent compounds with various metallic cations. One of the most significant characteristics of lichens is that they generate a range of organic compounds not found elsewhere in nature which are a powerful agent in extracting essential trace elements from rock and can produce small weathering pits on the upper surfaces of rocks like granite. Otherwise most humic acids produced by anaerobic fermentation are too weak to form stable complexes and despite their acid reaction supply little H^+ to solutions. Organic acids are more effective in transportation processes, particularly that of transporting iron as complexed ions. The complexes are sufficiently stable to be transported by percolating water but, if brought into contact with ferric oxide, they are sorbed and the ferrous iron is then readily re-oxidized. Nonetheless, more than half the total quantity of iron, and also manganese and copper, dissolved in the water of the Dneiper travels in metallic—organic compounds. The amount of dissolved iron in rivers flowing through swamp lowlands may be as much as 15 ppm instead of the more usual values about 0·5 ppm.

The aspect of decomposition of organic matter of greatest geomorphological significance is its eventual fate under aerobic conditions of complete oxidation to carbon dioxide, water, and other simple end products. As a result the carbon dioxide concentration, 0·03 per cent in the atmosphere is often in the range 0·2–3·0 per cent in soil air. If biological activity is depressed by averages of either temperature below 10° C or moisture contents less than 10 per cent, carbon dioxide concentrations are unlikely to be above 0·2 per cent. In aerobic conditions, but with ample soil moisture available, carbon dioxide concentration describes a smooth summer-maximum curve, following temperature closely (fig. III.15.A). However, mid-summer desiccation of a soil below 10 per cent moisture content may depress the carbon dioxide peak, but under favourable conditions bacteria and other micro-organisms involved in the oxidation of organic matter may generate 1·3 times their own weight of carbon dioxide in 24 hours. Microflora are able to do this work because they have the necessary energy, transmitted to them by the sun. In habitats with the same temperature and moisture conditions, carbon dioxide is highest under oak and birch, intermediate under larch and lowest under spruce and pine, although both carbon dioxide and vegetation type are probably both variables dependent on soil fertility. Carbon dioxide output is essentially the product of microbial activity, root respiration itself accounting for perhaps only about 20 per cent of the total soil air. However, microbial activity is probably closely linked with root activity and its exudations on which the organisms flourish. Seasonal changes of the amount of calcium carbonate in solution in seepage water in Poole's Cavern, Buxton (fig. III.15.B), and at several points in other nearby caves, show a well-defined early autumn peak believed to reflect the carbon dioxide conditions in the soil, allowing for a time-lag of some weeks or a few months for water flow-through time. In fig.

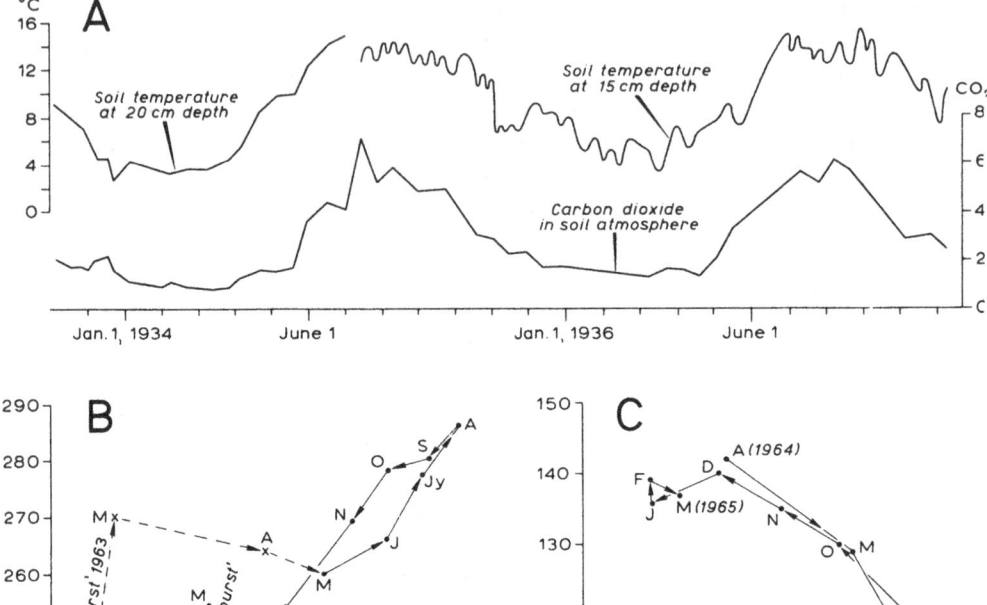

Figure III.15 Relationship between temperature and biochemical processes in soil and in water.

A. Soil carbon dioxide output at 20 cm depth, Wicken Fen, Cambridgeshire (*from V. M. Conway, 1936, New Phytologist, Vol. 35*).

B. Calcium carbonate in seepage water, Poole's Cavern, Buxton (*from Pitty, 1966*).

C. Calcium carbonate in a freshwater lake, Malham Tarn, Yorkshire (*from A. F. Pitty, 1967, Brit. Assoc. Lecture, Section E*).

III.15.B an unusual feature, attributed to the degree and abruptness of the thaw following the exceptionally severe 1962 winter, is high values of dissolved calcium carbonate that might reflect the 'spring burst' of organic activity that follows such a thaw. Although the significance of carbon dioxide, when dissolved in water to form carbonic acid, is seen particularly clearly in the solution of limestone, carbon dioxide has a profound and general significance in rock-weathering. The role of carbon dioxide in increasing the solubility of iron, for instance, has also been stressed, and indeed traditionally the theoretical significance of atmospheric carbon dioxide in rock-weathering has frequently been indicated. This emphasis, however, has been misplaced where the comparatively colossal output of biologically produced carbon dioxide in the soil has been passed over.

There are many forms on large and small scales indicating accelerated weathering that are associated with organic activity. Fig. III.16.A shows how

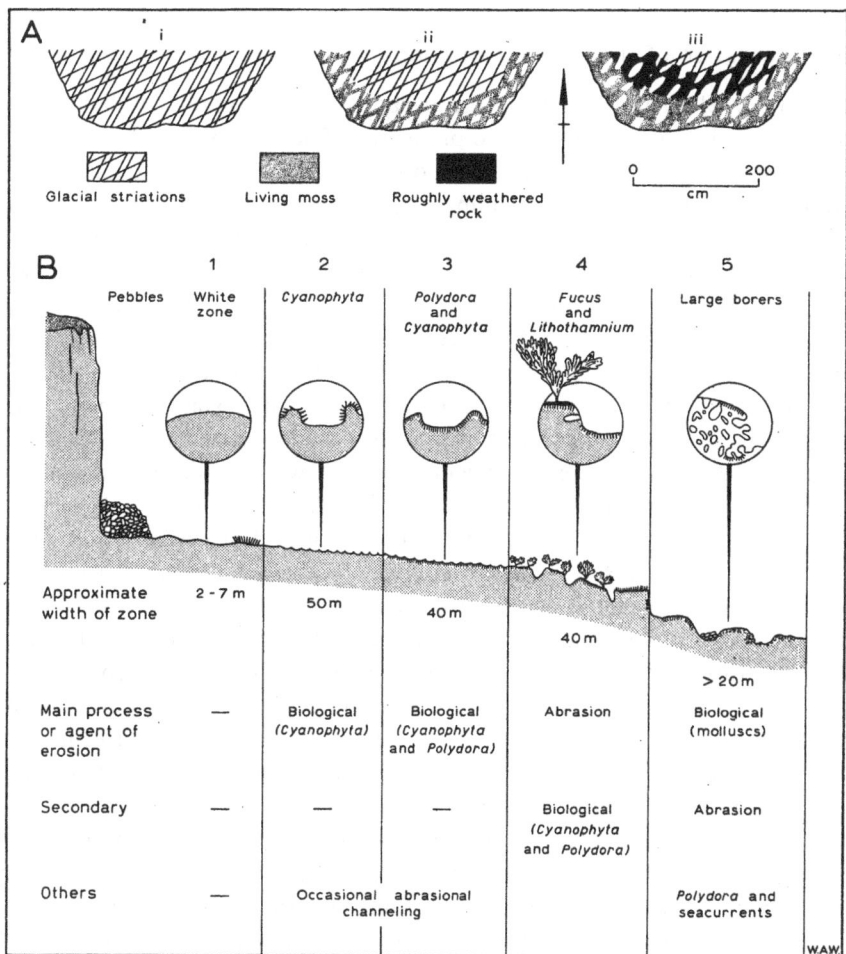

Figure III.16 Biochemical weathering on rocky shores.

A. Pseudo-pillow structures in greenstone on Sakami Lake shore, Quebec (*from J. E. Anderson and J. P. Mills, 1968,* Bull. geol. Soc. Am., *Vol. 79*): (i) outcrop glacially striated at N 30° E and N 65° E, (ii) mosses growing along striations where outcrop is wet, and (iii) mosses outlining pseudo-pillows where outcrop is wet, weathered rock outlining pseudo-pillows where outcrop no longer stays wet, and little-weathered striations on outcrop well above lake level.

B. Erosional zones on a shore platform in chalk at Életot, Pays de Caux (*after W. D. Nesteroff and F. Mélières, 1967,* Bull. Soc. géol. de France, Ser. 7, *Vol. 9*).

colonization of lines of weakness in a rock surface by vegetation can produce low elliptical domes in the intervening area. In favourable conditions in the shore zone, biochemical processes can consume rock at least as quickly as physical and purely chemical erosion.

5. Retardation of chemical weathering

In still or slow-moving water the carbon dioxide supply in the water is diminished by the photosynthetic processes of aquatic plants. This may lead to the precipitation of part of the dissolved load and some plants, notably the stonewort *Chara*, absorb calcium bicarbonate in order to extract carbon dioxide, leading to further precipitation. Like the increased intensity of solutional processes, the amount of precipitation in such situations increases with the more vigorous biological activity at higher temperatures. Water draining out of Malham Tarn in the Carboniferous Limestone area of north-west Yorkshire shows the results of this process, and the contrast between figs. III.15.C and III.15.B epitomizes the antagonistic roles of organisms in chemical weathering. Due to a different environmental change ferric hydroxides may be precipitated by the action of specific bacteria which utilize all the oxygen in the water for their respiration. In soils, particularly in lower latitudes and especially under savanna conditions, desiccation leads to the precipitation of iron solutions particularly close to tree roots due to the high osmotic pressures. Thus elements like aluminium and calcium, as well as iron and organic compounds taken into solution in the wet season are immobilized almost entirely in the sub-soil horizons, which accordingly becomes ferruginized. Similarly on coasts in a warmer environment, algae and other organisms probably play an important part in creating the conditions for intensive chemical precipitation in the formation of concretionary deposits like beachrock, not to mention coral islands.

Plants may appear to retard chemical weathering by incorporating a range of mineral elements into the biomass. Even in the tundra, where the decomposition of plant remains follows the same pattern as in many of the coniferous forest communities of the temperate zone, and is accompanied by relative accumulation of silicon, alumina, and iron, considerably less accumulation of calcium and magnesium and losses of the mobile substances like potassium, chlorine, and phosphorous. The retarding effect in these cool environments is that minerals remain locked up in plant residues for long periods. There may be more than 4000 kg/ha of minerals immobilized. In areas of active peat accumulation rates of addition are of the order of 20–35 cm/1000 years. In the B horizon of a soil at Garpenberg, central Sweden, the residence time of soil organic matter is 370 years ± 100. Even in temperate latitudes it may take two centuries to transform a fallen tree-trunk into humus and the organic mass on the forest floor is two to five times that of litter fall. By contrast, in tropical rainforest the very rapidity of biological processes is involved in the retardation of chemical weathering. The accumulation of organic matter, although litter fall is four times as great, is rarely more than a third or a half that of deciduous forests of the temperate zone and contains only between 80 and 300 kg/ha of minerals. This is a very small reserve in comparison with the annual re-uptake of about 2000 kg/ha, of which about 800 kg/ha may be silicon, with calcium and potassium about 200 kg/ha each. There would therefore appear to be little chance of minerals released from humified organic matter entering drainage waters. In fact some tropical plants accumulate significant percentages of certain minerals, particularly silica, the cycling of which distinguishes tropical forests from those of other zones. Five per cent of certain savanna grasses is silica by ash weight. As long ago as 1896 A. Grob found clots of amorphous silica in the internal cavities of bamboo. In tropical forests of China silicon contents may be more than 7 per cent. Although the degree to which minerals are irreversibly lost from the biological cycle could be of profound geomorphological significance if it were substantial, the problem has received little attention. Lovering (1959), however, has suggested that a forest of silica accumulator plants averaging 2·5 per cent silica and 16 tons dry weight annual new growth would extract 2000 tons/acre/5000 years of silica, and this calculation merits closer consideration. To the contrary, in fact, it seems that the balance between uptake and return of mineral nutrients approaches the ideal steady state. Measurements of uptake in arctic tundra and forest tundra were 38, 111, and 166 kg/ha respectively, compared with returns of 37, 108, and 157 kg/ha. In deciduous forest communities with quantities of 200–400 kg/ha of minerals involved, there is a very small imbalance with accumulation exceeding losses slightly. In the dry savanna of India measurements include an uptake of 1000 kg/ha and a return of 800. These and similar figures show two things. First, that for a given reserve of nutrients, the biological cycle is closed with most minerals, on release fall, being immediately reutilized by growing plants. There is therefore in a natural situation, with roots tapping solutions at depths of several metres and the restricted or non-existent surface runoff in the litter layer, little chance of erosional loss of mineral elements released from litter fall. In the high oak forests of Voronezh province, U.S.S.R., measurements have shown that of the elements which experience some irreversible loss, this proportion of the total of these elements retained with litter fall is less that 1 per cent. Secondly, even if all such elements were removed due to human interference, they would constitute for a few years only a scarcely noticeable addition to the total denudational losses. It seems therefore that the role of plants in cycling nutrients is essentially to retard denudational losses and that the basis of Lovering's calculations is an overestimate of the degree to which the litter fall is swept off the forest floor, and an underestimate of the degree to which the biological cycle is closed.

A final consideration is the possibility that organic acids, far from being a significant agent in weathering as is perhaps too readily assumed, might act in a protective role. From laboratory experiments which showed that amorphous silica solubility in several humic acid solutions was much less than that shown in distilled water, R. Siever (1962) concluded that the colloidal organic compounds were absorbed on the free silica gel surfaces and thus prevented the solution of that surface.

6. Organisms as aids in landform study

The stratigraphical use of organisms and other ways in which biological evidence may help in establishing chronologies will be illustrated in a later section concerned with methods of dating stages in landform evolution. Vegetation is an equally useful indicator in a dynamic depositional geomorphological environment tending towards some equilibrium condition as it is of indicating stages of sequential development. In the latter case plants, being exceedingly sensitive to variations in soil quality and development, closely reflect the uni-directional change in their environment. However, in a continually changing environment, like a delta, the one factor which in itself gives the closest approximation to the continued average effect of the other processes may be the vegetation. No plant community is in a static constant condition and the stands continue to evolve according to channel activity and may not approach a climax. On the landward fringe of small deltas in temperate environments woods, meadows, and sedge fens do not normally receive sediment, and in willow thickets there may be a very thin veneer at the most. Conversely if a delta is growing rapidly, due perhaps to greater sediment transport or to shallow basin depth, pioneer communities will occupy a large part of the land areas. Reeds cover 29 per cent of the 9000 km^2 Ili river delta, U.S.S.R., and their average height declines from 4·2 m in semi-flooded areas to either 3·0 m in wet valley conditions and 2·1 m in dry valley conditions. Stems on the left bank are 0·1–0·2 m higher than on the right bank. In a dynamic geomorphological environment, the longer a plant lives, the greater are its cumulative chances of being undercut and swept away. B. L. Everitt (1968) in the study of part of a valley occupied by the actively meandering stream, found the valley area of any age to be decreasing exponentially, 69 years being the time involved for an area of any age to decrease by half (fig. III.17).

Secondly, again in addition to its value in establishing times of change is the use of biological evidence in calculating amounts of change. The position of the base of trunks of older trees in relation to the present land surface

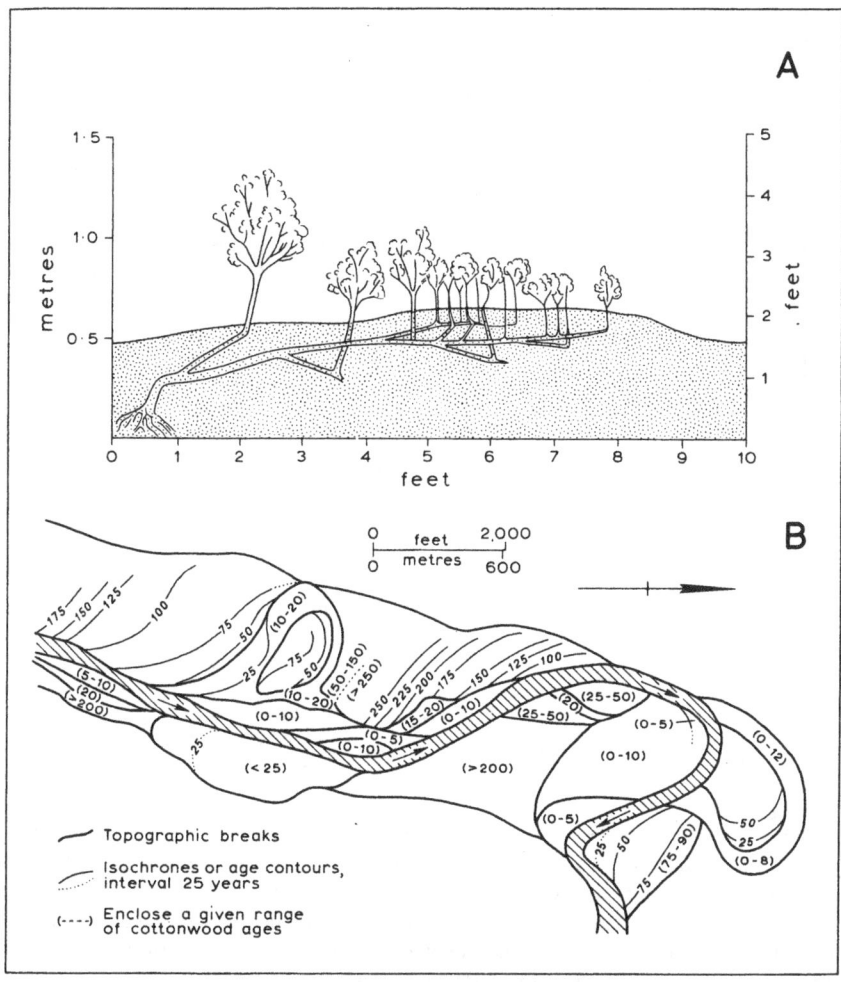

Figure III.17 The process of flood-training and the use of its results in dating flood-plain features (*after Everitt, 1968*).

A. Supple saplings less than 10 years old are bent over and buried with only shoots, now vertical, rising above the sand.

B. Age map of part of the Little Missouri valley floor compiled from age data obtained by coring cottonwoods.

(fig. III.18.B) may be useful in calculating amounts of pronounced erosion or deposition. One of the more spectacular examples is the bristlecone pine (*Pinus auistata*) in the dolomite areas of the White Mountains, California. Here, on living trees more than 4500 years old, the stubs of ancient side roots may project into the air 0·9–1·2 m above the present ground surface. The depth of movement of a solifluction feature is often indicated by a con-

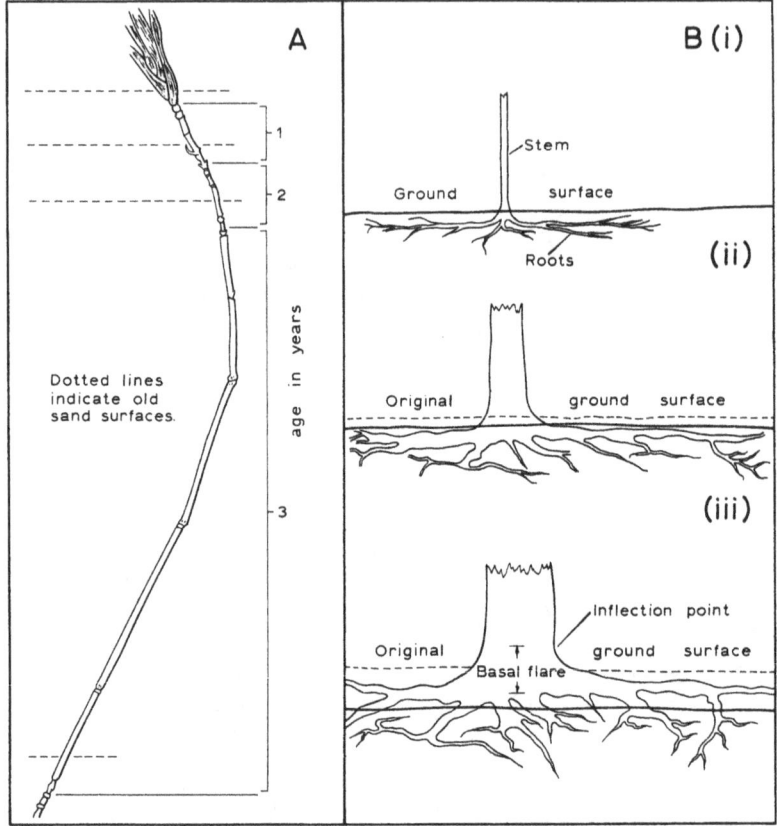

Figure III.18 Botanical evidence of amounts of land surface change. (A) Accumulation, as recorded by nodes on a buried marram grass stem (*from J. S. Olson, 1958*, Jour. Geol., *Vol. 66*). Wide spacing of internodes occurs in years following rapid deposition of sand in winter. Sudden decreases in annual grass elongation indicate stabilization. (B) Degradation, as shown by progressive root exposure (*from V. C. La Marche, 1968*, U.S. Geol. Surv. Prof. Paper 352-I).

spicuous layer of organic matter in the soil profile resulting from overrun vegetation.

G. Human Activity

'A change amounting to but little less than a revolution in the long established processes by which the features of the earth's surface are modified and developed, accompanied the advancement of man from a state of barbarism to one of civilisation.' I. C. Russell, 1904.

About 2 million years ago the ancestors of man emerged, with Early man (*Homo erectus*) standing up straight about half a million years ago. Human society and the deliberate organized work of Modern man began no more than 10 000–15 000 years ago, and has since disturbed biological and erosional balances over two-thirds of the land surface of the earth. The first major study of this phenomenon, C. P. Marsh's *Man and Nature; or Physical Geography as modified by human action* was published in 1864 and by this time large scale plantations were started in areas like the Aigoual area in France, where intensive erosion by floods in the Tarn basin were understood to reflect a link between deforestation and erosion. Despite the precedents, Economics man has during the past century pressed a vigorous claim to being one of the main converters of energy into erosional work.

One source of profound change has been man's development of water resources and drainage networks. Even on the smallest scales, like drainage ditches, the increase in surface drainage density may be substantial; in southern Scotland natural drainage densities of about 2 km/km^2 have increased twenty-fold. On a larger scale, water levels in the interior basins of south-west United States have steadily declined over the past 50 years. For example drops of more than 10 m in water levels beneath playa surfaces are very common. Farther south the ground level itself in the Mexico City basin first began to subside as water was removed from pore spaces by drainage channels installed in Pre-Columbian and Colonial times and accelerated after 1879 when the basin was systematically drained. In the last half-century intensified pumping lowered water levels even farther, and with the added factor of the sheer load of a city of 4 million inhabitants, some parts of the land surface have subsided at rates of up to 15 cm/year. Reservoirs not only alter flow regimes but the impounding slopes may become unstable with saturation at their base. Landslides occurred with 'unusually great and unexpected frequency' in the bordering Pleistocene deposits as the Grand Coulee dam reservoir filled. The trapping of sediment in reservoirs may accelerate erosion downstream in the river channel or on a coastline deprived of sediment. The total weight of stream-borne sediment reaching the Gulf of California would have been about 370 million m tons/year. Since the interception of the Colorado river by the Hoover Dam its contribution to Gulf sediment, formerly half the total, has been reduced to only some 8 million m. tons/year.

Man's removal of the natural vegetation, apart from his ploughing of the cleared areas, has had such a profound effect on erosional processes because of the variety of changes which all combine to increase surface runoff and at the same time reduce the soil's resistance to erosion. Apart from reduced interception and evapotranspiration losses, the porous litter and root layer vanishes and the soil, depleted in humus, has less power to absorb water and to create a fine porous crumb structure. Fine soil particles no longer filtered out in the litter layer are washed into lower horizons in the soil resulting in reduced permeability. Table III.3 shows how the hydrological regime in the Chernozem regions has changed in the past millennium. The simultaneous effect of increased amounts and intensities of surface runoff acting on soils with reduced cohesion is to increase erosion rates. In the Midwest of the United States, sediment yields became as much as seventy-five times greater than those under non-cultivated conditions. In India, deforestation in the Godavari drainage basin has exposed vast areas which have subsequently been eroded on a massive scale. In humid tropical areas the deeply weathered mantles were vulnerable to even the smallest changes. The lavakas of Madagascar are steep gullies 150–250 m long, 100–120 m broad, and 30–40 m deep cut in the plateau edge, probably over the last 500–1000 years. In Hong Kong gullies 150–200 m long and 2–30 m deep are due to progressive deforestation over the last 2000 years or more, and in many more upland areas of tropical forests, like Cuba, severe erosion followed deforestation.

Table III.3 Transformation of the water balance due to human interference over the past millennium in The Central Chernozem regions, U.S.S.R. *(from H. I. L'Vovich, Soviet Soil Science, 1966).*
    Elements of the water balance (in mm) are P = precipitation, R = full river flow, S = ground-surface runoff, U = underground flow to rivers, E = evapotranspiration, and W = soil water

	ELEMENTS OF THE WATER BALANCE						
	P	R	S	U	E	W	U/R%
Natural (some time before ninth–tenth centuries)	480	50	18	32	430	462	64
Elementally transformed nineteenth–twentieth centuries	480	100	72	28	380	408	28
Partly transformed by socialist agricultural measures 1925–50	480	95	65	30	385	415	31
1960s	480	85	55	30	395	425	36

In other environments accelerated erosion has been produced in different ways. In the savanna areas of Africa, the widespread firing of vegetation may have removed half the biomass in a single sweep, prevented the addition of humus to the soil and made the surface less permeable by baking irreversible changes into the clay colloids. More recently engineers have been surprised by the delicacy of balances in areas of frozen ground. Within a few weeks of the bulldozing of 30–60 cms of gravel and sand off an arctic beach, a hummocky and pitted surface developed because the underlying ice lenses came within reach of the zone of seasonal thaw. The removal of natural vegetation in such areas, scant though it may be, produces the same effect by allowing deeper penetration of thawing temperatures. Around Anadyr in Siberia, the removal of the tundra vegetation led to gullying, which, by improving drainage, facilitated deeper penetration of the thaw, in turn favouring intensified gullying. In drier climates removal of vegetation or lowering of a watertable reduces the resistance of soil to wind erosion. Farmers in most sandy areas, like those in the sandveld of South Africa, have found that ploughing greatly increased wind transport of sand, a problem recently accelerated in eastern England by the removal of hedge boundaries. Agricultural engineers suggest that in the Kara-Kum desert the uncontrollable grazing of livestock is sufficient to initiate wind erosion and that as a result severely wind-eroded soils develop at villages and at nomad camp sites. Again, as in the case of gullying in the humid tropics, the immense power of Technological man is not a prerequisite in creating the relatively small disturbance required to trigger off accelerated erosion in the more delicately balanced environments.

Human activities in modifying land form have in many ways assisted in an understanding of their development, particularly by the uncovering of fossil evidence and by the exposure of artificial sections revealing the contacts between biosphere, soil, landform, and subjacent rock. In fact it might be suggested that one of the handicaps in the study of tropical weathering processes is the paucity of artificial exposures in an area where natural exposures are almost non-existent.

IV

Inter-relationships between processes and landforms

The complexity of the factors involved in determining how the physical, chemical, and biological forces decompose or disintegrate rock, and carry away or deposit detritus, makes generalization about landform development much more difficult than is sometimes appreciated. Rock breakdown transportation and deposition are inextricably interlinked, but in order to study their wide variety of inter-relationships in the development, perpetuation or modification of landforms, an initial arbitrary separation to focus attention on the main characteristics of each of these three phases is required, together with discussion of aspects of landform that are largely the product of one only of these phases.

A. Rock Breakdown

There are three aspects of rock breakdown. First, denudation is a general term describing the breakdown of rock and the removal of part or all of the weathering and erosional products from the site of their detachment. As denudation may involve merely the transfer of weathering and erosional products from one part of the land surface to another, the term net denudation or net removal describes the amount of material lost from the land surface to the sea. Secondly, in its most restricted sense, rock breakdown may, due either to chemical weathering or to mechanical pressures building up *in situ*, involve little movement. A third aspect of rock breakdown, erosion, is also due to mechanical pressure, that of the abrasive action of weathered fragments during transportation and other stresses due to the motion of running water and moving ice.

In considering rock breakdown *in situ* it is vital to differentiate between situations in which removal of material is also involved and those in which parent material is merely reorganized rather than removed. The latter becomes increasingly the case in hotter climates where the ratio widens between water entering the soil from above and water leaving the weathering zone as runoff. Materials may be added to a weathering zone from an external, allogenic source, although detecting this may be possible only by detailed mineralogical analysis.

As all weathering reactions involve water either as a reactant or in the transportation of the reaction products, more than one chemical process is often involved during decomposition. For instance, the weathering of feldspar to clay involves both hydration and carbonation. Frequently the dominant process changes within short distances and even within the span of a soil profile. For instance, weathering of a calcareous boulder clay will involve rapid changes due to oxidation and hydration near the surface, the carbonation of limestone fragments at greater depth, and thirdly, near the base of the weathering zone, the comparatively slow dissociation of silicates and their formation into clays.

Although the course of weathering is related to the physical and chemical properties of water, the amount of denudation is equally related to the amounts of water leaving the weathering zone. This is obvious from the appearance of limestone surfaces in arid areas where few of the solution features of karsts in humid climates develop. Conversely one of the factors associated with areas of more intense bedrock corrosion in the gritstones of the southern Pennines is an abundant supply of slow-moving seepage moisture giving continuous wetting of the bedrock. Undoubtedly the most important single factor controlling denudation is the quantity of water leaching *through* the weathering profiles. Another factor which influences the amount of chemical denudation is the time during which water is in contact with the rock. For instance, the mineralization of soil water and superficial ground water may be two to three times that of slope surface water. Fig. IV.26.B shows how water in the limestone areas in the Pennines, after acquiring 140 ppm calcium carbonate relatively instantly, gradually increases its content of calcium carbonate with progressively longer spells underground.

Chemical and mechanical weathering processes are often closely inter-related. For instance, as chemical reactions occur on the surface of materials, areas containing loose, mechanically fragmented rocks and minerals will tend to yield far more solutes and solids than areas devoid or stripped of comminuted fragments. Another factor which makes it difficult to distinguish between chemical and mechanical weathering is the increase in volume which accompanies the weathering reaction, hydrolysis, which has a disruptive force greatly in excess of that exerted by most physical mechanisms.

1. Erosion

Erosion is a term perhaps best reserved for rock breakdown by the dynamic action of an agent like moving water or ice. Erosion may include abrasion if there are collisions between debris transported by water, ice, wind, or gravity fall and the adjacent bedrock. It is of vital importance to distinguish between erosion, implying the breakdown of coherent rock, and the mere rew orking of unconsolidated material produced by a previous phase of wea' iering or erosion. An immense volume of literature, particularly in fluvial g comorpho-logy, is devoted to fluvial 'erosion' where discussion relates only t) reworking of unconsolidated material and almost entirely ignores the problems of describing how running water might erode solid rock and of d ;ciding on the relative significance of this process in denudation as a whole. Discussion of the significance of glacial erosion is also scant with many examples of neat circular arguments inferring the scale of glacial erosion solely from the size of features believed to be the product of its action. In both instances there are several known circumstances which could theoretically contribute to erosion but whether they play a major, minor or negligible role in rock breakdown is as yet unknown.

Bedrock erosion in stream channels requires the operation of much larger forces than those which cause soil erosion on slopes or the reworking of channel alluvium. Abrasion of channel bedrock by the impact of bedload moving at flood velocities is believed to have some erosional significance in streams in middle and higher latitudes. Despite the probable significance of reduced abrasion in pebble-deficient tropical areas, the absence of ice could also be significant. Even by its weight alone ice could aid lateral pressure-release jointing on the lips of waterfalls in cool temperate and arctic environ-ments. Abrasion by the saltating load of sand is also possible and the smooth-ness of some bedrock surfaces in channels may be due to wet sand-blasting. In humid tropical rivers where the process may have increased significance in relation to other bedrock channel processes, J. Tricart has observed that some bedrock surfaces are in consequence smoothed and like roches moutonnées in appearance. A distinctive product of abrasion associated particularly with the cutting through of rock barriers are potholes where initial hollows are drilled deeper by eddies, swirling pebbles, and sand in spiral paths. On steeper slopes the loosening of blocks by the force of impact of falling material is a distinc-tive type of abrasion, and in a periglacial environment grooves up to 20 m long have been ploughed by gliding boulders.

Water erosion may involve two purely hydraulic forces. The first, hydraulic lift, is due to suction forces of eddies in turbulent flow which, on a small scale, keep sand grains in suspension and can in larger flows lift boulder-sized blocks. If lines of weakness in bedrock have developed due to pressure release, or by chemical or mechanical weathering, it is conceivable that hydraulic lift could detach such blocks. In fact as many observations of flood conditions reveal, like recent flash floods in Vermont, erosion is essentially in a horizontal direction, and degradation down through bedrock is the result of long-term weathering. The frequent discovery of solutional widening of limestone joints to 30–60 m below some river courses would support this view. Dam engineers have found that rock resistances in valley floors like the granite base of the Dours gorge may be fifty to a hundred times less than that of laboratory samples.

A second hydraulic force, observed in turbulent flow confined in artificial conduits and moving at speeds faster than 7·5 m/sec, is cavitation. Here water molecules are pulled apart by momentary shocks to form airless bubbles. The collapse of such bubbles involves pressures exceeding 30 000 atmospheres and the rapid hollowing of conduit walls, even if made of metal, testifies to the erosional potential of this explosive hydraulic action. Although cavitation is not to be expected in the unconfined condition of a surface stream, it may well be involved in producing cavities along phreatic cave passages and sub-glacial streams.

One of the most widely recognized focal points for stream erosion appears to be immediately below rock ledges or cliffs, particularly where the underlying stratum is relatively weak. Undermining proceeds due to the action of deepening and widening in the plunge pool, but the problem of exactly which factors exert the dominant forces has not been examined in detail. Nonetheless, huge gorges have been developed at rapid rates by waterfall recession, particularly along the lines adopted by large streams after glacial diversions of drainage. The classic example is the postglacial Niagara gorge where measurements over the past century show an average rate of retreat of 1·2 m/year for the Canadian Falls. Indications of rates of downcutting through bedrock by streams can only be approximate due to the problem of defining an initial reference point. Also, as the present example of the Niagara gorge shows, apparent downcutting might be due essentially to headward retreat of a waterfall position rather than to slow downwearing along an extended line. Several examples suggest that an average rate of bedrock channel lowering in middle latitudes might be roughly of the order of 15 cm/1000 years. In the Front Range, Colorado, downcutting of 60–120 m appears to have occurred within Quaternary time and probably started in pre-Wisconsin times or earlier. In the Beartooth Mountains, Montana, streams are now 90–180 m below gravel-covered benches which may be early Pleistocene or even younger. In the humid tropics even the larger rivers, in crossing a resistant outcrop, scarcely cut a gorge through it. There may be waterfall or rapids recession, but rarely does this extend more than a few 100 m upstream from the outcrop.

If a river course, instead of flowing over a resistant stratum, flows parallel with its edge, the underlying less resistant stratum might be excavated by lateral erosion. If lateral erosion in inclined strata is dip-controlled it may also involve a certain vertical component in the erosion of the channel bedrock. The contribution of a stream to lateral erosion is not easily separated from the activity of weathering and rillwash on the undercut slope. R. P. Sharp (1940) concluded that about 40 per cent of the mountain front retreat in the Ruby–East Humboldt Range, Nevada, was due to lateral erosion by streams.

The same hydraulic and abrasional processes that operate on the bedrock of a stream channel are active on abrasional platforms and cliffs along coasts. There are several isolated records comparable with that of a 60-kg rock hurled 30 m above sea-level on the Oregon coast. It is impossible to say if abrasion is more important than hydrolysis, salt crystallization, biological activity, or mass movement, and whether their relative significance varies with latitude. However, in the absence of abrasive materials the most powerful storm waves are relatively impotent against walls of massive well-indurated rocks. R. J. Russell (1963) observes that there are instances where sharp and complex changes in hydraulic pressure, frequent alteration between wetting and drying, and the activities of water-level organisms have produced inconsequential changes in the past 4000 years. For instance, granitic cliffed coasts near Rio de Janeiro are somewhat stained but display few abrasional features along the strandline.

Where it occurs cliff recession might be largely the result of landslipping rather than erosion, and depends on some seaward gradient on the erosional platform for the evacuation of debris from the cliff foot. Therefore, marine erosion tends to be self-arresting after a wave-cut platform some tens to hundreds of metres in width has been developed. Abrasional platforms on the Israel shore are continuous for long stretches and may be 30 m or more wide. On the north coast of Jamaica, west of Rio Bueno, the reef flat bench is 0·4 km wide. Probably due to the importance of biochemical weathering, platforms on coral limestone are often wider and flatter than average. One example is the Kenya to Tanga coast where the inclination of a coral platform 45–750 m wide is less than 1:200. In New Zealand a gradient of about 1:100 is often observed, closely approaching the suggested theoretical equilibrium value. However, slopes of erosional platforms along the south coast of England range from 1:14 at Eastbourne to 1:90 at Studlands. The influence of fetch is well illustrated on the shores of small islands. The slope off the south-east shore of Visingsö island in Lake Vättern, Sweden, is 1:19 to 1:11 (3–5 degrees) whereas the north-east shore slope is 1:70 (0·8 degrees).

Although some sea-cliffs may have remained in essentially the same position for 100 000 years, postglacial rise in sea-level or local circumstances have led to measurable rates of retreat of others. On Gotland, built exclusively of Silurian limestones, marls, and sandstones, wave-cut platforms are 100–200 m wide and cliffs retreat at 0·4–0·6 cm/year. On Barbados, a boulder from the Krakatoa eruption indicates 30 cm of marine erosion in 60 years, suggesting an average rate of 0·5 cm/year. Rates of marine erosion are often considerable in unconsolidated materials. Many stretches along the west Louisiana coast were retreating at rates of up to 1·8 m/year prior to the arrival of Hurricane Audrey in 1957. Cape Cod cliffs retreat by approximately 0·9 m/year. Boulder clays and cliffs on the Yorkshire coast from Flamborough Head to Spurn Point have retreated 3 km since Roman times. On the southern coast of the Lleyn peninsula, Caernarvonshire, cliffs 1·5–9 m high are cut back at 0·15–0·60 m/year.

These figures suggest that retreats of boulder-clay cliffs of a few kilometres will have occurred along many coasts since the last glacial retreat, and that during the repeated sea-level rises in Pleistocene times it is conceivable that cliff retreat in even moderately resistant rocks might have also amounted to a few kilometres during the span of Pleistocene time. However, for cliffs like those of the English Channel in south-east England and Normandy lining sea breaches through narrow necks of land, initial rates of erosion comparable with average rates of erosion in boulder clay seem possible. The degree to which coastal erosion in coherent strata is a self-regulating mechanism has not been established. As a wave-cut platform develops, increasingly the energy of wave advance is expended in crossing the platform and dissipated in reworking sediment veneers before it reaches the base of the cliff. In such circumstances cliff erosion depends on the possibility of little understood processes lowering the platform. Perhaps the intense organic activity in the intertidal zone should not be overlooked in studies attempting to assess whether appreciable lowering of wave-cut platforms takes place.

To an observer a glacier or ice-sheet may convey an overpowering impression, yet to any obstacle it may transmit a maximum pressure of 2 kg/cm² only. The physicists concerned with the mechanisms of ice movement have not given the same consideration to the problem of specifying the actual way in which ice might, with or without the aid of abrasion by bed material, erode rock. In consequence there is a range of opinion on the efficiency of ice as an agent of erosion. Large boulders, regardless of the force by which they are transported may by their own weight be sufficient to crush much smaller obstacles and to scratch smooth surfaces. For instance, striations on quartzites in the Fort Churchill area, Manitoba, occur once every 12 cm with lengths varying from 3 to 280 cm with breadths between 0·2 and 4·5 cm. Depths usually a fraction of a millimetre rarely exceed 2 mm. Striae and larger sinuous grooves and similar scouring forms often occur on the lee-side of resistant protuberances, indicating that the erosional agent was not associated with ice deformations due to pressure but rather to a separate scouring sub-stance. J. Gjessing (1967) suggests that scouring forms might be asso-ciated with soaked ground moraine or a compound of water, ice particles, and rock material, flowing in these lee-side positions between the bedrock surface and the underside of the ice. On a larger scale, testimonies to the efficiency of erosion by glaciers rests on the scale of forms assumed to be the product of glacial erosion. Glacial action appears to remove irregularities in valley sides. For instance, sharp ridges, pyramidal peaks and many ravines characterize the land surface above 1850 m along the south-west wall of the Shakwak valley in the St Elias Range, Yukon. The 1850 m level is an upper limit above the lower valley sides, smoothed apparently by glaciation. As present-day large-scale landslides in mountainous areas show, many steep slopes have been weathered to an extent that little internal cohesion remains. These features suggest that ice might therefore truncate obstacles like valley-side spurs, thus straightening valleys of the preglacial topography, and also that ice failed to dislodge parts of slopes where lack of internal cohesion was approaching critical limits. Conversely there are examples in glaciated areas of valley constrictions corresponding to resistant rock outcrops like those in the Aberdeenshire River Dee upvalley from Banchory. Neither is a glaciated valley invariably U-shaped. Like Glen Nevis they may vary from a U-shape, narrow to a V-shape, and then open out into a broad U-shape. Other instances that complicate the specification of landform characteristics uniquely asso-ciated with glacial erosion include the truncated spurs of recently faulted relief and the valley cross-sections of the U-shape, that may be pronounced along the sides of fluvial valleys, like the lower Driesam in the Black Forest. Just as there is an inter-relationship between pressure-release jointing and the dome shape of hills in areas of crystalline rocks a similar association for the inverted dome shape of the U-shaped valley is easily envisaged. In more open lowland situations the resistance of certain rocks is important, seen particu-larly well in areas where small igneous intrusions resist erosion and in the lee of which less resistant rocks escape erosion. The many such 'crag and tail' features in the Central Lowlands have been discussed and illustrated for nearly a century. In other lowland situations, straightened sides of an adjacent upland mass may give the impression of ice activity. In moving north, the Keewatin ice-sheet is supposed to have straightened and smoothed off the eastern scarp of the mountain overlooking the Mackenzie delta.

Although it is usually supposed that glaciers have straightened out valley irregularities in plan, the argument is reversed for irregularities in longitu-dinal section. Valleys at present or formerly occupied by glaciers often have a longitudinal profile like a staircase (fig. II.15D). Each step or tread is rela-tively flat or even overdeepened with a steep riser at its upvalley end and a riegel, a knob of resistant rock, at its downvalley end which constitutes the

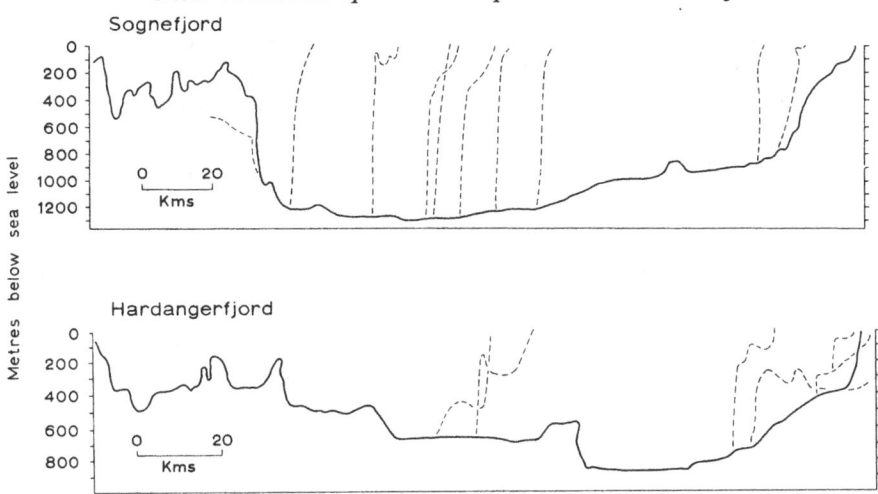

Figure IV.1 Longitudinal profiles of Norwegian fjords (*after Holtedahl, 1967*).

riser for the upvalley end of the adjacent, lower tread. Again, however, this feature of a staircase longitudinal profile is not uniquely associated with areas of glaciation. In valleys attributed largely to fluvial erosion analogous series of levels have in the past often been linked with former, higher sea-levels and the steeper sections regarded as knickpoints introduced by rejuvenation as successively lower levels encroach on those above. In Corsica the longitudinal profile of the upper Vecchio river valley in the Monte d'Oro massif is a series of steps only a few hundred metres long, separated by a series of subvertical cliffs about 20–40 degrees in inclination. Observations in areas where there is no transported load to fill up hollows show that rivers, like the Göta Älv in Sweden, may erode hollows as deep as 20 m below the level of a downstream rock barrier.

In fiords, long narrow arms of the sea with parallel, steep-sided walls, often rectilinear in plan and common on higher latitude coasts, the longitudinal section is distinctively basin-shaped overdeepened inland and with raised rock thresholds near the outlets. The overdeepening of these basins in relation to the thresholds, often cited as the clearest indication of the erosive power of ice, may be substantial. For the Skelton inlet, an arm of the Ross Sea, Antarctica, a maximum figure is 1933 m. This compares with depths like 1288 m for Messier Channel, Chile, 998 m for Scoresby Sound, East Greenland, 764 m for Finlayson Channel, British Columbia, or 527 m for Breaksea Sound, New Zealand. Along the Norwegian fiord coast, Sognefiord is overdeepened by 1308 m, although neighbouring fiords are not usually more than half this depth (fig. IV.1). Not all fiords show a systematic relationship between fracture pattern and their orientation, as seen in the Glomfiord–Melfiord area farther north. For the formation of fiords, R. F. Flint suggests that the importance of various factors must vary substantially from one fiord to the next. A. P. Crary (1966) considers the significance of possible inter-relationships between fiord ice and the ocean. He suggests that once floating ice is formed downvalley, erosion is limited to an area near the junction with the grounded ice (fig. IV.2). As the depth of such ice thickens inland, the erosion of bedrock could therefore lead to overdeepening inland, particularly if land margins were rising slowly through the order of 1000 m that occurs during isostatic rebound. In contrast to ice floated out of an inlet he considers grounded ice as most efficient in clearing away obstacles to gravity flow above sea-level. The spreading of an ice-sheet on leaving the constriction of valley walls also involves a thinning of ice and this effect therefore has some similarities to the thinning produced by the floating off at coastal margin.

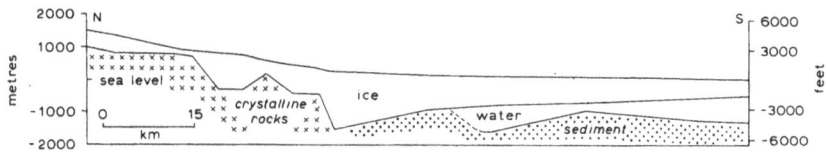

Figure IV.2 Glacier ice floating on to sea-water in Skelton Inlet, Antarctica (*after Crary, 1966*).

There is also the view that ice erodes comparatively little. In 1890, A. C. Lawson considered that there was no evidence to suggest that the surface of the Canadian Shield had undergone any material reduction in level due to glaciation. Around Flin Flon, 650 km south-west of Hudson Bay, an accidented pre-Ordovician topography, with a relief of 20 m where it was covered by the Quaternary ice-sheet, is essentially the same as where it remained fossilized by an Ordovician cover. Similar features have been recorded in Lower Cambrian strata in Finland. In the Fall Zone of Connecticut, R. F. Flint considers that glacial erosion, and preglacial erosion too, appear to have altered structural alignments very little. A. Godard (1965) suggests that much of the landscape of north-west Scotland appears to be largely the product of preglacial weathering. In areas of valley glaciation like Alaska, it is observed that glaciers can advance and recede without greatly modifying the terrain. In Iceland, lava flows indicate that a large portion of the landscape is of pre-Pleistocene or very early Pleistocene age, and in British Columbia show that the late-Tertiary relief of the Interior Plateau has been little modified by Pleistocene glaciation. In the Alps, some estimates, including the possible effect of long periods of interglacial sub-aerial erosion, suggest 300–400 m of overdeepening, while other investigators suggest that almost no overdeepening by ice took place. Evidence on a smaller scale of inefficiency in ice erosion includes the weathering pits that remained unerased by the Farmdale glaciation in the Rock River area of northern Illinois. On a much larger scale are huge overdeepened troughs lying at right-angles to the direction of ice movement and as deep as those lying in the direction of its general movement. J. K. Gilbert (1904) remarked on this unexpected pattern in Alaska. Earlier J. Geikie (1878) discussed a similar pattern in north-west Scotland, where basins along the whole of the east side of the Outer Hebrides reach about 200 m below sea-level near Barra Head. However, he did not stress that the direction of ice-sheet movement was almost perpendicular to the alignment of the basins. J. H. Winslow (1966) discusses the problem posed by deepwater continuations of some fiords, like Storfiord in Norway, extending beyond the known limits of glacial advance.

Although the necessity for some breakdown of rock prior to glaciation is increasingly recognized as a prerequisite for removal of rock debris by ice, there are several examples of loosened materials that have remained unaffected by the passage of ice. In the lowland area from the Orkneys south to Central Scotland deeply weathered rock remains beneath boulder clay deposits. In Swedish Norrland around Junsele, soft peaty lacustrine deposits are similarly buried, and a thick bed of kaolin at Ivö in Skåne survived vigorous ice movement. In northern Baffin Island a body of ice, admittedly less than 30 m thick, receded recently to reveal undisturbed patterned ground features and vegetation.

Cirques are another overdeepened feature, sunk into the higher parts of areas where valley glaciers formerly existed or persist today. Their distinctive hollow form, remarkably regular for an erosional feature, appears to show little modification with differing structures. In the Lake District, cirques unmodified in form cut across complex structures like the faulted contact between the Skiddaw slates and the Borrowdale volcanic series. In higher latitudes cirques become progressively few as the area occupied by major continental ice-sheets increases. Their development appears to be greatest near the limit of permanent snows and the importance of oscillations through freezing point is a significant consideration in many hypotheses. Originally, following the work of Willard Johnson who in 1904 was lowered down the gap between a cirque headwall and the ice of a cirque glacier, freeze–thaw at the base of this gap, or bergschrund, was believed to be significant. Other hypotheses have depended on the assumed erosional activity of ice. In the last of a series of hypotheses proposed by W. V. Lewis, rotational slip was suggested. However, this hypothesis appears to depend on the pre-existence of a hollow shape and a further objection is that measured ice velocities are not greatest at the base of a cirque glacier near its outer threshold end as the phrase 'rotational slip' implies. Many areas may be like the Lake District with cirques located in structurally weak zones where pre-glacial hollows might have readily developed. Once again distinctive landforms are not easily causally linked with areally correlated glacial phenomenon.

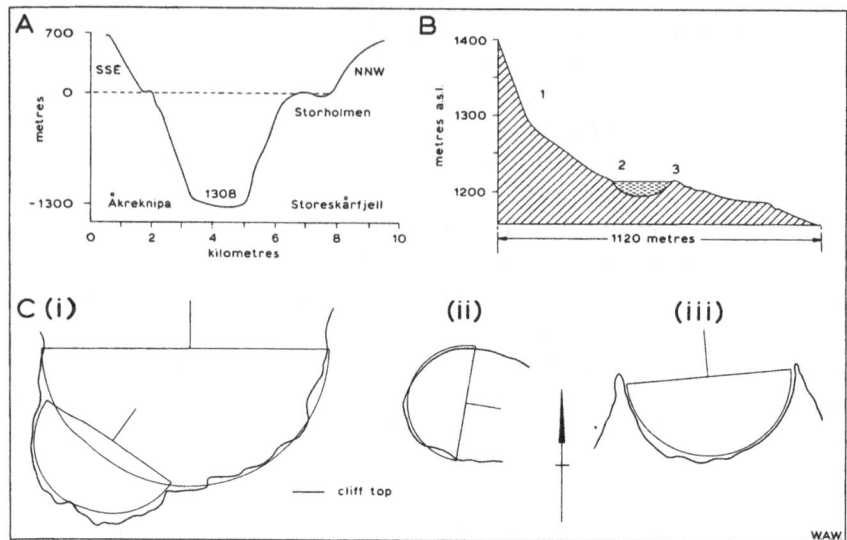

Figure IV.3 Forms typical of middle and high latitude mountains formerly occupied by ice.

A. The U-shaped valley cross-profile, as illustrated by the middle part of the Sognefiord (*J. Gjessing, 1965–1966*, Norsk Geog. Tidsskr., *Vol. 20*).

B. Cross-section of a cirque, Mount Olympus, central Tasmania (*E. Derbyshire in Fairbridge, 1968*). This section across the Lake Enone cirque typically displays: (1) a steep headwall, (2) a concave floor, in this case a rock basin modified by talus accumulation at the foot of the headwall, and (3) a threshold of bedrock covered by a thin veneer of glacial drift.

C. Plan views of corries (cirques) in the Cairngorms (*D. E. Sugden, 1969*, Scot. geog. Mag., *Vol. 85*): (i) Corrie an-t-Sneachda, (ii) Corrie-an-t-Sabhail, and (iii) Corrie Ruadh, emphasizing the symmetry of these forms by showing their approximation to the arc of a semicircle. The headwall of Corrie an-t-Sneachda includes a secondary corrie.

It is not easy to draw conclusions about the importance of glacial erosion. Even without considering the mechanisms involved it is difficult to observe points at which ice might erode rock and unsafe to consider that the amount of debris in melt water represents material actually eroded by ice. For land-form evolution in many areas once occupied by ice, contemporary opinion is that tectonic and preglacial erosion may well have been at least as important if not dominant factors. For instance, H. Holtedahl (1967) considers that the presence of well-developed, typical fiords along the west coast of Norway is a natural consequence of the oblique Tertiary uplift of the Norwegian land-mass, leading to increased fluvial erosion which was especially active in cutting deep preglacial valleys on the seaward slope. It is suggested that a glacier is, above all, an agent of transport with ice too plastic and too weak to attach itself firmly to a coherent rock. In contrast there are other areas for which interpretations are based largely on the appearance of the forms themselves and some investigators record that they are unable to avoid the conclusion that deep excavation by ice has taken place on a vast scale. J. B. Sissons (1967) suggests that 'there is no doubt that glacial erosion has been extremely effective in many parts of Scotland.' .

Although many workers have increasing doubts about the efficiency of ice in the erosion of unweathered or unloosened rock, their views should not be confused with the 'Protectionists', a school of anti-glacialists who at the beginning of the twentieth century went beyond the belief that ice had little erosive power and claimed that an ice cover protected the underlying rock from erosion, notably from frost-shattering. Many present-day workers, whose beliefs in the efficiency of ice erosion may vary widely, agree that in certain localities there is substantial erosion beneath ice due to the pressurized flow of meltwater confined by the ice cover. The heavy loads of fine material in meltwater from valley glaciers indicate the importance of comminution of transported material in sub-glacial meltwaters if not to actual degradation of the channel floor. Giant potholes appear to be associated with former sub-glacial environments (fig. IV.4). In the Flåm valley, a tributary of Sognefjord, organic material in the bottom of one of the large potholes close to the summit of Furuberget was dated as 9350 ± 300 years B.P. suggesting that active erosion took place in pre-Boreal and earlier times, presumably in

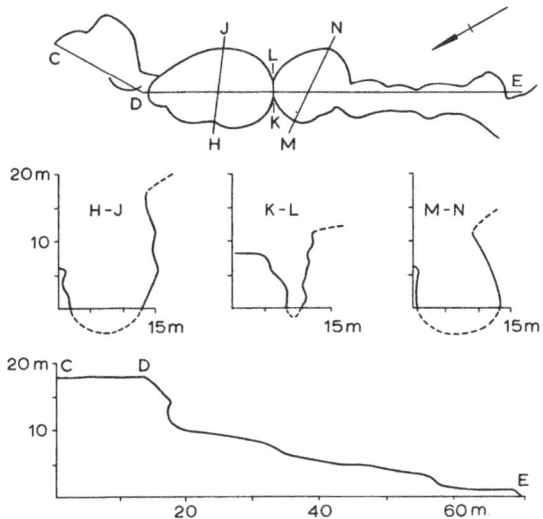

Figure IV.4 Sections through giant potholes (*from Holtedahl, 1967*). The potholes were surveyed by I. Klovning at Furuberget, in the Flåm valley, Norway. Erosion by sub-glacial streams under hydrostatic pressure is suggested.

a sub-glacial environment. For comparison, potholes at Taylor's Falls, Minnesota are up to 4 m in diameter. In addition there are smoothly sculptured micro-forms on rock surfaces including channels, troughs, and hollows, termed *p*-forms by Scandinavian geomorphologists and ascribed to the flow of sub-glacial melt water. Hjulström (1935) was the first to suggest that cavitation might be involved in their formation. A large number of localities containing *p*-forms are situated on mountain plateaux in broad interfluve zones. In many parts of the British Isles formerly covered by ice, channels cut across interfluves with a characteristic 'up-and-down' longitudinal profile distinguishing them from present-day stream occupied valleys, are now widely attributed to sub-glacial melt-water erosion. In some parts channels descending at the gradient of the local slope have, following the work of C. M. Mannerfelt (1945), been described as sub-glacial chutes and attributed to accelerated drainage in this direction following a supposed release of water held under hydrostatic pressure in the lowest parts beneath an ice-sheet.

In addition to meltwaters flowing under hydrostatic pressure beneath ice, there is the possibility in stagnant ice of melt waters flowing freely at the base of crevasses reaching through the ice to the ground beneath. M. M. Leighton and J. A. Brophy (1961) have termed straight and shallow channels believed to have been cut in this manner as crevasse traces and have described many examples showing striking alignment in western Illinois. Many are 5–10 km in length and one is over 50 km long. Melt-water channels in areas formerly covered by stagnant ice are among the more spectacular glacial features of the Plains region of the United States.

2. Frost action

The degree to which frost action is involved in general weathering processes, as opposed to the production of a specific form, is not clear. An initial difficulty is that for freezing of water in microfissures to be effective the crystals must possess microfissures already enlarged sufficiently for the freezing of contained water. There are reports that particle size in arctic soils is largely controlled by the grain size of the parent rock and that in the Antarctic, soil-forming processes are barely discernible. Generally, in arctic and sub-arctic areas, diurnal freeze–thaw can cause only differential movement within a surface layer of a few centimetres and may not make a significant contribution to the amount of weathered material. However, frost breakdown of mineral particles may be significant on glacial outwash plains where saturated conditions, seasonal changes in freezing conditions and fluvial reworking of the sands appears to reduce mineral particles to sizes predominantly in the 10–100-micron range, indicated by C. Troll (1944) as the size-range typical of frost-rived particles, and subsequently observed in high latitudes both in North America and in Russia. Many attempts to evaluate, by controlled experiment, the importance of freeze–thaw action in the initial disintegration of rocks have been attempted since A. G. Högbom made some preliminary experiments in 1899. In one experiment a simulation of 16 years of freeze–thaw on rock fragments increased the proportion of the 0·1–1·0 mm range by 10–20 per cent. S. Wiman (1963) used 36 'Icelandic-type' − 7° to 6° C cycles and 9 'Siberian-type' − 30° to 15° C cycles, but produced debris which was only a small percentage of the original weights. Samples in dry plastic containers showed no weathering, but it is difficult to decide whether this shows that water is a prerequisite for frost action or whether it illustrates hydration producing a similar disruptive effect. Table IV.1 provides a further illustration of materials subjected to laboratory freeze–thaw cycles. Blümche, cited by Tricart and Cailleux (1967), showed that the number of cycles needed to develop a fissure in a sandstone of 25 per cent porosity was 3 whereas in a sandstone of 5 per cent porosity 43 cycles were necessary. In a chalk sample with 30 per cent porosity 1 cycle was sufficient and several experiments have observed how porous chalk disintegrates completely with a few freeze–thaw cycles.

Table IV.1 Artificially frost-weathered products as a percentage of the original weights of the rock samples (*from Wiman, 1963*).

ROCK TYPE	ICELANDIC	SIBERIAN
Slate	1·16	0·16
Mica-schist	0·25	0·04
Granite	0·15	0·07
Porphoritic granite	0·29	0·19
Quartzite	0·02–0·19	0·007
Gneiss	0·65	0·01

Any conclusions on the role of freezing and thawing in the disintegration of coherent rock material must emphasize the difficulty of separating this from the role of other processes and point to the likelihood of its effective operation being restricted to a certain number of rocks like schists or porous sandstones where pre-existing openings are large enough to permit ice-formation. R. J. St. Arnaud and E. P. Whiteside's (1963) experiment shows the range of results between duplicate samples (Table IV.2). This reduces the significance of the contrast between their reduction of a granite sample by 0·48 per cent compared with the 0·15 per cent reduction of 36 'Icelandic' cycles by S. Wiman (1963). Table IV.2 also shows a decline in debris produced in the second 100-cycle treatment, perhaps because the initial stages in artificial weathering exploits pre-existing weaknesses. A more fundamental reason for believing that the results of laboratory experiments cannot be applied to natural situations is that where frost-shattering is demonstrably effective the process, if the weathered material is not removed, becomes self-arresting by producing a layer of material through which freezing temperatures fail to penetrate to bedrock.

Table IV.2 Effect of repeated artificial freeze–thaw cycles on the breakdown of mineral grains (*from Arnaud and Whiteside, 1963*).

		FIRST 100 CYCLES		SECOND 100 CYCLES	
Initial material	Size range, millimetres	First sample	Second sample	First sample	Second sample
A	0·05– 2·0	1·30	1·93	0·64	0·44
Horizon	2·0 –10·0	0·61	1·58	0·44	0·32
B	0·5 – 2·0	3·49	2·84	0·32	0·35
Horizon	2·0 –10·0	1·68	0·66	1·40	0·60
Crushed	1·0 – 2·0	0·16	—	—	—
Quartz	2·0 –10·0	0·13	0·18	—	—
Granite cube	60	0·48	—	—	—

3. Other mechanical processes of rock breakdown

As early as 1925 E. Blanck and S. Passarge attributed the crumbling of rocks in the Egyptian desert to the crystallization of salts. In recent years workers have reported increasingly on the apparent importance of salt crystal growth in rock breakdown. One of the most important weathering agents in the semi-arid zone may be the salty dust which lodges in the openings in the rock and swells when light rain falls. Basalts near Laguna Pueblo, New Mexico, have been weathered in this manner. Similarly this mode of mechanical disintegration is possibly significant on tropical coasts where evaporation is intense and the salt in sea-water spray crystallizes in fine cracks in the rocks. In dry areas in Antarctica salt-weathering may be the dominant process of rock breakdown and appears responsible, at least in part, for cavernous weathering features up to 2 m in diameter, and for the reduction of quartz diorite in McMurdo Sound to a poorly sorted sand with diameters as small as 3 microns. Some workers believe that an essential characteristic of salt-weathering is its more rapid action on the lower rather than the upper side of rock surfaces. A somewhat similar mechanical–chemical weathering process may be the crystallizing of caliche. It is conceivable that over long periods of time this might be a sufficiently disruptive force to disintegrate both superficial deposits and bedrock in semi-arid regions. Temperature changes can theoretically induce some expansion and shrinkage of rocks. During the past 30 years many investigators have concluded that this process is ineffective.

The role of wind in the disintegration of solid bedrock, as with that of stream action, is easily exaggerated if the scale of transportation of debris usually ascribed to 'wind erosion' is not carefully evaluated. Again by far the larger part of material removed and transported by wind represents the reworking of previously disintegrated, loose material with little contribution from the breakdown of solid rock.

Only very soft rocks will disintegrate under the blast of wind-blown sand at some distance above the ground. However, as ventifacts show, in arid climates the comminution of debris and the cutting of resistant rocks is possible in a shallow layer close to the ground surface. Near Marble Point, Antarctica, the amounts removed are usually less than 3 mm of rock. P. Kuenen (1960) suggests that such modification might take a dozen or even a hundred years. In an experimental plot set up in a windy pass in the Western Coachella valley, California, R. P. Sharp (1964) found maximum wear at the discrete height of 23 cm above the ground, which is above the height of 13 cm below which half the saltating material travelled. In this area the critical wind speed of 18 km/hour is exceeded 16·5 per cent of the time. The height of maximum wear probably represents a level at which grain size, number, and velocity combine to give greatest impact energy. Using bricks of a 2·5–3 hardness on Moh's scale in a natural setting, he first observed pitting on the 228th day of the experiment, and after 11 years found a maximum cutting of 5·5 cm. Two other features were noted. The first was the development of a residual sill at ground level on windward sides; the second, fluting on the top surfaces of bricks produced by descending grains. In desert areas the effect of wind abrasion is now generally thought to be restricted to the slight under-cutting of low-standing outcrops. The drying effect of wind, in assisting evaporation and thus accelerating salt crystallization in arid areas, is probably a more significant aspect of the role of wind in erosional processes than abrasion by wind-driven particles.

4. Chemical weathering

However difficult it may be to judge the relative importance of chemical weathering from the appearance of disintegrated rock, and regardless of the degree to which a separation of the interlinked processes of chemical and mechanical breakdown may be arbitrary, one fact is clear; the scale of the invisible operation of chemical denudation has probably been underestimated in a large number of landform interpretations. Some theoretical schemes of landform development are remarkable in the degree to which inferences about processes are restricted to mechanical weathering and erosion. Measurements of the actual amounts lost show increasingly how widespread and substantial is the scale of chemical weathering. The measurements made by A. Rapp (1960) during his nine seasons' study of mechanical weathering and mass movement in a mica-schist area of arctic Sweden have, in themselves, aroused great interest. The fact that he demonstrated also that this contribution to net denudation was no greater than chemical denudation of 26 m. tons/km^2/year, even at a latitude of $68\frac{1}{2}°$ N, receives less attention. In the adjacent Skojem district, R. Dahl (1967), from the weathered appearance of granite surfaces concluded that chemical weathering was possibly an important factor in the formation of cavernous recesses. In most arctic areas the precise nature of physical and chemical weathering has never been well-established although for more than 30 years there has been a suspicion, first expressed by Glinka (1914), Taber (1943), and others, that the importance of mechanical disintegration could be over-emphasized. In well-drained sites it appears that hydration is quite active, although the presence of feldspar in the clay-size fraction suggests that hydrolysis is not strongly developed.

Although little is known about the changes in the chemical character of water as it passes from the surface through soil and rock to a zone of saturation below, some measurements are available. In general, concentrations tend to drop where precipitation is heavy and to increase where it is slight. Figures may be very much a reflection of other local conditions, difficult to place within broader zonal generalizations. For example, concentrations of calcium carbonate in limestone areas in Britain commonly fall between 150 and 250 ppm, yet in the distinctive environment of a shallow, seasonally dry lake in the McMurdo Sound area, Antarctica, concentrations

Table IV.3 Amounts of calcium and magnesium in solution in contrasted weathering environments ranged across European U.S.S.R. (*from B. G. Skakalskiy*, Soviet Hydrology, *1966*).

Values are in mg/l. and represent the range of several measurements made in each environment.

	CALCIUM		MAGNESIUM	
	Slope water	Ground water	Slope water	Ground water
Tundra	1– 5	5– 63	1– 3	2– 23
Taiga	2–17	14–165	1– 4	4– 21
Mixed forest	5–17	23– 82	1– 4	8– 36
Northern steppe	26–54	85–156	1–11	15– 81
Southern steppe	4–81	112–303	3–53	71–161
Forest steppe	17–33	65–433	1–12	8– 59

of 3630 ppm have been observed. Table IV.3 lists the range of values observed for calcium and magnesium concentrations from north to south across the Russian plains. Due to rapid reprecipitation, values from tropical areas are difficult to interpret. At Koullon in the Congo a range of 90–270 ppm calcium carbonate has been recorded. It would take several years if not centuries for terrestrial waters to approach equilibrium in silica, even in humid tropical environments. Amounts of silica dissolved in superficial groundwater as well as in river water are low. Calculations suggest a weighted mean concentration of silica in all rivers of the world to be 13 ppm. Even though removal of silica is an intrinsic part of humid tropical weathering, rivers in the tropics have only a slightly greater load of colloidal silica than those of the ancient massifs of Europe. Concentrations in waters draining from tropical forests are 10–40 ppm compared with the 14 ppm average for North America. In fact silica is the least variable of the major constituents in ground water. Values for the River Po range from 4 to 8 ppm. In Honolulu, the range in superficial groundwater is 8–16 ppm silica with a mean of 12 ppm, values for large springs average 20 ppm, and for deep wells, 38 ppm. In springs draining from Pre-Cambrian ferruginous cherts in Cerro Bolivar the silica concentration is 10·5 ppm and 15 ppm in drill holes. In weathered epidiorite in British Guiana, the figure for springs is 12·5 ppm and 33 ppm for wells 50 m deep. For the Mahadanz area in India, the amount is 26 ppm. In the 35 years after the 1883 Krakatoa eruption the silica content of the ash decreased from 67 to 61 per cent. At present rates of removal in forested areas of the Ivory coast, at 0·7–2·5 mg/cm^2/year all the silica in the silicates would be removed in 44 000 years. On Oahu Island, Hawaii, silica removal is 3·6 mg/cm^2/year.

Some controlled experiments on weathering show substantial losses. In 1930 the French workers Demolon and Bastisse broke down 800 kg of granite into sizes in the 2–4-mm size range. Natural processes achieved the further break down of sizes between 1935 and 1945 as shown in Table IV.4. Over the period 1931–45 elements in the drainage water were removed in substantial amounts (Table IV. 5).

Table IV.4 Percentage change in various sizes of granite fragments after 10 years' natural weathering (*from A. Demolon and É. Bastisse, 1946*, C.r. Acad. Sci., *Vol. 223*).

	FRAGMENT SIZES IN MICRONS				
	<0·75	0·75–2	2–20	20–200	>200
Percentage in 1935	0·32	0·49	2·96	25·20	71·0
Percentage in 1945	0·45	1·00	3·85	29·50	64·75

General weathering rates are relevant to landform studies but not necessarily easy to interpret. The nearly uniform and complete weathering shown by the first 4·5 m of the soil on Kaui in the Hawaiis would require about 70 000 years. The weathering of dacitic ash to a depth of 0·9 m on the El Salvador would take at least 5000 years. Volcanic rocks erupted in recent times in Indonesia and the West Indies show that the establishment of a complete weathered zone could take 1 million years, and in Venezuela data from springs suggest that the present iron-rich formations could be the result of weathering processes operating over a span of 20 million years. In a

Table IV.5 Amount of dissolved solids removed from granite fragments after 10 years' natural weathering (*after A. Demolon and É. Bastisse, 1946*, C.r. Acad. Sci., *Vol. 223*).

	CaO	MgO	K_2O	P_2O_5	SiO_2
Initial weights (kg)	29·680	7·52	1·28	1·20	1·16
Amount removed (kg)	0·276	0·043	0·059	0·003	0·038
Annual average removal in solution (gm)	19·8	3·0	4·2	0·2	2·7

forested area in the Ivory Coast, all the calcium could be removed in 3000 years and all the potassium in 117 000 years. In loess deposited in the Slims river valley, south-west Yukon, 9780 years ago decalcification is complete in the upper 22 cm and partial in the subjacent 20 cm, whereas Neoglacial loess is unweathered. Soil profiles on moraines left by the receding Mendenhall Glacier, near Juneau, Alaska, show that podzol profile formation is slight after 250 years in this environment and that the establishment of an equilibrium condition between environment and soil profile would take 500–1000 years. In deposits of the ancient early Quaternary or late Tertiary River Teays weathering has destroyed all but the most resistant original materials. Little remains other than quartz sand and siliceous gravel and boulders.

In appearance, changes of colour often indicate the progress of weathering. Weathering by oxidation is typically indicated by a red or yellow surface layer on weathered rock. Because of its abundance and easy oxidation, iron shows the general progress of this type of weathering. Most frequently described is the deep-weathering of acid crystalline rocks, on which the signs of weathering are, first, a whitening of the rock apparently due to the development of fine fractures in the feldspars. When the plagioclase is partly decomposed and as the attack on orthoclase begins, the rock breaks down to platey fragments of decomposed granite called grus. As corestones develop, the most prominent fractures are parallel with their boundaries, and zones of weathering in the grus are concentric around the corestones (fig. IV.5). Corestones rarely remain coherent at sizes less than 1 m in diameter. Of the visual impressions of rock breakdown the depth of weathering often attracts notice. In Table IV.6 the fissuring of granite and gneiss in the Kola peninsula, to depths comparable with many areas of deep-weathering is particularly noteworthy.

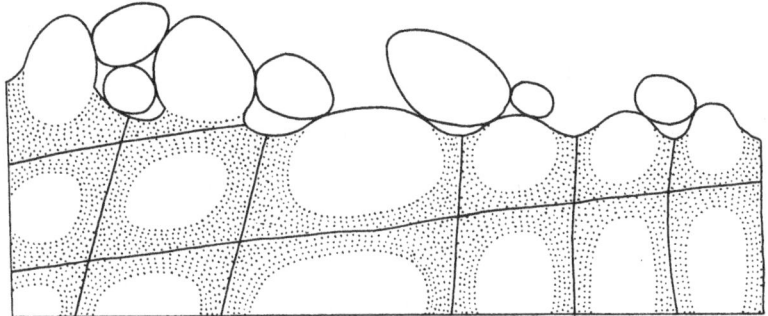

Figure IV.5 Development of granitic boulders by sub-soil weathering (*from W. M. Davis, 1938*, Bull. geol. Soc. Am., *Vol. 49*).

In certain favoured localities it is possible to observe the effects of denudation by the amount of ground surface lowering in relation to some reference datum. These tend to be greatest in areas of relatively soluble rocks like those rich in calcium where precipitation is moderate to heavy and where evapotranspiration losses are relatively slight. In a granite area near Narvik in northern Norway, R. Dahl (1967) calculates a lowering of 1·1 mm/1000 years at 85–95 m above sea-level and 1·3 mm/1000 years at 120 m above sea-level. This suggests an order of a negligible metre of lowering if present climatic conditions prevailed for a million years. For the humid tropical areas, P. Birot (1965) states the amounts of dissolved material removed from crystalline rocks into approximate equivalent of 1–3 cm/1000 years surface lowering. For the Hawaiian basalts, the theoretical lowering would be 13 cm/1000 years. In total, the Koolau extinct volcano could have been lowered by 650 m in a period starting 1·3–5 million years ago. The rates of denudation of the Hydrographers volcano, north-east Papua, increase inland from 8 cm/1000 years near the coast to 52 cm/1000 years at an elevation of 533 m. Bermuda's precipitation, now about 1400 mm a year, has been sufficient to lower the land surface by 125 cm since Sangamon time. At 2230 m altitude in the Mären mountains, lowering of limestone surface takes place at 1·5 cm/1000 years. In Kentucky, lowering rates are about 15 cm/1000 years. In chalk areas in southern England, open fissures and voids only partly filled with clay indicate the solution active at the chalk surface. Differences of level in localities where the chalk surface is locally protected by prehistoric earthworks suggest a lowering of the chalk surface at a rate of 10–12·5 cm/1000 years. For most areas it is possible to calculate a rate of weathering indirectly from amounts of material transported from a drainage basin. Illustrations of these rates and difficulties in their interpretation will be discussed in the next section.

Table IV.6 Some examples of type and depth of weathering.

LOCALITY	LATITUDE	ROCK TYPE	DESCRIPTION OF NATURE OF WEATHERING AND DEPTH
Ballangen, 38 km west of Narvik	69° N	Mica schist and phyllite	Grus 2–3 m
Kola peninsula	68° N	Granite and gneiss	Fissured to 10–30 m
Moraineless Buchan	57½° N		Weathered to 0·5–10 m
Aberdeen	57° N	Granite	Weathered to 9 m and even to 12 m
Scotland	56° N	Granite, syenite, norite, felsite, schist, quartzite, Old Red Sandstone, gneiss	Chemically weathered to at least 12 m
Shap Fell, Westmorland	54½° N	Granite	Average depth of weathering 0·6m
Banbury	52° N	Ironstone	Partial decalcification and oxidation, averaging 3–6 m
Boulder, Colorado	40° N	Granodiorite	Maximum depth 25 m
Sierra Nevada	38° N	Granite	Disintegration to 15–30 m
Teays river area	39° N	Sandstone	Weathering to more than 12 m
Japan	36° N	Granite	In places, decomposition of 20–30 m
Wiluna–Meekatharra area, Western Australia	26–27° S	Granite and gneiss	Weathering to more than 15 m
Sao Paulo, Brazil	24° S		Weathering of open joints 100–130 m
Hong Kong	22° N	Granite	Weathered to more than 30 m
Hawaii	20° N	Basalts	In places, partly weathered to 30 m
Minais Gerais	19° S		Average weathering about 100 m
Salisbury, Rhodesia	18° N	Granite	Decomposition to more than 10 m
Malacca	2½° N	Granite	Average depth of weathering 4 m
Singapore	2° N	Gabbro and granodiorite	Regolith depth 10–20 m

5. Weathering forms

A variety of rock surface features have been described and attributed to some form of weathering. However, a great deal of uncertainty surrounds some of the explanations linking certain forms with an areally associated weathering process. The main difficulty is that the association of a weathering form with a given process does not necessarily imply that the latter can exert the forces required to produce the given form.

Many features are too small to be classed as landforms but assume an importance in landform study out of proportion to their size because they illustrate, at manageable scales of time and size, the interaction between land surface and weathering process. Widely observed are localities where small-scale recesses develop by 'cavernous weathering'. These may give a pitted appearance to certain vertical rock or boulder surfaces, or form weathering pits on horizontal surfaces. The dry Victoria valley, Antarctica, provides an example of the first mode of weathering where the hollows on boulders (fig. I.4.A) on the desert pavement surface and on moraines tend to coalesce to leave mushroom-like forms rather than honeycomb surfaces. The hollows have no apparent preferred orientation or position in the boulders. For these features and for similar forms developed on coasts where higher temperatures involve periodic drying of sea spray, salt-weathering appears to provide a satisfactory explanation. These pits are frequently noted in tropical and subtropical regions, particularly on the flat surfaces at high levels on inselbergs of acid igneous rocks. They even occur on hematite outcrops in Minas Gerais, Brazil. Their diameter is about a metre and their depth may be 10–20 cm. Those on the huge inselbergs in South Australia may be up to 10 m in diameter and a metre deep. In middle latitudes they occur on the upper surfaces of isolated tor rocks, blocks, and on other flat surfaces at relatively high levels and on shore platforms. Diameters of about 0·3 m are often equalled by their depths. In Tasmania, pits are about 75 cm across and less than 15 cm deep. They are well known on the Tertiary plateau basalt at Gobbins Island, County Antrim, and have been observed as far north as block-field zones in north-east Baffin Island. Weathering pits may develop in sandstones and schists as well as in granites, lavas, and limestones. Although H. Wilhelmy (1958) has pointed out that weathering pits form sub-aerially without covering layers of soil their interpretation is uncertain. Possibly different processes can produce similar self-accelerating mechanisms that lead to the concentration of weathering activity at sites where initially some chance irregularity existed. The presence of a pool of water if periodically dry seems to be essential. In fact, drying out is a feature of at least some pits; those on tors on Monte Montarone near Lake Maggiore in Italy, dry out a few hours after heavy summer rainstorms. Pits in limestones, karst wells, are often several times deeper than their diameters, suggesting the importance of chemical weathering and the importance of ease of removal of weathered products for deepening to proceed. In addition to the removal of soluble elements by biochemical and chemical weathering, it seems that hydration will exert a mechanical effect in disrupting the grains or mineral crystals in damp spots and that on shore platforms in lower latitudes, salt-weathering may be involved. The surface of massive limestone monoliths of a prehistoric monument in Derbyshire, receiving perhaps 1000–1125 mm of precipitation annually, suggest rates of karst-well deepening of about 15 cm/1000 years. On adjacent gritstone escarpments weathering pits are often filled with algae and the pH of the water may be as as acid as 3–3·5. One self-accelerating factor might be that the initial depression favours not only wafer accumulation but also provides a slightly less harsh environment in which algae colonies might survive or provide natural flower pots in which plants become established. The bottom of pits in the Narvik area is lined with detritus consisting of a mixture of gravel and mud-like silt suggesting the decomposition of organic material. In many situations high winds perhaps provide the most effective force in removing resistant mineral particles from the base of the pits and this may explain their siting on exposed tor summits and account for why periodic drying out is a characteristic of the hollows.

Some depressions are too large in scale or too broad in relation to their shallow depth to be termed weathering pits. Some differ by occupying the floors of ill-drained basins rather than level summits or isolated peaks. In tropical rainforest deep weathering can introduce reversed gradients in the underlying rock surface by the more rapid and deeper decomposition of less resistant material. Again the process is most clearly displayed in limestone areas, where steep-sided depressions develop into the distinctive cockpit karst landscape well seen in West Indian islands like Jamaica. In more open limestone depressions, the distinctive feature is the seasonally flooded level floor bevelled across the limestone structures between steep cliffs in the humid tropics, moderate slopes in less hot environments and with sides as gentle as 4 degrees in the basins or turloughs in western Ireland. Enclosed depressions possibly due to biochemical weathering may even occur on quartzite, as observed in the Blue Ridge, west of Morganton, North Carolina. The depressions range from a few metres to as much as 60 m in diameter with a deepest point of a metre. The dip is nearly horizontal which is probably an important factor and a thin layer of algae and aquatic moss which floats on the surface of the ponds and covers the bottom materials when the ponds are dry may be involved in the weathering process.

In South Africa basins known as pans are widespread. Sub-circular, oval or irregular in shape, they range in diameter from a few tens of metres to several kilometres, and in depth from a metre to as much as 60 m. Although many appear to be wholly or partly the result of chemical weathering, the deepening of many results from the removal of weathered debris by wind transport. These pans are very much like the basins which occur by the thousand on the Great Plains of the United States, ranging south from Montana to New Mexico in a belt where precipitation is between 250 and 500 mm/year. The piping of silt into underlying coarse-grained strata is another possible explanation for these basins.

In cooler environments enclosed depressions have long been recognized. However, as the semicircular walls around a cirque and the depression itself have always been linked with erosion by ice, it would be unwise to discuss these forms in the context of weathering however much they may in appearance resemble the small-scale weathering pits. The case of nivation hollows (fig. I.4.D) is somewhat different. In form they may not involve a reverse gradient. The floors of those near Resolute Bay, North West Territories, have a downhill slope of $3\frac{1}{2}$–7 degrees and range in width from 10 to 60 m and in length from 10 to 65 m. In occurrence they are associated with snowdrifts, but the way in which the melt water from the snow acts in weathering a nivation hollow is obscure. In addition to the problem of shallow depth of freeze–thaw penetration there is the complication that snow cover if anything blankets the underlying ground from sub-zero temperatures and from oscillations through freezing point. Abundant water supply and unfrozen soils might conceivably favour localized acceleration in rates of chemical decomposition.

On moderately to steeply inclined bare rock surfaces, particularly in tropical latitudes and on some limestone surfaces in most humid areas, channels or flutings develop, following the slope of the rock surface. They often occur on the steep slopes of sugar-loaf-shaped domes. One of the earlier descriptions of their development on igneous rocks was that of H. S. Palmer (1927) in discussion of the flutings on joint faces of basalts in Oahu, Hawaii. In southern Malaya they are usually found on surfaces steeper than 60 degrees, very few on inclinations less than this and never when below 23 degrees. A shallow groove may be only 1 cm in depth, the deeper ones about 0·5 m but most are less than 30 cm deep. On acid crystalline rocks in Liberia, the channels may be 1 m in depth. On the surfaces of inselbergs in South Australia, precipitous and even slightly overhanging faces are scored by narrow grooves 20–30 cm deep and occasionally as deep as 60 cm. These do not reach the foot of the wall but either fade out gradually or end abruptly in a small hollow set into the rockface. This last feature suggests that these channels cannot be regarded simply as originating as drainage channels, even if subsequently this is the role they perform.

Just as weathering pits on vertical surfaces have counterparts in exposed horizontal areas, so also do flutings develop on horizontal as well as on steeply inclined surfaces. These are usually associated with joint widening and

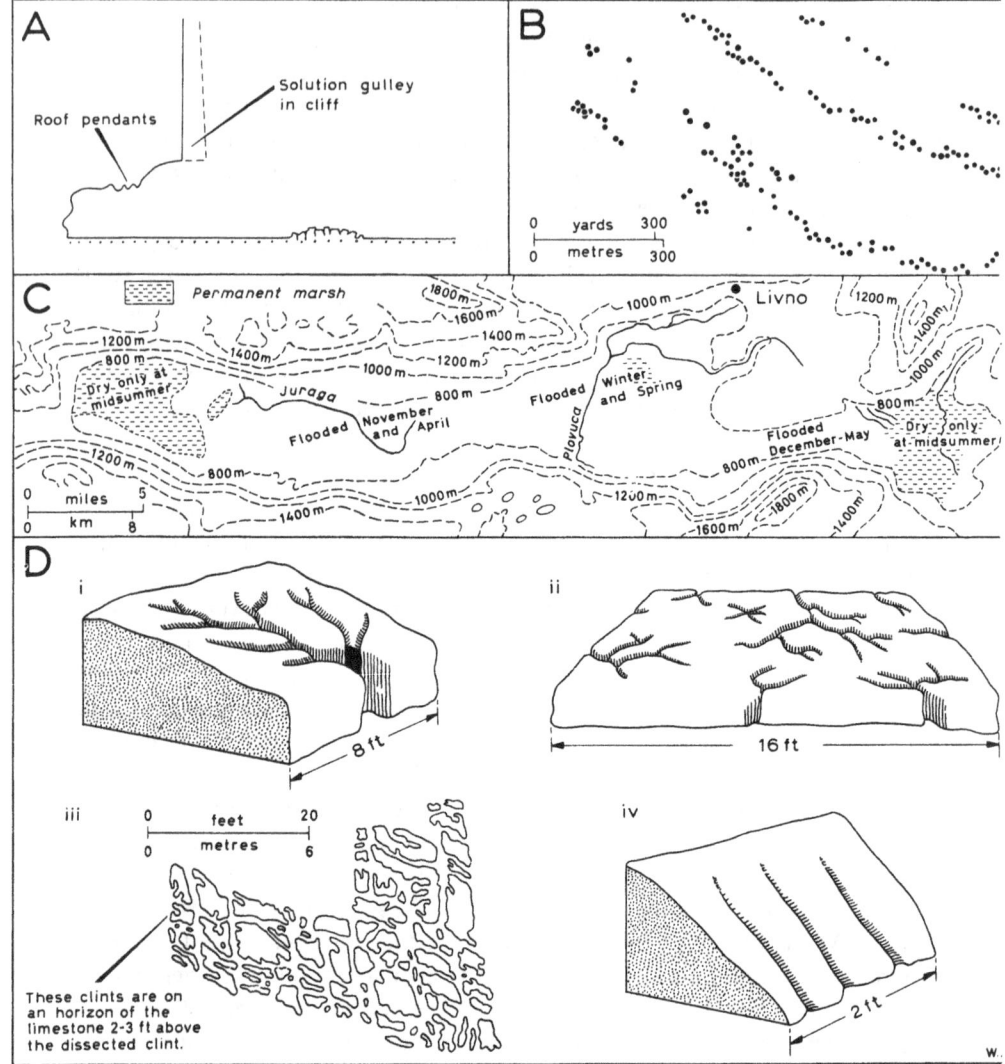

Figure IV.6 Weathering forms in limestone.

A. Steepening and undercutting at the base of a cliff in a tropical environment (*J. N. Jennings and M. M. Sweeting, 1963*, Bonner Geog. Abh., *Vol. 32*), a cliff-foot cave near Barnet Spring, West Kimberley, Western Australia.

B. Small enclosed depressions on Cow Close Fell, Littondale, in north-west Yorkshire (*from K. M. Clayton, 1966*, Field Studies, *Vol. 2*). The depressions follow contacts in the alternating limestone–shale sequence of the Carboniferous Yoredale beds and are formed largely in a mantle of boulder clay due to piping down open joints in the underlying limestone.

C. A polje, the largest limestone enclosed depression form in middle and low the rectilinear pattern of channels reflects the appearance of the joint directions. In addition to the enlargement of joints on limestone pavements, the same process operates on joints in gneisses and schists and other jointed rocks containing silicates of calcium. However, in situations where joint expansion could be due largely to lateral expansion due to pressure release or to slight failure in an incompetent underlying substratum, the amount of channel widening along joints to be attributed to solution must be evaluated with caution.

Just as some processes combine to promote deepening of hollows in certain situations, others combine to produce upstanding residual eminences. Tors are residuals of bedrock isolated on all sides by cliffs a few metres high. Those in Tasmania consist of a number of blocks, each one 2–4 m high but with most tors totalling less than 6 m in height. Such isolated rock pedestals and pinnacles are found in all massive coherent rocks like limestones, sandstones, granites, and basalts. They tend to occur in sites high up in relation to their immediate surroundings. They may represent the remnants of the nuclei of more massive portions of rock formerly surrounded by less resistant material which was more thoroughly weathered and more readily removed. There is a possibility that in massive rocks some little-understood combination of self-enhancing processes leads to the progressive reduction of weathering on tor summits in comparison with surrounding areas, whereas at a different scale other processes combine to produce depressions by focusing weathering on a few specific points. On some flat summits the absence of any appreciable downslope gradient may be a factor contributing to their preservation because an important consideration is the removal of the weathered material assumed to have previously surrounded the residuals of sounder rock. Tors are usually attributed to a phase of differential weathering followed by a phase dominated by the removal of the loosened material by mass movements. D. L. Linton (1955) emphasizes the importance of how a fall in base-level could produce this change. For arctic environments the differential weathering was attributed to frost action by Högbom (1912), Eakin (1916), and J. Palmer (1956). In tropical areas workers have developed on the early

latitudes (*after Jugoslav official 1 : 100 000 map series*). This map of the Livno polje also shows the seasonal flooding regime on the flat, enclosed floor of the polje.

D. Weathering forms on bare limestone pavements (*from Jones, 1965*; D iii from Sweeting, *1966*): (i) the effect of gentle dip on solutional channelling, (ii) channelling on a horizontal pavement (clint) block, including (*top centre*) an instance of centripetal channelling, (iii) the dominance of joint control on a less weathered clint, Chapel-le-Dale, and (iv) the dominance of solutional weathering on steeply inclined surfaces; the German term *karren* is sometimes used to describe such runnels.

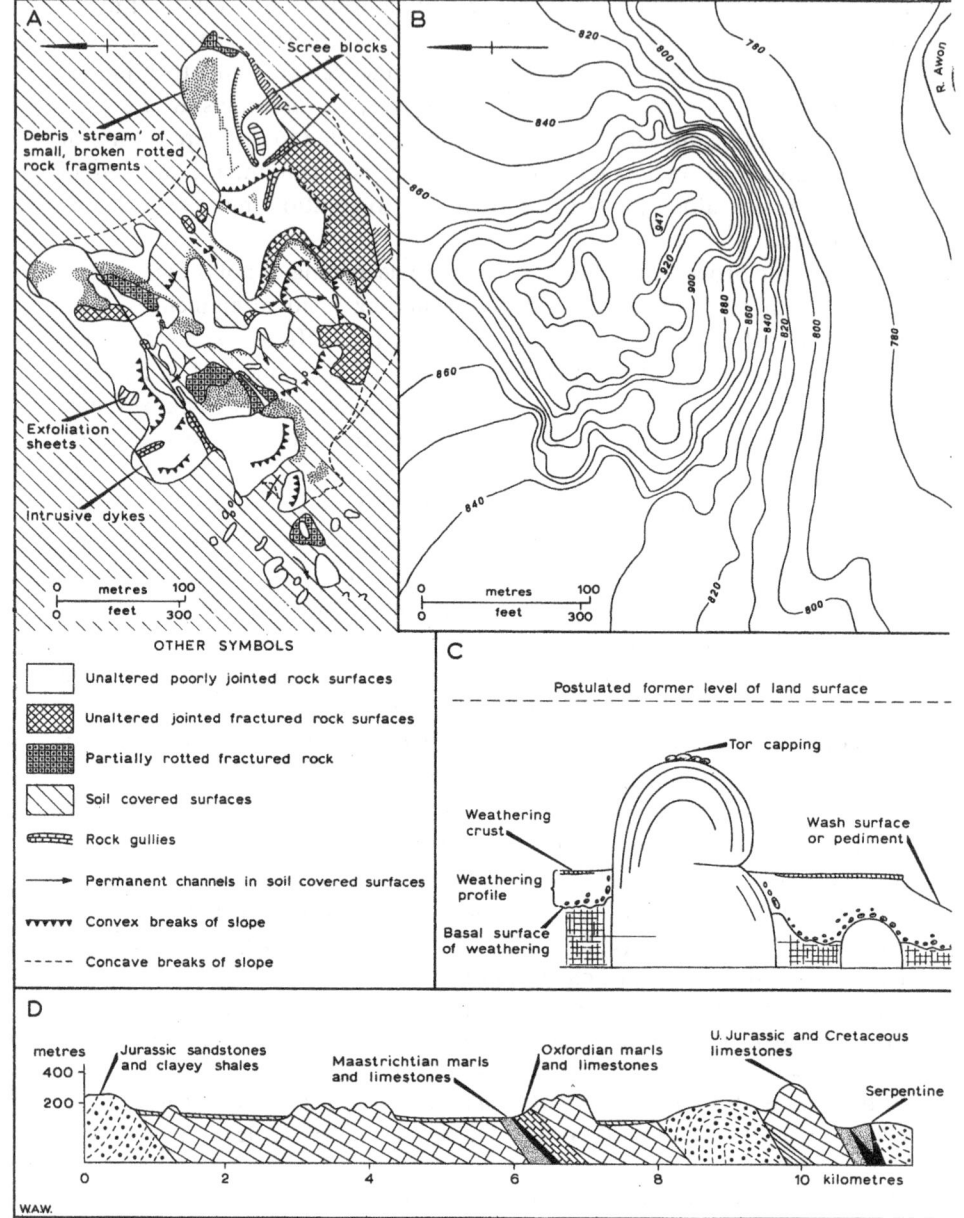

OTHER SYMBOLS

☐ Unaltered poorly jointed rock surfaces

▨ Unaltered jointed fractured rock surfaces

▦ Partially rotted fractured rock

◪ Soil covered surfaces

⬭ Rock gullies

→ Permanent channels in soil covered surfaces

ᴠᴠᴠᴠ Convex breaks of slope

---- Concave breaks of slope

Figure IV.7 Isolated rock domes in tropical environments.

A. Morphological map of the Oyo bornhardt, western Nigeria (*after M. F. Thomas, 1967*).

B. Contour map of the Oyo bornhardt (*after Thomas, 1967*); contour interval 10 feet.

ideas of J. D. Falconer (1911) with the initial phase being one of chemical action prior to mass movement in a removal phase. With chemical denudation known to equal that of physical denudation in some arctic localities, the mild temperatures of interglacials, higher temperatures in Tertiary times, and the shallowness of freeze–thaw penetration in solid rock, the initial phase of rock breakdown probably involved some chemical weathering even in arctic environments. The corners of loosened and residual blocks, however, remained essentially angular. In mountainous areas, particularly near cliff edges in jointed sedimentary strata, open joints are sometimes observed and may reach depths of about 30 m and be as wide as 1·5 m at the top. Any freeze–thaw activity at such depths would have an amplitude of only a degree or two and free drainage in open joints would limit the amount of moisture present. This point increases the possiblity that pressure-release jointing or expansion due to slight failure in an underlying incompetent stratum keeps near-surface rocks in a continually fractured state, open equally to mechanical, chemical, and biological agents of rock breakdown.

Of other processes possibly involved in the development of tors, wind abrasion, despite the appearance of pedestal rocks, is not now believed to have sufficient force to undercut steep sides of tors. The possible significance of salt weathering is now considered in some areas. In the arid Central Otago area in New Zealand, there is widespread cavernous weathering near the base of schist tors with admixtures of decomposed rock and salts within the hollows. Fig. III.11 is a reminder of the possible significance of tree roots in this context.

B. Transportation

Some land surfaces inclined across bevelled structural features are now referred to as transportational surfaces rather than as surfaces of erosion or as erosional slopes. This usage reflects an increasing awareness of the intimate link between transportation processes and many landforms. In studying the transportation of debris, therefore, the geomorphologist gains not just an insight into the mechanisms involved in the removal of weathered material and in influencing depositional characteristics but also, in many instances, into influences which directly or indirectly are an integral part of the

C. Postulated development of a bornhardt by the removal of the deep chemically weathered profile around the dome and etching at its base (*Thomas in Fairbridge, 1968*).

D. Kegelkarst (cone-karst) in Cuba (*from H. Lehmann, 1960,* Ann. de Univ. Lyon, Spec. No. 11) along a north–south section through the Sierra de los Orgánós, with flat polje floors terminating abruptly at the foot of the residual masses.

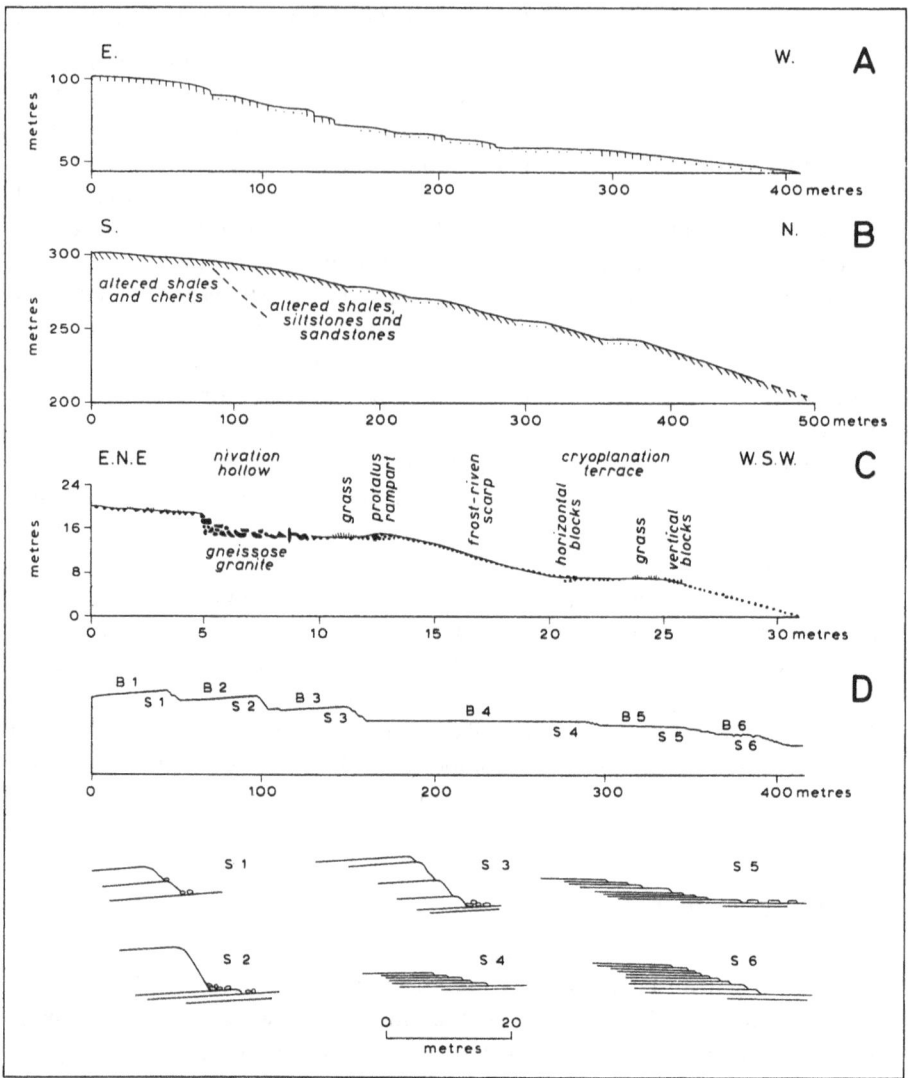

Figure IV.8 Illustrations of difficulties in interpreting hillslopes as erosional forms. (1) The role of differential resistance of rocks in the development of benched slopes in past and present periglacial and glacial environments.

A. Profile of a stepped slope on Blomesletta in Vestspitsbergen *(from R. S. Waters, 1962,* Biuletyn Peryglacjalny, *Vol. 11).*

B. Profile of a stepped slope on Dartmoor, Devonshire *(from R. S. Waters,* idem.*).*

C. Profile of a stepped slope in the Bohemian–Moravian Highlands with periglacial processes inferred *(from J. Demek, 1968,* Przeglad Geograficzny, *Vol. 40).*

D. Profiles of a stepped limestone pavement in Glencolumbkille South, County Clare for which glacial scouring is often inferred *(after Williams, 1966):* 'B' indicates a bench, 'S' a scar.

processes shaping the landforms themselves. However, it is essential to note at the outset, as hydraulic engineers and sedimentologists stress, that the heterogeneity of sizes in natural sediments are not readily susceptible to theoretical treatment, and in the absence of basic measurements, even the expert's understanding of the mechanisms involved in the entrainment and transport of sediments is incomplete.

1. Transportation of particles in a fluid

The transportation of particles in a fluid involves three forces. These are the velocity of the fluid motion, the force of gravity acting downwards and the fluid resistance acting in a direction opposite to that of the motion of the particle. With other factors equal, the force of gravity is proportional to the cube of the grain diameter and therefore, while the settling velocity due to gravity becomes increasingly high for particles of larger size, it is reduced to negligible proportions in comparison with fluid forces for very small particles. Although it seems obvious that particles of diminishing size from gravel or sand-size scales are progressively easier to move, F. Hjulström in 1939 recognized a critical factor in the transport of sediment in his studies of fluvial processes by noting that the smallest size of material which can be transported as bedload is 180 microns. This effect is due to the settling velocity of material with a specific gravity of 2·65 being higher than the frictional drag or threshold velocity above this size and lower for sizes less than 180 microns (fig. IV.10). The comparable turning-point on R. A. Bagnold's graph for wind transport is about 80 microns, close to the upper median grain sizes recognized in loess material. As smaller sizes become progressively more difficult to move, the velocities required to dislodge particles are still greater than those needed to transport them. It is as a result of this phenomenon that dunes are not normally found in loess deposits because the fine-grained nature of the material presents a smooth surface texture, and once particles finer than 80 microns have settled they cannot be swept up again individually because they sink into a viscid surface layer and are out of reach of turbulence. In water too the laminar sub-layer shields the finer grains in interstices of coarser material. Also there may be much greater cohesion among smaller particles due to inter-particle forces, an effect dominant at about 100 microns. Fig. IV.10 shows this very important feature of transportation. This necessity for a greater mean velocity to entrain material than that needed to carry the load is sometimes termed the Hjulström effect. Thus in streams where velocities fall to a rate at which particles begin to settle out, such velocities are inadequate to dislodge particles of a similar size on the channel floor. On beaches cusp growth depends mainly on the fact that velocities of flow must be higher to entrain than to carry load. In a gross

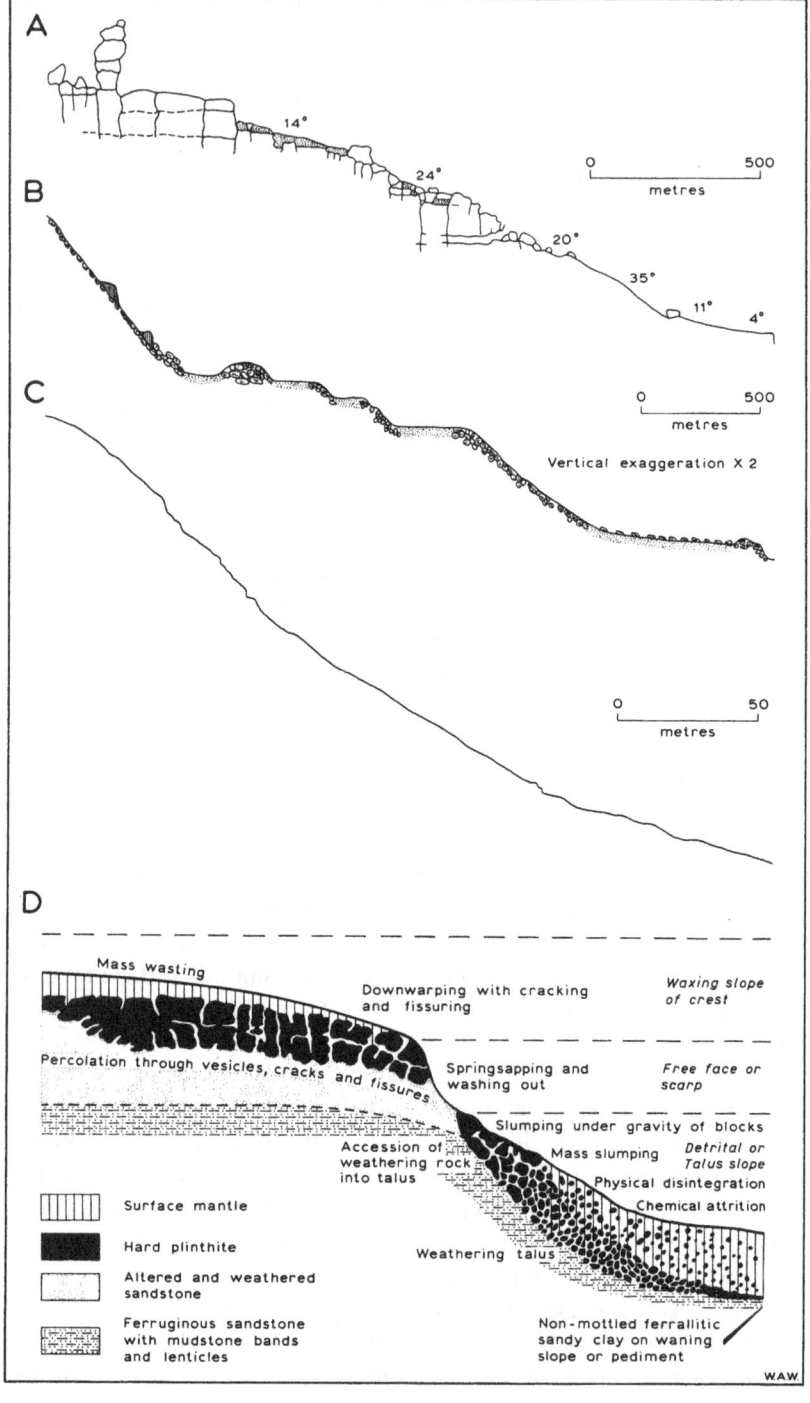

A

14°

24°

20°

35°

11°

4°

0   500

metres

B

0   500

metres

Vertical exaggeration X 2

C

0   50

metres

D

Mass wasting

Downwarping with cracking
and fissuring

*Waxing slope
of crest*

Percolation through vesicles, cracks and fissures

Springsapping and
washing out

*Free face or
scarp*

Slumping under gravity of blocks

Accession of
weathering rock
into talus

Mass slumping

*Detrital or
Talus slope*

Physical disintegration

Chemical attrition

Weathering talus

	Surface mantle
	Hard plinthite
	Altered and weathered sandstone
	Ferruginous sandstone with mudstone bands and lenticles

Non-mottled ferrallitic
sandy clay on waning
slope or pediment

W.A.W.

statistical sense the diminution of particle sizes coarser than fine silts away from the bed follows a logarithmic law of decrease in both air and in streams. In breakers, as far as sampling difficulties permit, it seems that there is a gradation upwards from coarse sizes and larger volumes near the bottom. It is not easy to observe other effects. Fines cannot settle out within the length of laboratory flumes or tunnels and the observation of the movement of coarse material in natural situations is difficult and hazardous. There is also the problem of currents moving at different velocities and operating in different directions, like hot desert winds. On a beach, transport up a beach in turbulent swash may tend to be essentially by suspension, whereas that down the beach is by sliding and rolling. Thus material with a high settling velocity tends to move downbeach while that of low settling velocities moves up a beach. A further complication is that particles of the same mean-size category but of different shape may differ by a factor of 2 or more in their settling velocities. However, hypotheses are as yet inadequate to explain the transport of debris involving a wide range of particle sizes under the wide variety of conditions found in nature.

Bedload is usually coarser than 200 microns. Typical of the bedload of major streams is the Ob at Barnaul of which 70 per cent falls in the 200–500-micron range. Although there is some sliding in surface creep of bedload material the movement is largely that of rolling particularly for particles larger than 500 microns because the increase of current velocity upward from the bed, by exerting greater force on the upper half of a pebble, tends to cause rotation. Pebbles with one long dimension will tend to roll into a position with

---

Figure IV.9 Illustrations of difficulties in interpreting hillslopes as erosional forms. (2) The influence of a distinctive lithology, the problem of homologies, indications of the operation of more than one process, and the chemical recombination of weathered material.

A. The influence of the lithological characteristics of a massive sandstone on a slope in the Javorniby Mountains in the Moravian Carpathians (*J. Demek, 1966,* Geographia Polonica, *Vol. 10*).

B. A slope south-south-west of Tahat, typical of many cases in the Atakor mountains in the central Sahara (*from Rognon, 1967*). The distribution of blocks and the stepped profile are similar to that observed in areas of arctic mollisol, but in this case may be due to block-glide on silty soils during very wet phases of climate.

C. A slope-profile in Glen Etive, Argyllshire, showing the general shape attributed to glacial overdeepening, but with its midslope third showing a constant declivity suggesting debris control at the scree-repose angle (*from Pitty, 1969*).

D. (*from R. P. Moss, 1965,* Jour. Soil Sci., *Vol. 16*). The diagram shows how slope processes in south-west Nigeria become closely interrelated with the presence of a ferricrete (plinthite) crust 0·7–2·0 m thick and its mode of retreat.

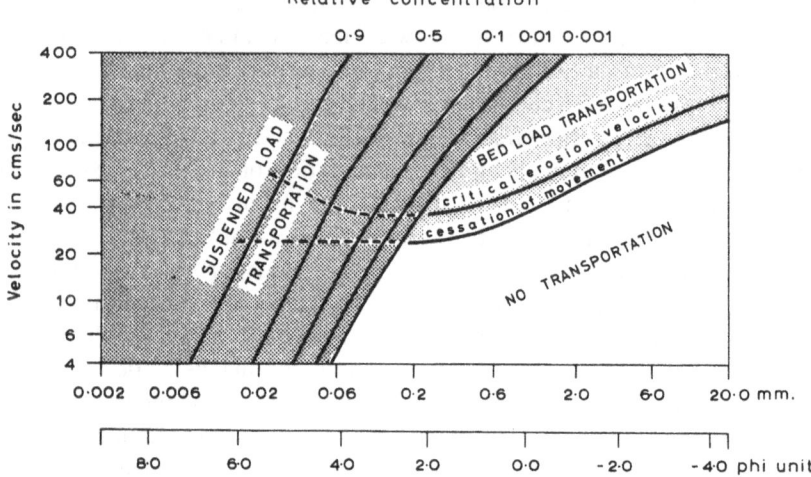

Figure IV.10 Approximate relations between flow velocity, grain size and state of sediment movement (*after Sundborg, 1967*). The sediment is uniform with a density of 2·65 gm/cm^3; the velocities, critical for erosion or for cessation of bedload movement, refer to a level 1 m above the flume bottom; the relative concentration of suspended load is the ratio between the concentrations at half the water depth and at a level close to the bottom.

this axis perpendicular to the current. Currents may raise discoidal pebbles on edge and roll them along like wheels, although when packed into a beach deposit discoidal fragments may shuffle along as a downbeach member of a pack is displaced. Usually bedload material travelling at about half the velocity of the floodwater moves in a series of discrete steps, involving distances of the order of 100-grain diameters for particles of average sphericity. In traction, spherical and rod-shaped particles tend to move faster than the discoidal. However, due to its relatively slow settling velocity, discoidal shingle is transported farther and piles up at the base of cliffs near the summit of beaches whereas spheroidal ones remain seaward, as confirmed by numerous field observations. The ease with which currents roll a pebble may depend on whether the relative sizes of adjacent particles support the pebble or leave it exposed to fluid-dynamic forces. With a wide range in particle size the larger particles may roll readily over a surface made up of finer particles. Once a coarse particle stops rolling on a bed of fines, up-current scour starts immediately and the particle settles in the scour pocket. The deeper the depth of burial of larger particles the more immovable they become, whereas the largest particles will remain in movable positions for longer periods of time. Therefore there are certain situations where the transport of the larger particles remaining as bedload, perhaps surprisingly, is favoured. For example, the selective removal of coarser grains from beaches might be attributed to

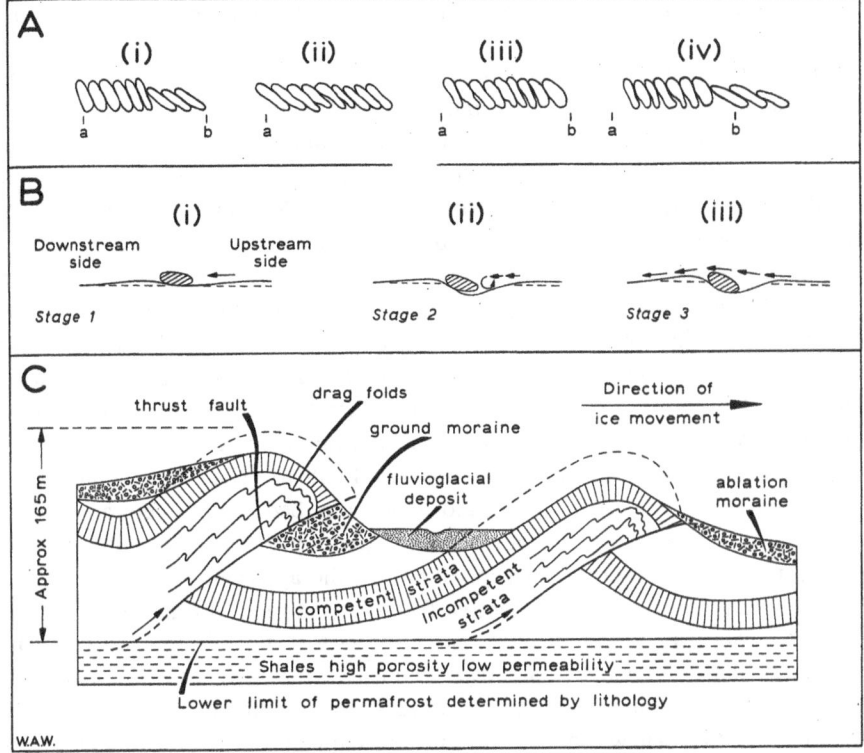

Figure IV.11 Some transportation mechanisms.

A. Disc migration on a gravel beach (*from Bluck, 1967*). By displacing a few of the discs at the seaward end of the column, and with the others conforming to this new dip, the gravel discs move past the stationary points *a* and *b*.

B. Pebble tilting on its upstream side to adjust to the groove made by undercutting (*from S. Sengupta, 1966*, Jour. Sed. Petrol., *Vol. 36*).

C. Ridges formed by ice-push (*from Kupsch, 1962*).

large diameters projecting upward from the bedload into higher seaward velocities. While surfaces of fines may facilitate the removal of exposed large particles, dominantly pebbly surfaces tend to discourage sand accumulations because of the increased trajectories of rebounding grains. With wind transport, grains too large to be lifted may be impelled forward by smaller saltating grains which may move stationary particles up to six times their diameter or two hundred times their weight by high-velocity impact. In water, the impact momentum of a descending grain is sufficient to raise a surface grain only by a very small fraction of a diameter. The degree to which the threshold drag velocity is exceeded has not been studied, but for the Blue Creek, California, at least, the value appears to be equalled or exceeded 5 per cent of the time for the coarse bed material in this mountain stream. A further point about bedload is the maximum size that might be moved by the occasional extreme flood. Wolman and Eiler record a 3-m-deep flood travelling at a mean velocity of 2·7 m/sec, with a peak possibly of 4·5–5·5 m/sec, moving a boulder 2·7 × 1·5 × 1·2 m. A mean current velocity of 7–21 m/sec bedload velocity has been known to move a 3-m-diameter boulder. From calculations for the Truckee River near the California–Nevada border, Pleistocene flood velocities, apparently sufficient to move boulders up to 12 × 6 × 3 m in size, might have been 9 m/sec on a 0·007 grade. Bedload carried by exceptional melt-water floods released from the amount of the Nisqually glacier in 1955 included boulders greater than 1·8 m in diameter. Floods in the Chiapas river in southern Mexico can carry boulders at least 2·4 m in diameter. In Colombia, sudden violent floods entrain much coarse material including boulders as much as 1·2 m in diameter, and in general, tropical mountain torrents can carry a boulder a metre in diameter.

McGee introduced the term saltation in 1908 to describe the manner in which particles appear to be lifted from the bed at a steep angle, are accelerated forwards by the current and then settle back on a descending gradient which is much less steep. In water, the apparent lift may be a few particle diameters and most effective for sizes between 500 and 62 microns (1–4 ø), whereas in air it may be hundreds if not thousands of grain diameters. Particles somewhat larger than 125 microns may attain heights of 20 m during saltation, and grains up to 2 mm in diameter have been found wedged into cracks in wooden power line poles in semi-desert areas at heights up to 6 m. In the saltation of particles in water it is difficult to dissociate apparent particle lift from turbulent movements of the water itself. In plan, however, the paths of saltating grains are straight lines parallel to the general current direction, and differ significantly from those suspended by turbulent eddies. In airborne grains, it is much clearer how the energy of forward motion of wind is converted into an upward one by impact with the floor. This is also seen on steep slopes where saltation of rockfall fragments may be important. In this case a snow or ice cover facilitates motion and the occurrence of depressions caused by this process suggests its relative frequency. On a 200-m-high scree in Insigsuin fiord, Baffin Island, beneath a rockwall several hundreds of metres high, there is evidence of free-falling blocks arriving with considerable impact approximately two-thirds of the way up the scree. G. K. Gilbert (1914) noted that the particles in saltation in streams occupy a definite sheet like zone above the bed. Inter-particle collisions maintain the density of the traction carpet as few rebounding or lifted grains escape collision to attain a height in proportion to their initial rebounding energy. This traction carpet may have a transitional status between the bedload, always in contact with the bottom, and the particles carried along well above the bed in a state of suspension. In wind-blown material 50 per cent of that observed by R. P. Sharp (1964) travelled at heights from 7·5—15 cm above the ground. A similar smooth-topped traction carpet accompanies backwash on beaches. However, the depth of this layer may only equal a few grain diameters outside the high energy zones. On the Santa Monica beach, California, a measurement of the traction carpet depth is 2·5 mm, compared with 0·5 mm on La Jolla beach. Also in contrast to the saltating motion of unidirectional currents, the oscillatory motion of waves tends alternately to lift and to deposit material as an intermittently suspended load.

Finer particles up to a certain size limit, which is related to the amount of turbulent energy available to support the sediments, may be maintained in more or less permanent suspension by the turbulent exchange of particle-rich fluid from near the bottom with fluid from higher levels less dense in particles. Suspension in water affects sizes less than 140 microns (2·8ø) and particularly sizes less than 62 microns. R. A. Bagnold (1937) observed that the velocity of upward air movements near the ground were on average approximately one-fifth that of the mean forward velocity of a wind. In rivers the material in suspension is normally almost completely independent of that in a traction carpet and turbulent exchanges within the suspended load are balanced. In some circumstances it is possible, however, that water moving upward may contain a greater silt content than that moving downward. Finer sizes which can be held indefinitely in colloidal suspension with the very fine-grained particles kept dispersed by ionic-electrostatic forces are sometimes termed the washload. Unlike the suspended load the washload is relatively unrelated to the hydraulic conditions and their changes. Occasionally, but from a range of environments, come reports of material being transported by flotation. In 1926 J. S. Behre estimated that in an hour the River Ohio transported 130 000 kg of material by flotation.

There is no realistic definition for critical velocities just sufficient to suspend the sediment. Instead there is continual transition as progressively more particles of a given size are entrained. The fluid velocity theoretically required to dislodge grains of a given diameter is sometimes termed the threshold drag velocity. The velocity for moderately efficient suspension of sand-sized particles might be 2—2·5 times the threshold drag velocity. On A. Sundborg's (1967) modification of Hjulström's well-known illustration of the orders of size and velocities involved (fig. IV.10), four different areas may be noted. The area of high velocities and small particle sizes describes conditions of net scour. The bottom left part of the diagram represents conditions of transport in suspension, but of net deposition of bedload. The grey shaded area indicates the condition of entrainment of coarser sizes from the bottom and the transport of bedload while, with lower velocities, the unshaded area indicates the absence of prolonged transport and the deposition of bedload and possibly suspended load too. Recently it has been shown that the critical factor is not the average velocity of flow but the bottom shear velocity. In consequence, the height of the column of water above the bedload is critical and in deeper water at normal flows a larger average velocity is needed to transport particles of a given grain size than in shallow water. For this reason water flow in a beach swash zone is more competent. That the usefulness of fig. IV.10 is notional should be stressed also because it applies to uniform-sized material only. Where a mixture of particle sizes or discoidal shape are involved there is as yet little precise theory or observation available. Tentatively A. Jopling (1966) suggests that velocities might have to be increased by perhaps 20 per cent for poorly sorted sediments. Another complication is that for a given flow past a fixed point neither the fluid velocity nor the shear stress which varies approximately as the square of the velocity are constant, but fluctuate considerably due to turbulence and momentary increases in velocity. Inevitably contrasts are far more pronounced on shores where there is a high, momentary velocity under the crests of waves which are about to break, and under breakers in association with the development of vortices, which can lift particles far above the bottom. Another related factor which again applies indirectly to a beach environment, is the concept of the abruptness of flood. Experiments in Russia with surges released down laboratory streams revealed intense dragging and suspending of sediment associated with the passage of a surge. It seems that in an accelerating current, velocity at the bottom approaches velocity at the surface. In a natural situation in small Scottish streams in high flood gravel disturbance appears to be associated with smaller sharp increases of velocity than with the slower rises of larger floods.

Where surface water flows over a ground surface instead of concentrating in channels some additional factors have to be considered. Often the entrainment of particles is the result of rain splash which in badland gullies may raise fine sands and silts by at least 5 cm. In Turkmenia, comparable heights of 30 cm have been observed involving lateral displacements of 40–50 cm, but where vegetation covers more than 60 per cent of the area, there is practically no removal of solid material. Experiments have shown that raindrop splash may carry a higher percentage of large particles and aggregates than surface flow. Particle sizes in the 50–400-micron range are most susceptible to movement. Coarser sizes are too large to be dislodged and smaller sizes seal the surface so that a water film on the surface dissipates the energy of raindrop impact. Sheetflow or sheetwash involve shifting rills a couple of centimetres to half a metre in depth, dying out with distance from the hill-front. On fans, a maximum water depth of 15 cm has been observed, with channels continually filling up with debris before shifting to another line. The process is essentially a transportational one, moving loose material of sizes up to coarse sand. In all situations experiencing sheetwash there are extensive areas unprotected by a permanent vegetation cover and where there is a seasonal or sporadic supply of water. Rainwash, however, may be important on the scantily covered forest floor in tropical forests as rainwater pouring down tree-trunks is sufficiently concentrated to expose roots downslope.

Piping is a transportational process which produces tubular sub-surface drainage channels in insoluble clastic rocks by the grain-by-grain removal in suspension by moving ground water which opens into free drainage. Broad subsidence may follow. This essentially mechanical process termed piping is particularly well-developed beneath weathering crusts where cavities several metres or even 100 m in length may develop. The process occurs in granite areas in Hong Kong. In dry environments the coarser material that fills desiccation cracks favours the development of sub-surface seepage lines. In cool environments the transportation of fine material in blockfields may take place below the ground. On beaches fine materials may move through a framework of coarser material.

The critical general factor in the movement of particles by wind is a limitation on the amount of effective precipitation rather than wind velocities although the limits of 17·5 km/hr needed to suspend dry sand and the inability of wind to turn particles larger than 5–7 cm are clearly defined. Otherwise, in middle and low latitudes vegetation may provide insufficient cover on sands in continental interiors with precipitation as high as 650 mm/year. Areas of active longitudinal dunes in south-west United States and in central Australia are limited approximately by the 250-mm isohyet. In arctic deserts the necessary surface dryness for sand to be movable may be realized only when precipitation is less than 75 mm/year.

2. Amounts of sediment transport

Apart from the mechanisms of transport, the amounts moved are another vitally significant aspect of landform studies. In streams there are the concentration of the suspended load, the percentage contribution of the bedload and the net amounts removed to consider. Sediment concentration or turbidity is important because it gives a clear indication of the rates of supply of detritus to the channel system. Exceptionally high concentrations arise momentarily, due to mass movements or to an intense phase of slopewash at certain vulnerable points in a catchment. High concentrations occur frequently in semi-arid areas where supply may include a large proportion of fine-grained wind-transported material which fails to cross a perennial stream or the depression of an ephemeral stream channel. In these circumstances amounts may exceptionally reach 50 per cent or 500 000 mg/l. In contrast in winter in high latitudes, measured concentrations may be less than 1 mg/l. As examples of intermediate values some results from northern Sweden might be noted. Upstream from the Laitaure delta the highest concentration is 1260 mg/l., and near the water surface at bankfull stage is about 950 mg/l. Nearby the highest concentrations are of the order of 310 mg/l. in the Kanajokk and 275 mg/l. in the Tarraädno rivers. The Rhine, upstream from Lake Constance, with an average flow of 224 m³/sec has an average sediment concentration of approximately 825 mg/l. In the relatively flat European U.S.S.R., the turbidity of the Polomet river, with a catchment area of 631 km², at approximately 250 mg/l is higher than most rivers. Here values may rise to 7000 mg/l in high water due essentially to bank erosion and channel reworking. Of the larger streams, the mean turbidity of the Ob, at approximately 500 mg/l, is somewhat low. In the Negev where discharges range from 2 to 1000 m³/sec turbidities observed range from 60 to 680 mg/l.

A relatively unknown factor in the calculation of amounts transported is the contribution of bedload movement which is usually estimated from formulae. Russian scientists have estimated that of the sediment discharged from continents, 9 per cent is bedload. From mountainous areas the proportion is 11 per cent, from plains 8 per cent. Individual examples may fall within or outside this framework, usually for reasonably clear reasons. In the lower Mississippi and the Amazon bedload is only 5 per cent. An estimate for the Congo is 6 per cent. For the Rhône near Villeneuve, the proportion is 12 per cent, for the Linth near Walensee, 22 per cent. In a drainage basin including the most ice-covered high mountain area in Sweden, the proportion is 14 per cent. Fifty-three kilometres downstream in Rapaälven, bedload amounts to about 16 per cent of the suspended-load discharge. One feature of coarse bedload is the relative slowness of travel. J. Tricart observed that during the 1957 flood on the Guil, unprecedented in postglacial times, 10-cm pebbles rarely travelled more than 1–2 km. Within a rocky gorge in Hérault province subject to violent floods he estimates an average progression of 0·2 km/year. In wind transport about 25 per cent of the load moves by surface creep, perhaps reflecting the fact that if sizes equivalent to the washload of streams were initially present they could soon be swept a few kilometres above the ground.

G. K. Gilbert defined 'capacity' as the maximum load that a stream can carry. However, a natural stream cannot be saturated with sediment as it might be with a salt. Again it is the wide range of particle sizes that may be carried which introduces a complication as does the range of transport mechanisms. Thus a stream unable to move a boulder a few kilograms in weight may be competent to shift thousands of tons of finer silt, which in the case of semi-arid floods, may exceed one-quarter of the total weight of the moving suspension. Another problem is that concentrations, if exceeding 5 per cent, begin to exert profound changes on the hydraulic laws that apply for pure liquids. In fact the increased density of the fluid mass means that the relative density of particles diminishes, thus facilitating transportation. Also, shearing forces on the bed increase. The submerged weight of a rock in a mudflow is reduced by perhaps more than 60 per cent compared with its submerged weight in water. Thus, although there must be a theoretical limit to the amount of material a stream can transport in suspension, the practical limit is nearly always set by a deficiency in the supply of suitable material available for transportation. By far the most geomorphologically significant statement on the amount of stream load is its actual measurement. Nonetheless, the significance of such a measurement in landform interpretation may be limited, as it is impossible to infer the relative importance of processes which dislodged or eroded the material at the sediment source area within the catchment, the degree to which the sediment is being reworked or evacuated from the river network and the degree to which human activity has accelerated the previous rate. As these complications are added to the diverse results produced by local factors, only the broadest outline is given in Table IV.7. Examples of more specific local observations might include the transportation of loessial areas in China, equivalent to an overall surface lowering of 7–9 mm/year. In the Italian provinces of Emilia, Romania, and Marci, comparable figures are 0·2–1·4 mm/year. Reflecting in part the fact that the erosiveness of the same amount of runoff might be one or two dozen times that of humid subtropics, in the arid sub-tropics the theoretical lowering in Central Asian mountain drainage basins range from 1 to 6 mm/year in the south to 0·003 mm/year in the north.

On beaches, even the slightest current will cause some horizontal displacement during the gravitational fall of suspended sediment. Thus in the breaker zone where particles move to and fro under the waves, they are laterally displaced from the net direction of this current by even the slowest longshore current. Within the inner breaker zone the oblique uprush and more perpendicular downwash will lead to particles including pebbles following the well-known zig-zag pattern of longshore drift. The velocities of these movements on the California coast average about 30 cm/sec with a maximum recorded of 1·3 m/sec. Individual grains move at 2·4–3·6 m/min on Californian beaches, at 1·1–2·0 m/min on average on Florida beaches, and at 0·60–0·75 m/min as a maximum rate on Sandy Hook foreshore, New Jersey. There are reports of grain movements at 0·15 m/min on Baltic coast beaches, rising to 1·65 m/min under wind forces of 2–6. On the shore of Ikroavik Lake, Alaska, littoral drift may reach 0·30 m/sec during 30 km/hr winds, moving pea-sized gravels at a rate of 1·28 m/min. Velocities of 0·8–1·0 m/sec are strong enough to divert most sands alongshore and even moderate longshore velocities are sufficient to deter offshore movement. Thus volumes of alongshore transport are substantial. At Vridi, south of Abidjan, Ivory Coast, the average annual displacement from artificial obstructions is 800 000 m³. Along the coast of British Guiana longshore drift transports 150–200 million m. tons annually as suspended sediment and in vast migrating mudbanks. In the Baltic

Table IV.7 Transported load and turbidity of selected rivers (*from F. Fournier, 1969, Bull. Inst. Ass. Sci. Hydrol., Vol. 14*).

For each order of drainage basin size the lowest. L, the median M, and the highest. H. values have been selected from Fournier's data. The number of rivers in each basin size is given in brackets. The next-to-the-lowest value in each group is on average 2·7 times the lowest for load and 2·8 times the lowest for turbidity. The highest value in each group is on average 1·7 times the next-to-the-highest for load and 2·7 times for turbidity.

ORDER OF SIZE OF DRAINAGE BASIN, KM²	RIVER	COUNTRY	LOCALITY	AREA, KM²	LOAD, TONS/ KM²
10–100	L. Mangami	New Zealand	Tariki Road	82	8
(11)	M. Wanganui	New Zealand	headwaters	80	50
	H. Mangatepopo	New Zealand	Katetahi	15·5	727
100–1000	L. Thames	Canada	Ingersoll	518	15
(37)	M. Neveri	Venezuela	Batalon	976	301
	H. Marecchia	Italy	Pietracuta	357	4 570
1000–10 000	L. Pembina	Canada	Windygates	7 800	4
(50)	M. Iller	W. Germany	Krugzell	1 118	292
	H. Waipaoa	New Zealand	Kanakanaia	1 580	6 983
10 000–100 000	L. Meuse	Belgium	Hedel	29 000	24
(22)	M. Tone	Japan	Matsudo and Toride	12 000	273
	H. Yesilirmak	Turkey	Ayvacik	36 000	1 228
100 000–1 m.	L. Oder	Poland	Gozdowice	109 400	1
(18)	M. Waal	Belgium	Hulhuizen	160 000	13
	H. Chao Phya	Thailand	Nakorn-Sauau	103 470	106

					Turbidity mg/l
10–100	L. Melito	Italy	Olivella	41	23
(9)	M. Main	W. Germany	Marktbreit	27	74
	H. Eleuterio	Italy	Risalaimi	79	971
100–1000	L. Thames	Canada	Ingersoll	518	23
(35)	M. Tronto	Italy	Tolignano di Marino	911	1 350
	H. Hii	Japan	Nadabun	924	28 619
1000–10 000	L. Yodo	Japan	Hirakata	7 120	23
(51)	M. Me Wang	Thailand	Mew Lom	2 708	559
	H. Maticora	Venezuela	Don Pancho	2 490	23 500
10 000–100 000	L. Meuse	Belgium	Hedel	29 000	68
(20)	M. Me Ping	Thailand	Wang Kra Chao	26 386	283
	H. Yesilirmak	Turkey	Ayvacik	36 000	4 893
100 000–1 m.	L. Oder	Poland	Gozdowice	109 400	9
(18)	M. Saskatchewan	Canada	Saskatoon	140 000	149
	H. Euphrates	Syria	Tabqa	120 650	1 709

an annual transport of 1·9 million m. tons/year moved from south to north past Klaypeda. On the Californian coast figures range from 76 500 m³/year to 215 000 m³/year at Santa Barbara.

In contrast to the very turbulent swash there is very little turbulence in the downwash, with the exception of rip currents which transport a lot of sand beyond the breaker zone by their stream-like flows. Similarly there is not usually much movement seaward of the breaker zone. Off the East Anglian coast tracer movement suggests a drift of perhaps 6·4 km/year with a peak rate of 3 km/hr where a sand stream moves from the north end of the Norfolk sand banks 160 km north as far as Flamborough Head and spreads across 100 km. At the southern corner of the North Sea, sediment from the English Channel enters at a rate of 600 m³/year. In contrast the current through the Bering Strait is perhaps unusual in being the main source of sediment in the area. Such slow offshore currents are of direct geomorphological significance where the sea-floor material drifts shoreward, as from depths of 10 m or less it can be moved on to the beach. Sand moving on to Mediterranean beaches in France comes from depths down to 9 m. Where foreshore bars develop, deposition on the landward face may favour an onshore movement. On the Georgia coast, the rate of this movement is generally about 10–30 cm/day.

In coastal areas, dune sands more than 100 m above sea-level illustrate qualitatively the transporting power of wind in some environments. Measurements include those of H. J. L. Beadnell (1910) who observed a year's dune advance at the Kharga Oasis, the highest dunes (20 m) moving at 10·9 m/year, with the smaller ones (4·10 m) moving at 18·4–18·8 m/year. For coastal sand dunes of Guerrero Negro, Baja California, mean travel velocity of dunes is 4·9 cm/day or 18 m/year. Nearby, inland from Bahia Sebastian Vizcaino, a dune field has travelled approximately 12 km across a lagoon in the last 1800 years. Between 1956 and 1963 the Salton dunes in Imperial Valley moved 14–40 m/year, compared with 7–25 m/year in the preceding 15 years. However, in observations over 12 years of the Kelso transverse dunes, R. P. Sharp while observing a maximum rate of movement of 15 cm/day found that, despite great activity, this involved so much shifting back and forth that net displacements were small indicating 30–45 cm/year over the 12-year period. In southern Peru, there is some reversal throughout the year in the mornings and during the four winter months this reversal predominates. In south-west France, the 100-m high Pyla dune advances from an eroding shore at a rate of 1 m/year. The volumes involved include measurements of 41 kg/hr/m for the Kharga oasis. In January in southern Peru, barchan advance involved transport of 79 kg/hr/m, and for the Kelso sand ripples R. P. Sharp estimated 89 kg/hr/m under a 50 km/hr wind.

The degree to which very strong winds on isolated rocky summits are effective in reworking material has not been widely investigated, although a lowering of 2–3 m on the summit of Mount Nussbaum, Taylor Valley, Antarctica, has been attributed to deflation. It is also difficult to see how the possibility of the action of wind in moving sediments downhill could be evaluated, although the importance of this factor is obviously related to the high transportation rates of semi-arid streams.

The significance of wind transportation of finer silts can usually only be inferred from deposits. From the amounts of fine dust added to the land surface on desert or high dry plateau margins the significance of the process appears appreciable. In 1863, 4 million m^3 of dust fell on the Canary Islands, westward of the Sahara. In the vast expanses of Turkmenia, wind accounts for 60 per cent of sediment transport, compared with 20 per cent by water and 20 per cent for the combined action of wind and water. In the periglacial environment wind, as a transportational agent, is often more significant than rillwash, particularly along the margins of arctic outwash plains. Loess deposits in Alaska following the last (Kluane) glaciation are 35–100 cm thick.

If it has so far proved impracticable to measure actual amounts of bedload transport by streams, it follows that similar measurements for a glacier would be even more difficult. G. Østrem (1965) measured the amount of englacial material along a 40-m stretch of the Isfalls glacier snout, but the delivery of 1·3 m. tons/year appears negligible compared with 100 000 m. tons in the same 40 m of the adjacent end moraine. Transport on a glacier surface may occur where lateral moraines of valley glaciers lead on to the ice surface as medial moraines below glacier junctions; examples on the Siward glacier, on the east coast of Baffin Island are about 12 m high. However, in general, it seems that dominant bottom transport should be assumed for ice, and that the competence of glaciers is undoubtedly due to the compactness of ice and, compared with water, its relative resistance to compression. The huge size of some ice-moved boulders testifies to the efficiency of ice as a transportational process. One of the largest erratics in Scotland, on the Arran coast near Corrie Burn, is 3 m^2 at its base, 3·7 m high and weighs approximately 400 tons. Ice may also transport material up reversed gradients even in some terminal zones. Erratics were carried obliquely up the southern slopes of the Kilsyth hills to 210 m above their outcrop within a distance of 3 km. As far south as the Illinois Ozarks, the Illinoian glacier, after radiating from the Labrador dispersion area, surmounted obstacles 90 m high. Another testimony to the efficiency of ice transport is the considerable distance over which some large erratic blocks have been carried. The mechanisms in continental ice-caps are even more obscure than valley and piedmont glaciers. However, from the results of recent investigations in Antarctica, it appears that some basal transport occurs, but that the rates could be unimportant. The ice bottom 5 km down at Byrd Station contains a lot of debris, including beds of silt and sand and larger fragments up to 5 cm across. Erratics suggest a distance of transport of 360 km.

In valley glaciers where sub-glacial water exists it is difficult to apportion the transportation achieved between the ice and the sub-glacial stream. The volume of material in outwash plains compared with the size of moraines suggests the dominance of fluvioglacial transport in some areas. They owe part of their efficiency to flow under hydrostatic pressure. Subglacial streams entering the Slims river fountain 2·4 m into the air with maximum reported heights of 6 m. Around the openings are large accumulations of well-rounded boulders 0·6–0·9 m in diameter, suggesting the efficient removal of all sizes finer than this.

In coastal areas drift ice can transport pebbles of considerable size. An ice-front on disintegration becomes a powerful transportation agent as beach material frozen to the underside is rafted away. Drifting ice seems to be the main agent of transport of the large quantity of coarse sediments found on the muddy flats along the St Lawrence.

Ice can also be an agent in deforming land surfaces over which it passes, causing movements *en masse* but without entraining the material. As these deformations have some resemblance to those of gravity mass movements and because original sedimentary structures remain essentially intact, these ice-drag or ice-push features are perhaps appropriately discussed together with other aspects of glacier transport rather than in the context of glacial erosion. Because of the preservation of sedimentary structures, ice-pushed ridges appear to depend on the existence of a thick permafrost tract in front of the advancing ice-sheet. After postglacial thawing the pushed-up blocks remain largely intact due to the internal friction of their constituent materials. Similar ice-pushed ridges are a feature of frozen shores. They are common along the north Alaskan coast, but the ridges, 0·6–4·5 m high, probably amount to only 1–2 per cent of the sediment above sea-level. In some places ice-pushed boulders are so regularly arranged along the shore that the impression is that of artificial features modified by erosion. A wall of turf and stones pressed up by ice is a common feature around many Swedish lakes. In ice-drag features movement may take place on gliding surfaces often parallel to the bedding. An example studied in some detail in the north side of Beaver Creek Valley, near Elkton, Ohio, is a rift 10·5–18 m wide and 9 m deep cutting across a bulbous spur. This rift appears to have been pulled about 15 m apart by the passage of ice, filled with glacial debris, then capped with sandstone and shale displaced 45 m by a second ice-drag mass movement.

Another transportation mechanism in which ice is probably involved is the rock glacier. The tendency for rock glaciers to be composed of equiangular debris which provides for the greatest interstitial accumulation of ice, without itself possessing any tendency towards mobility may add to the reasons for supposing that rock glaciers owe their mobility to interstitial ice. The present-day activity of some rock glaciers is evident where they override trees or saplings.

Apart from ice-covers, ice-drag, and interstitial ice, underlying ice, where it exists, may have some influence on transportation. In Antarctica, ice-cored talus aprons have been observed below cliffs on which the movement and the sorting of the debris has been attributed to vibrations caused by the cracking of the underlying ice, induced by rapid temperature changes accompanying the passage of shadows across the aprons.

The transportation of material by moving snow may be significant in colder areas if snow accumulations are appreciable, particularly if the movement is that of a slush avalanche. The site for these movements is the upper parts of slopes where a collecting basin exists. In winter ice may block the gully and hold back wet snow and water which on release in the spring thaw may carry much debris in the mass of water-saturated snow and melt water. In northern Sweden a boulder $5 \times 3 \times 2$ m in size and 75 tons in weight was moved 120 m down a slope of only 5 degrees in this manner. Total volumes moved may be as much as 2000–4000 m^3, slush avalanches moving ten times the total load of other avalanches. However, avalanches and snowblock falls from hanging cliffs are important in clearing away smaller sizes of rock waste.

3. Mass movement

There is a range of mechanisms involved in transporting material down slope gradients in which the amounts of water involved become progressively less. At one extreme is the mudflow, its liquidity and its channel floor position making it transitional between muddy stream flow and fluid mass movements. The water content involved in soilflow tends to be less than the theoretical liquid limit with debris yielding to continuous plastic deformation above a shear plane. Soil creep may operate in unconsolidated materials on slopes less than approximately 35 degrees. Materials again tend to be drier than the theoretical liquid limit but, unlike soilflow, the irregular motion of loosened individual particles or aggregates, characteristic of soil creep, does not involve a basal shear plane. Landslips describe mass movements of solid material over a sliding plane with perhaps some changes in structures but without continuous deformation. Finally falls, at the opposite end of the scale to the mudflow, involve the free fall of dislodged solids alone. The distinctiveness of mass movements is not, however, invariably closely related to water contents as gravity is often the overriding factor. It is not easy to categorize types of movements and it may be misleading to specify distinctions too rigorously. The susceptibility of a slope surface to mass movement and the landforms which result are influenced by a wide range of factors continually changing in space and time and even within the same slip area. Geological factors which influence the shape of shear or tension fractures may play little part in the form assumed downslope by the transported material. The momentum and effects downslope of mass movement depend not merely on the geological conditions favouring its release but also on the infiltration capacity of the overwhelmed ground. Surface weathering conditions range from the frost and snow melt of cooler environments to the deep chemical weathering in humid tropical areas. The depth of decomposition, the weathered clays adding to their weight with their capacity to absorb the large quantities of water and expanding in the process, the abrupt contact between weathered soil and subjacent rock providing an ideal sliding plane, all make the humid tropics the environment in which mass movements are most characteristic.

Mudflows are fluid masses moving in surges down stream channels. They may develop from steep scantily vegetated slopes made up of unconsolidated material containing enough clay to make the mass slippery if wetted by abundant water received in a short period of time. The pressure of water in pore spaces also reduces intergranular friction. The weight of water is a critical factor also, because flow starts when the weight component parallel to the slope exceeds the internal cohesion of the mud and the friction at its base. At some point a distinction between a heavily loaded stream and a mudflow becomes arbitrary. The water content may be 20–60 per cent by weight or by volume, and densities may be as high as 2·0–2·4. The high density of mud exerts forces on stream bedload substantially different from those of running water and quantities of channel alluvium are incorporated into the fluid mass as it passes. An example of a larger area of mudflows covers an area of more than 40 km^2 in Chile, at 27°S on the southern slope of the Carro Cadillal. Mudflows also occur near the south Crimean coast. In 1949 when 178 mm of precipitation fell in 12 hours, a mud- and rockflow involving at least 1·5 million m^3 of material moved down the Uchau-Su river valley into the sea within a day. Some mudflows are so viscous that they come to a halt in the stream channel.

On slopes, soilflow or earthflow depend less on the water content of the mineral material and more on gravity as the major factor, and are common on declivities between 5 and 30 degrees. In the tropics, much steeper slopes are often involved. The movement is essentially one of plastic deformation, but initially, either at the beginning of seasonal melting or following an abrupt failure, flow may be more affected by liquidity. The shortness of the lobes which develop downslope are due to the loss of water by drainage and infiltration which eliminates any element of liquidity in the flow mechanism and tends to extinguish rapid movements. Where sufficient volumes of debris are involved, boulders several decimetres in diameter can be moved.

Many workers use the term solifluction non-genetically to describe the viscous flow of soil under saturated conditions, but as the term was first used in the context of an arctic environment, others imply soilflow only where freeze–thaw conditions prevail or even only where a permanently frozen sub-soil exists.

The similarity between soilflow in the humid tropics and in periglacial environments arises because the moisture changes in the top 10–15 cm of the clayey tropical soils produce the effects of alternating expansion and contraction which also accompanies freeze–thaw activity. In damper periglacial areas solifluction is the main agent of transport favoured where permafrost is present as this provides an impermeable layer below the seasonally thawed soil which easily becomes waterlogged. Where frost penetrates to depths sufficient to induce large-scale solifluction mean annual air temperatures are not usually above 1°C. Material in the lobes is often only 0·25–1·5 m thick on steeper 20–25-degree slopes. Measurements of solifluction lobe movement include rates of 2–6 cm/year in Alaska, 4–5 cm/year in Spitzbergen, and on a gentle 5-degree slope in Sweden, 0·9–3·5 cm/year. On steeper slopes rates increase as shown by S. Rudberg's measurement of a 12-cm/year rate in northern Sweden in 1958 on a 20-degree declivity. In an adjacent locality A. Rapp (1960) found that the centre of lobes moved at 25–30 cm/year but that the sides moved only at 4–7 cm/year. Also, at times of initial release, movements may be more rapid. In an area of active retreat of supersaturated frozen ground in the Mackenzie delta area, the enlargement of active scarps at 1·5–4·5 m/year may give some indication of such rates. In many areas recognizable solifluction features are no longer active. Measurements on Mount Northcote in the Snowy Mountains of Australia, made 60–220 m below the summit, while indicating a rate of 0·4 cm/year until 1540 ± 160 years B.P., suggest little subsequent movement.

Soil creep is another slope process encountered in most latitudes occurring on slopes as gentle as 2–4 degrees with again more than one mechanism producing similar effects. In cooler damp conditions, freezing temperatures, although they do not penetrate sufficiently to produce a solid mass of ice, convert interstitial water in superficial soil layers into segregated ice crystals overnight. These crystals of needle ice, called pipkrakes, develop below individual particles or aggregates of soil which are better conductors than the surrounding soil. On a slope the subsequent melting of the segregated ice crystals and the vertical drop of the raised soil particles involves a slight downward displacement. On a bare 55 per cent slope near Wenatchee, where annual precipitation of 525 mm falls mainly as snow, 25 per cent of particles labelled with a radioactive tracer moved 0–7·5 cm, 36–41 per cent moved 7·5–15 cm, and 20 per cent 15–30 cm during a winter spell. Frost creep was the main process involved. In the Tatra Mountains, Poland, T. Gerlach

(1959) has recorded particle displacements of 2–6 cm in one freeze–thaw cycle. A critical factor, the height of pipkrakes, varies with local conditions, diminishing as diurnal changes through freezing point at the soil surface become less frequent. In mountainous areas in central Germany, pipkrakes may be 10–15 cm long, and J. Schmidt (1955) calculated that weights of material lifted in the Black Forest and the Taunus areas were 650–7200 gm/m². The weight of individual particles lifted is also appreciable. Pipkrakes 3–4 cm long have been observed on the Chambarrau plateau. Records show pebbles 0·1–2 kg in weight lifted 1·5–3 cm and that one 9·5 kg fragment was raised 1 cm.

Displacement of particles in accumulations of steeply inclined coarse material is also pronounced. In the English Lakeland hills, marked stones moved 12–15 cm during a winter period. In Spitzbergen stones in areas of stone stripes moved 3–4 cm/year on 6–18-degree slopes, and on 3–6-degree slopes by 1–2 cm/year. In Sweden there are records of coarse material moving at 0·2–1·6 cm/year.

As environments become warmer, and where moisture supplies are adequate, organisms become increasingly important in soil creep processes. Examples of the scale of disturbance in superficial soil layers by organisms were given in the preceding chapter. Rates may range between 0·1 and 10·0 mm/year.

Movements discussed so far relate to those at the surface or involving superficial layers. In landslides entire sections of slopes are removed to depths of several metres. The most common type of landsliding is the slump in which a long slice of cap-rock or surface material separates from a scarp face and, without being greatly deformed, moves downslope on a curved slip surface (fig. IV.12.A). The strike of the slump blocks is controlled by the trend of the cliff. The backward tilt of the rotating block's ground surface creates a depression at the headward end of the tilted block which becomes marshy or the site of a pond. As in solifluction a crucial factor may be the intake of large volumes of water, which by displacing air from the pore spaces and building up pore-water pressure, imparts buoyancy to constituent materials if these are sufficiently loosened or weathered. In 1952 when there was the largest and most spectacular of landslides in the upper Columbia river valley, springs in the area dried up 10 days before the failure occurred. Perhaps initially, therefore, some minor displacement, by damming back these underground springs, triggered off the much larger movement. Where a massive cap-rock overlies less competent strata rotational movements may also occur. However, some blocks glide along intact, tipped outward away from their source, particularly where cambering precedes mass movements. These blocks then load contiguous segments of the slide paths. In some instances the detached blocks are displaced vertically without tilting, accompanied by lateral

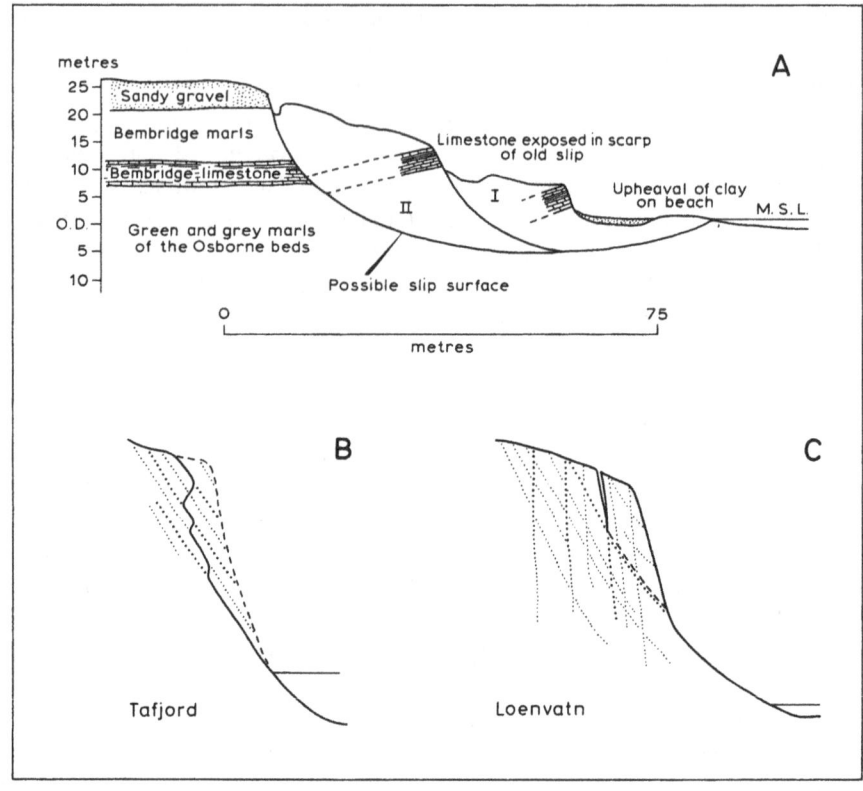

Figure IV.12 Types of mass-movement slides.

A. Rotational landslides on Oligocene clay induced by increased water pressure in the sedimentary strata, Isle of Wight (*W. H. Ward* in *Steers, 1962*).

B and C. Rockslides in Norway, illustrating the control of jointing (*O. Holtedahl, 1960*, Geology of Norway). $1\frac{1}{2}$ million m³ were involved in the 1934 rockslide at Tafjord; (B) Waves in the narrow fjord were more than 62 m high.

spreading of the incompetent strata. This mechanism, which might be described as block glide, has been significant in the lowering, by some 180–240 m of sandstone slabs from the Chuska Mountain scarp, north-west New Mexico. If the slumped material contains sufficient water and if structural controls are not rigorous, the displaced material flows out from a pear-shaped hollow to form a tongue-shaped mass downslope, these phases resembling those in the creation of a solifluction lobe. However, it is quite common for mechanical failures upslope to be rotational slump involving essentially slippage, whereas removal downslope is essentially by flowage. Quick-clay slides demonstrate these features particularly clearly. These involve much flowage and they can move on slopes as shallow as 1 degree. Slumps in tropical areas, where heavy downpours fall on deeply weathered ground, are characteristic of slopes of between 42 and 48 degrees. If the slumped material is essentially dry it forms a jumbled and disaggregated mass of individual fragments. The scars left by the departure of sliding material may be more conspicuous landscape features than the large areas covered by slumped debris. The main scar of the 1964 Sherman slide in Alaska is 700 m wide and over 1 km long in the direction of the slip.

The amounts of material removed in some landslides are huge. Debris amounting to 1800 million m^3 from the Tin Mountain landslide in south-east California covers 14 km^2. The 1925 Gros Ventre slide in Wyoming was one-forty-fifth this size. In Chile in 1960, earthquakes released slides with volumes as great as 100 million m^3 in the vicinity of Lago Riñihue. The largest slip of several on the Upper Columbia river occurred in 1944, involving 3–3.8 million m^3. Even quick-clay slides may involve large quantities of material. One of the larger, at St Alban, Quebec, incorporated 19 million m^3. East of Ottawa city a series of retrogressive slips, occurring about 1000 years ago, removed some 14 ha of a terrace made of quick-clay. By inference, rates of movements of quick-clay are much more rapid than for most slides. They may take only 2–3 minutes, as at Surte in Sweden, or as much as half an hour. The largest 1944 slip in the upper Columbia River valley moved at 90 m/day. Rates in terms of recurrence are most striking for tropical areas. For every square kilometre in a basalt outcrop in the Hawaiis, landslipping recurs once in less.than 2 years.

Rockslides are defined by C. F. S. Sharpe (1938) as the downward and usually rapid movement of newly detached segments of bedrock, sliding on joint, fault, bedding, or similar plane of separation. The structural control of the sliding plane is a distinguishing feature of the rockslide compared with the rotational slip. The slide shown in fig. II.10.C is believed to have moved about 300 m away from an anticlinal axis at places along a surface inclined at less than 10 degrees. A bentonite horizon probably acted as a lubricant and localized the position of the sliding surfaces. Again the amounts of material displaced represent volumes that would take thousands of years to remove by slow everyday processes. The 1756 rockslide at Jjelle in western Norway comprised 12 million m^3 of gneiss and granite. On a smaller scale rockslides occur on the line dividing relatively thin weathered material from solid rock, or even where a narrow weathering horizon at shallow depth underlies comparatively sound rock above, as happens frequently in sheeting along granite joints.

Rockfalls involve the relatively free fall of newly detached segments of bedrock. A distinctive feature is the disintegration of the detached mass during its descent. If such material is heterogeneous the falling material, due to air trapped in the interstices, may cover surprising lateral distances at rapid speeds.

4. Transportation of dissolved material

In higher latitudes there is little dissolved material transported in streams partly because much of the small quantities initially taken into solution is captured by organisms. In the tropics the synthesis of clay minerals means that in the transportation of weathered products the suspended load may be three to four times that of the dissolved load, even though the initial weathering processes were largely solutional. In mountainous areas mechanical transport removes up to seven times the quantities carried in solution. By contrast in many areas the amount of material transported in solution approaches that of mechanical transport, as in much of Europe, and in some lowland areas transport in solution may account for nearly twice the amount removed by mechanical transport.

The scantiness of information on how the dissolved load is transported, compared with the detailed attention to mechanical transportational processes, is clearly out of all proportion to its relative significance. In fact this imbalance may have placed one of the greatest limitations on progress in the understanding of landform development. Its significance relates not simply to the appreciable if not preponderant quantity of materials removed in solution but to the fact that transport by solution is as rapid as the rate of waterflow and is effective on and beneath all land surfaces where water moves, however low the gradient on the surface or the degree to which gradients are reversed beneath land and water surfaces.

The main element transported in solution is calcium which, in bicarbonate form, is chemically stable over times involved in river transport although it may be reduced by the activity of organisms in slow-moving reaches. Next may come either silica or iron, their proportions varying according to how climatic conditions influence weathering processes and the other substances present in the solution. As an example of the latter circumstance, the solubility of silica decreases appreciably when alumina is also in solution, posing an obstacle for large-scale transport of kaolinite in solution. K. B. Krauskopf (1956) found that little or no silica was removed when kaolinite, montmorillonite, calcite, or Fe_2O_3 were added to silica solutions made up in distilled water. The mode of transportation of iron is also problematic. As ground water enters a stream, oxygen from the atmosphere oxides iron bicarbonate to ferric oxide. Because of low solubility of the oxide, a greater part precipitates from solution as a gel, but the remainder is protected by organic colloids or in the form of iron-organic compounds. For the most part these gels join the suspended material and are evacuated from the drainage basin. Because of either hydrolysis, adsorption, or both by alumina ions, the transportation of significant amounts of this element even through a weathering zone may be possible only in the form of complex ions or as protected sols. Stream sediments may also affect the chemical composition of the dissolved load. Some minerals will dissolve, some may cause precipitation of certain dissolved ions, and because of their exchange capacity, others may stabilize the chemical composition of the dissolved load. In the eastern United States, where the washload clays less than 4 microns in size include kaolinite, illite, and vermiculite, their exchange capacity is less than 20 m.e.g./100 gm. In western states, montmorillonite increases and in consequence so does the exchange capacity. For washload clays in the Crooked river, near Post, Oregon, the figure is 63 m.e.g./100 gm. It is significant that during periods of very high concentrations of suspended sediment, the ratio of cations adsorbed on colloidal sediments to those in solution probably reaches a maximum. Of the maximum flows observed in the United States, the smallest ratio of adsorbed cations to cations in solution, 0·1, was on the Juniata river at Newport, Pennsylvania and the highest on the Rio Grande, in New Mexico, where there were three times as many cations adsorbed on the suspended sediments as there were in solution. Intermediate ratios at maximum flow include 0·2 for the Crooked river at Post in Oregon, and for two rivers well-known in geomorphological literature, the Brandywine Creek, Delaware, ratio was also 0·2, and for the Green river at Munfordville, Kentucky, 0·3. These and other figures suggest that during periods of very high sediment concentration, cations adsorbed on suspended sediment may approach or even exceed the cations carried in solution. This conclusion is of profound significance for geomorphological studies as it shows that measurements of transported solid and dissolved loads in such circumstances could give the impression that the importance of mechanical weathering in relation to solutional weathering was three times its actual significance. There is also the risk of further exaggeration because there is also the possibility that organic

Table IV.8 Average figures for mechanical and chemical denudation *(from N. M. Strakhov, 1967,* Principles of lithogenesis) based on calculations by G. V. Lopatin.

CONTINENT	AREA, 10^6 KM^2	DISCHARGE OF SUSPENDED LOAD, 10^6 TONS/YEAR	DISCHARGE OF DISSOLVED LOAD, 10^6 TONS/YEAR	INTENSITY OF DENUDATION, TONS/KM^2	
				Mechanical	Chemical
Europe	9·7	420	305	43	32
Asia	44·9	7445	1916	166	42
Africa	29·8	1395	757	47	25
N. and Central America	20·4	1503	809	73	40
South America	18·0	1676	993	93	55
Australia	8·0	257	88	32	11

matter may be responsible for an appreciable proportion of the exchange capacity of stream sediments. Under average to low flow conditions, suspended sediment concentrations are usually very low, and with the ratio of adsorbed cations to cations in solution being negligibly low and probably less than 0·001 there is no chance of confusion in equating solid/dissolved load ratios with the relative importance of mechanical and solutional weathering.

In addition to some general overall calculations of the amount of material removed in solution shown in Table IV.8, some specific examples are worth mentioning. The loss in the main rivers of the northern half of European U.S.S.R involves 21 million m. tons of calcium and 15 million m. tons of magnesium.

C. Deposition

Modes of deposition and origins of depositional landforms are often controversial issues. One of the problems, apart from the practical problem of observing the process in operation, is that usually more than one factor is involved, either simultaneously in one environment or in effecting similar results in different environments. For instance, some boulder clays laid down on land are sometimes difficult to distinguish from glaciomarine drift deposited from floating ice. Trough-shaped sets of cross-strata are produced in different media in different environments, the products of nearshore marine, beach, river, and eolian processes wherever the depositing currents flow primarily in one direction.

The most widely applicable generalization is that deposition follows a decline in the competency of a transporting agency. In a stream this is when the velocity which dislodged a particle of a given size slackens by about one-third. There is a time-lag between the moment at which a decelerating current can no longer suspend a particle and the moment at which this particle reaches the bottom. This settling lag is an important factor in interpreting the accumulation of coarser material on beaches. Deposition may initially be self-enhancing because the deposition of the coarser particles increases bottom roughness. This, in slowing the stream down, leads to the deposition of less coarse material. This mechanism may, however, become self-arresting with the deposition of sizes which are sufficiently small to increase the bottom smoothess. However, over a large size range, from 70 microns to 30 mm at least, the balance of forces acting on a particle does not change in such a way as to cause a major change in the mechanism of particle selection during deposition. Once deposited, a particle may remain in a given position for a time that could range from a few seconds to tens of millions of years.

Decrease in competence may not be constant throughout a transporting medium. Deposition takes place in areas that are beneath, or to the side of, some thread of maximum velocity and its flanking threads of maximum turbulence. Enough material may accumulate at a given place to give rise to a distinctive form, starting with two particles coming to rest so that one leaning on the upstream side of the second is itself in part blocked and in part shields the downstream particle from the full force of the current. As this is a· self-accentuating process, a traction clog may develop which could eventually become a central bar or even an island in a braided stream (fig. IV.15.A).

It is possible to recognize three types of depositional landform. A form resulting essentially from deposition in the 'past', where the definition of the past reflects a subjective preference on the part of an observer, is only one of these. The description of the characteristics of this static form, if the depositional material remains little unaltered, is relatively straightforward. A second type occurs where a given volume of material moving more or less continually, the definition of time-spans again involving a subjective decision, may assume a distinctive form during its transportation. Thirdly some depositional forms depend essentially on a quasi-equilibrium between amounts of material removed and continued renewals in supply. In this case the amount of material passing through the system may be of much greater volume than that temporarily incorporated in the form.

The purpose of the following discussion is to describe static depositional landforms and the unchanging characteristics of dynamic depositional landforms.

1. Depositional landforms on slopes

The most distinctive depositional forms on slopes are those assumed by material released and transported downslope by abrupt mass movements. The chaotic arrangement of several irregular ridges, mounds, and depressions is characteristic of many landslide areas. The more liquid the moving mass the more regular and lobe-like the form assumed by the deposited material. On slopes in cool environments where higher vegetation is absent, material moving slowly downslope may assume distinctive small terrace shapes, fronted by a bank of turf, or of coarse debris (fig. IV.13). In general, the form assumed by solifluction deposits is that of a smooth sheet, usually 0·5–1 m deep on slopes and thinning on summits but often thickening to several metres in hollows and valley bottoms. The term 'head' describes such crudely stratified debris produced by solifluction. The extent of these deposits is much greater in many areas than was formerly believed. J. M. Ragg and J. S. Bibby (1966) have observed the stony mantle in the higher parts of the Southern Uplands. The rubble layer, where examined, is well sorted and has a uni-

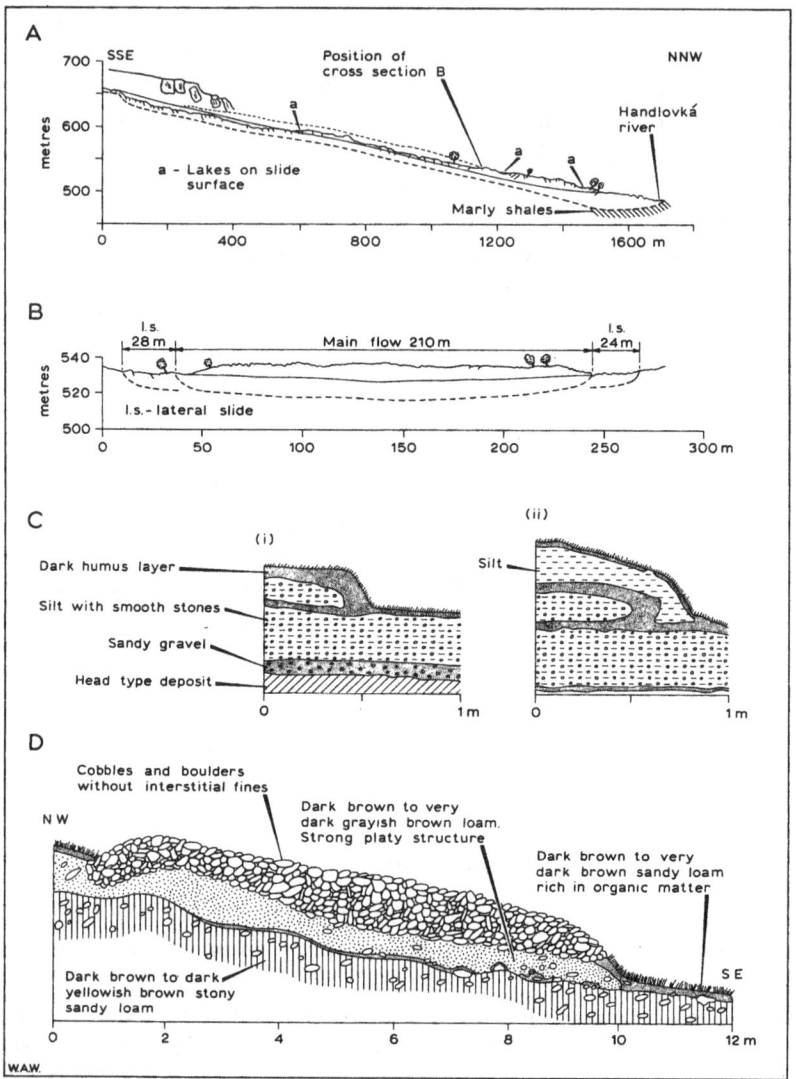

Figure IV.13 Slope deposits.

A and B. (A) Earth-flow deposit (*from Zaruba and Mench, 1969*), one of the largest in Czechoslovakia, produced in 1960 near Handlova. The movement involved debris of volcanic rocks and clayey and silty sediments of Sarmation (Neogene) age which had already been entrained downslope by previous movements. Factors involved were probably changes in clay consistency by periglacial climatic effects and also possibly the squeezing out of a plastic substratum by the load of the overlying volcanic sheets. (A) Longitudinal section. (B) Cross-section near the base of the tongue of debris.

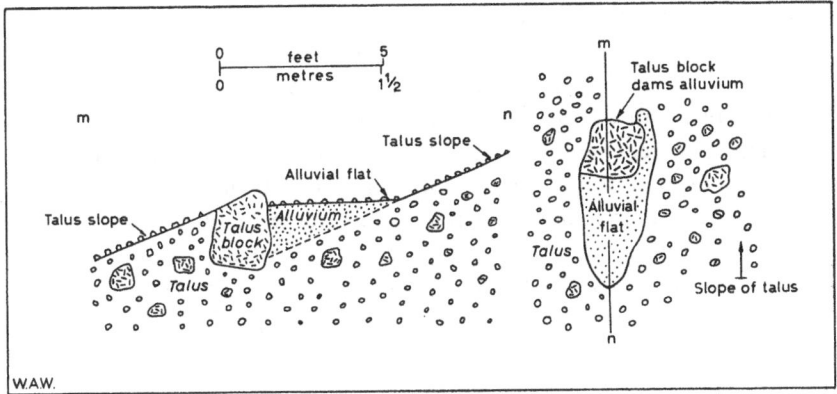

Figure IV.14 Alluvium deposited on a talus slope (*from Nichols, 1960*). Alluvium on a slope is often distinguished by the term colluvium if it has been transported by slope processes.

modal peak. The mode of the surface rubble layer is generally less than 32 mm whereas the coarse modal peak in the deeper layer is greater than 32 mm. The subjacent layers are often bimodal and contain considerably more fines in the coarse silt to fine sand range (31–250 microns). There is a coating of sand and silt on the upper surfaces of stones whereas these lower surfaces are clean and often rest on coarser material and nests of small stones.

The superimposition of soil layers on hillslopes commonly results in stone fabric differences between surface and buried layers. The arrangement of rock fragments in the overlying layer may be chaotic whereas undisturbed bedding planes of the parent rock may show through the buried layer. In central Poland, rhythmically bedded slope deposits, due to the rillwash, dominant during phases when ground ice melted, were formed during the climax of the last cold stage.

Where free fall of individual particles occurs, screes may accumulate. In cool and arctic environments, as on Nevy Island, Antarctica, screes may be a hundred or more metres high. Average angles for entire screes are usually between 32 and 36 degrees. Tinkler (1966) found that 40 per cent of measurements made on limestone screes in the Eglwyseg valley near Llangollen, North Wales, were 35 degrees. Other measurements on some steep-fronted moraines and on the lee-side face of sand-dunes have averaged 34 degrees.

C. Turf-fronted solifluction terraces on Slaetharatindur, the Faeroes (*from C. A. Lewis and G. M. Lass, 1965*, Geog. Jour., *Vol. 131*). Species of *Rhacomitrium* form the riser of these terraces; both are at 400 m: (i) south-facing, and (ii) north-facing.

D. A stone-banked terrace on Niwot Ridge, Colorado Rocky Mountains (*from Benedict, 1966*). The terrace is at 1065 m on a 10–15-degree slope.

Angles steeper than 38 degrees are not usually maintained for more than a few metres. An example at the other extreme is the declivity of scree slopes in the vicinity of the Kalagarh landslip, Uttar Pradesh, where quartzite blocks accumulate at an average angle of 31 degrees. One factor influencing the steepness of screes may be the tendency, as observed in experiments by van Burkalow (1945), for more angular material to bank up at slightly steeper gradients. Another major influence is the shape in plan of the supporting valley wall, with screes banked up more steeply where contours are concave in plan. Many screes, like those on Palaeozoic rocks in parts of Snowdonia, which average 34 degrees, are slightly concave in profile.

2. Depositional landforms in fluvial environments

Although they are of comparatively rare occurrence, there are reasons for discussing mudflows as the first depositional form in fluvial channels. Mud-flows have abrupt, well-defined margins, are convex in profile and decrease regularly in thickness downvalley or down-fan. At the toe there may be a radial arrangement of surface ridges composed of rounded boulders. Mudflow material is poorly sorted due to the presence of large boulders and abundant smaller pebbles embedded in the silt and clay matrix. Occasionally debris includes organic material, which, because of its light weight, would be carried away in water flows. Some indistinct stratification in beds 1·5–10 m thick, sometimes outlined by layers of water-laid material, may be visible in large exposures. A fluid mudflow shows graded bedding with the larger rock fragments sometimes orientated in vertical as well as in other positions. Apart from the surface form of the mudflow and its valley-floor position there is often little to distinguish the material from boulder clay.

The shape of an alluvial fan is a result of deposition by a stream that swings back and forth over the accumulating material but is fixed at the apex by the position of its bedrock valley upstream. Deposition may result in some cases from a decrease in slope where a stream enters the apex of the fan, but more generally might be related to a decrease in depth and velocity of flow as a stream spreads out downvalley from the confines of its headwater valley. In the initial growth of a fan a powerful self-enhancing factor is reduction of surface runoff by infiltration into the porous accumulation of debris. There are two types of water-laid sediments on alluvial fans, for reasons somewhat similar to those which explain the contrast between river channel and flood-plain deposits. On fans most of the water-laid sediments consist of sheets of sand and silt deposited by a network of braided streams which are rarely more than 30 cm deep. These sands may be well-sorted, with a quartile deviation of 1·6 *phi* units being reported from fans in western Fresno county, and contrast with the interstitial sand and gravel deposited in the more permanent beds of the main stream channels. Efficiency of sorting and some rudimentary stratification may increase down the fan. Fig. IV.28E and F illustrates the downslope decrease in size which accompanies an increase in pebble wear and the decrease in gradient. On some fans there may be a contrast between mudflow deposits descending from the fan head and water-sorted material deposited nearer the toe.

Fans are widespread and are most common in arid and semi-arid areas of the world. About one-fifth of California is covered by alluvial-fan deposits. Coarse gravel fans on the western front of the Black Mountains in Death Valley vary from a few hundred metres to 1 or 2 km in radius with inclinations ranging from 3 to 25 degrees. The steepest fan in the Tucson area is about 10 degrees. On a strip of coalescing fans 20–30 km wide in the foothills of the San Sanquin valley the inclination of fans decline from 1:35 to 1:530. Of much gentler inclination than valley-fed fans are those which imperceptibly extend downslope the surface inclination of gravel- or sand-veneered pediments. These alluvial fills, made up of confluent fans of poorly sorted clay, silt and sand, known as bahadas, may thicken downslope to depths of 5–60 m. Their declivities in the Tucson area vary from 0·5 to 2 degrees. In the Ajo region, declivities are about 0·5 degrees. Fans are not confined to semi-arid areas. Coalescing fans, for instance, form a tilted plain on the west side of the Mackenzie delta in north-west Canada. Here the average frost-free period is only 66 days but there are annually twenty-five freeze–thaw cycles of a 34°–28° F amplitude which might be an important factor in the disintegration of bedrock. Two fans that have been studied in detail are of similar dimensions being 2·8 km long and about 4·0 km wide at the base. The inclination of the first fan starts at 1 degree declining to 0·6 degrees at the toe; the second from 3 degrees to 1·1 degree. There are sands and pebbles at the apex, passing down into material less than 2 mm at the toe where silts predominate. Faulting is often an important factor leading to sedimentation of a fan surface.

Downstream from the steeper headwater reaches where fans may be present small but significant features are the traction clogs which may form spool-bars. These diminish in frequency downstream as the critical concentration of bedload particles for jamming to occur is seldom reached.

Farther downvalleys a distinctive feature of deposition on the level floodplains is that, even over spans of thousands of years, building up the floodplain may be a progressive process. It is difficult to decide whether long-continued accretion is a response lagging behind postglacial sea-level rise, or is due to man-accelerated erosion or whether some other factor is involved. In part at least continuous alluviation appears to result from deposition of sediment trapped by vegetation at occasional high flows. Parts of the Nile floodplain build up at a rate of 10 cm/100 years. An average of 45 cm/100 years overbank accumulation seems possible in the Chemung valley, a tributary of the Susquehanna. In the Little Missouri valley, where a flood-water height over 6 m may occur once every 50 years, floodplain build-up may be by 1–2 m/100 years. In the Tesuque valley, which rises in the Sangre de Cristo Range in north-central New Mexico, deposition averages between 19 and 29 cm/100 years. In Eastern Europe, parts of the Danube floodplain are 4·5 m above average low water, and comparable figures for the Dneister, Southern Bug, and Ingul are 3·0 m, 12·0 m, and 21·0 m. In humid tropical areas the clays decanted into hollows may build up rapidly, particularly in the stagnant water of mangrove swamp areas where accretion rates may be several centimetres a year. In cooler areas, floodplains like the Russian plain are in part relict features produced during thousands of years of flooding by fluvio-glacial melt water from a receding ice-sheet and settling in huge temporary backwater lakes at the edge of the ice. In fact, in higher latitudes and in areas near the margin of former ice-front, alluvial plains are described as outwash plains. Where they are confined as a valley train their scale may make these flat depositional features one of the most striking landform characteristics of these areas. Scandinavian workers use the term sandur to describe these depositional plains flooded by melt water and distinguish between plain and valley sandurs. In Sweden, one of the thickest Quaternary deposits fills the Petsaure outwash plain to depths exceeding 100 m. Even as far south as the headwaters of the Mississippi, valley fills of 60 m or more are outwash gravels, relics of their former marginal position to an ice-sheet. Similarly, in the Forth valley, south-east of Stirling, there are extensive near-level spreads of gravel which become very coarse towards the north-west upvalley end. As this surface leads upvalley into areas of ice-contact forms and has a down-valley gradient of 5·5 m/km, which is ten times that for raised beaches with which it merges, J. B. Sissons (1963) recognizes a relict glacial outwash plain. Outwash plain material is generally coarse sand and gravel, contrasting with floodplain fine silts and clays, except where fine silts, settling in lakes, escape wind transport. In some former lake basins, as in Østerdal in central Norway, an infill of lacustine silts and sands may eventually be covered by coarser outwash material.

Another characteristic of the floodplain environment in areas where flow is irregular is the scale of accretion that may follow the extreme flood. In 1965 in eastern Colorado, when most of a 200–300 mm downpour fell in an hour and the subsequent flood was nearly twice the volume of the previously recorded maximum flow, the floodplain sands deposited within hours were many hundreds of metres in width and up to at least 3·6 m in thickness.

In a sinuous channel, deposition takes place on the inside of the river curves of material coarser than that of the flanking depositional plain. The ridges, or point bars, that form, consist mainly of fine-grained, well-sorted sands with interbedded silts. Their mode of deposition is not well understood, although it is possibly related to a Hjulström effect where the slackening flood velocities, at which deposition begins with traction clogs building up, is insufficient to entertain material of the same size. Gravelly sands, with coarse fractions ranging from 2 to 11 mm in diameter were found in deeper parts of the bar shown in fig. IV.15.D, where trough cross-stratification is characteristic. Each trough-shaped set consists of an erosional scour, its long axis parallel to the local stream direction, 0·5–1 m wide and filled with scoop-shaped layers less than 0·3 m deep. Horizontally laminated sand, although not abundant, is found in layers 0·3–1 m thick at various locations and at different levels in the point-bar deposit. Also trough cross-bedding is not characteristic of all point bars. From the study of a 320-m-long point bar on a Mississippi meander where finer materials in silt and silty clay sizes ranges are inevitably involved, the channel on the toe of the slip-off slope appeared to be related to a portion of the current returning upstream rather than to a thread of slackening floodwater crossing the toe of the slip-off slope. As point bars grow through accretion of sediment outward from the bank and as river undercutting leads to continuous shifts in the position of the outer side of a river meander, a series of point bars, or scroll, may be stranded on the slip-off slope. With higher floods than those which can form point bars, overbank flows may form levees due to the accumulation of sediment on top of the channel banks. These asymmetrical ridges are often lense-shaped in cross-section and may decrease in size downstream. The crest is close to the channel because the quantity and calibre of sediment deposited is normally greatest near the channel and decreases in a lateral direction (fig. IV.15.E). In general, levee deposits are on the whole finer than most channel deposits and may contain plant remains. Their structures show evidence of their sub-aerial exposure during the greater part of the year, particularly disturbances due to organic activity, and may be reddish brown in colour due to oxidation.

One final aspect of those forms and features, due wholly to deposition from rivers, involves chemical precipitations. The floodplain and its sediments, particularly if these include colloids, form a geochemical barrier to the migration of chemical elements which does not exist higher upstream. Of those most commonly found in the natural environment the Al, Fe, and Si compounds are the least soluble and are the first to be precipitated when the water moves. As temperatures become sufficient to induce evapotranspiration losses from the floodplain water-table, deposition of these compounds becomes increasingly marked. The soils of the Vikhra river floodplain, for instance, contain five to eight times more iron and five times more manganese than do soils of the adjacent watershed. The deposition of chemical elements in floodplains or swampy ground adjacent to streams is an important factor in landform study as a whole and of crucial significance in humid tropical environments. However, it is probably only the deposition of carbonates

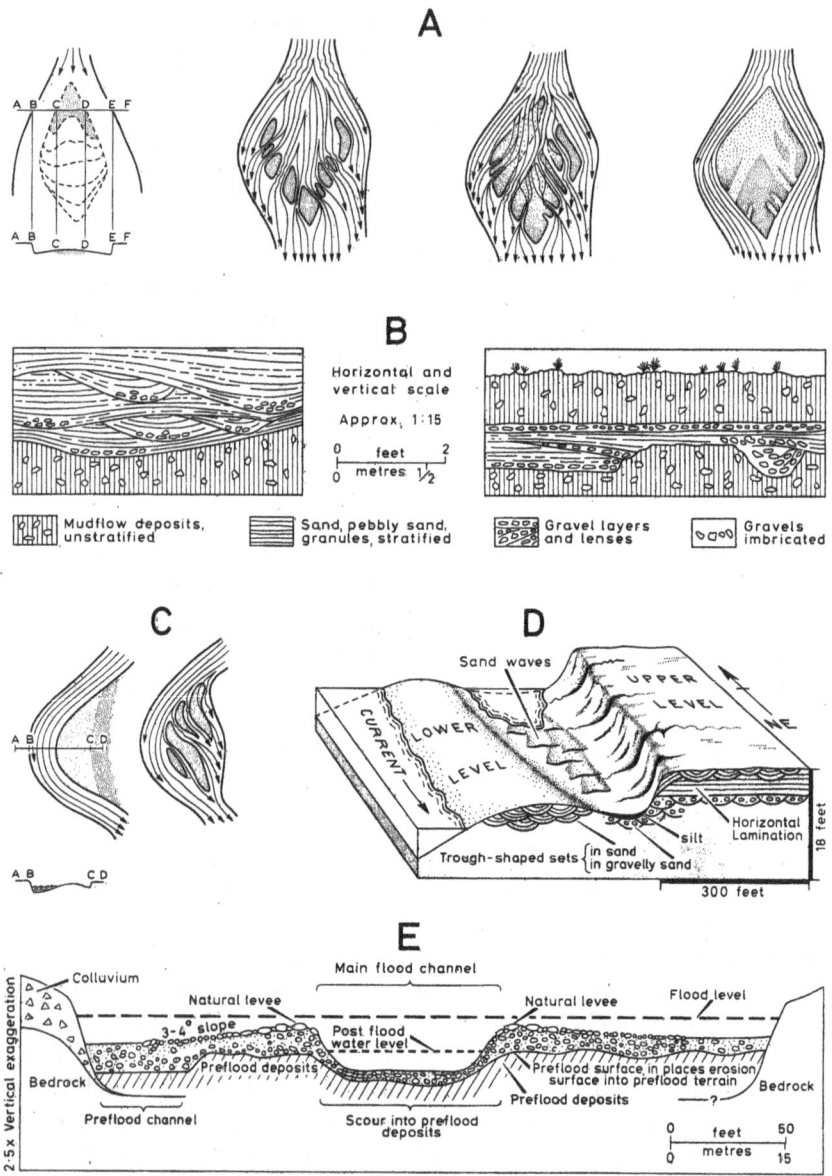

A

B

Horizontal and vertical scale

Approx. 1:15

0   feet   2
0   metres   1/2

Mudflow deposits, unstratified

Sand, pebbly sand, granules, stratified

Gravel layers and lenses

Gravels imbricated

C

D

Sand waves

UPPER LEVEL

CURRENT LOWER LEVEL

NE

Trough-shaped sets { in sand, in gravelly sand }

silt

Horizontal Lamination

18 feet

300 feet

E

Colluvium

Main flood channel

Natural levee

Natural levee

Flood level

3-4° slope

Post flood water level

Preflood deposits

Preflood surface in places erosion surface into preflood terrain

Bedrock

Preflood deposits

Bedrock

Preflood channel

Scour into preflood deposits

—?

0   feet   50
0   metres   15

2·5 x Vertical exaggeration

Figure IV.15 Fluvial deposition forms and their sedimentary structures.

A. Formation of spool-bars and spool-like islands in sandur streams (*from Krigström, 1962*). From left to right the diagrams indicate – the embryo (C–D) and hypothetical growth of a spool-bar; an almost submerged spool-bar; the same spool-bar at a lower water level; and a spool-bar transformed into a spool-like island.

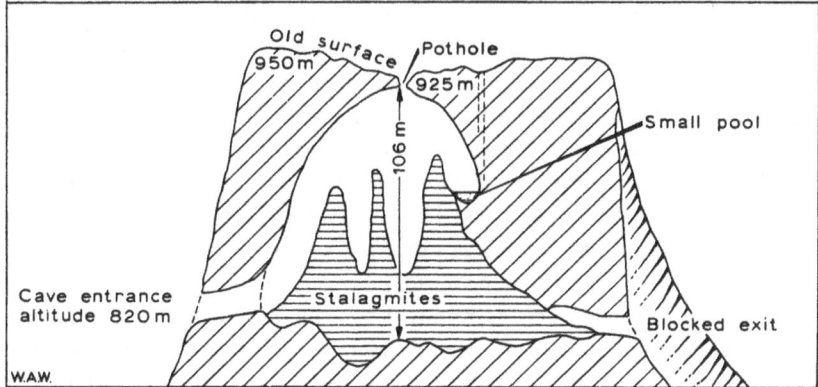

Figure IV.16 The Garcia cave, north-west of Monterey, Mexico, two-thirds filled by stalagmite (*after H. Enjalbert, 1968,* Mém. et Doc. Centre Réch. Doc. Cartog. Geog., *Ed. C. N. R. S., Vol. 4*).

which assumes distinctive forms. Organisms like algae, by extracting carbon dioxide from the water or from the dissolved bicarbonate, may create large volumes of calcareous tufa which forms dams across streams and smooths the outline of waterfalls. Calcareous material may also be deposited in floodplain soils. In a 60-m section of a stream channel below a spring in Virginia, with a flow rate of 390 l/min approximately 2900 kg of calcium carbonate were deposited annually as tufa. Along the Kaap escarpment, north-west of Kimberley, there are remarkable accumulations of travertine.

3. Deltas

While some transported material may remain at the base of slopes for long periods, much of that entering rivers may similarly accumulate at the river mouths where deltas form. Approximately half the amount of the material

---

B. Sedimentary structures in an alluvial fan (*from Blissenbach, 1954*). Main direction of transport from right to left; based on fan deposits in the San Catalina Mountains, Arizona.

C and D. Point-bars. Because of the convex cross profile on the spur of the 'slip-off' slope (C *below*), currents at rising water are deflected towards the inner bank (*from Krigström, 1962*). (D) shows the sedimentary structures of the Beene point bar, on the Red River near Shreveport, Louisiana (*from J. C. Harms* et al., *1963,* Jour.-Geol., *Vol. 71*).

E. Form and structure of flood deposits in the channel of a small mountain stream (*from J. H. Stewart and V. C. LeMarch, 1967,* U.S. Geol. Surv. Prof. Paper 422-K). The deposits are the product of a catastrophic 100-year interval flood in Coffee Creek in Trinity County, Northern California.

transported to the Yangtze-Kiang or Nile deltas is deposited there. However, the loss from the clay fraction may be substantial. A delta is a partly sub-aerial deposit built by a river when spreading of flow into a body of little-disturbed water checks its velocity. Postglacial rise of sea-level has accentuated this braking effect. However, a small tidal range is not a pre-requisite for deltaic accumulation, as a broad shallow area can diminish wave action sufficiently. Thus the Ganges and Irrawaddy deltas continue to advance into coastal areas where the respective tidal ranges are 4·5 and 5·5 m. In general, delta shorelines continue to prograde as long as the supply of sediment, mainly from the river, exceeds removal by wave and current action, particularly in shallow seas. The Volga delta's spectacular advance is by 170 m/year. The huge volume of the Mississippi's load is a major factor contributing to the 400 m/year mean growth of the south-west part of its delta. The Orinoco delta progrades by 200 m/year. Less striking are figures of 4 and 4·5 m/year for the River Meander delta in western Turkey and the Gulf of Thailand respectively, but which still represent huge rates of change in the context of millennia. Examples of intermediate rates of advance include 10 m/year for the Don, 20 m/year for the Po, and 27 m/year for the Kilia delta of the Danube. Extrapolation of rates of accumulation suggest that the bulk of deltaic series is the product of postglacial times. For instance, the Krikkjokk delta in Sweden could have accumulated in 7200 years if the rate of growth for the last 80 years had been constant over this period. Conversely abandoned portions of a delta mouth may be cut back. Since the outflow through the Rhône delta has been through the Grand Rhône, the banks of the Petit Rhône have receded by 2 km since the seventeenth century. In relation to its mode of accumulation the delta is the last downstream in a series of subaerial lobate depositional forms produced by the shifting and splitting of streams supplied from a relatively fixed source. However, the possibility of shifts in the supply point, restrictions imposed by the position of adjacent uplands, makes the plan of deltas the least regularized of the fan-shaped depositional forms. The slope of the shelf offshore is a further diversifying factor and it would be as realistic to enumerate the unique characteristics of individual deltas as to search for generalisations. The Orinoco resembles the Mississippi delta in the predominance of fine sediments and similarity of facies types. However, the processes of delta growth and distribution of facies types are fundamentally different. Whereas the Mississippi delta advances by accretion of bar and channel sands and by levee construction at the radiating distributary mouths, a strong marine current deflects material from Orinoco distributaries abruptly north-westward and the delta advances by littoral accretion. Turbulence of Atlantic swells is another factor preventing sub-delta development. In some situations delta growth is spasmodic rather than continuous; half the total annual growth may take place during a single week in the

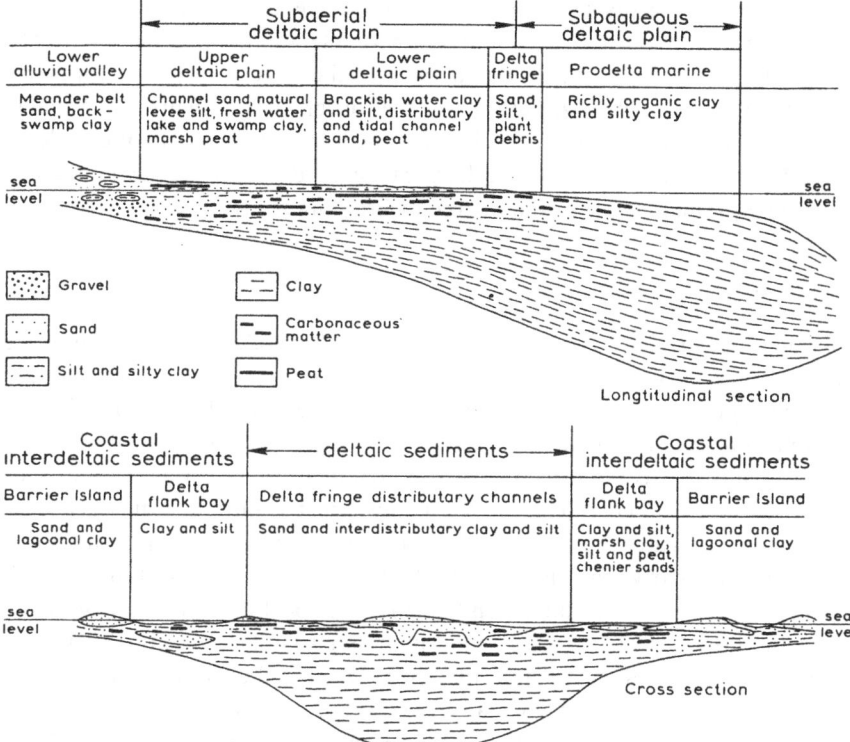

	Subaerial deltaic plain			Subaqueous deltaic plain	
Lower alluvial valley	Upper deltaic plain	Lower deltaic plain	Delta fringe	Prodelta marine	
Meander belt sand, back-swamp clay	Channel sand, natural levee silt, fresh water lake and swamp clay, marsh peat	Brackish water clay and silt, distributary and tidal channel sand, peat	Sand, silt, plant debris	Richly organic clay and silty clay	

Longtitudinal section

Gravel

Sand

Silt and silty clay

Clay

Carbonaceous matter

Peat

Coastal interdeltaic sediments		deltaic sediments		Coastal interdeltaic sediments	
Barrier Island	Delta flank bay	Delta fringe distributary channels	Delta flank bay	Barrier Island	
Sand and lagoonal clay	Clay and silt	Sand and interdistributary clay and silt	Clay and silt, marsh clay, silt and peat, chenier sands	Sand and lagoonal clay	

Cross section

Figure IV.17 Section through a deltaic mass of sediments (*after E. H. Rainwater* in *Shirley and Ragsdale, 1966*).

Laitaure delta in northern Sweden. One general rule, however, is that the thickest section of a delta tends to be in the proximal pro-delta region where silt and clay are dominant. Landward, seaward, and coastward, the deltaic mass tends to thin. Also the grain size of the bed material tends to decrease slowly downstream in active deltaic channels. In the lowest sub-aerial parts of the Godavari delta, India, sizes decrease by about 1 *phi* unit/25 km. There are lateral and vertical gradational changes and sharply delimited scour and fill structures. Organic matter may accumulate in stretches of stagnant water and water-logged depressions which form substantial portions of the area of the sub-aerial deltas. The Ila river delta, which widens to 210 km at its outer edge along the Lake Balkash shoreline, includes 12 per cent of open water in its total 9000 km². The principal structural features of coarse-grained deltas differ considerably from those of fine-grained deltas. Where bedload is carried into the delta area this material gives rise to more rapid changes in deltaic channel pattern and if it reaches the delta front may be deposited as foreset beds. In fine-grained deltas, where accumutions are essentially deposited from the suspended load, distinctive sets do not develop. There are also contrasts in the average inclination of the sub-aerial parts of a delta, those on small coarse-grained deltas being rather steep, up to several metres per kilometre and overlapping with the order of gradients on bahadas. On large fine-grained deltas the inclination is much lower, of the order of 5 cm/100 m. Also the emerging bars of bedload origin are less numerous in the fine-grained delta, so the degree of braiding is low.

4. Estuaries

On some coasts tides and other factors like river velocity may combine to prevent large-scale deposition in river mouths where many well-known estuaries funnel out to the sea. Within estuaries the electrolytic effect of sodium chloride in sea-water flocculates the clay suspended in fresh water. The neutralized clay colloids by drawing together into larger aggregates settle faster than individual particles. Velocities greater than 0·7 m/sec tend to break up aggregates but at velocities less than 0·28 m/sec flocculation is rapid. It is primarily the rising tide which carries this material, first, up the channels of the estuarine flats, then like a river at bankfull stage, spreading out on to the bordering flats, which may include parts of deltas and river flood-plains as well as estuarine salt marshes. Due to the Hjulström effect the receding ebb tide does not normally have velocities sufficient to dislodge the deposited clay aggregates and rates of accretion may be appreciable. In the Wash and other shores of the North Sea and Baltic, like the north German coast, accretion may be up to 3 m/year. Accretion of the substantial clay fractions in the loads of tropical rivers may also be rapid. Following the deposition of a sandy marine horizon, homogeneous clay layers some 20 cm thick have been observed to form within weeks on the east coast of Malaya. Indeed, tidal inlets are among the most variable and mutable of land surface features. Of the features created by estuarine deposition, the marsh or swamps may have a negligible expression in cross-section but introduce profound and rapid modifications to the shape in plan of the river mouth and adjacent coast. The shores of the Wash embayment on the east coast of England where tidal range is 6·7 m in spring tides and 3·4 m during neap tides, provide a good example of a salt marsh environment in mid-latitudes. The gradual decrease in the velocity of tidal currents as they move over inter-tidal flats causes a reduction in their competence and results in size differentiation of the deposited materials which, because of the Hjulström effect and the braking and trapping action of vegetation, tends to accumulate. However, the ways in which ebb of water draining back down the salt marsh creeks might contribute to a redistribution of fines are not clearly understood, partly because of the hazards of field work in such areas. The marsh in the Wash is flooded only at high water spring tides when silty clays, clayey silts and small amounts of sand are laid down in well-defined laminae. Seaward are the higher mudflats which in turn give way to broad sand flats, a small cliff a few inches high often demarcating the transition. Lower down the sand flat burrowing organisms like *Arenicola marina* destroy initial stratification, but the sand surface is usually a distinctive pattern of straight-crested ripples, of 0·5–1·0 cm amplitude and spaced some centimetres apart. In section the ripples may be rounded, truncated by wave action, or remain sharp. A fourth zone,. the lower mudflats, coincides approximately with the more steeply sloping part of the intertidal zone, and occupying a flatter zone between this and the low-water mark is a fifth zone, the lower sand flats. Here material is the coarsest in the area and megaripples are 25–35 cm high and spaced 0·5–3·5 m apart. In tropical areas or in cool or cold latitudes, estuaries are not well-developed; this is possibly related in part to absence of extreme flood discharges in tropical areas and unfavourable conditions for vegetation growth and clay weathering in cooler environments. In the latter context an unusual feature characteristic of the St Lawrence salt marshes is the abundance of boulders 0·5–2·5 m in diameter deposited from floating ice and forming ridges or pavements on the mud surface. These areas are also pitted with numerous irregular hollows 0·5–3 m in diameter and 25–60 cm deep. These are due to lifting of grass rafts frozen on to the underside of ice blocks.

5. Depositional landforms along coasts

Although it is difficult to ignore the fact that the shore is a highly dynamic environment with the continual shifts, readjustments, and replacements of beach materials scarcely separable from the unceasing motion of the waves, certain aspects are sufficiently unchanging to be described as for those of static depositional landforms. One such feature is the highly characteristic 'step' which occurs in the surf zone close to the mean low-water line, and for many workers marks the seaward limit of the beach. Spring-sapping by groundwater draining out of the beach at low water during times of maximum tidal range is an important agent in steepening the base of the beach face. Immediately upbeach from the step there is a small convexity but the cross-section of much of the remainder of the beach is somewhat concave upward. Backshore beaches do not occur on coasts where a wave-cut cliff occurs directly upbeach from the foreshore. Frequently the summer berm forms a noticeable near-level crest just above the upper limit of the zone of swash and backwash on the beach face, landward of which there may be sand-dunes.

Despite the continual shifting of beach sands some sedimentary properties remain relatively unchanging. The step is usually characterized by the

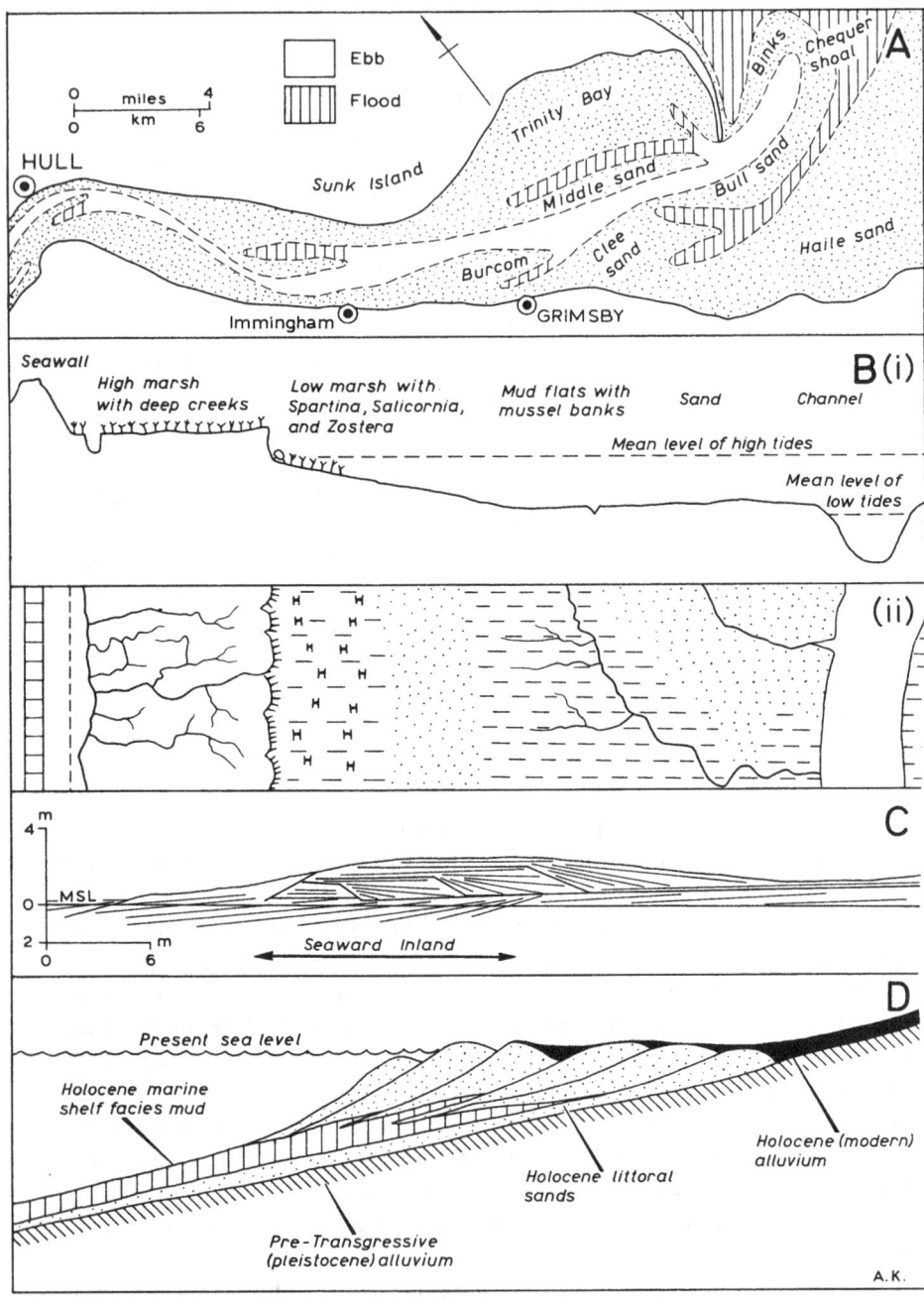

A

Ebb

Flood

miles 4

km 6

HULL

Sunk Island

Trinity Bay

Middle sand

Clee sand

Bull sand

Binks

Chequer shoal

Haile sand

Burcom

Immingham

GRIMSBY

B(i)

Seawall

High marsh with deep creeks

Low marsh with Spartina, Salicornia, and Zostera

Mud flats with mussel banks

Sand

Channel

Mean level of high tides

Mean level of low tides

(ii)

C

m

4

MSL

0

2

m

0   6

Seaward Inland

D

Present sea level

Holocene marine shelf facies mud

Holocene littoral sands

Holocene (modern) alluvium

Pre-Transgressive (pleistocene) alluvium

A.K.

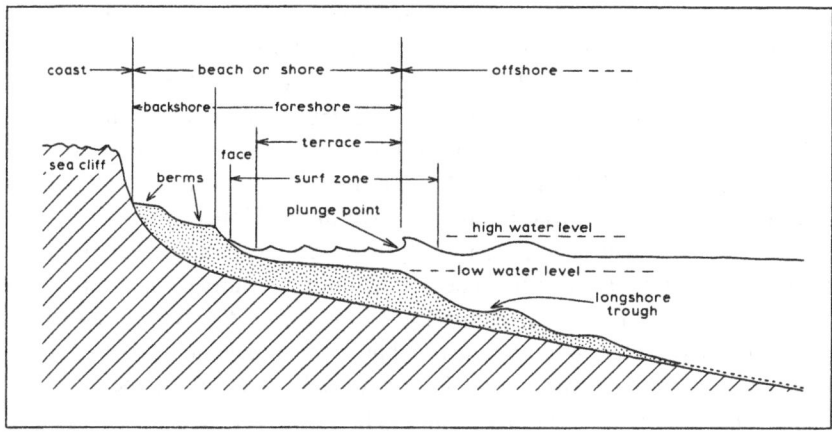

Figure IV.19 Profile of a beach and nearshore region (*after D. Inman, 1962*, in McGraw-Hill encyclopedia of science and technology).

presence of coarser sediment such as gravel, in rather sharp contrast to sand above and below, with the best sorting frequently occurring on the beach side of the surf zone. On Sandy Hook beach, New Jersey, the material at the step rapidly approaches a maximum diameter in the 4–6-mm range. Here the gravel is rather well-sorted. Immediately seaward of the step base the sands are about 0·5 mm in diameter and decrease in size offshore. A narrow range of grain sizes in the zone of the swash is a characteristic feature on most large

---

Figure IV.18 Estuaries and beaches.

A. The shape of the lower Humber estuary and the ebb-flood channel pattern of the bottom (*from A. H. W. Robinson, 1968*, Zeit. f. Geomorph., Suppbd. 7). Spurn Point spit modifies somewhat the funnel shape typical of many estuaries.

B. Section and plan of an estuary bank, Wadden Sea area (*after L. M. J. V. Van Straaten, 1954, in Guilcher, 1958*).

C. Cross-section of a Tabascan beach ridge (*from Putsÿ, 1965*). At the seaward end of the section there is a berm on the foreshore. To the right of the centre of the diagram there is landward-dipping foreset stratification with gently inclined components merging inland with a more steeply inclined component. At the inland end of the section swale stratification with complex near-surface component overlying seaward-dipping stratification common to the beach.

D. Cross-section through coastal plain and inner continental shelf (*from J. R. Curray and D. G. Moore in Van Straaten, 1964*) based on observations along the regressive Costa de Nayant, Mexico. Successive accretion of submerged longshore bars creates ridges. During periods of low waves, a bar is built to the surface and becomes the new beach, isolating the former beach and creating a narrow lagoon.

sand beaches, predominantly in the medium- to fine-sand range. This is because sand of approximately 0·2 mm is generally too fine to roll along the bottom and too coarse to be kept steadily in suspension. A median figure of 0·165 mm has been suggested as representative of Californian beaches, 0·27 mm from the Scanian coast in the Baltic and a median of 0·18– 0·30 mm at Recife. Offshore from the step there is only transport and deposition of fine particles removed from shallower areas. Upbeach from the top of the step the grain size decreases rapidly towards the midpoint of the swash action. The sorting may also become better, but the grain size again increases before gradually and steadily decreasing to the top of the zone of swash and backwash where sizes are often least. The occurrence of coarser particles within the swash–backwash zone is particularly marked on beaches with plunging breakers where sediments may in fact be coarsest at the plunge point seaward of the mean-tide level. Immediately above, coarse material on the top of the summer berm may constitute a secondary size maximum. From this point sizes diminish towards the dunes.

Although sedimentary characteristics on each part of a beach are distinctive because of particular combinations of factors that control its development at these points, the advance or retreat of a beach with time may develop the essentially static depositional characteristic of sheet-like stratification. For instance, the seaward inclination of the foreshore beach on Mustang Island near Corpus Christi, Texas is 2–5 degrees. The inclination of backshore slopes is as much as 2 degrees inland. Inclination of stratification is 5 degrees on the upper foreshore, and on the backshore is irregular and characterized locally by beds inclined at 6–12 degrees.

One of the beach characteristics of most immediate geomorphological significance is the actual slope of the beach which tends to be steeper the coarser the median grain size of the beach material. Along the Californian coast beach, declivities may be up to 10 degrees, commonly are 7–8 degrees and contrast with Texan beaches where declivities are usually 5 degrees or less. On the east African coast between Kenya and Tanga, the narrow coral sand beach at the base of coastal cliffs has moderately steep inclinations ranging from 1 : 10 to 1 : 20.

A second significant form-characteristic of beaches is bars. These include in a genetic sense, those that may form seaward of the plunge-point of breakers from the accumulation of coarse material moving landward outside the plunge-point together with that of sand moving seaward from points upbeach from the plunge-point. The identification of a sub-aerial ridge as a former plunge-point bar, however, is very difficult, partly because both the plunge-point bar and the berm formed at the upper end of the swash zone may be driven inland by rising water to produce ridges, like the Tabascan beach ridges which attain heights of 1–2 m above the swales. Spacing varies between 20 and 90 m. Conversely it is possible to mistake sub-aerial dunes that have been submerged by a progressive sea-level rise for plunge-point bars. The interpretation of beach bars or related relict features on the shore is complicated by landward movement of barrier islands. The development of any form of bar, barrier or berm is favoured by abundant sediment supplies from offshore or transferred from rivers mouths by longshore drift, as suggested by G. K. Gilbert in 1885. On broad, gently inclined coastal plains, where the occasional high-water level is superimposed on a sustained gradual rise of the relative level of sea in relation to the land, long ridges may develop parallel to the shore and attain a substantial size. Barriers are best developed where the tidal range is relatively low, as on the south-east coast of Australia. A large tidal range, by generating strong ebb and flow currents, prevents accumulation in a gap through a barrier. Some elongate barrier islands may range from a few kilometres to more than 150 km in length, separated from the mainland by a bay, lagooon, or marsh area; the Pleistocene and Holocene barrier islands which border the Georgia coast area are 11–25 km long and 3–6 km wide and separated from the mainland by 6–10 km of salt marsh. The predominance of fine-grained horizontally stratified sediments landward of a barrier may indicate that, provided an origin as a submerged sub-aerial dune appears unlikely, barrier development was essentially an onshore movement, whereas lagoonal conditions would develop landward of a barrier extending longshore only in a comparatively late stage. The main effect of migration on sediment properties is the modification of stratification. The littoral and nearshore neritic sediments deposited along the fronts of barrier islands of the Georgia coast have low depositional slopes, rarely steeper than 6 degrees. Because of postglacial sea-level rise enormous Pleistocene sand ridges with smaller, present-day longitudinal dunes in the inter-ridge areas are a feature of many coasts with gently inclined shelves, particularly in lower latitudes; such a zone on the Senegal coast is about 92 km wide.

So far this discussion of coastal depositional forms, although stressing the complications introduced into the interpretations and classifications by sea-level changes, has not yet included changes alongshore. Deposition on a shore with rocky headlands often occurs as in many sub-aerial environments, with a divergence in the flow lines of the transporting agent, reducing its competence. On the shore this occurs as waves enter the relatively concave sectors of a coastline with most flow lines focused on the headlands by refraction, and the form of bay-head beaches, concave-seaward in plan, is well known. With prevailing winds blowing obliquely onshore, reduction of energy down the

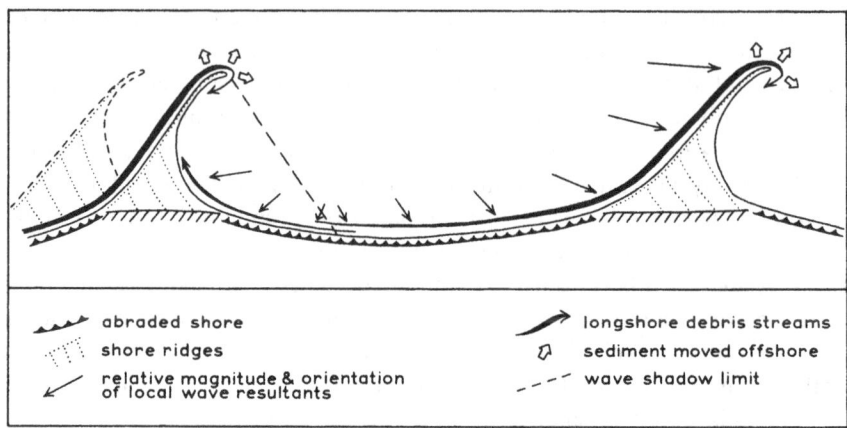

Figure IV.20 Coastal features involved in cuspate spit formation (*from Zenkovitch, 1959*). A previous position of the spit appears on the left of the diagram, as spits may be dynamic depositional landforms (see Chapter IV. D.6).

length of a beach usually results in deposition of the longshore drift load. Where sediment supply is abundant and the declining energy of longshore drift the dominant process in coastal deposition, and the trend of the coastline changes, a depositional ridge will continue the updrift trend of the coastline as a spit, to a point where the depth of water in the bay or estuary is too great for waves to be effective. The strength of currents in or out of an estuary may also be a limiting factor. The five examples on the northern shore of the Asov Sea have lengths of up to 40 km. In areas of active spit formation rates of advance may be by several metres per year, as at the mouth of the Courantyne river on the British Guiana—Surinam border. The spit, exposed to the same north-west drift that dominates the form of the Orinoco delta, advances at 62 m/year. North of Benguela, the Lobito spit advances by 20 m/year.

If the end of a spit is limited and if material continues to arrive there by longshore drift, a hook may form. The direction of the hook will depend on whether the force acting at an angle to the longshore drift is dominantly seaward or onshore. Usually, refraction at their distal ends produces a recurve which doubles back towards the coast. On the northernmost tip of the United States a spit extends 8 km north-eastwards from Barrow and then hooks sharply and extends south-east for 5 km. In width it varies from about 10 m to almost 1050 m at Point Barrow, it is 4·5 m high and 1000–4000 years old. If a recurve joins the coast, the landform, now triangular in plan, is termed a cuspate foreland, which recurves may continue to join on its down-

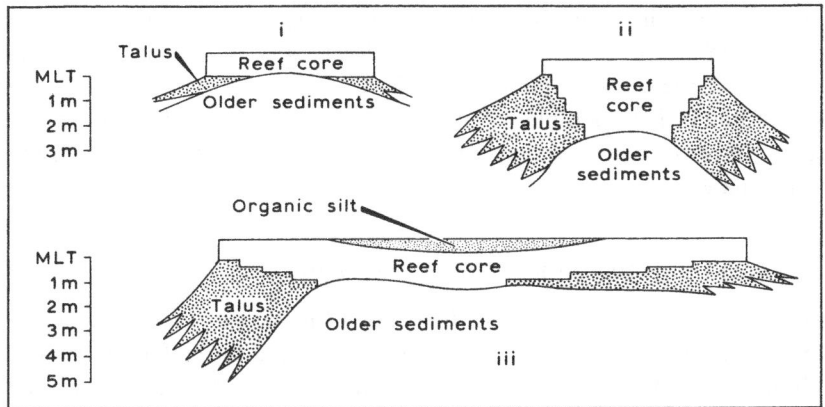

Figure IV.21 Representative reef structures, Ten Thousand Islands, Florida (*from D. E. Shier, 1969,* Bull, geol. Soc. Am., *Vol. 80*). (i) The reef has rapidly expanded to cover a shoal area during a period of little or no sea-level rise. (ii) The reef has built upward and expanded slowly in area during a period of sea-level rise. (iii) Growth on the periphery of this large reef has kept pace with sea-level rise and increased the area of the reef. Poor water circulation and the settling of fine sediment have prevented growth in the central part of the reef.

drift side. The tip enclosing Lake Bumbe on the east shore of Lake Albert turns through 110 degrees. In Dungeness the highest levels consist of unimodal pebbles. 1–1·5 m below the surface the closely packed pebble frame, with larger particles 7·5–40 mm in size, is filled with sands. Bedding planes dip constantly 8–10 degrees in the direction of inclination of the shore face at the time of deposition. Advance by accretion on the eastern side may be about 0·75 m/year.

In the formation of reefs, chemical and biochemical processes produce one of the most distinctive depositional forms of the marine environment. They are found between 25° N and 25° S, particularly in the Pacific and Indian Oceans. Shapes which may be altered by structural weaknesses range from circular to very irregular or ribbon-like. A ring-shaped reef enclosing a lagoon is termed an atoll. Biohermal reefs consist of a growing network of interlinked organisms, mainly corals and coralline algae, and contrast with detrital reefs; fragments of older consolidated reef rock, eroded from the growing biohermal reef on its windward side, are a common constituent of many reefs. The largest detrital deposits accumulate along the leeward side. Substantial vertical thicknesses of reefs, which may be up to 1500 m in deep oceans, are due to the foundering of the submerged volcanic cones which provide the basement from which most deep ocean reefs grow. Reefs contrasting in origin with the atolls of the deep oceans rise from continental shelves in the tropical zone. The Great Barrier Reef off the Queensland coast of Australia is the most extensive example. It consists of an outer barrier made up of a line of innumerable 'ribbon' reefs separated by shallow passages less than 1 km wide, with countless 'platform' reefs between the outer barrier and the mainland.

6. Depositional landforms in eolian environments

Although sand covers only a small portion of most deserts the areas of wind-blown sand deposits can be huge. The area of sand in the Sahara is about 7 million km^2, in the Gobi desert about 2 million km^2. The extent of some vast sand spreads, like the Kalahari in interior South Africa, was considerably greater in the Quaternary and in other areas formerly active tracts of sand movement are now largely stabilized like a 20 × 115-km strip in the Las Bela valley, inland from the West Pakistan coast. Here the maximum depth of sand is 120 m; in parts of the Libyan sand sea depths exceed 300 m. Eolian deposition is also a feature in many localities in humid areas where an unvegetated surface is made up of unconsolidated sands. In sub-aerial parts of deltas, fluvial deposits are reworked and redeposited by wind and identifiable by bedding characteristic of wind-laid sediments, like the fore-dunes built up at the back of a beach or on the crest of a beach ridge. On the plains adjacent to Pleistocene outwash areas in Poland and northern Germany, long belts of wind-deposited sands remain today as strips on the present ground surface up to 30 km in length and as much as 1 km in width. They are characteristically unstratified and not as well-sorted as many wind-blown sands due to their mixture with snow during winter blizzards. These sand covers are usually less than 20 m deep.

The distinctive feature of wind-deposited sands is cross-bedding stratification with some sets inclined at 30 degrees or more if they formed by gravity sliding on a steep lee-side slope. Angles on some beach dune slip-faces, formed by grain by grain deposition and accompanied by partial cementation by salt and water, are stable at 42 degrees. Individual laminae in most dune deposits average 2–3 mm in thickness, but extremes range from 1 to 100 mm, with single lengths of up to 15 m or more. In some unusual localities, but including examples from five continents, where clay includes salt, clay aggregates may be transported, stratified by wind and even form conspicuous elongated ridges. The south part of Bonneville Flats, U.S.A., is one example, the north and west side of Oso Creek estuary near Corpus Christi, Texas, another, and other instances were first recorded in Western Australia, 130 years ago. Downwind from sandy areas on desert margins wind-entrained fine silt-sized particles are eventually deposited as loess mantles. Those widespread on the plateau and upper valley slopes of the Dneiper uplands, being 5–35 m thick, give an indication of the blanketing effect of a loess mantle. As vegetation cover is vital for creating calm ground surface conditions for fine silt to settle permanently, the removal of vegetation has had particularly severe repercussions in these areas. As early as 1922 the resetting rate of dust in Idaho was 10 cm/100 years.

In Antarctica, where wind is the only transporting agent over much of the non-glacier area, dunes demonstrate their azonal characteristics. Those in the Victoria valley are 200–600 m long and 10–15 m high. In some areas the pattern of sand-dunes is not necessarily regular. In the eastern Mojave desert, for instance, where sand-moving winds blow from all directions in addition to that of the prevailing westerly winds, large parallel ridges have developed up to 6 km long and 170 m high, with short reaches diverging by as much as 35 degrees from the prevailing trend. R. P. Sharp (1966) considers that these dunes, now showing little systematic movement, might be relict features dating back several thousands of years, perhaps even 20 000 years. In other areas sand ridges and intervening troughs are remarkably straight and nearly parallel over broad areas. In Egypt they are hundreds of kilometres long; some Australian examples are longer. The remarkably clearly defined corrugations in southern Arabia are at least 30 km long and are commonly 200–300 m from crest to crest. Stabilization of areas of relict dunes has occurred in some sub-humid areas, as in the High Plains or in Rhodesia where ridges may approach 130 km with many lengths unbroken for 50 km. Spacing is about 1·6–2·6 km and occasionally as much as 5 km. Although these dunes have preserved their pattern in plan they are characteristically very low due to sheetwash reducing the ridges to heights a mere 2–4 m above the intervening troughs. The sands here are characteristically very well sorted with a median size of 0·2 mm.

Barchan dunes, due to the smoothness of curvature in their crescentic shape in plan and the regularity of their asymmetric cross-section tapering to two horns downwind (fig. I.4.C), are one of the landforms approaching most closely a geometrically ideal shape. In some instances the length and width are approximately the same and the height about one-tenth this size. Measurements of height, width, and length average 3–12, 40–250, and 60–210 m, respectively, for the larger Salton dunes in Imperial Valley, California, 1–6, 37, and 120 m for dunes in southern Peru, and 30, 400, and 400 m for the maximum sizes in the Egyptian sand sea. Almost invariably slip-faces average 32–33 degrees in inclination, with a maximum of 34·5 degrees just beneath the crest. Barchans occur on desert floors downwind from fields of more amorphous dunes. Downwind summits are aligned *en echelon* with the upwind horns. In southern Peru, where the mean density of barchans is 1/7·1 ha, their distribution is relatively even. There is sometimes a tendency for barchan field patterns to diverge downwind perhaps related to reduced velocity of the wind due to lateral spreading of the current after wind-funnelling through a constriction. The downwind spread forms a pattern similar to that of some drumlin fields. Only coarser material remains in the horns and near the crest, the sites from which material is most readily removed.

Seif dunes are remarkable single continuous ridges which run straight across relatively featureless desert areas. Leeward they dwindle to a sharp point and appear to follow the direction of the long-period wind regime. According to local conditions, however, the overall plan of seif-dune patterns may vary considerably (fig. IV.22). Individual dunes may be perhaps 60–100 km in length. In the Egyptian sand sea many dune summits are 100 m high and in southern Iran over 200 m high. The transverse width at their base is about six times their height. In cross-section the seif dune stratification reflects the important distinction between deposition by accretion, at

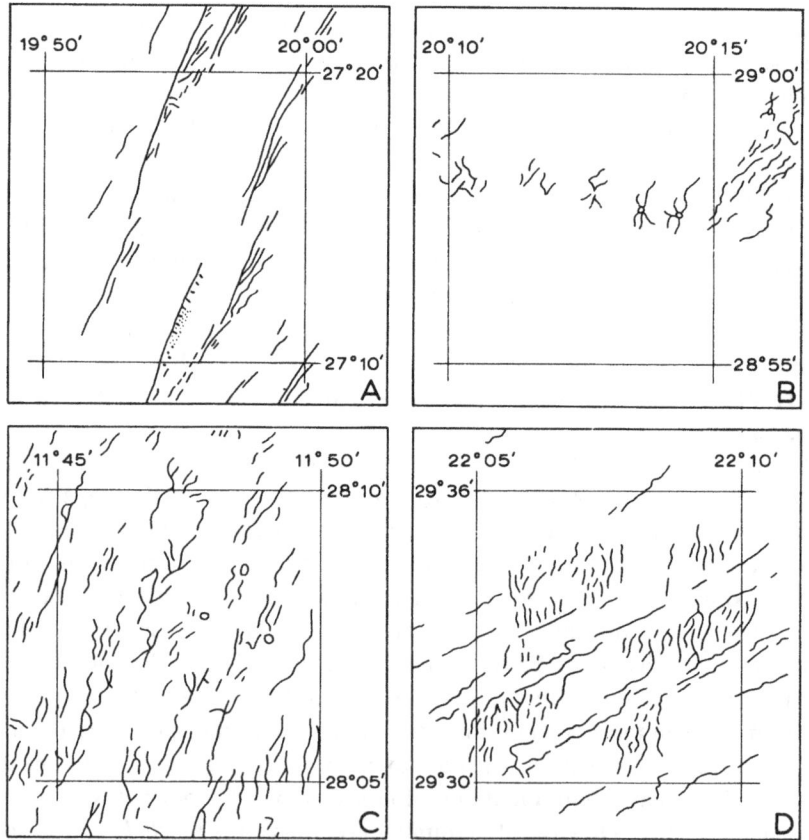

Figure IV.22 Some seif dune trends (*from McKee and Tibbitts, 1964*). The patterns were traced from air photos of Libya by L. C. Conant.

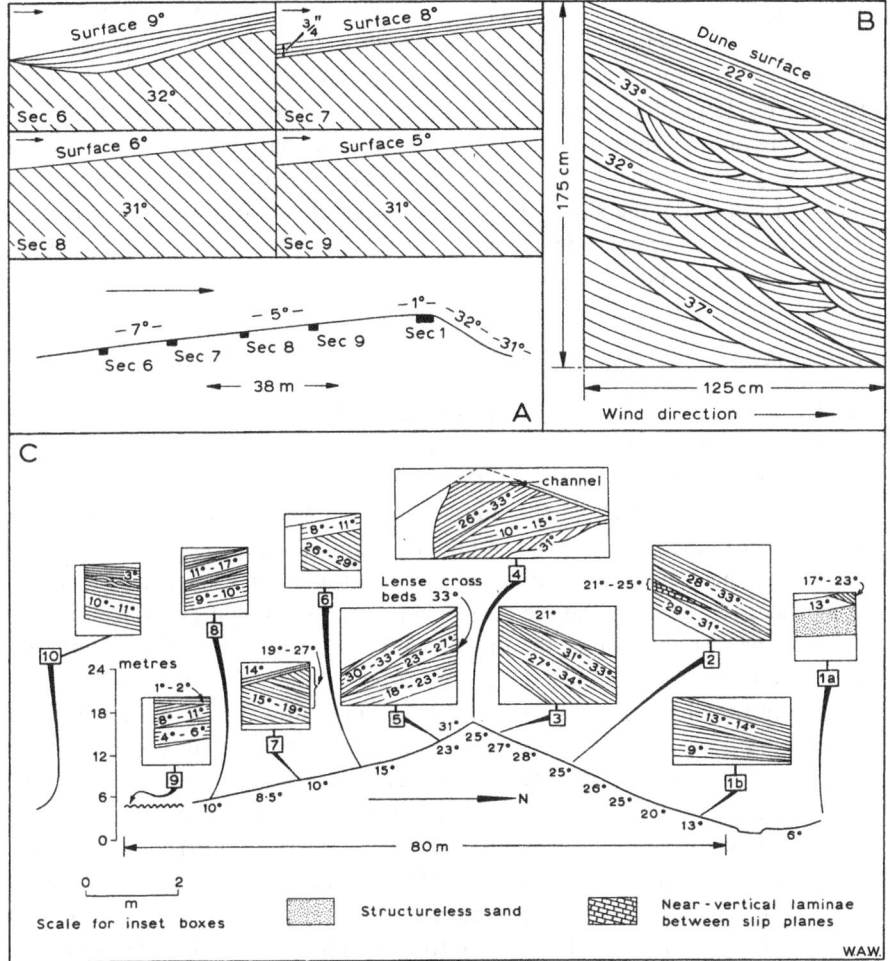

Figure IV.23 Stratification in sand-dunes.

A. Profile of a barchan dune near Leupp, Arizona, taken through the crest in direction of dominant wind (*after E. D. McKee, 1957*, Bull. Am. Ass. Petrol. Geols., *Vol. 41*).

B. Cross-section of a high transverse dune, Mustang Island, Texas, anchored by vegetation (*from McKee, 1957*).

C. Cross-section of a seif dune, west of Sebhah, Libya (*from McKee and Tibbitts, 1964*).

angles of 15 degrees or less, and by gravity flow creating dips of 24–34 degrees. These inclinations may flatten out near the base of the deposit and bottom zones may represent the accumulation of relatively coarse sand from the nearby desert floor (fig. IV.23.C). The slip-face, instead of lying transverse to the prevailing wind as in a barchan dune, runs parallel with it and on the side away from that of the prevailing cross-wind. In the case of the symmetrical seif dune, the slip-face is a temporary phenomenon which is formed and reformed on either side according to the direction from which the cross-wind happens to blow.

7. Landforms due to glacial deposition

A characteristic of glacial deposition, particularly of basal tills, is the large amount of clay involved, for many workers the product of interglacial and preglacial chemical weathering rather than the result of bedrock erosion by the ice itself. Nonetheless, the amount of clay that accumulates in glacial deposits still poses the problem of understanding the depositional mechanism involved. One suggestion is that intergranular melt water, arising from frictional heat of abrasion and ice-movement processes might possibly facilitate the downward migration of the fine products and their deposition as the sub-glacial matrix of the boulder clay. Preferred pebble orientations are usually aligned in the direction of ice movement, and the flow of a fluid or semi-fluid till appears necessary for the production of such patterns. Thus in addition to the resemblance of the material of sub-aerial mudflow to that of sub-glacial tills, the condition of two materials at the time of deposition may also have certain similarities. Most of the development of preferred orientation in clay compacted artificially under pressures of $0–100 \, kg/cm^2$ takes place very early, at pressures near $1 \, kg/cm^2$, the most critical factor possibly being the amount of water held by the clay. In addition to lack of any precise knowledge on the mechanisms by which ice lays down its load, there are problems of differentiating and interpreting landforms produced by glacial deposition. Distinctions between forms in the field are often more difficult to grasp than between idealized-type examples. Generalization about typical dimensions and shapes is not easy, and the description of some ice-pushed ridges as end moraines is just one instance where mistaken identifications have arisen.

The main contribution of ground moraine is to the appearance of the land surface. The smoothing effect is due more to the lodgement or basal till which was plastered on to the bedrock contours. The coarser ablation till settles on to the ground surface with little regard for the relief beneath the ice and with thicknesses of $1 \cdot 5–3 \, m$ is usually an inconsequential aspect of the shapes of morainic terrain. In some areas an additional factor in the laying down of till is sub-glacial processes beneath active ice which produce aligned ridges, and grooves parallel to ice movement may corrugate the till surface. In higher mountainous areas of Scotland, such features may occur on cirque and valley floors, ranging in height from 0·3 to 6·0 m and in length from 20 to 300 m.

An end moraine may be one of the most clearly defined landforms made up of till, usually regarded as marking an extreme position reached by a glacier. It is not simple to decide if the ice remains in this position for any length of time, maintained by an equilibrium between accumulation and ablation and continuously supplies a load to pile up as a ridge of increasing size at its margin; whether ice could exert pre-existing force to push material forward or would tend to override it—if the inward-facing or proximal slope is still as steep as 30–35 degrees, push might be indicated whereas a long proximal slope might indicate overriding; or to what extent an end moraine once formed provides an outer limit for subsequent minor readvances. Regardless of exact mode of origin, a characteristic usually observed in well-defined end moraines is a contrast in morainic material between the inner and outer slopes. On the inward-facing slope the material is similar to ground moraine. It is unsorted and with rare or indistinct stratifications, but may include irregular lenses of well-washed pebbles laid in melt-water channels cut during previous retreat stages of the ice. On the gentler outward-inclined slope the former action of running water, including sheetwash, means that bedding is common, and fragments are more rounded and better sorted. Sizes may decrease down the slope. An end moraine is not necessarily elongate and narrow but may be a complicated mosaic of round and broad hummocks, perhaps 150–300 m in length, and lacking any systematic pattern. Of various possible origins for such complex end moraines, deposition by stagnant ice is often favoured. Nor is a distinct end moraine necessarily continuous but when traced laterally may become more poorly defined or may even temporarily disappear. For instance, the Illinoian drift sheet has no continuous end moraine, and the margin of many drift areas is without definite morainic ridges. If the discontinuity is due to washing of till by melt water during and after deposition, rather than non-deposition, a trace of large boulders without fine material may remain. In valleys end moraines are usually higher and broader in cross-section on or near valley floors, but they often extend a hundred or so metres up valley slopes. Amplitudes of 400 m have been reported from Greenland. In general, heights are very variable. Where several are closely spaced, individual heights tend to be less, as along sections of the Labrador coast where the heights are 4·5–9 m. Some end moraines in north-west Scotland are up to 40 m high, some in Iceland are about 20 m high. A distinctive type of morainic deposition is the cross-valley, washboard, or de Geer moraines which occur where deglaciation took place in a subaqueous environment like south-central Sweden. In Baffin Island they are 3–20 m high. In 70 km of the Isortoq valley, Labrador, there are at least 2000 moraines spaced on average

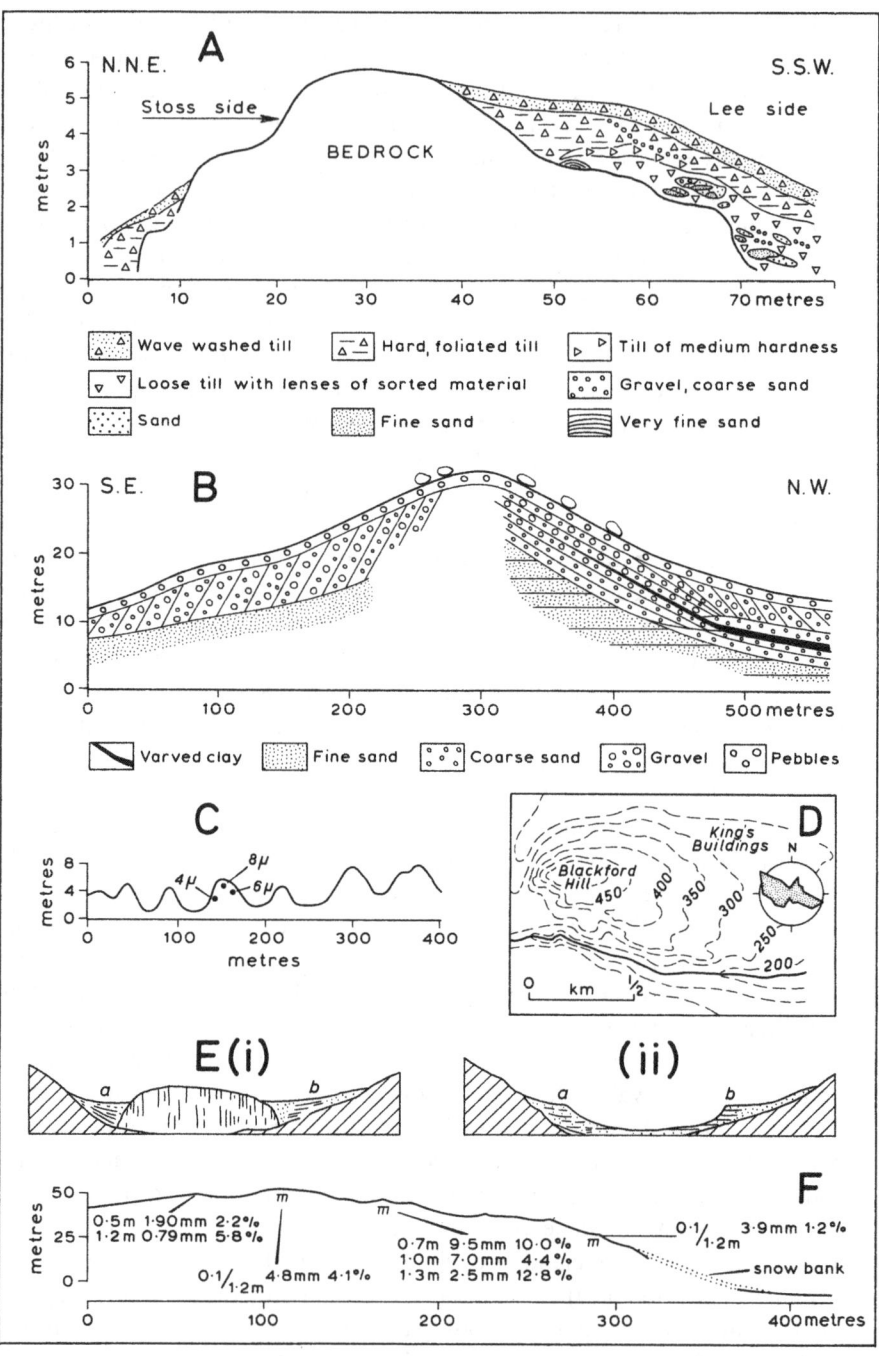

A

N.N.E.   S.S.W.

Stoss side   BEDROCK   Lee side

Wave washed till   Hard, foliated till   Till of medium hardness

Loose till with lenses of sorted material   Gravel, coarse sand

Sand   Fine sand   Very fine sand

B

S.E.   N.W.

Varved clay   Fine sand   Coarse sand   Gravel   Pebbles

C

8μ

4μ 6μ

metres

D

King's Buildings

Blackford Hill   450   400   350   300   250

200

0   km   ½

E(i)   **(ii)**

a   b   a   b

F

0.5m 1.90mm 2.2%
1.2m 0.79mm 5.8%

$0.1/_{1.2m}$ 3.9mm 1.2%

0.7m 9.5mm 10.0%
1.0m 7.0mm 4.4%
1.3m 2.5mm 12.8%

$0.1/_{1.2m}$ 4.8mm 4.1%

snow bank

50 m apart. In Manitoba examples are 5 m high, about 100–150 m apart and may be 1·5 km in length. J. T. Andrews (1966) suggests that cross-valley or washboard moraines may be formed by the squeezing of plastic water-soaked ground moraine into basal crevasses or to the ice-front at the edge of a proglacial lake. The weight of the ice itself could exert the necessary pressure, but some slight ice-flow, possible along basal thrust planes might have contributed in certain places.

The terminus of a stagnant glacier is quite different from that of a receding glacier because a large loss of volume results from vertical shrinkage rather than from terminal recession and because material is no longer transported to the ice-front. The surface becomes covered with englacial debris and a chaotic surface of stagnating ice with debris knolls and ridges separated by ponds and streams emerges. During stagnant stages of glacial lobes linear ridges may develop superficially moraine-like in appearance but composed essentially of fluvioglacial material. Those on the Plains of North America range up to 10 m in height, may be 7·5–100 m broad and from a few metres to 10 km or more long. The most prominent features of these ice-disintegration ridges is that they are essentially straight and lie normal, parallel or at an angle of 45 degrees to the direction of flow and therefore suggest a genetic relationship to crevasses. The ridges may be related to the filling of a massive crevasse

---

Figure IV.24 Landforms due to glacial deposition.

A. Section through a typical lee-side lense of till from central Sweden (*after H. Möller*, 1960, Geol Fören. Stockholm Förh., *Vol. 82*). On the stoss side till is hard and foliated. On the lee-side a loose till, with many lenses is covered with foliated till. A layer of very fine sand often occurs between till and bedrock.

B. Cross-section through an ice-marginal ridge (Salpausselkä I) in southwest Finland (*after K. Virkkala, 1963*, Bull. Comm. geol. Finlande, *No. 210*), observed between Ojakkala and Otalampi. It seems probable that parts of the ridge were deposited in lakes at the edge of the continental ice-sheet while other parts were deposited sub-aerially, according to local differences in relief.

C. Profile levelled across some small moraines in Little Cataraqui Creek, southeastern Ontario (*after O. H. Løken and E. J. Leahy, 1964*, Canadian Geogr., *Vol. 8*). Figures indicate median particle size in microns.

D. Plan of Blackford Hill 'crag-and-tail', Edinburgh, with till fabric (*after R. P. Kirby, 1969*, Geografiska Annaler, *Vol. 51A*). Compare with A, above.

E. The conversion of gravel deposits, *a* and *b*, laid down against ice, into kame terraces after ice-melt (*from T. F. Jamieson, 1874*, Quart. Jour. geol. Soc., *Vol. 30*). The removal of lateral support often induces slumps in the terraces.

F. Cross-section through an ice-cored end-moraine at Gràsubreen, Norway (*after Østrem, 1964*). For each group of figures the three columns represent, from left to right, depth at which soil sample was recovered, median particle size in millimetres, and moisture content as a percentage of dry soil.

complex and associated knolls 20–30 m high to the infilling of melt-holes at crevasse intersections.

In many glaciated lowlands, particularly where ice recession was rapid, like much of lowland Sweden, eskers are the most common type of depositional landform. Whether they are better regarded as ice-contact features or as essentially fluvioglacial in origin may depend on local conditions. They consist of well-worn, sorted, and stratified material with pebbles in the 5–20-cm range and often containing maximum sizes larger than a metre, although some eskers consist entirely of sand. The inclination of bedding is usually away from the axis of the esker at dips of 10–20 degrees or even coinciding with the surface slope. Like crevasse fillings eskers can ascend reverse gradients and the disregard of their trace for underlying relief is a noted feature. Unlike the linear crevasse fillings, eskers are characteristically sinuous in plan and always narrow in cross-section, 3–15 m in height and up to 50 m as a maximum. Angle of slope of esker sides is usually 25–30 degrees. An esker is not necessarily a continuous ridge. It may have gaps in some places and in others expand into a form with a more bulbous ground plan, the beaded esker, with many kettle holes in the wider section. Lengths of single continuous ridges of 1·5 km are reported from both the Kennebec river area, Maine, and the proglacial area of the Casement glacier in Alaska. The fact that meltwater under hydrostatic pressure could flow up reverse gradients has favoured an explanation of eskers as the deposits of the flow of sub-glacial streams with competence reduced as they emerge from the confines of an open crevasse or tunnel. Tunnels are a feature of stagnating ice as actively moving ice would tend to close them up. However, in some areas the beginnings of eskers appear to develop on the surface of the ice. The preservation of steep sides of eskers is a somewhat enigmatic feature in easily reworked sands and gravel. Perhaps the binding effect of roots of the vegetation which rapidly colonizes the debris could be sufficient to preserve even the form of a superglacial debris ridge. In many areas like southern Norway and the Alps forests grow close to the glaciers. R. J. Price (1966) records the rapid and considerable colonisation by alders of eskers at the Casement glacier ice-front.

A feature of glacial deposits when chiefly composed of water-sorted gravel and sand is that they may have been deposited against ice or on ice. When the ice melts away 'ice-contact' landforms remain. As with the melting of buried ice, the loss of lateral support in ice-contact landforms leads to collapse and the associated deformation and contortion of structures. Mounds produced in this way are sometimes described as kames. Material deposited in an ice-marginal lake and banked against the ice is termed a kame terrace, relict forms being distinguishable from fluvial terraces by their irregular altitude and irregular surface. In higher latitudes the coarser debris may similarly accumulate to form a block terrace.

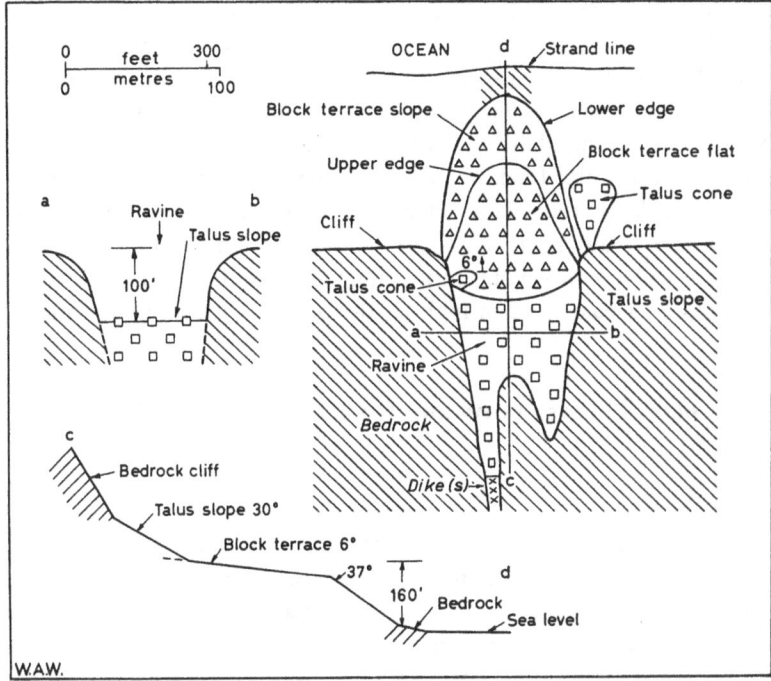

Figure IV.25 Block terrace, Marguerite Bay area, Antarctica (*from Nichols, 1960*).

Another feature of glacial deposits is that in areas adjacent to present-day glaciers they may still contain blocks of buried ice and that in areas of Pleistocene glacial deposition the subsequent melting of such buried ice blocks has given rise to a distinctively irregular land surface due to numerous 'kettle-holes' depressions created. Ice may also persist in the cores of some moraines as in the Jotunheim area of Norway where the distribution appears to be related to the altitude of the temporary snowline. A characteristic of these ice-cored moraines is a rounded surface and outline with minor ridges superimposed on the top of the larger feature.

In contrast to ice-marginal features the emplacement of drumlins appears to be the work of active ice, but the nature of the mechanisms involved is obscure. Whatever they may be, drumlins appear to be favoured by flatter, more open terrain where ice-flow is spreading out, possibly losing some competence in developing longitudinal crevasses in the process. Drumlins are low elongate domes usually strikingly symmetrical in appearance. They are not usually more than 50 m high or 0·8 km in length. Widths vary and some drumlins are so narrow as to be virtually linear ridges. There are many contrasts in the ratios of the three main dimensions from one drumlin field to another. The symmetrical drumlins in the Wadena field, Minnesota, are only

6–15 m high, yet are 1·5–6 km long and even 11 km on occasions, widths are 0·1–0·6 km. More typical are the Glasgow drumlins which are about 30 m high and 0·6 km long. Those on North Uist, are as small as 45 m long and 8 m high. Forms may change within one drumlin field. Near Long Prairie, in the Wadena field, the characteristic elongated form gives way to a broader oval or almost circular form towards the terminus of the former ice-sheet. In some drumlin fields there is a tendency for two size modes; many drumlins near Syracuse, New York State, are 30–45 m high, 600–900 m long, and 200–300 m wide, with smaller more elongate drumlins commonly occupying the intervening areas, 3–12 m high, 250–750 m long, and 60–90 m wide.

D. Inter-relationships between Rock Breakdown, Transport, and Deposition

1. Time-lags

A fundamental characteristic of inter-relationships between landforms, land-forming processes and time is the time-lag between some antecedent perturbation and subsequent adjustments to the changed condition. This makes the view of simultaneous interactions unrealistic in many geomorphological situations and the multiple correlation of several associated factors often unsatisfactory. For instance, even on the shortest time scales, a relatively high level of unassigned variation in the multiple correlations in the beach environment of Brodie Island, Carolina, was thought to be partly due to the time-lag between the changes in process intensity and resulting beach-face adjustments. One complication is that shoreward flow transports largest and heaviest material to high beach levels early in the declining phase of wave heights. In cool environments delayed thawing explains some apparently anomalous late summer or autumn features, like the occurrence of rockfalls observed at that time of year in Scandinavia and elsewhere (Table IV.9). In streams, although correlations between discharge and sediment load are always very highly correlated because of the statistical effect of sediment load being itself the product of sediment concentration and discharge, the higher discharges, despite their greater competence, may be only poorly correlated with sediment concentrations. One reason for this is that quite often, particularly in spring in temperate regions, the maximum sediment concentration precedes the peak of the flood. F. Hjulström (1935) attributed this to the silt reaching rivers from the interfluves more rapidly than the water-level in the river rises. A more likely explanation is that the first sediment entrained during a rising stage is that loose channel bed material which was deposited during the preceding falling stage. During summer the lag between sediment concentration peak and flood peak may be less marked as the interfluves and banks are better protected by vegetation. In glacial outwash some observations indicate that the peak of silt concentration may be 2–3 hours before the daily runoff maximum. Similar time-lags between the initial loosening of material and its subsequent removal may span millennia, where the weathering of bedrock is involved. Much of the load of many mid-latitude streams is cut mainly from river banks of glaciofluvial material and till, which before its movement by ice might have taken millions of years of weathering to become detached from solid bedrock.

Table IV.9 Frequency of mass movements in Norway (*from Rapp, 1960 and O. Holtedahl, 1960,* The geology of Norway).

	J	F	M	A	M	J	J	A	S	O	N	D
Frequency of rockfalls near Bergen, 1920–58	13	11	11	38	9	8	11	6	3	13	7	10
Frequency of quick-clay slides, southern Norway, 1900–60	5	1	4	9	9	5	4	3	4	9	6	5

The time taken for changes to be transmitted through the ground is another significant source of a lag between an antecedent condition and a subsequent change. Because of slowness of penetration of thermal waves, residual permafrost layers can remain for a long time. The characteristics of spring water on emergence may reflect the influence of soil conditions at some previous time when the water entered the soil. Fig. IV.26.A shows how water temperature and amount of dissolved calcium carbonate, the latter reflecting the temperature-controlled biological production of soil carbon dioxide, in a cave pool show an early autumn peak and general seasonal lag of about 2 months behind the controlling temperatures that would exist at the surface.

There is also the possibility of a geographical distribution in the degree to which some points lag behind the time of change at others, even though the lagged responses may be a reflection of the same perturbation. In some cases runoff peaks from tributaries on either side of a valley do not coincide in time after precipitation. On a broader scale, it is possible that the date of the transition from Hypsithermal to late postglacial conditions in north-west North America lagged according to latitude, due to the greater influence in the north of the south-flowing cold polar air. This change may have occurred about 4000 years ago along the western Gulf of Alaska, 3500 years ago in south-east Alaska, 3000 years ago in southern British Columbia, and 2500 years ago in California. There is also the possibility of different responses to the same perturbation lagging one behind the other; one of the major difficulties in correlating glaciations with pluvial episodes is the possibility of pluviation lagging behind glaciation.

Finally, as rates of tectonic movements may greatly exceed erosional rates

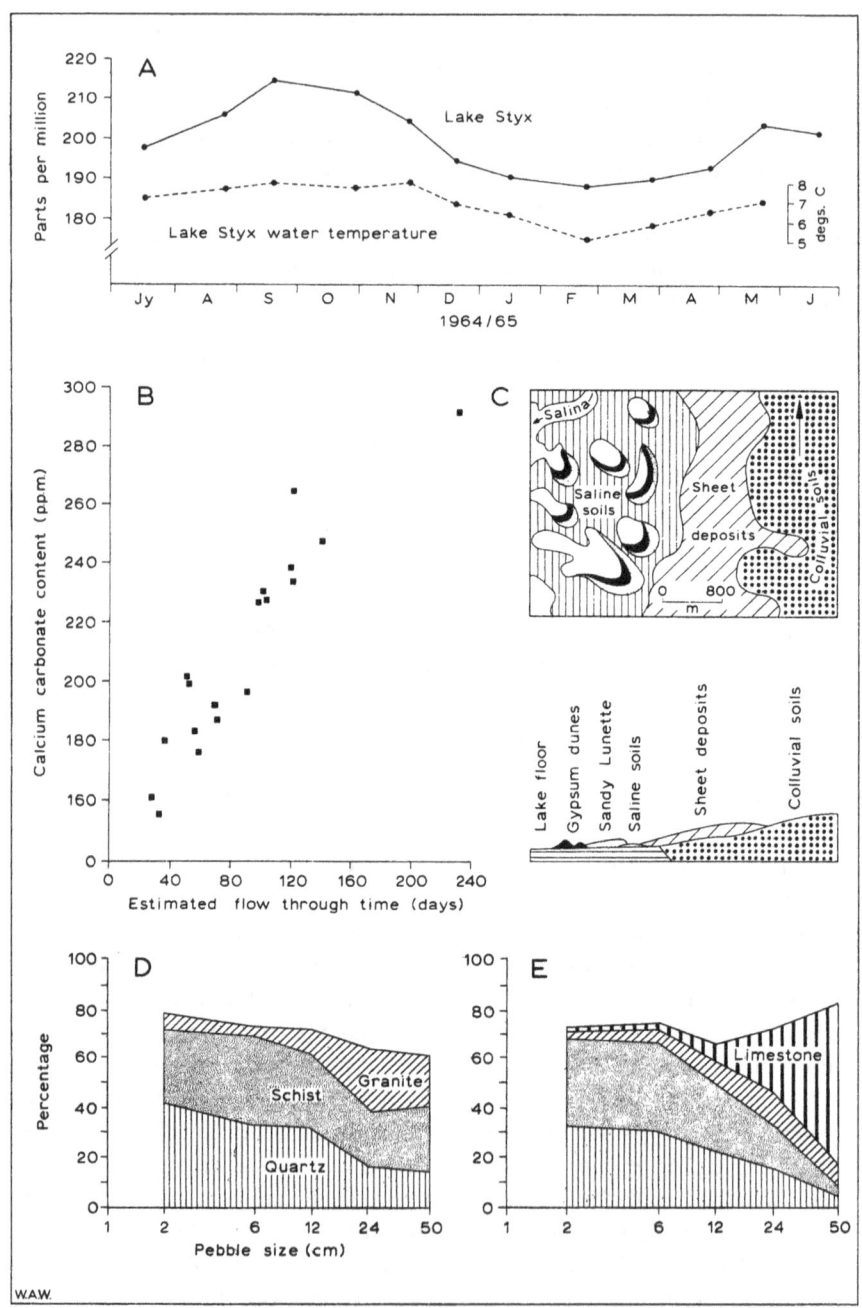

Figure IV.26 Lagged effects in time and space.
A. Calcium carbonate content and water temperature of Lake Styx, Peak Cavern, Derbyshire, lagging two months behind air temperatures (*from A. F. Pitty, 1968,* Trans. Cave. Res. Group Gt. Britain, *Vol. 10*).

and with an approach towards some equilibrium therefore unlikely, much of the broader-scale erosional history of landforms reflects the comparatively slow response of erosional processes lagging behind tectonic movements.

Therefore in landform development there is a tendency for rock breakdown, transportation, and deposition to succeed each other in distinct phases, with measurable time-lags between the antecedent conditions and the subsequent change. This is no more than a tendency and the definition of what would constitute a measurable time-lag varies according to the interactions under investigation and the subjective judgement, if not merely the taste, of the investigator. Therefore the consideration of the interactions between these phases and with landform in the following section is as much a change in viewpoint as a specific consideration of those areas and environments where the interaction between rock breakdown, transport, deposition, and landform is essentially inseparable.

There are five contexts in which it is essential to consider the inter-relationships between two or more of these four variables. The first is situations where certain local features put inevitable and distinctive constraints on the way in which some response might operate. Secondly, there are co-variances where two or more phenomena occur together in either space or time but where the mechanism of their interaction is often obscure. Thirdly, there is the rearrangement of weathered products which may involve essentially a reworking or recombination of weathered products *in situ*. Fourthly, there are dynamic depositional landforms, and lastly, there are situations where the characteristics of some previously weathered material appear to influence the way in which subsequent weathering processes operate.

2. Influence of source

In many ways rock breakdown influences transportation, deposition and further rock breakdown simply because the results of its action are the characteristic products from distinctive environments. The supply of abrasives is an obvious example. For instance, in northern Alaska and Saskatchewan, areas of sand-blast striations on outcropping sandstones are

B. Greater lengths of estimated flow-through time of karst water in the southern and central Pennines leading to greater solute concentration (*from A. F. Pitty, 1968*, Nature, *Vol. 217*).

C. Deposits downwind from the salt lake system at Hines Hill, Western Australia (*after Bettenay, 1962*). The cross-section (*below*) is not to scale.

D and E. Petrographic changes downstream from changes in bedrock (*from Tricart, 1965*). These observations on the Ceze river by J. Capolini show that the persistence of granite near the ancient massif at Peyremale, D, falls off downstream at St Ambroix, E, as do the schists and quartz.

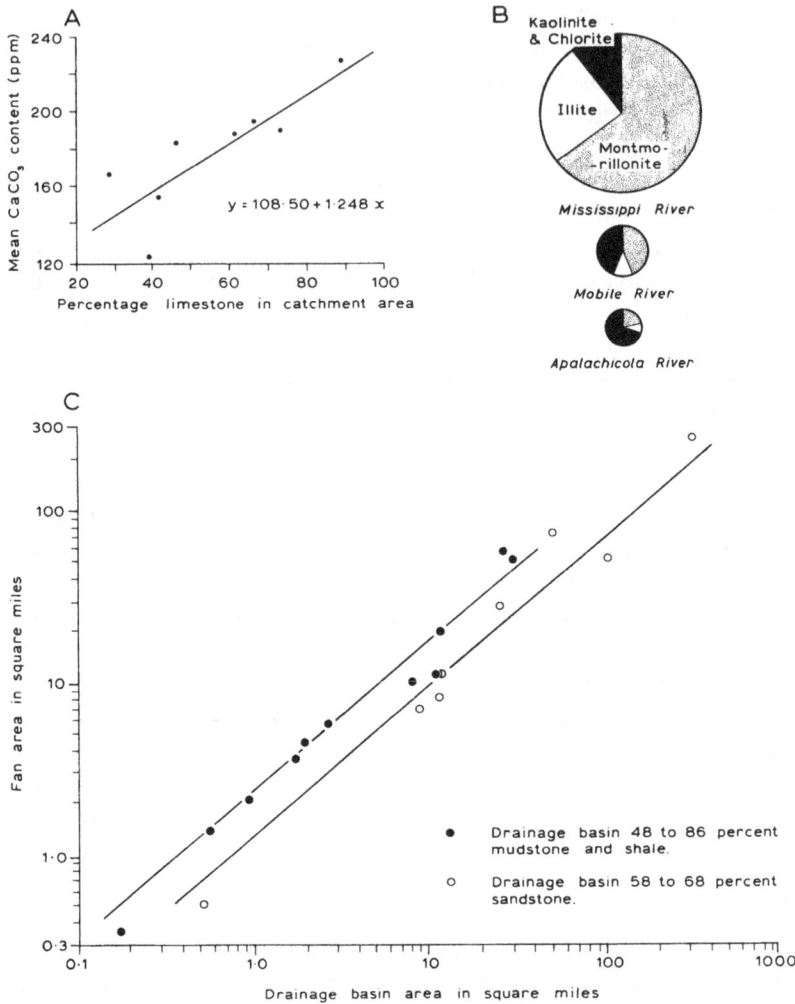

Figure IV.27 Influence of drainage basin characteristics on geomorphological responses.

A. Increased concentration of calcium carbonate in streams draining from catchments made up of greater proportions of limestone (*from Pitty, 1968*) as observed in the southern Pennines.

B. Composition of three suspended clay loads reflecting an easterly increase in kaolinite relative to montmorillonite in the soils of their river basins (*from G. M. Griffen, 1962*, Bull. geol. Soc. Am., *Vol. 73*).

C. Greater alluvial fan areas at the outlets from larger drainage basins (*from Bull, 1964*), observed in western Fresno County, California.

well displayed perhaps because of the abundant abrasive sand in glacial deposits about 60 km to the north and north-east. In Sweden in sediments on the bottom of the Göta älv river there are erosional grooves downstream from a moraine, whereas none exists upstream where there is almost no bedload. The paucity of material coarser than sand in tropical rivers, due to rapid chemical decomposition and the limitations imposed on erosional activity in streams, has been much emphasized in recent years. In fact more generally, and in a negative sense, one of the most significant controls on transportation is that of very slow weathering providing little load. Rivers draining from the west and south slopes of the Ukrainian Carpathians carry little suspended sediment, despite their turbulent flood regime because few fines are available for transport. In Western Australia, gypsum dunes, lunettes, and sheet deposits are almost entirely restricted to semi-arid areas where the extensive bare desiccation-cracked surfaces of salt lakes are ideally suited to wind deflation and provide a ready source of gypsum, sand, silt, and clay. Along the eastern coast of South America, the highest concentrations of kaolinite occurs at the mouth of the small rivers which drain in the coastal areas where lateritic weathering is well developed. There is relatively less kaolinite in the clays at the mouths of the large rivers which drain from mountainous Andean areas where chemical weathering is less significant in processes of rock breakdown, and the absence of a pebble strand is typical of this as of other tropical coasts. From fig. IV.27.B it is apparent that the dominantly montmorillonite clay from the Mississippi river will dominate the clay mineralogy of the eastern Gulf of Mexico, whereas the Mobile and Apalachicola rivers will introduce only local modifications.

The effectiveness of transportation depends on the degree of prior size reduction and on the properties of the parent material which affects the internal friction of debris. Deeply weathered or poorly consolidated material if falling from a free face might disintegrate on impact and like the 'rotten' rockfalls in south-west United States add little coarse material to lower slopes. The development of screes, and as a result, that of scree-controlled bedrock features too, is in consequence reduced. Conversely, in cooler environments forms like rock glaciers develop only where rocks weather to blocks. In the central Alaska Range, rock glaciers developed in granodiorite are absent on higher walls of the canyon where schist outcrops. The proportion of mudflow material in and on alluvial fans varies widely and may be absent on one fan yet predominant on an adjacent one. These differences depend on the control of local lithology which limits the amount of fine material that might be available for transport. Initial size is also a factor in the amount of subsequent chemical solution of weathered particles, the rate of attainment of equilibrium being theoretically strongly dependent on particle size. A final point, so obvious that departures are perhaps more intriguing than examples following the general rule, is that the size of a source area will be a dominant factor on the size of depositional landforms accumulating down current. Thus, on the shores of Lake Albert, the size of spits on the Bukobya coast is small in comparison with those at Kaiso and Tonga, possibly because of the short length of coast from which they can draw loose material. More clearly established is the correlation between size of alluvial fans and source area in south-west United States.

3. Influence of transportation on rate of supply

It is sometimes assumed or implied that in certain circumstances a transportation system could, according to the amount of material it removes, have some controlling effect on the rate of supply. One of the main reasons why approaches to this state of balance are often not realized is that more than one transporting agent, in conjunction or in sequence may be involved in the movement of weathered debris. This situation is most clearly seen in sub-humid and in semi-arid areas where inter-relationships between wind and water transport may be highly unpredictable. On the one hand river alluvium may provide a source for wind-blown material as appears to have been the case in many parts of the ancient Mississippi valley. Alternatively wind in semi-arid areas may supply to streams huge quantities of sand. In fact, supplies might be voluminous enough to fill up valleys and cause stream diversions. If intermittent streams flow infrequently, sand may continue wind-blown across valleys hundreds of metres deep and several kilometres wide. In the south-west Palouse, Washington–Oregon, there is a clear example of how a balance might be struck with smaller streams tending to be aligned with the lineations of wind-blown deposits, whereas the more sizeable streams, unrelated to these trends appear to have maintained their courses. In semi-arid western Australia, 'Wanderrie' country covers many thousands of square kilometres, consisting of alluvial flats alternating with low banks made up of wind-blown sand from the drainage ways and aligned longitudinally with the prevailing wind direction.

   More generally, it should be noted that the size of particles or volume of material supplied to a river from its valley-side slopes may not be determined by the river but be semi-independently produced by the rocks, relief, vegetation, and other properties of a drainage basin. This is seen to be so most clearly in the extreme case where the stream is incised into a rock-walled gorge in the valley floor.

   Similarly concepts of balance do not apply if a river introduces a huge load into a nearshore environment, such as along the south-western Louisiana coast, which the available energy is unable to move. Beach sands in this area have in consequence the textural parameters of a fluvial sediment.

4. Covariances between phenomena

One of the more obvious examples of concurrent processes is the rock breakdown resulting in changes in size and shape of debris during transportation. Results obtained by simulating these conditions in an abrasion mill are difficult to compare with observations from natural situations, partly because of the paucity of the latter, particularly from high-gradient mountain streams, partly because the initial characteristics of the debris are determined by characteristics of the drainage basin and its environment, and partly because of the trapping or the influx of debris, from slope, channel floor, or tributary, during transport. In some cases lakes trap all the bedload and part of the suspended load of a stream and in many cases tributaries modifying flow regimes in the main channel may introduce large volumes of additional sediment load with different characteristics to that entrained in the main channel upstream from the tributary junction. On a local scale there is sometimes an inverse relationship between debris size and channel width. In a stream an ideal case of a steady exponential decrease in particle size has rarely been observed. Conversely in the south Canadian river, textural parameters and composition remain essentially constant over a downstream distance of 1000 km. Another discrepancy between laboratory and field data is that, where sizes do decrease downstream, the rate of reduction is greater than might be predicted. This probably reflects a lag-deposit effect, with selective transportation of finer materials adding to the impression of size reduction of material downstream. The separation of these two processes contributing to size reduction downstream defies generalization. In his study of three streams on the eastern flank of the Black Hills, J. Plumley (1948) concluded that only about 25 per cent of size reduction was due to abrasion and attributed the remaining 75 per cent to selective transportation. To some extent the formation of traction clogs in the upper reaches of rivers causes any excess of saltatory particles to be passed on downstream. Also winter frost action on sediments exposed at low water flows and chemical weathering, such as the penetration of micro-fissures by hydroxides, are probably more significant than mechanical collisions under water. In semi-arid areas salt weathering may be significant in damp localities, like the unweathered gravel from alluvial fans which disintegrates in the saline marshes surrounding playas in northern Iran. Whatever may be the contribution of mechanical collisions to the breakdown of coarser debris in transport, sand grains smaller than 1 mm have few mechanical markings. The abrasion involved in an average distance of transport in one sedimentary cycle is apparently insufficient to account for any significant rounding or size decrease. However, as with the breakdown of stationary fragments, there is the possibility of mechanical and chemical processes working in conjunction. An important mechanism may be

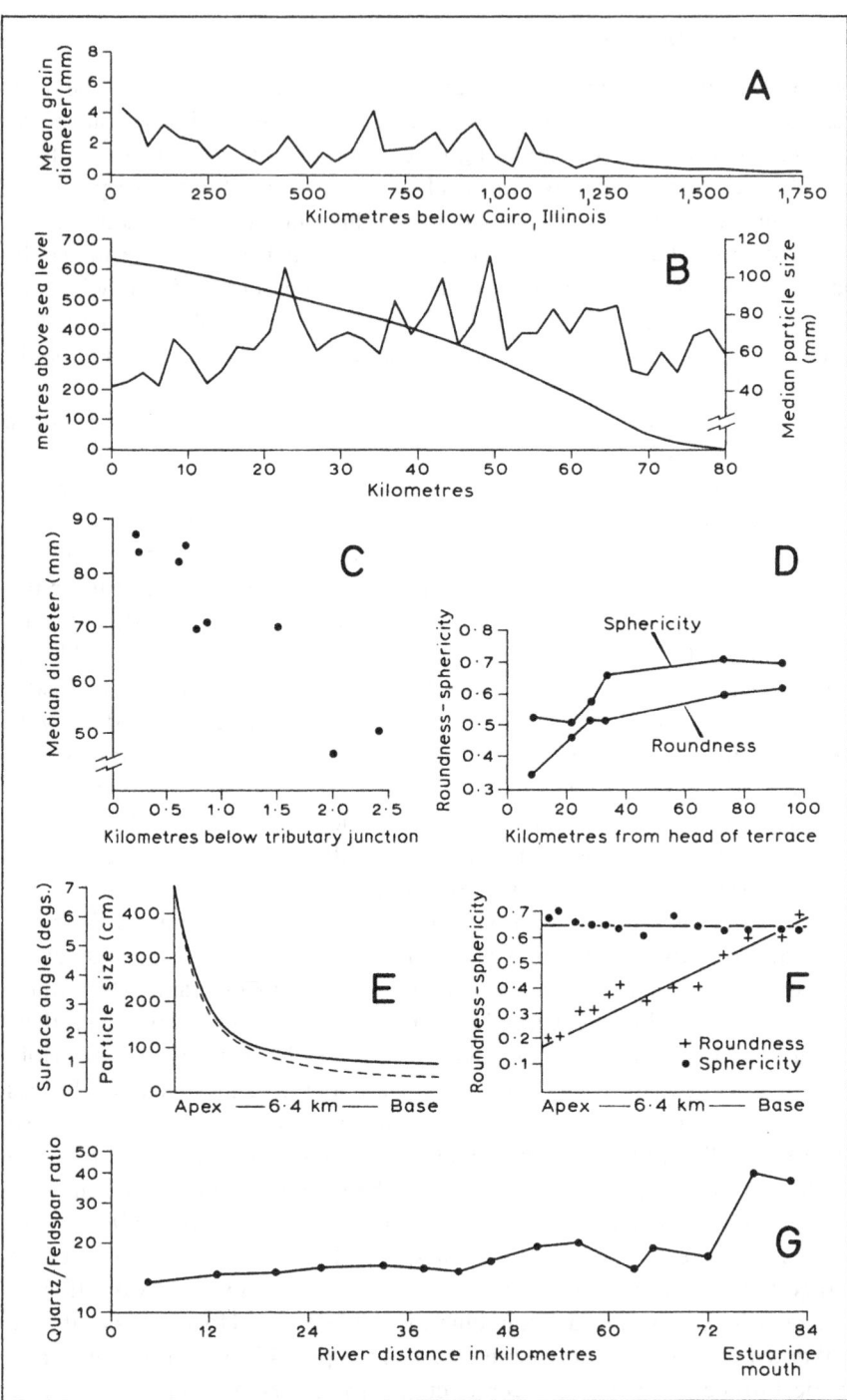

that of abrasion solution which involves disordering and hydrating the surface layers of quartz by continued impact of other grains to produce a finely divided amorphous silica, and results in the probable rounding of particles. Similarly, solution may follow abrasion, as on the hydrated silica crystals and newly exposed metal cations on the separating surfaces of silicate particles in the rock flour suspended in glacial melt waters. Hydrogen ions are immediately available for replacement of the newly exposed cations on the broken surfaces of the silicate minerals. Conversely, in sand-laden swash, size reduction is predominantly an abrasional process and any rounding of river gravels might similarly be the result of wet sand-blasting. The importance of crushing and abrasion of stones during glacial transport has long been recognized. Much of the abrasion occurs on flat surfaces originally related to stratification, jointing, and fracture planes, but amounts bevelled off corners and edges are more significant.

Downstream there may be progressive changes in mineralogical composition, with climatic conditions playing an important part in determining the initial composition as well as their progressive changes during transit. As with size changes, composition changes below tributary junctions or points where rocky outcrops are eroded. J. Tricart (1965) records that in many humid tropical rivers as in Colombia, Brazil, and West Africa, the proportion of siliceous pebbles is always very high in coarse alluvium, from 85 to 100 per cent. Individual quartz fragments 4–6 cm in size may travel 10–15 km

---

Figure IV.28 Downstream changes in properties of fluvial sediments.

A. Mean particle sizes of bed-material along the Mississippi *(from Leopold Wolman and Miller, 1964)*. The data, collected by the Mississippi River Commission, runs from Cairo, Illinois, to the river mouth.

B. Median particle sizes of bed-material in a North African wadi *(from I. Douglas, J. D. Leatherdale, and A. F. Pitty, 1960, unpubl.)*. The example is the Wadi Kuf in Cyrenaica, Libya, along which flow is intermittent. The surveyed longitudinal profile is also shown.

C. Diminution in median particle size with increasing distance below tributary valley junctions in the middle reaches of the Wadi Kuf (ibid.).

D. Relation of roundness and sphericity to distance from head of terrace *(from R. F. Hadley, 1960, U.S. Geol. Surv. Prof. Paper 352A)* in Fivemile Creek, Fremont County, Wyoming; pebbles are in 16–32-mm-size class.

E and F. Morphometric properties down an alluvial fan in the Santa Catalina Mountains, Arizona *(from Blissenbach, 1954)*. (E) shows the distribution of maximum particle sizes (dashed line) and surface angles (solid line), and (F) the distribution of roundness and sphericity indices.

G. Downstream changes in the quartz/felspar ratios in the 0.25–0.50-mm particles sizes in the lowest part of the Godavari river, India *(from A. S. Naidu, in Shirley and Ragsdale, 1966)*.

downstream before penetration of hydroxides along fissures during low water leads to their fragmentation. Crystalline debris may disappear 1–2 km downstream from its source in a rocky bar. In Dutch Guyane, a distance of 5–6 km has been observed. In contrast, in the Lower Colorado river, Texas, a granite proportion of 24 per cent of bedload pebbles, decreasing to 2 per cent 250 km downstream only disappears completely after 400 km. Rocks like gneiss, schist, and sandstone decreased rapidly from a combined proportion of 23 per cent to about 1 per cent 150 km farther downstream. Limestone, as in the Black Hills, appeared to wear at about the same rate as the granite, suggesting that in these environments felspars and limestone tend to be equally durable. As such studies of changes in pebble composition yield a natural index of the relative resistance of a range of rock types in a given drainage basin, and an index independent of the rock's role as a relief former, the investigation of the sedimentary properties of present-day streams should repay more detailed geomorphological study. These include changes in individual minerals as well as in the composition of pebble suites. Felspar breaks down along twin composition surfaces. Disintegration is fairly rapid in high gradient streams but lost more slowly in large streams of low gradient. In the Findhorn river, Morayshire, a feldspar content of 42 per cent is halved 50–65 km downstream, whereas in the Mississippi feldspars are about 25 per cent of the sandy material at Cairo, Illinois, but still 20 per cent 1750 km downstream in the Gulf of Mexico. It is possible that heavy minerals that increase in roundness downstream may be less than a specific hardness depending on the abrasional rigour of the stream. In the south Canadian river, only heavy mineral particles of hardnesses less than 6 were rounded. Quartz and feldspar show a downstream decrease in roundness in larger size grades, perhaps due to chipping and cleaving respectively.

Changes in debris moving by gravity downslope are as intricate and variable as those involving other transportational agencies. One distinctive feature brought out by a study of nearly 500 soil samples from varying lithologies in the Oxford area is that the sorting of mineral particles was very similar regardless of whether the underlying rock was homogeneous, like the Oxford Clay or whether it was heterogeneous, like the contrasted beds included in the Inferior Oolite. This result demonstrates how downslope movement of soil is of fundamental and widespread importance in its influence on the nature of the debris mantle. On a granite slope in Rhodesia, sorting appeared absent in the summit area (fig. IV.29). Downslope sorting in the upper 5 cm was better than in underlying horizons at most sites, but there was little systematic change downslope until the lowest site. As in stream channels, debris tends to move more rapidly on steeper gradients and the likelihood of breakdown of coarse debris during transportation diminishes. Where free falls are involved size gradation on the scree may be related to the height

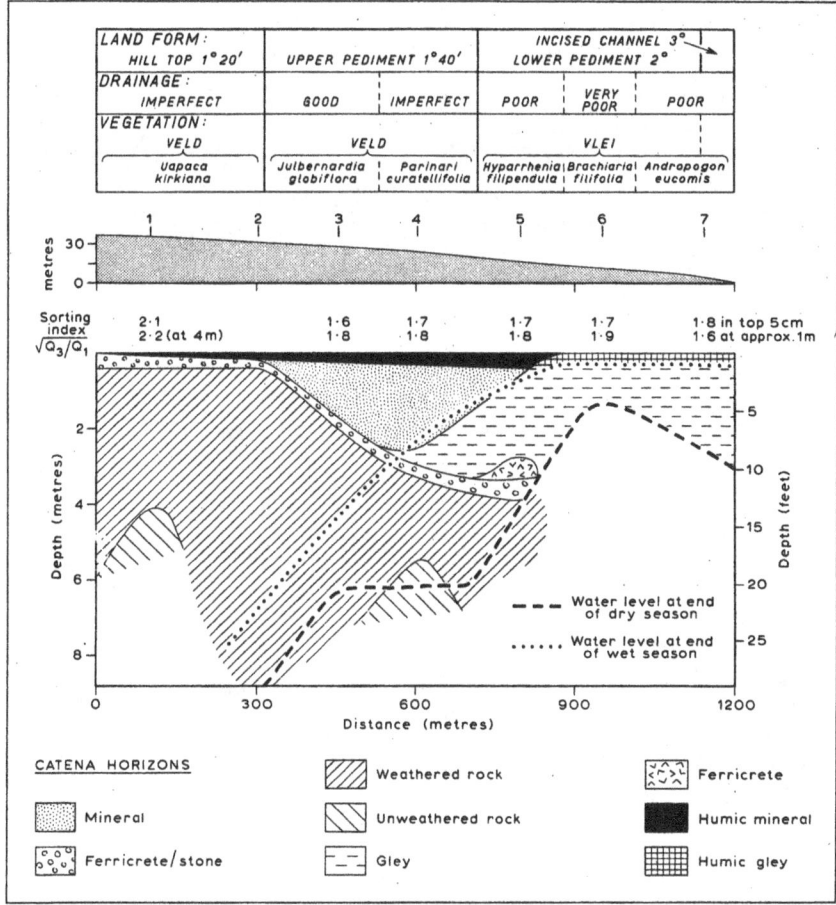

Figure IV.29 Changes downslope in a granite area in Rhodesia (*after Watson, 1964*).

of the cliff-face. If large boulders have a substantial initial fall their range is greatest, whereas with little critical momentum in falling off a low free face they may instead remain near the top of a scree. Quarter of the way down a scree beneath the high rockwall in Insigsuin fiord, mean dimensions of fragments are 49 × 34 × 22 cm, with a smaller set averaging 41 × 28 × 20 cm slightly lower down the cone. At the base blocks are up to 10 m in diameter. By contrast the fragments on some screes in Tasmania do not change significantly in size downslope. Once again conditions in intertropical areas appear different, with the absence of scree at the base of cliffs being striking. Blocks may disintegrate within 200–300 m beyond their source outcrop, with only a few small pieces of quartz remaining. With the more rapid breakdown in material during transport on less steep slopes a diminution of sizes downslope

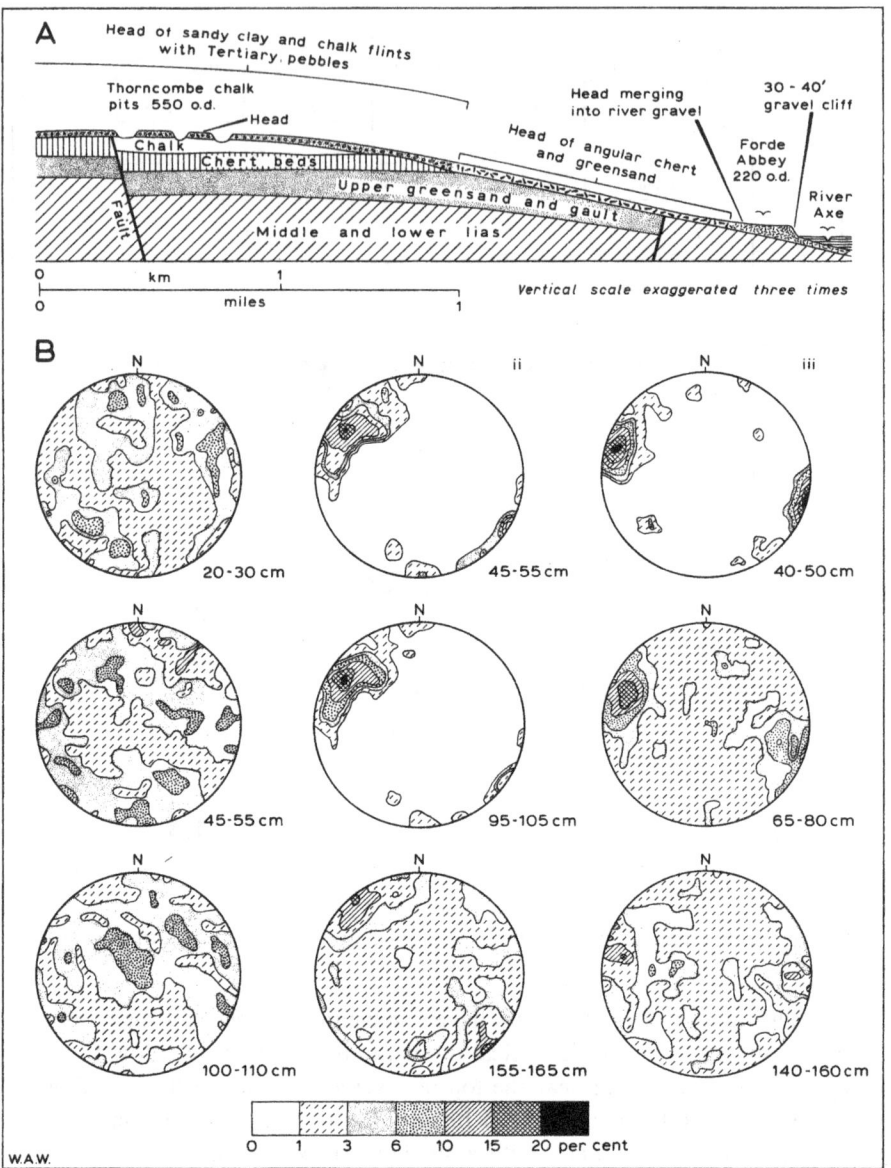

Figure IV.30 Downslope changes in solifluction products.

A. Types and distribution of head on slopes above the Axe valley, Devon (*from Dines* et al., *1940*).

B. Fabric diagrams on a Schmidt net showing changes in strength of particle orientation down a slope in the Southern Uplands of Scotland (*Ragg and Bibby, 1966*): (i) near the summit, 1 degree, (ii) 16 degrees, and (iii) 9 degrees, near the base.

may be marked. This is the case on many alluvial fans and on some slopes. On slopes in stony deserts of Central Australia, material downslope from boulder-strewn areas was less than 5 cm in size when it reached the footslope pediment. Similarly, around the steep isolated rocky hills which characterize the savanna landscape in central Sudan, the surface sequence of solid rock, boulders, sand, and clay corresponds with the topographic sequence of buttress, hillslope, footslope and plain. In south-west United States, J. Mammerickx (1964) observed sizes grading from boulders 3–6 m in diameter on mountain slopes, which were absent at the foot of the slope, to coarse sand and granules containing fragments 10–15 cm in diameter. Talus, above pediment slopes in dissected Triassic sandstone tablelands in New South Wales, also showed a similar pattern, but across the pediment surfaces the 200–600 μ weathered material was relatively uniform in its textural parameters. The degree of orientation of coarse material is another physical property that may change downslope. On the north west flank of Broad Law in southern Scotland, petrofabric analysis at site I (fig. IV.30), where the slope declivity is only 1 degree, showed that the degree of orientation was very low at all three of the depths examined, with a tendency for stones in the subjacent layer to be aligned vertically. At site II, on a 16-degree slope, the alignment in the rubble layer was marked with subjacent layers less strongly orientated. Where the steepness of the slope decreased to 9 degrees at site 4, the orientation was still strong in the rubble layer but fell away more rapidly with depth. A number of steep axial dips were recorded, a feature uncommon at the surface. Most subjacent layers in this area have a secondary concentration of stones dipping into and upslope. Mineralogical changes in the detritus may occur downslope due to the weathering out of less stable compounds (fig. IV.31). Changes in soil water chemistry may also occur downslope but these have scarcely been studied and so their possible significance in landform studies is not well known. However, springs at the base of some savanna slopes are known to contain enough iron to form crusts on the basal portions of the slope, and similarly, caliche is a prominent feature of pediments in the limestone (fig. IV.30) ranges of the Fitzroy basin in semi-arid tropical Western Australia.

Longshore drift may grade beach and nearshore sediments laterally, as a general decrease in grain size can be expected from areas of high energy towards areas of low energy, seen most clearly in the contrast of material between exposed headlands compared with sheltered bays. As in other environments residual material only remains in the most exposed areas. On shingle beaches differentiation of particle shapes may take place downbeach. Spherical particles, on some beaches at least, move more quickly through the interstices than do other shapes. In fig. IV.32 the large disc zone has an apron of spherical pebbles a few feet seaward of the bands of seaweed marking the turning-points of the flood-tides. The spherical pebbles may have moved

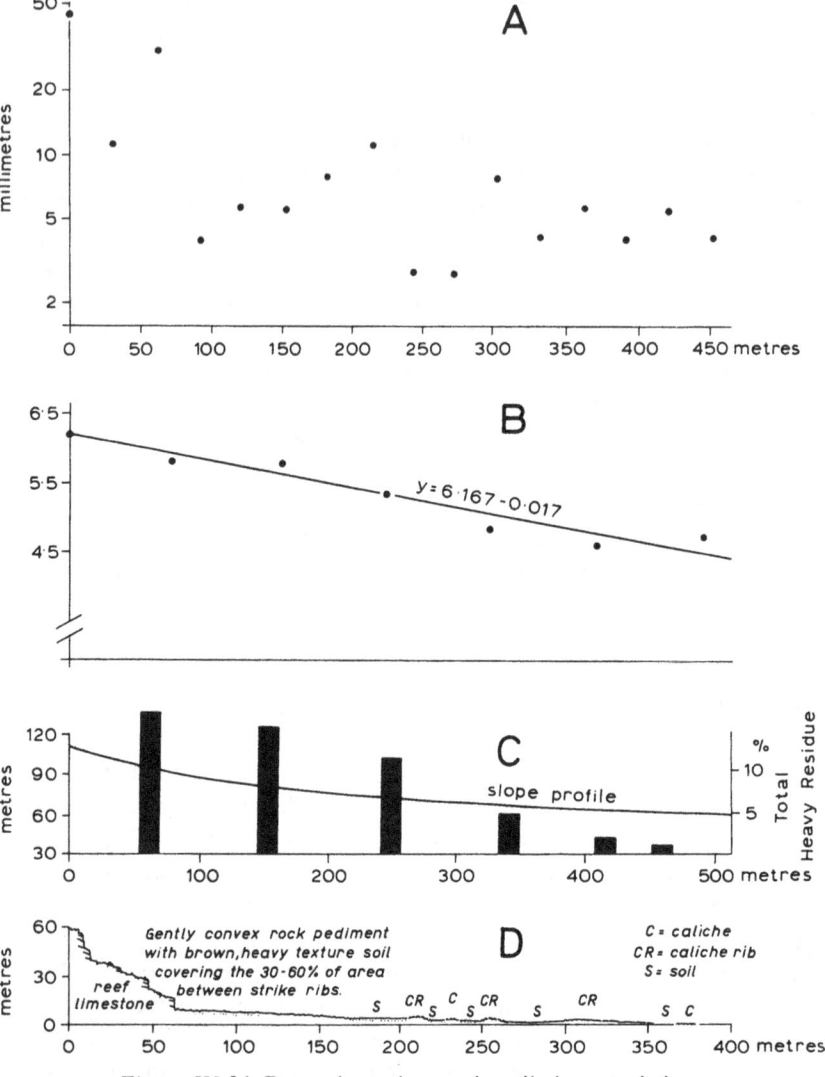

Figure IV.31 Downslope changes in soil characteristics.

A. Median particle size on a pediment slope at Broken Hill, Australia (*after G. H. Dury, 1966*, Aust. Geog. Studies, *Vol. 4*).

B. Decrease in percentage of felspar in the sand-sized fraction in the surface soil below a gritstone escarpment caprock (*from P. Zalasinski, 1970, unpubl. B.Sc. dissertation, Univ. of Hull*). Samples were taken at 4·5-m intervals, the first being detritus from a bedding plane in the Upper Carboniferous Rivelin Grit outcrop on Curbar Edge, Derbyshire.

C. Decrease in percentage heavy mineral residue in the fine-sand fraction of the top 15 cm of soil on a slope at Prospect Hill, New South Wales (*from R. Brewer,*

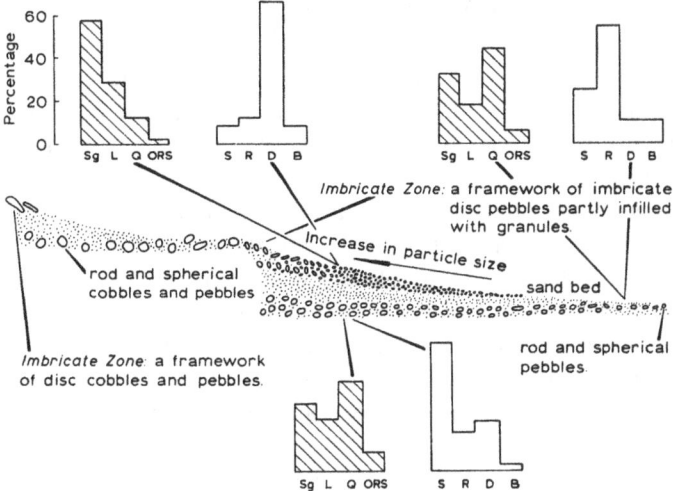

Figure IV.32 Inter-relationships between lithological composition, shape and sorting of material on a gravel beach (*after Bluck, 1967*). The section was obtained from trenches at the Newton beach, near the mouth of the Ogmore river, South Wales. Sg = subgreywacke, L = limestone, Q = quartzite and ORS = Old Red Sandstone; S = spherical, R = rod, D = disc, and B = blade.

through the cobble frame of the large disc zone. The same lateral filtering of particles according to their shape also takes place in the outer frame with the spherical particles moving more readily through the cobble frame than the rod-shaped pebbles.

Changes occur downwind in the size of wind-blown materials. Wind-blown silts become finer away from source areas at the edge of glaciers, temporary lake floors or river beds. Similarly, in dunes climbing over an irregular bedrock surface, fine-grained sand is winnowed out on the windward slope, making the remaining material relatively coarse, particularly on the lower slope. In dunes descending on the lee-side, relatively constant finer sizes are likely to predominate. However, the size of wind-blown sands in some situations, as in the Peruvian desert, may become coarser downwind possibly due to additional coarser material being supplied from the desert floor.

There are not many studies of changes in debris sizes and shapes in glacial

---

1950, Jour. and Proc. Roy. Soc. N.S.W., *Vol. 82*). Basalt is the bedrock at the summit of the slope, the underlying sandstone becoming the soil parent material for the lowest third of the profile.

D. Caliche accumulations on the lower part of a pediment slope in the Fitzroy basin, Western Australia (*from J. M. Jennings and M. M. Sweeting, 1963,* Bonner Geog. Abh., *Vol. 32*).

material. Some end moraines contain sharp-edged boulders and stones, but the general tendency is for glacial transport to reduce somewhat initial irregularities by abrasion. Southward in the direction of ice-movement in central New York State, rhombohedral-shaped fragments decrease by about one-fifth and wedge-forms by about one-sixth, with increases in ovoid and discoidal tabular forms.

In flat areas a lag of coarse material may develop as a significant feature, involving weathering, removal of fines, with the residue of coarser material, although it has undergone negligible transport, remaining *in situ* as a distinctive type of 'deposit'. Commonly, on summit plateaux in cooler environments, boulder fields develop where gradients are too slight for the evacuation of coarse material. If the rubbly material is thought to be a residue resulting from weathering *in situ*, the term 'mountaintop detritus' describes its autochthonous character. In many formerly glaciated upland areas weathering processes may have started only in postglacial times, and the depth of mountaintop detritus may be very variable. The blockfield cover in the Skojem area, Norway, is 10–20 cm and up to 2 m in thickness. In north-east Scotland it may be as much as 45 cm deep, with stone sizes of 5–10 cm, and best developed at altitudes between 600 and 900 m. Rocks which, unlike granite, resist disintegration into gritty particles, provide the most typical layers. The cover is thickest on norite in the Skojem area and less on granite. However, stone-pavements of frost-terraced ground near the summit of Hallival, on the island of Rhum, west Scotland, have the stability of a relict feature. In Scandinavia, frost-splitting of similar material is certainly not rapid. On gentle slopes or on valley floors blockfields, superficially similar to mountaintop detritus, may reflect the selective removal of fines. In 1906, Andersson observed the development of blockfields from solifluction deposits in this way on the Falkland Islands, but there, as in similar cases where blocks and boulders accumulate on a slight slope due to the washing-out of fines and hollows, sometimes a crude branching pattern may appear on the blockfield surface. In tropical areas stones and ferricrete fragments may accumulate in similar sites to the mountaintop detritus of cooler areas, although the rounding of material, as observed in Rhodesia, may indicate their accumulation during millions of years of erosion. In the absence of frost, insects may be responsible for bringing stony material to the surface. Desert pavements are smooth, gently sloping surfaces composed of closely packed rock fragments, commonly coated with desert varnish. The stones range in diameter from a centimetre or less to a metre or more. Many coastal sediments are also lags. Even a beach sand can be considered as a lag deposit with the fines removed by waves and currents and the sand remaining as a relict sediment.

Unlike the covariances considered so far which illustrate changes with distance, a second type of correlation is sometimes recorded between gradient

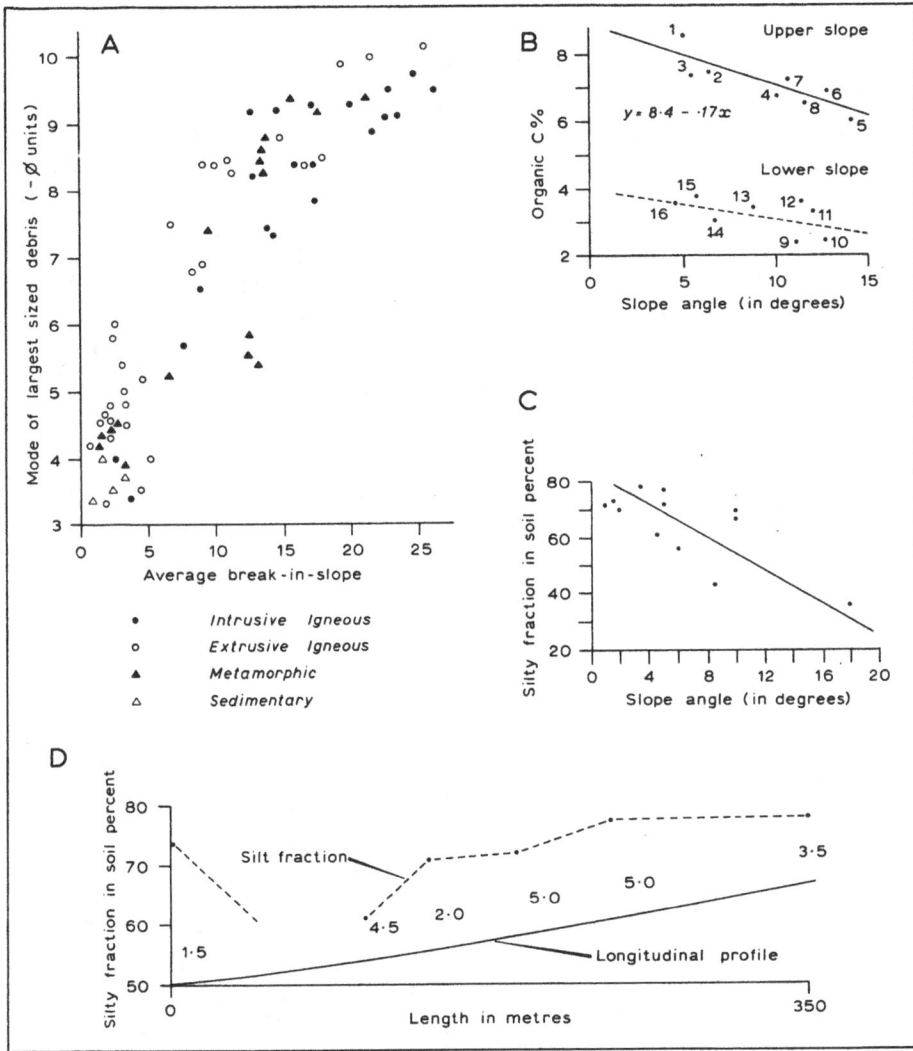

Figure IV.33 Relationships between slope-angle and characteristics of debris and soil.

A. Relationship between the mode of largest-sized debris and the average break-in-slope around an inselberg in south-western Arizona (*after Rahn, 1966*).

B. Relationship between organic carbon percentage and slope-angle on a Cotswold scarp (*from Furley, 1968*).

C. Relationship between silty fraction in the soil and slope angle (*from A. Jahn in Macar, 1966*) in a plateau area made up of moraine clays and fluvioglacial sands in the Upper Odra basin, Poland.

D. Longitudinal profile of a dry trough-like denudational valley in the Upper Odra basin and the silty fraction in the soil (*from A. Jahn in Macar, 1966*). Angles in degrees.

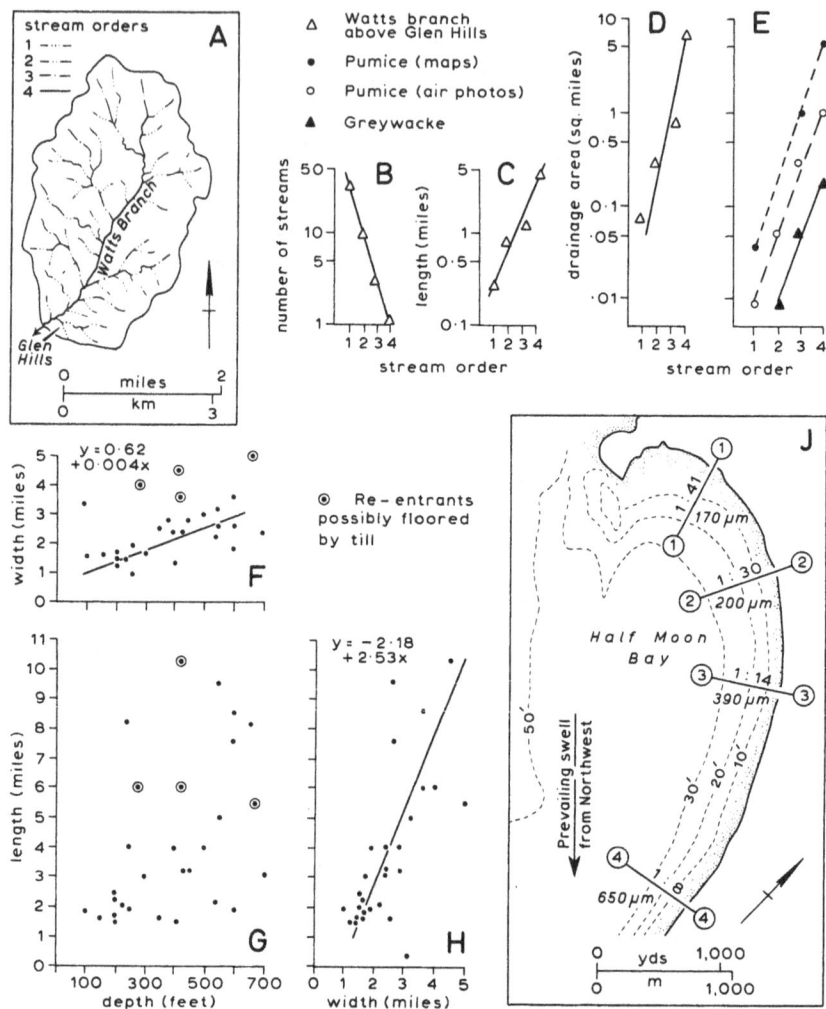

Figure IV.34 Inter-relationships between morphometric properties of drainage basins and of other landforms.

A–D. Hortonian analysis of Watts Branch above Glen Hills, Rockville Quadrangle, Maryland *(from Leopold, Wolman and Miller, 1964)*. (A) Map showing stream orders. (B) Relation of stream order to number of streams. (C) Relation of stream order to stream length. (D) Relation of stream order to drainage area. (E) Relation of channel order to drainage basin area *(from M. J. Selby, 1967,* Proc. 5th New Zealand Geog. Conf.) showing contrasts between rock types and between the same basins using different basic data. To the left are pumice drainage basins according to map-based data (dots) and air photos (open circles); triangles are greywacke basins.

and associated variables. These results are often less useful than those describing the first type of change in which the order of occurrence is fundamental. Being removed from their spatial context, they are often difficult if not impossible to interpret. Fig. IV.33 illustrates three ways in which this problem has been reduced. In A, four different rock types are distinguished. In B, where there is no significant relation between organic content and slope angle for an Upper Chalk slope as a whole, separation of the observations reveals that there is a clear relationship between slope-angle and organic content in the upper part of the slope and the numbers indicate the sample sequence downslope. In C, the increased coarseness of soil on steeper slopes is shown to be less clearly developed in a sequence of observations along a longitudinal profile, D.

A third group of geomorphological covariances are derived from the geometry of river basins, rather than from the study of processes and their inter-relationships with landform. However, some of those who have followed R. E. Horton's (1945) lead have concluded that the degree of covariance among the geometrical properties of 'normal' basins indicates that a balance exists among the associated dynamic processes in the drainage basins. Therefore some illustration of these relationships is relevant in the present context. Fig. IV.34 shows how, in a Hortonian analysis, stream-order plots on semilogarithmic graph paper against number of streams, channel length, and drainage area as a straight line. The slope of the line in fig. IV.34 is Horton's bifurcation ratio, expressed as the mean of the ratios of the number of streams of any given order to the number in the next lower order. However, because of the statistical effect due simply to the way in which a hierarchy of stream orders is built up, the bifurcation ratio has little discriminatory power and inevitably tends towards a constant value for widely contrasted types of drainage basin. Furthermore, the position of the line on the graph, as opposed to its slope, is of little value as it fluctuates widely depending on the detail available on the map or photograph of headstream areas. Fig. IV.34 illustrates the difference obtained in Hortonian analyses of the same drainage basin in an area of pumice in New Zealand, using maps and also air photographs. The degree of covariance between some other measures is exaggerated in instances where both include the measurements of the same variable. For instance, parameters of both drainage area and drainage basin volume include basin length in their calculation, and part of a covariance of any pair of these three is due simply to this statistical effect. Some workers consider these covariances as possibly trivial and probably inadequate and

F–H. Relations between length, width and depth of re-entrants in the Niagara escarpment (*from Straw, 1968*).

J. Relations between degree of protection of a beach from littoral currents, beach slope and mean particle size (*after W. Bascom, 1960*, Sci. American, *Vol. 203*).

inappropriate as a basis for concluding that process and form are in equilibrium, because when translated into qualitative terms they merely state intuitively obvious relationships. For instance, the bifurcation ratio describes how the larger a stream, the greater will be the number of its headstreams and tributaries where the number of headstreams and tributaries has in fact been the sole basis for describing the larger streams as 'large'.

Other geometrical covariances in which the dimensions are not statistically independent also often guarantee some statistical relationship. If the basic data is open to fewer sources of error than those to which stream ordering is exposed, the slopes of a series of regression lines may provide useful information, provided that the difficulty of interpreting correlations is fully stressed and that the misleading misuse of the word 'law' avoided. Fig. IV.34 illustrates the relationships between the length, breadth, and depth of valleys used by A. Straw (1968) in his study of the erosion of the Niagara escarpment.

The most valuable geometrical relationships are those between variables represented by statistically independent sets of data, established in the confirmation of careful qualitative observation, or in the testing of shrewdly construed working hypotheses, by measurements. Fig. IV.34 shows the effect of a headland on the slope of a beach eroded by a littoral current. The increased slope of the beach on the lee-side is attributed to the increased winnowing on the increasingly exposed parts of the beach, leaving progressively coarser material down-current. The example, Half Moon Bay, is on the Californian coast.

5. Reworking of sediments in stream channels

The inter-relationships in streams between erosion, transportation, deposition, load characteristics, reworking, and further erosion produce distinctive features on the land surface. These features can be viewed in plan, in cross-section, or in longitudinal section.

In plan, one of the organized patterns to appear nearest to a stream's headwaters is associated with the small sandy flats that build up behind boulder clogs, spanning the channel. Eventually floods may notch the boulder clog at a weaker point with a narrow scour channel, with the sandy level remaining at a slightly higher elevation. Farther downstream similar diversions may occur, but the drop in elevation is much less than in a boulder-clog channel, and a channel fork develops with water continuing to flow through both the old as well as the new channel. Both branches may rejoin farther downstream, creating the anastomosing or double-branching pattern of the braided stream because of the likelihood of the creation of a further bar at the confluences. If the efficiency of one channel decreases in relation to the other branch channel, the channel will tend to diverge at a greater angle from the

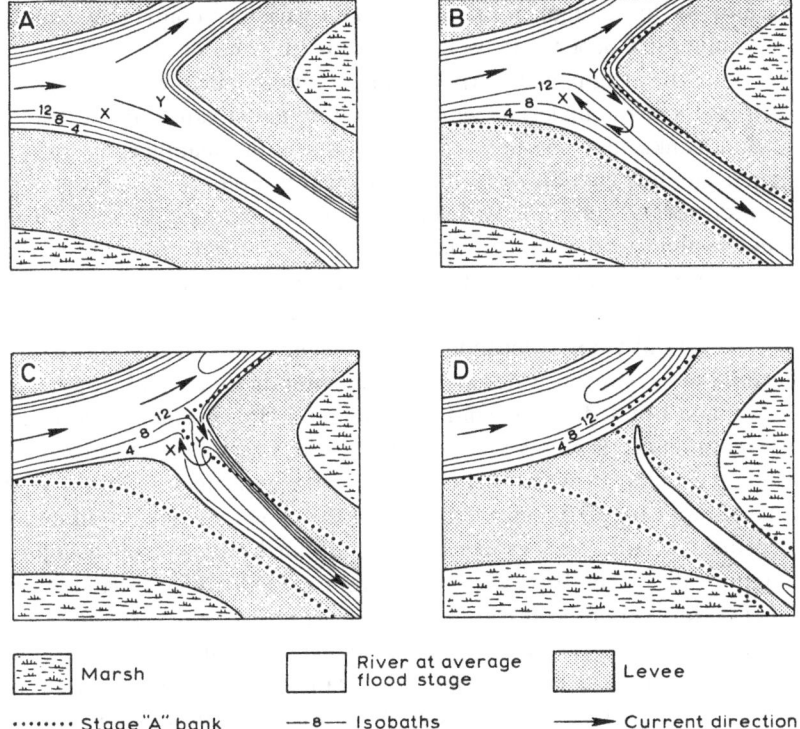

Figure IV.35 Stages in the closing of a distributary channel (*after F. A. Welder, 1959, Coastal Studies Inst., Louisiana State Univ. Contrib. 59–7*).

trunk stream than the more efficient branch, and with backwater conditions developing, to become the site of deposition. This process involves a reverse eddy forming in the side channel opposite the point of bifurcation (fig. IV.35.B) which, by reducing the effective area of the section in the main channel, accelerates its flow. The water surface of the constricted flow becomes super elevated at the point of bifurcation and a very high percentage of the bedload is diverted into the side channel. Continual splitting, formation, oscillation, and closing of distributaries is a characteristic of dynamic depositional landforms other than the braided stream, like the outwash plain, the alluvial fan, intertidal flats or the sub-aerial part of a delta. For any of these features bar formation and resultant channel splitting tends to be more frequent where comparatively coarser material is involved, and where there is strong aggradation raising depositional levels and thereby increasing the chances of bank overflow. In dry inter-channel areas there are abandoned channels tending to be above the ones active at any given time.

In many downstream reaches of rivers the pattern associated with the reworking of alluvial material is the meander, not the braided stream. Although it is difficult to distinguish the independent variables controlling this change from numerous dependent variables, evidently one of the most important factors in the origin and perpetuation of floodplain meanders seem to be the firmness of the materials which constitute the stream bed and banks. The bed and banks must be firm enough to allow the maintenance of an open channel but soft enough to facilitate channel migration. It is possible that if this intermediate condition varies towards either extreme, braiding or incised meanders tend to develop. In the latter case there must also be, however, a certain amount of coarser material for the development of point bars along the convex bank. In areas where banks are firm but such a supply is deficient, as in the humid tropics, meander development is less prominent. Even in an intertidal environment, watercourses traversing the intertidal zone tend to be shallow and braided where sand is abundant and to meander where clay and silt predominate. It seems that where banks are easily eroded broad shallow channels develop in which helical flow is too weak to impose the rhythmic order of point bar and concave bank and that it is the bank erodibility which determines whether a meandering or a braided pattern will develop. It is difficult to dissociate bank erodability from the tendency for sand and gravel shoals to develop in the stream bed which encourage braiding, as the presence of coarser material favours both developments. Stream meanders show very consistent geometric relationships from laboratory streams a foot wide to those of the Mississippi nearly one mile wide. The meander length is consistently between seven and ten times the channel width. Also, as shown on the River Elbe, relative depth of water at a bend increases as an inverse function of the relative radius of the bend. However, for meanders in unstable material the relative depth may increase as the relative radius increases.

The point of inflection in a river bend is closely associated with a shallow portion of the reach or a depositional bar on the bed. Material eroded from one bank tends to be deposited on a point bar downstream on the same bank. The bar tends to concentrate scouring of the concave bank downstream from the axis of the bend, the material thus loosened being added to the next point bar. In this way the meander bends move progressively downvalley. Forward movement of meanders on the Russian plain is commonly as much as 10–15 m/year. Between 1880 and 1938 bends on the Oka river above Dzerzhinsk moved at 3·6 m/year, with a maximum value of 7–8 m/year. This compares with a movement of 3 m/year on the Chemung, a tributary of the Susquehanna, between 1938 and 1955. The degree to which bank recession is actually the product of mechanial scour or hydraulic drag is debatable. During rising-flood stages the thread of maximum velocity tends to shift away from the outside of the curve towards the centre of the channel, and during bankfull stage tends to straighten out, scouring across the point-bar deposits as its stream flow approximates to the line of the downvalley slope. Clearly, therefore, the main process causing bank collapse cannot be associated directly with high flows but with the instability of water-saturated banks after a flood subsides, and with winter freeze—thaw when flood discharges are less than those in summer. On the lower Mississippi, banks steepen until they are steeper than 1 in 2, the ensuing caving producing a more stable slope of 1 in 3 or 4. According to C. W. Carlston (1965) meander wave length is controlled by the range of flows equalled or exceeded between 10 and 40 per cent of the year when the processes of transportation and point-bar deposition are most effective. The significance of sand-dunes moving down the channel floor is unknown, but those observed in Russian rivers are often about eight times the channel width and therefore have lengths similar to the meander length. Their movement appears to be related to the sudden local increase in gradient and velocity that accompanies the cutting through of a narrow meander neck. Observations on the River Ob suggest that a meander neck might be broken through once every 30 years.

It is difficult to establish how the observed hydraulic characteristics of meandering streams are inter-related with the associated geometrically shaped forms. Although helical flow is characteristic, due to water piling up on the outside of the bend and initiating a bottom channel flow towards the inside of the bend, no single parcel of water crosses much more than two-thirds of the stream width. Regardless of genesis an important fact is that meanders dissipate more energy than a straight channel of similar dimensions and boundary roughness. Because of the increased length of channel per unit length of valley, a meandering stream is associated with a low gradient and a straight stream with a high gradient. S. A. Schumm (1963) suggests that the maintenance of a straight channel during valley aggradation resulted from the need to utilize all the stream's energy in overcoming frictional resistance to flow and in transport of sediment through the channel.

The functional difference between meandering streams and braided streams might be overemphasized as the former, with its channels scoured across the point bar deposits and its periodic cut-off diversions is essentially a braided pattern for certain short-lived periods. Also, when discharges exceed bankfull stage, both patterns of flow are submerged beneath a sheet-like flow down the axis of the valley floor. Nor should the possible significance of downstream change in sediment type as the controlling variable be overlooked. It is readily observable that watercourses traversing the intertidal zone tend to be shallow and braided where sand is abundant and to meander where clay and silt predominate and it has long been recognized that these two materials, in areas of glacial deposition, give rise to contrasted landforms.

Another aspect of fluvial deposition which influences the characteristics of its reworking has already been mentioned in comparisons between braided and meandering streams and relates to channel cross-sections. S. A. Schumm

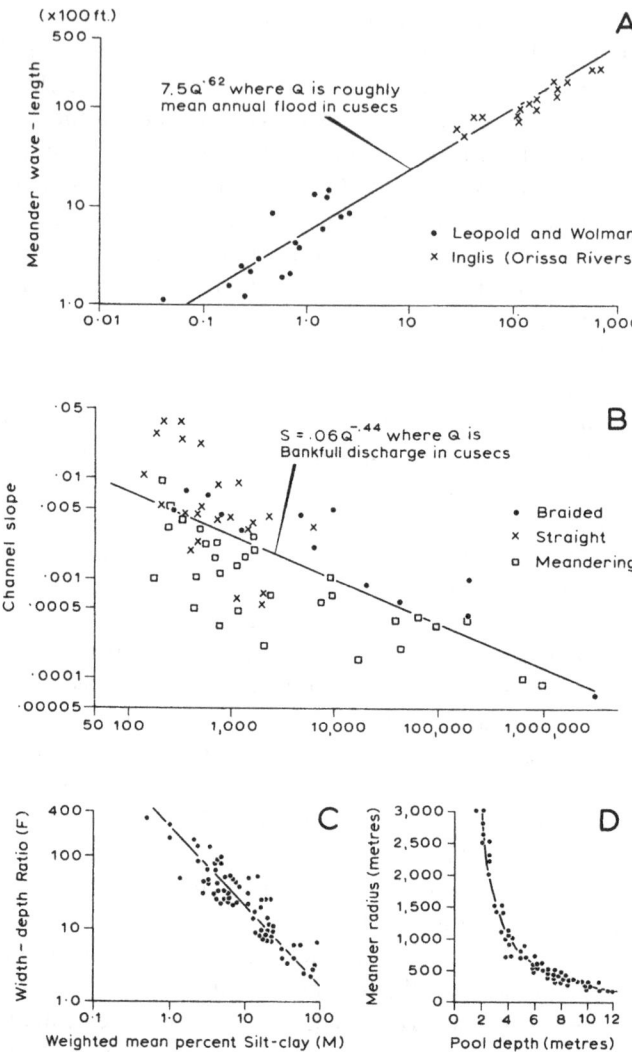

Figure IV.36 Some inter-relationships observed in alluvial channels.

A. Increase of meander wave length with greater discharge (*from Carlston, 1965*).

B. Relation of discharge to slope, and a line which separates data from meandering and braided channels (*from Leopold, Wolman and Miller, 1964*).

C. Relation between a width–depth ratio, F, and weighted mean per cent silt-clay, M (*from Schumm, 1960*).

D. An instance of pool depth decreasing with increased meander radius (*from E. A. Kondiferova and I. V. Popov, 1966*, in Soviet Hydrology, *1966*). This graph by V. A. Orlyankin shows a relationship inverse to that observed in some other situations.

found that relatively broad channels are associated with bank and channel deposits high in silt and clay with little material coarser than 74 microns. It seems possible that where coarse material is available it could, as in so many other situations, protect the underlying material, thus favouring bank erosion and channel widening rather than channel deepening.

The mechanism of scour and fill is not known at present, but appears to take place as one continuous process beneath the water surface in channels which are relatively shallow. One suggestion is that scouring of troughs takes place by eddies generated at the front of shoals or of advancing sand waves. In headstream areas floods leave a basic stepped nature to the channel profile with more level areas of fines between irregularly distributed boulder clogs. This pattern is best seen in the waterless channels of ephemeral streams. In higher reaches scour and fill may be inter-related with the channel cross-section, with scouring by the water moving faster through constrictions and the building up of riffles and bars where water decelerates into broader cross-sections. An initial deepening favours further deepening just as an initial depositional structure, with flow-lines tending to diverge from its summit, may favour further deposition. Ultimately both processes become self-arresting. In straight or non-meandering channels the spacing may become regular at a distance of 5–7-channel widths apart. As with many dynamic depositional forms, like barchans or point-bar deposits, a riffle is a set of sedimentary particles retaining a characteristic shape, while individual particles move intermittently downcurrent from one form to the next. After devastating floods scoured channels in mountainous areas may not be filled up and sizeable excavations remain. After a flood in 1947 in Kantz Creek Valley, the channel scoured out was 7 m wide and 3–20 m deep. In mid-latitudes a flood of 500 cusecs in a 10 000 km² catchment could scour a channel some 3 or more metres in depth.

6. Dynamic depositional landforms

Different from traction carpets or suspended loads are dynamic depositional landforms in which particles are, at least for part of their transportation time, incorporated in masses rather than moving as individual particles. The mass itself may move or remain stationary while some particles arrive, usually by surface creep, change their relative position within the mass, and others depart. Where the forms themselves move they are usually wave-like in appearance and may range in size upwards from ripple marks which may be only 1·5 cm apart. They arise, it seems, because a flat sand surface is unstable, any small chance deformation tending to become accentuated by the mode in which saltation moves sand. To the lee of a dune summit the reverse wind up the slip-face is below the threshold velocity for grain motion. The

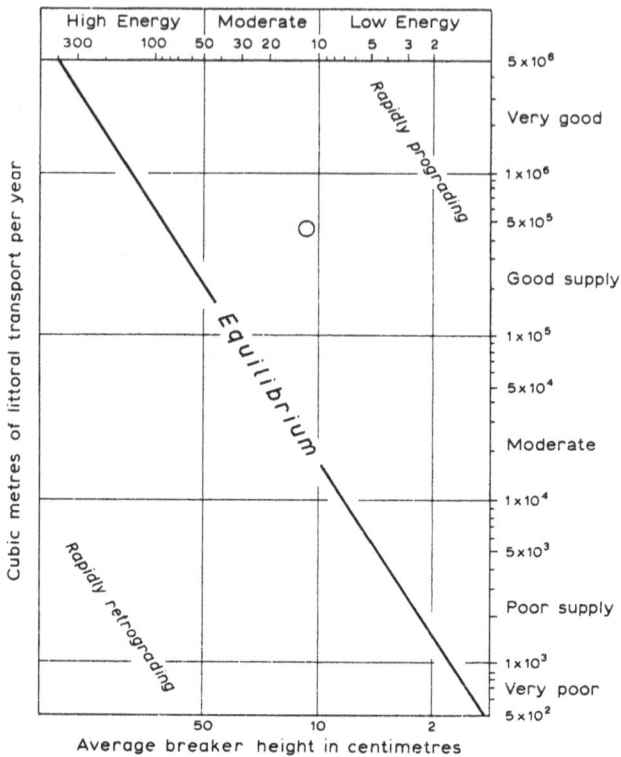

Figure IV.37 Equilibrium chart for evaluation of various shoreline features (*from Tanner, 1961*). Wave energy is the average breaker height in centimetres, and the sand supply units are cubic metres of littoral transport per year. The circle plots the Apalachicola river delta, suggesting that the amount of material supplied by the river is too much for the waves and currents to redistribute.

existence of the lee eddy and reverse wind causes sand blown over the summit to be deposited on the upper part of the slip-face. This mechanism keeps the sand from blowing from dune to dune, and thus is essential in conserving the size of dunes as they travel, and the slumping is probably the principal agent in shaping them. The surrealistic appearance of barchan dunes reflects a quasi-equilibrium between a dynamic depositional form and aerodynamic forces evident both in section and in plan. The sand coming to a steady-state dune from upwind serves principally to set the dune's sand in motion and to replace the loss from the tip of the horns. Sand flies rapidly across the bare stretches between barchans, especially to the leeward of the horns, accumulating in the next barchan which obstructs its path. Although at a given time an individual barchan can be considered as an aerodynamic equilibrium form this does not apply to a barchan field as a whole. For instance the size distribution of barchans from the upwind end of the dune field towards the

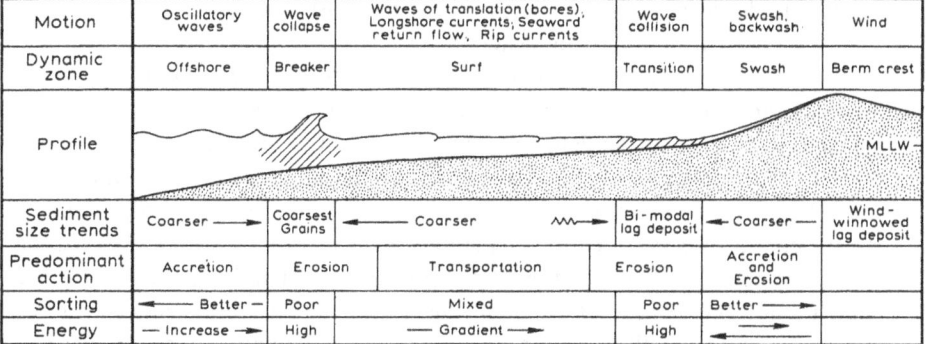

Motion	Oscillatory waves	Wave collapse	Waves of translation (bores), Longshore currents; Seaward return flow, Rip currents	Wave collision	Swash, backwash	Wind
Dynamic zone	Offshore	Breaker	Surf	Transition	Swash	Berm crest
Profile						MLLW
Sediment size trends	Coarser →	Coarsest Grains	← Coarser	Bi-modal lag deposit	← Coarser →	Wind-winnowed lag deposit
Predominant action	Accretion	Erosion	Transportation	Erosion	Accretion and Erosion	
Sorting	← Better →	Poor	Mixed	Poor	Better →	
Energy	— Increase →	High	— Gradient →	High	← →	

Figure IV.38 Summary of the effect of the four major dynamic zones in the beach environment *(from Ingle, 1966)*. MLLW = mean lower low water.

north-north west end of the Pampa de la Joya, Peru, suggests that dunes are formed in this area, grow to a maximum size, and then shrink again and vanish in the downwind part of the field. It appears that fine sand is imported into the area not only in the form of dunes but also that the desert floor represents an additional source. Also sand transport in the downwind part of the barchan field takes place increasingly in the form of streamers. Some investigators consider alluvial fans as stationary dynamic forms with their size and declivity reflecting an approach to a balance between rates of supply and rates of removal from the fan. If a fan receives an increased supply of debris at its head this will steepen the gradient part way down the fan where gullying may lead to accelerated removal. The concept could also be applied to spool-bars.

The best-known example is the sandy beach where according to local conditions material is provided by longshore drift of fluvial sediment, shoreward movement off the sea floor and perhaps a little is due to erosion of adjacent parts of the coast. Balanced against this supply are losses alongshore and offshore and landward removal of sand by wind. The alternation of scour by storm waves and berm-building in calm weather is known as 'cut and fill'. A distinctive feature is that losses may take place in a matter of hours, whereas accretion to replace some former volume takes several days to several months.

A critical factor is the 9-m limit on the depth from which waves can supply sand. Shoreward, the sand lense of the beach is a profile of equilibrium that is balanced against gravity and wave energy. Despite some difficulties in interpretation and conflicts between laboratory and field data, average breaker height at any period appears to be the most reliable index of accretion and loss on natural beaches.

As a specific example, on Brodie Island beaches, Carolina, waves higher than 70 cm occurring with water levels 80 cm or more above mean sea-level resulted in decreases of beach thickness of up to 60 cm and of widths by up to 7·5 m, with a flattening of the beach-face slope.

Considering a beach in plan, W. F. Tanner (1962) suggests that the curvature and sand prism characteristics of the hypothetical equilibrium beach are adjusted to each other so delicately that the potential littoral motion provides precisely the energy needed to transport the detritus supplied at the upcurrent end. The point of maximum net erosion tends to be close to the updrift end of the beach, its precise location depending on the rate at which the drift is moved, on the ratio between the fetch at the downdrift end and the fetch at the updrift end, the length of time during which the present processes have operated and on whether at present there is any systematic change in average energy levels. Beach slope is an important inter-related feature because longshore-current velocity increases with increasing volume of water shoreward of the breaker zone. Although material passes through many beaches and other dynamic forms of coastal deposition, other depositional forms on coasts may be essentially closed systems like Cape Cod, made up of material derived from a source immediately offshore or from the erosion of fluvioglacial material of the Cape itself. Other beaches can be regarded as reservoirs of sand and gravel slowly being released to the sea.

7. Reworking of sediments into distinctive patterns

In periglacial areas the reduced efficiency of fines as thermal conductors appears to be involved in the development of distinctive patterns in loose debris where an ample supply of water is available, as on former lake floors. Because thawing temperatures arrive sooner at a given depth below or close to stones than through finer material, stones on surfaces where vegetation cover is discontinuous may tend to migrate into thaw depressions. From these zones finer materials might be removed by eluviation in water which may flow in the fissure network. As well as frost-heaving, drying and shrinking are involved in the production of sorted patterns. Absence of continous snow cover is also necessary and in mountains the upper limit of structural soils tends to follow the lower limit of permanent snows. Most polygons have from three to six sides formed by straight to gently curving, orthogonally intersecting cracks. An initial random outline becomes more regular with increasing age. One- to five-metre diameter polygons occur on 0–5 degree slopes in north-east Greenland. Polygons of a much larger order of size develop in permanently frozen ground. In north Baffin Island, examples are up to 50 m in diameter. Comparable in size are some large ice-wedge polygons still discernible from the air or in vertical sections in former permafrost areas, as in south-central England. On the Three Pigeons plateau, 13 km east of Oxford, the diameters of relict polygons range from 40 to 150 m.

Despite the fact that polygons may dominate areas of frozen soil, there is nothing unique about a polygonal fracture pattern. Apart from soils in periglacial environments it may be seen on mud, basalt, concrete, varnish, paint, and other media which undergo volume change, resulting from the contraction of a layer of homogeneous material perpendicular to the cooling or shrinking surface. Desiccation polygons in semi-arid playa floors develop in most clays apart from kaolinite, due to exceptionally great volume changes. Widths range from 15 to 100 m and those of Guano Lake, Oregon, are up to 300 m in width. Like frost and ice-wedge polygons there are two orders of size with giant polygons rarely less than 15 m in diameter. Smaller features appear on many types of surfaces, due to day to day changes, if drying conditions follow rain.

Unlike polygons on flat ground, surface patterns on slopes are much more restricted to periglacial environments. Downslope from upland plateaux polygons are drawn out on gentle 2–5-degree declivities until the sorted material lies essentially in stone stripes. There is no general critical declivity, the transition from polygons depending on the mobility of the thawed material. In the Antarctic, for instance, polygons develop on some essentially immobile 35-degree slopes. Pipkrakes are probably involved in the sorting process that produces stone stripes as these features are best developed where night temperatures are very cold. T. N. Caine (1963) considers that periods of more than 24 hours may be required before the differential frost heave of stripes becomes apparent and suggests that periods of freezing of more than 3–4 days are more effective as far as the mechanics of sorting are concerned. Artificially disturbed stripes have reformed within a winter in Japan. The longer the duration of movement the more distinct becomes the boundary between stripes made up of different types of material. In observations in New Zealand, at an altitude of 660–800 m, twenty-one of sixty-eight marked stones moved from fine to coarse stripes in 2 years, and comparable rates of sorting have been recorded in the Lake District Hills. Rillwash, utilizing the coarser bands as natural drains, is an associated process accentuating sorting by removing fines. In fact in drier environments or on hillcrest sites in periglacial areas where rillwash is not important it is the portions of polygons parallel to the slope which persist, termed soil guirlands.

In cooler areas where permafrost is absent, frost processes may still play an important part in the development of ground surface patterns. The earth hummocks or 'thufurs' are well-developed under relatively mild conditions in Iceland. Somewhat similar mounds have been identified in some higher areas in Europe, as on Hohneck in the Vosges, but according to K. F. Schreiber (1969), those in the Swiss Jura are anthills. A peat or grass mat covers the loamy or clay interior of the thufurs. They are about 25–30 cm high and 1–2 m in diameter, and if artificially levelled are known to re-form within a few

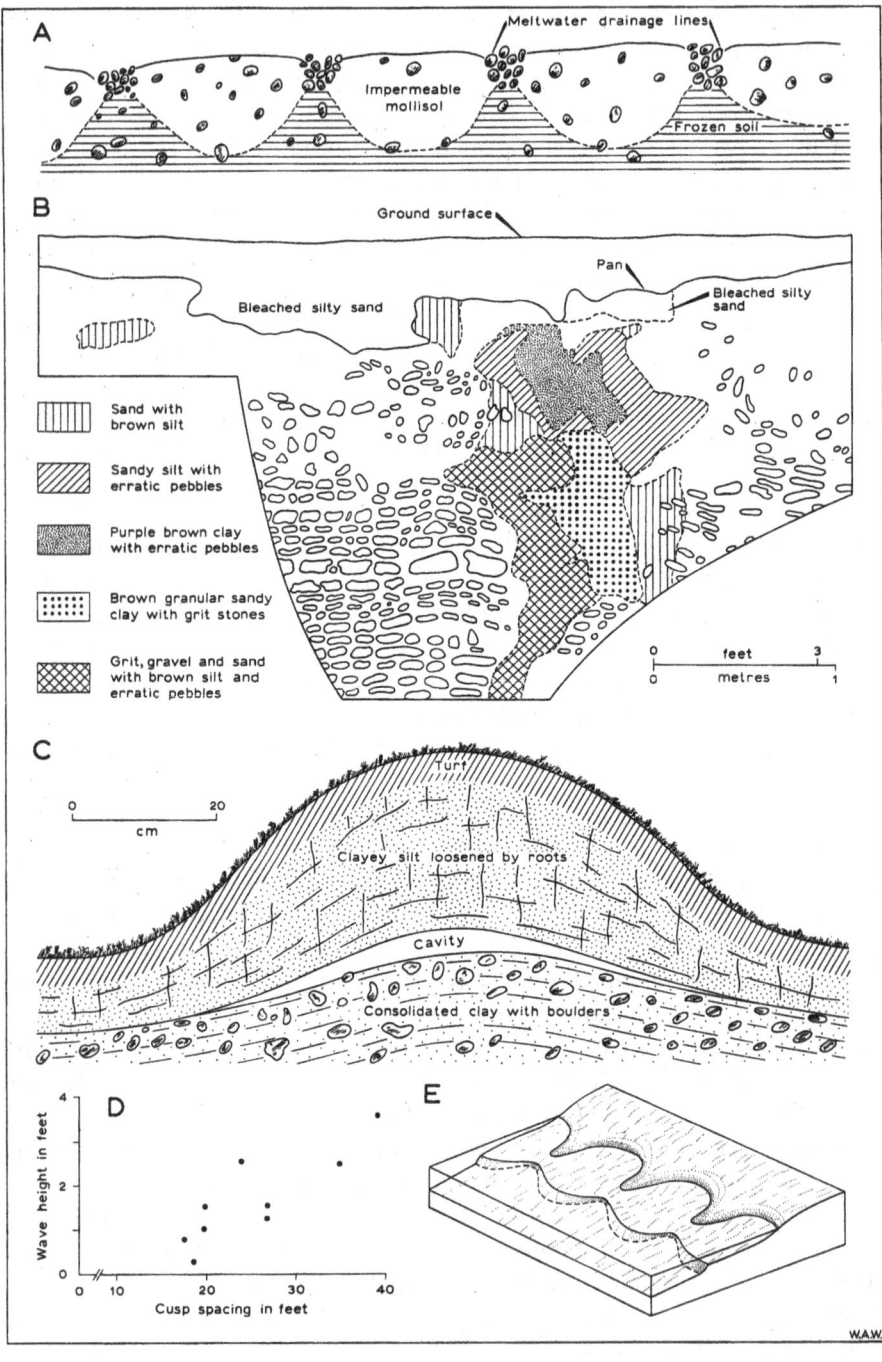

A

Meltwater drainage lines

Impermeable
mollisol

Frozen soil

B

Ground surface

Pan

Bleached silty sand

Bleached silty
sand

	Sand with brown silt
	Sandy silt with erratic pebbles
	Purple brown clay with erratic pebbles
	Brown granular sandy clay with grit stones
	Grit, gravel and sand with brown silt and erratic pebbles

0   feet   3
0   metres   1

C

Turf

0   20
cm

Clayey silt loosened by roots

Cavity

Consolidated clay with boulders

D

4

Wave height in feet

2

0
0   10   20   30   40
Cusp spacing in feet

E

W.A.W.

years, but by a process that remains obscure. In dry areas of Australia volume changes in clay soils, related to desiccation and occasional flooding produce the gilgai soil pattern which resembles the thufur in its external form. Pimpled plains in the western United States, consisting of low domes 60–120 cm high and 15–30 m across, may be related to the desiccation cracking of clay pans which they usually overlie.

If tidal range is small and if there is a balance between beach profile and wave characteristics, beach cusps may form a regular pattern on the seaward side of advancing and broadening berms. Initially patches of coarser material accumulate at irregularly spaced intervals along the upper limits of wave uprush and gradually develop flat tops. Regularization of form and spacing appears to be related to rotating vortices developed between the coarse patches by longshore drift, with water swirling into the gaps to pile up against streams from the developing cusp on their opposite sides. Occasionally, similar features of considerably larger size form in lagoons as cuspate spits of very regular and symmetrical outline, like those of the St. Lawrence Island in the Bering Sea or on the shores of the Black and Asov seas (fig. IV.40).

8. Recombination of weathered chemical elements

One of the most distinctive products of the combined action of rock breakdown and deposition is the development of weathering crusts and other products involving the recombination of elements released by weathering. Transportation, apart from vertical movements within the weathering profile, is not necessarily involved. However, the interpretations have involved some confusion because similar recombinations may be produced by solutions that have travelled some distance downslope or downstream as well as by dissolved elements reprecipitated essentially *in situ*. Another source of confusion is the development of a crust rich in one particular element either because it is precipitated from a solution rich in that element, a process termed secondary

---

Figure IV.39 Microforms and distinctive patterns produced by the reworking of unconsolidated surface materials.

A. Cross-section through striped soils (*from Lliboutry, 1965*) observed in the Santiago Andes at 3800 m on a 36-degree slope.

B. Cross-section through material filling a relict Pleistocene ice-wedge on the North Yorks Moors (*after G. W. Dimbleby, 1952*, Jour. Soil Sci., *Vol. 3*).

C. Cross-section through a thufur in the Tatra mountains, Poland (*from A. Jahn, 1958*, Biuletyn Peryglacjalny, *Vol. 6*).

D and E. Beach cusps. Relations between mean cusp spacing and wave height (*from M. S. Longuet-Higgins and D. W. Parkin, 1962*, Geog. Jour., *Vol. 128*) and (E), a block diagram of beach cusps (*from Kuenen, 1950*).

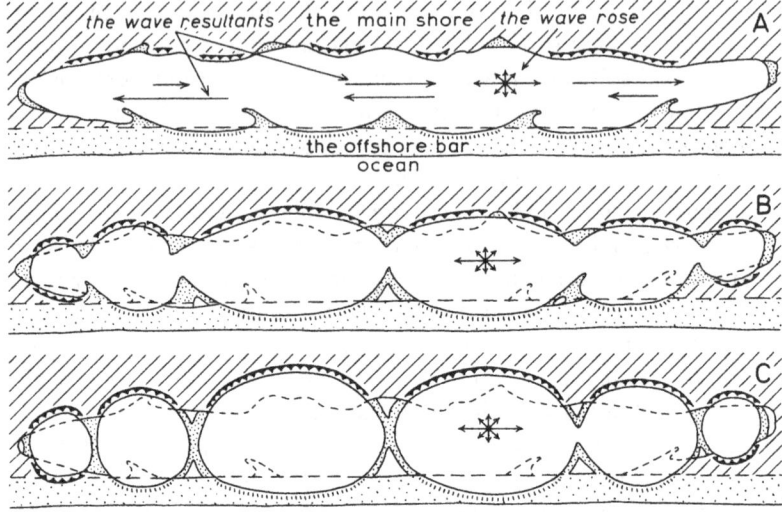

Figure IV.40 The development of stable spits and lagoons (*from Zenkovitch, 1959*). In the stable stage the wind rose is nearly round; key as for fig. IV.19.

enrichment, or alternatively essentially the same end-product might arise by residual enrichment where an element remains relatively undissolved while other elements are leached away. Another problem is that while organisms are known to be a predominant factor in the reprecipitation of some elements like calcium, their relative importance in other processes is unknown but suspected to be highly significant. A final problem is in deciding whether the weathering crust is in active formation under prevailing conditions.

One of the most immediate reasons for including the discussion of the complicated topic of weathering crusts within landform study is that many crusts have become distinctive landform features. Ferruginous crusts in Brazil, for instance, preserve ridges and plateaux because the underlying softened rock is readily eroded. Their resistance can be gauged from some areas where fragments from ancient crusts, like the early Tertiary siliceous layers in North Dakota, have been lowered 100 m or more, while the underlying rock has been eroded away. Similarly, some beaches in lower latitudes may be protected from erosion by beachrock encrustations. Secondly, crusts have been used as stratigraphic horizons marking stages in erosional histories in some tropical areas. In this context plateau top residual enrichments have sometimes been erroneously linked with valley floor secondary enrichments. Beachrock originating close to the strandline, is an excellent indicator of recent changes in the relative level of land and sea. Thirdly, some crusts, particularly several varieties of calcareous crust, have been used as paleoclimatic indicators. Gypseous crusts are characteristic of a desert climate, whereas others are associated with a humid climate supporting a forest vegetation.

Three factors are involved in the production of crusts which favour their development in lower latitudes. First, mechanical erosion of the surface must be less than rates of chemical weathering for residues to accumulate. Secondly, for cementation of the residues high temperatures are necessary to evaporate the mineral-bearing solutions while these are still in the weathering profile and for intense biological activity that might be involved in precipitation processes. Thirdly, for the preservation of the porous textured concretions, freezing temperatures must be absent. Thus in mid-latitudes it is only calcareous encrustations which may develop where flowing water, retaining some warmth from an underground source, always saturates the deposit.

In a semi-arid or arid climate where evaporation exceeds precipitation soil moisture may, having dissolved calcium compounds from soil and rock in its downward course during a rainy spell, move upward towards the surface during the ensuing dry period, where evaporation leaves behind as *caliche* a somewhat spongy calcium carbonate crust around pebbles or as layers just below the surface. An example of this type of duricrust or weathering crust is the 'caprock' of the High Plains in Texas and New Mexico. Relict caliche horizons are often better preserved today where they have remained beneath an alluvial cover. Recent caliche begins to form as isolated nodules around roots and pebbles beneath the surface. Eolianite is preponderantly sand of eolian origin also owing its degree of induration to the amount of cementing calcium carbonate present (fig. IV.41). Cementation might occur during the evaporation of rainwater and is favoured by the alternation of wet and dry

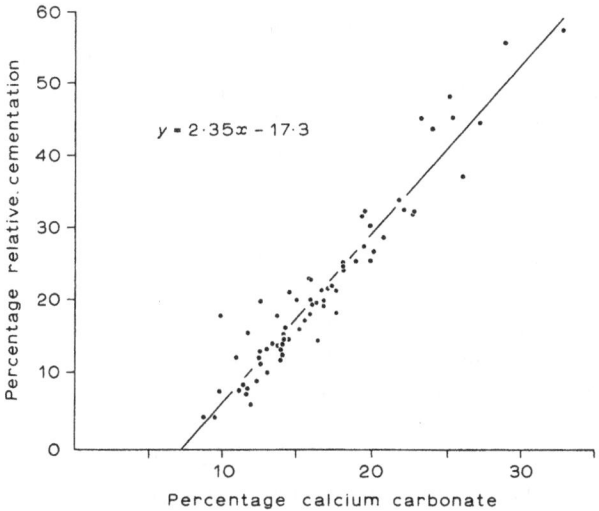

Figure IV.41 Cementation as a function of calcium carbonate content in partially consolidated eolianite (*from Yaalon, 1967*). The Wingate dune, Israel, provided this example of induration.

seasons which characterize the Mediterranean climate. Cementation is greatest near the surface where drying is quicker and more pronounced. Dune plants and shrubs might be involved in the process, but formerly held ideas on the capillary rise of lime-charged ground water are not now thought to be physically possible because of the coarse pores of sandy material. Induration is relatively rapid extending over tens or hundreds of years. Eolianite may be found close to shores with beachrock but is distinguished from the latter by better sorting and by distinctive aeolian bedding. Depths of 43 m have been proved along the Egyptian north coast, suggesting that formation started when sea-level was lower. Beachrock, incorporating sand, shells, and other residual materials not blown off the beaches may vary in composition over very short distances. The beds trend parallel to the coast and dip gently seaward. The cement is calcite and the beachrock forms near the shoreline where a dense algae population may play an important part in the chemical precipitation processes. The cementing calcite appears to be precipitated from percolating ground water encountering carbonate-saturated sea-water, although repeated wetting and drying as the water-table rises and falls with the tide may be a more significant factor. Where the cement is aragonite, as on British Honduras, the origin of beachrock is probably marine. A precise climatic control, possibly related to algal activity or to temperatures needed for evaporation limits active beachrock development to 33° N in California and within the Mediterranean. Some workers consider that beachrock is essentially a characteristic of retreating beaches.

Crusts due to the reprecipitation of iron are sometimes referred to as ferricretes. Boulders of this material on hilltops near Salisbury, Rhodesia, are probably related to a former more extensive spread of Kalahari sands although their nearest deposit is now 60 km distant. In Brazil, the superficial weathering product of an iron-rich rock formation is 89 per cent Fe_2O_3 and 2·9 per cent Al_2O_3, compared with 54 and 0·5 per cent respectively in the unleached rock. Estimates based on silica removed in spring water suggest that similar iron-rich crusts in Venezuela might have taken about 20 million years to form. Silica, once dissolved in percolating water is not easily reprecipitated, whereas the iron in hematite is readily reprecipitated as relatively insoluble hydrous ferric oxide. This is, therefore, not easily remobilized with the water-table rise in the succeeding wet season and the progressive enrichment in iron takes place. The source of renewed supply of iron appears to be solutions brought up to the surface. Examples of iron-rich weathering include crusts along the Niger, where quartz pebbles are cemented by iron oxide to a depth of 5 m at least. In the Llanos, older terrace deposits include a ferruginous hardpan at the base of gravels which cap the terraces. These examples show how widespread iron-rich crusts are in seasonally dry tropical environments where they form important landscape elements. They may also form

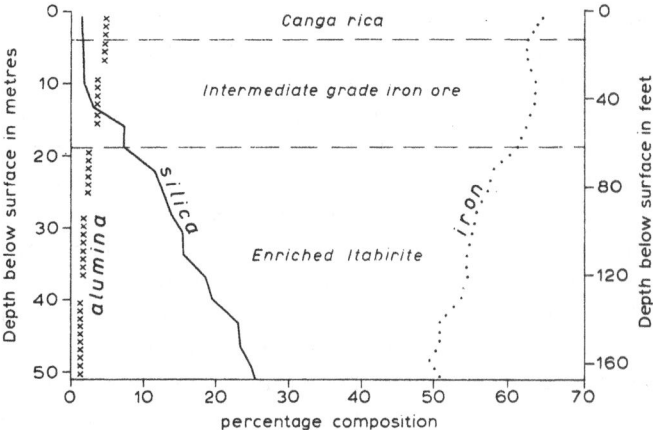

Figure IV.42 The effect on a tropical deeply weathered profile of an environment propitious for long-continued leaching of quartz and of hematite and for reprecipitation of iron (*J. V. N. Dorr, 1964*, Econ. Geol., *Vol. 59*). The profile is that of the Casa de Pedra itabirite deposit, Brazil. Itabirite is the local name for the residually enriched intermediate grade ore.

in more arid environments where the water-table is near the surface, but might equally be relics from a seasonal wet climatic phase in the past. For instance, in the desert in Upper Egypt, ferruginization of the Kharga oasis deposits is a characteristic feature, with the iron for the crusts and concretions believed to have come in part from ground water.

Bauxites are another partly residual product due to leaching out of other elements, but some are too thick to be produced by the solution of limestone alone and suggest the supply of siliceous and aluminous materials from allogenic sources. Bauxites are widespread in Jamaica. On resistant Oligocene and Miocene limestones in the centre of the island, the Al_2O_3 percentage is 43–45, Fe_2O_3 is 20–22 per cent, and SiO_2 3–4 per cent, the proportions of iron and alumina being rather higher than for many bauxites.

In cooler environments the differential downward movement of Al and Fe, termed podzolization, is less pronounced and occurs within even shallower depths, but leads to the gradual development of an indurated horizon as the clay is cemented by small amounts of translocated Al. At a Roman camp-site near Forfar in Central Scotland, a 15-cm indurated layer appears to have formed in about 1000 years. In Vermont, hardpan depths in a silt loam were largely restricted to depths of 30–40 cm below the surface. Although these cool environment concretions do not compare in extent, thickness, and resistance to the rock-like crusts of savanna areas, they do have some influence on land-forming processes by restricting drainage and tree-root penetration.

On bare rock surfaces in tropical areas thin veneers may appear, the result of processes similar to those which produce crusts in weathering profiles.

Like crusts, veneers are often allogenic rather than composed of elements drawn up from within the rock. For instance, the manganese oxide veneers a few millimetres thick which add greatly to the resistance of rapids to erosion are deposited from stream water where seasonal variations of flow exist. The rocky domes of inselbergs are often smoothed by a similar ferromagnesium veneer which has the significant effect of preventing the entry of water into fissures and aids rapid drying on the surface. Silica veneers on granite out-crops, as in Surinam, have been observed to have a similar protective role. Algae may be involved in the reprecipitation processes.

Clay mineral synthesis in tropical environments may accompany the formation of weathering crusts and is a geomorphologically significant process because of the depth to which soft weathered material is produced, the distinctiveness of slope form where it is eroded from beneath a weathering crust caprock, and the significance of differences between the properties of the various clay types. Clay mineral formation is also significant in that clays are usually the result of the same processes that produce weathering crusts in lower latitudes. Finally, an understanding of how dissolved elements are reincorporated in clay mineral synthesis is necessary in appreciating why net denudation rates in tropical areas are usually relatively low. The uptake of silica during the formation of new silicate minerals is one example.

Clay minerals of approximately similar compositions may be produced in more than one way. W. D. Keller (1964) recognizes two directions of approach towards a given end-point or equilibrium conditions between the quality of percolating waters and the residual clay minerals. On the one hand anamorphism describes the synthesis of clay minerals from smaller or simpler units such as alumina and silica gels. An example would be illite formation by the absorption of potassium and magnesium ions from drainage water derived from an allogenic source rich in these minerals. Conversely katamorphism describes the degradation of clays. This type of clay formation is widespread but not as significant as has been thought in the past. It is restricted to areas of free drainage and to where ground water does not participate in the soil-forming processes because it is essentially beyond the reach of plant roots. The relation between the exchangeable cations on the clay minerals and drainage water separates soils into two main groups, a contrast which may be as much in evidence between top and bottom of a tropical slope, or in sites adjacent or removed from a stream, as in a contrast between humid tropical and seasonally dry tropical zones. The two major tendencies in clay mineral formation are, first podsolic alterations in an acid environment, with the development of silica-rich superficial horizons. Secondly, in a near-neutral environment lateritic alterations produce a zone of aluminium and iron concentration.

The synthesis of crystalline alumina and silica into kaolinite occurs where landform, climate, or both produce an acid environment with the free removal of calcium, magnesium, iron, sodium, and potassium, and with hydrogen ions being supplied in abundance. Thus with a pH of 4–5 the soil clay minerals tend to become H clays. For iron to be removed in its soluble ferrous bicarbonate form, air must be excluded. In practice, some precipitation in the ferric iron form also takes place, and this association of iron oxides with kaolinite is one of the commonest features of tropical soils and accounts for their predominantly red colour. The iron oxide is precipitated as a coating on the surface of clay minerals formed in an acid environment, reducing the exchange capacity of these clays like kaolinite or halloysite, and protecting them from further decomposition in percolating waters. Aggregation of iron hydroxides may give the kaolinite a certain permeability. Desilication of kaolinite may yield gibbsite under conditions of strong leaching where good drainage makes the oxidation potential high. Gibbsite precipitates from percolating waters during drier periods within the weathering mass, fissures in the soil providing a drying surface. Root surfaces in addition to providing a drying surface also form a zone of relatively high carbon dioxide saturation which favours the precipitation of aluminium oxide and the formation of gibbsite.

In less severely leached situations than those in which kaolinite forms and where the water has an alkaline reaction, cations in the exchange position of clays may be Ca, Mg, Na, K, or others and favour the formation of montmorillonite. The concentration of hydrogen ions is correspondingly low. This may occur in a semi-arid environment where percolating waters carry little organically produced acid due to the very rapid oxidization of organic material and where hydrolysis of the silicate occurs in the presence of almost neutral water. Under such conditions of slight alkalinity the silica tends to be dissolved and removed, with an associated concentration of alumina and ferric oxide in the upper part of the weathering zone. During the drying which follows the seasonal rains the solutions originally dilute in cations become saturated with magnesium, calcium, iron, and sodium which combine with oxygen, silica, and alumina to produce the montmorillonite, often beneath an alumina-iron weathering crust. However, the same clay-mineral synthesis can occur in humid environments where poor drainage leads to a build-up of cations.

Illite is a term describing clay-mineral constituents belonging to the mica group where the magnesium content is low. Illite is structurally similar to montmorillonite except that 15 per cent of the Si^{+4} is replaced by Al^{+3} and the resulting charges balanced by K^+ ions. It tends to be characteristic of soils in drier environments. Combined depotassication and desilication of illite may yield kaolinite, the alumina content in the latter being relatively higher.

There are many variations of the three main types of clay mineral and even these are not clear-cut entities. The synthesis of elements in clay-minerals should be regarded as dynamic recombinations rather than static end-points, susceptible to continuous evolution according to the climatic conditions, site, and the composition of solutions moving through the weathering profile. The degree to which organisms are involved if only indirectly in these processes has not been established, but it is known, for instance, that most natural surface and near surface waters do not have enough silica dissolved to precipitate amorphous silica inorganically.

9. Influence of weathered material on subsequent weathering

One feature of processes of rock breakdown expressed in innumerable different ways is that the process becomes self-arresting by producing a sufficiently thick cover of material in which much of the energy reaching the ground surface is dissipated without reaching bedrock. In other areas vegetation cover may, in an analogous way to the debris mantle, intercept incoming energy. The peat cover of humid cool temperate latitudes in which the entire supply of incoming energy is absorbed in evaporating the excessive moisture content is one example that has already been considered. Similarly, in humid tropical areas weathering and either subsequent removal or recycling of elements is related to the intense biological activity and the degree to which this merely rapidly recycles the same mineral elements without appreciable losses to runoff water and significant gains from weathering bedrock. Such a state of biological equilibrium between soil vegetation and climate is termed biostasis. With a breakdown in equilibrium, rhexistasis, the forest progressively disappears, perhaps due to the climate becoming drier. In other areas a weathering crust affords immense protection from weathering to underlying materials, but in many cases, regardless of vegetation cover and without consolidation, the mantle of weathered debris arrests the penetration of rock-weathering processes. The significance of soil depth in protecting underlying bedrock is particularly marked in relation to frost penetration in cooler environments. On Kosciusko summit, the stony soil 1 m deep and an overlying stony layer appear to limit the frost-shattering destruction of bedrock outcrops today. In contrast to the self-enhancing immunity of bedrock outcrops in tropical environments, the exposure of bedrock above a debris mantle in cold climates hastens its disintegration by frost action. Depth of soil may also be significant in warm seasonally dry areas where there is an upward return of moisture before it reaches down to the bedrock. There is also the constancy of soil temperatures at depth in a soil compared with seasonally high temperatures near or at the surface, accelerating biochemical processes. J. A. Mabbutt (1966) describes how the mantle of debris on some Australian pediments imparts a general levelling to an otherwise uneven and heterogeneous bedrock surface directly through its base-level control of ground-level 'sapping'. Gilluly (1937) suggested that rocks yielding relatively coarse and resistant debris have steeper pediments than those yielding relatively coarse and friable materials. In cool temperate environments a residual surface-lag deposit, by making the soil more permeable, reduces mechanical erosion at the surface. Also weathering at depth, despite the dampness of the soil environment, is slow because of low temperatures. It protects the surface from puddling and the dislodging action of raindrop impact. Field experiments in Maine indicate the importance of this factor in a temperate climate (Table IV.10). In semi-arid environments the surface-lag deposit is more important. Studies in Israel showed that with the removal of stones from areas

Table IV.10 The effect of surface stones on soil erosion (*after E. Epstein et al., 1966,* Soil Sci. Soc. Am. Proc., *Vol. 30*).
  The soil is developed on till, Presque Isle, Maine. Averages are for the four years 1961–4, during which time annual soil loss was greatest in 1961.

	SOIL LOSS			RUNOFF		
	Natural	Stones removed Kg/ha	Stones crushed	Natural	Stones removed Millimetres	Stones crushed
1961	5954	7574	5711	205	245	220
1961–4	3224	4062	3224	153	191	167

where they made up 28–62 per cent of the surface led initially to a twelve-fold increase in erosion. As early as 1927, H. Mortensen described how such hamada covers stabilize surfaces in semi-arid areas by reducing the erosion on underlying soil to a minimum. In arid areas a lag gravel of small rounded pebbles accumulates which may constitute a continuous sheet or serir, protecting underlying sand from deflation. In many other areas, including some shingle beaches, there is a lag deposit cover due to the removal of the fines. Thus the beach material that would provide an abrasive for waves impinging on cliff faces may, if supplied in profusion, provide a protective ramp on which all storm wave energy is expended. Similarly, nearshore shoals have an important effect in protecting beaches from erosion by absorbing storm energy. The most rapid retreat of coasts is in areas where there is not enough residual material and where the abraded masses are continuously lost by beach drifting. On flat sites inland a relict sediment like mountaintop detritus, may develop and in warmer, damp areas certain mineral elements may slowly become more concentrated in surface layers due to the progressive leaching of more soluble compounds.

  Although a soil is in some senses the end-product of rock breakdown and may in certain circumstances arrest the effectiveness of some weathering processes in other situations, it may be the starting-point of further rock breakdown. One reason is that some weathering products, like an abundant supply of silts, will choke open pore spaces in the surface horizons of the soil and tend to favour sheetwash. Conversely, the sandy quartz residue produced on acid magmatic rocks increases permeability. This characteristic may become a self-enhancing element in weathering processes as soils of a certain depth, particularly in lower latitudes, will tend to retain moisture longer than exposed rock surfaces, and thus sub-soils at the periphery of bare rock surfaces may even tend to be especially moist due to drainage off the bare rock. This contrast in chemical weathering may explain the sharpness of the break at the head of some pediments and the steepening at the base of bare-rock domes. Accelerated chemical weathering in the moist sub-soil environment seems assured as long contact of moisture with rock is necessary for water to penetrate into capillary-sized spaces between crystals. Many writers have suggested that the chemical weathering of granitic rocks, for instance, is much slower on exposed outcrops which dry after each rain than on buried surfaces. In Hong Kong, granite corestones at the surface and at depth display this contrast and there are similar contrasts in granite and gneiss weathering in southern South Australia. In older tills a contrast between sound granite boulders on the surface and disintegrated boulders within the soil is frequently reported. In the Tahoe Till, 25–100 per cent of boulders have now disintegrated to grus, and although in this case and in others, such as the weathering of schist on Bjornefjell, near Narvik, there is no great demonstrable chemical change. On a much larger scale J. Büdel (1957) accounted for the contrast between broad plains and residual bornhardts in the tropics by emphasizing, in an hypothesis of a double surface of planation, the contrast between the essentially arid environment of exposed rocks and the warm humid sub-soil environment. Although Büdel was one of the first to appreciate the significance of different self-enhancing processes operating in adjacent areas to produce contrasted relief effects, both surfaces are usually too irregular for an hypothesis of a double surface of weathering to account for planed features.

Clays, apart from retaining moisture, may increase the destructiveness of a soil environment both by chemical and by mechanical means. An acid clay may by cation exchange act as an agent of weathering to produce more clay from silicate parent mineral and rock. Colloidal clay may fill fissures in some grains and the swelling of such colloidal infilling may accelerate the disruption of an already fissured crystal. The deposition of iron hydroxide in networks of microfissures smaller than 3 microns is possible and can exert a similar disruptive effect, as may the crystallization of caliche.

Finally, the influence which debris may exert on snow or ice can either tend to accentuate or to reduce some processes just as the surface soil influences weathering at greater depths. A very thin layer of debris on a glacier surface will increase melting owing to a lowered albedo. The minimum thickness of complete protection of underlying ice is 1–2 m. Similarly individual blocks, if thin enough to be warmed to their base, melt their way into the underlying ice and form a hole. Thicker blocks act as insulators and form rock tables as adjacent exposed ice melts down. After nearly one complete season of ablation the lower end of the Sherman glacier avalanche rested on a platform of ice 7 m high. The effect of this mantle will be to cause the glacier to advance beyond the point where the climate of the region can maintain it. The insulating effect is reversed if subjacent material is unfrozen soil and pipkrakes develop beneath and lift up particles small enough to transmit freezing temperatures to the subjacent soil. The presence of an impervious shield of moisture-saturated ground in a frozen condition may also have very different effects according to local circumstances. On the one hand there is a very limited infiltration capacity for surface water, yet if ground below is unfrozen a river at high stage may build up substantial pressures within the unfrozen material, particularly if this is sandy favouring the lateral transmission of the pressure. In low-lying periglacial areas surrounded by poorly drained flats, points on the surface may be domed up by the trapped water under hydrostatic pressure. Uplift rates of small domes in Russia have been measured at 1·5–2 m/20 years. Mounds entirely in arctic peats are a few metres high and termed palsas. Pingos are similar domed-up forms of sediments, usually at least 95 per cent sand and rarely more than 1–2 per cent silt, which may include ice layers. The degree to which the formation of an ice core precedes, accompanies or follows doming is not clear. The overburden is usually half to a third of the total pingo height. Their sides are usually as steep as 20–30 degrees but of different orders of size. Some are only 2–4 m high like palsas, but commonly they are 12–25 m high and 30–60 m long. If the summit of a pingo bursts the form gradually sags back to form a ridge encircling a small pool.

10. The concept of quasi-equilibrium

The concept of a quasi-equilibrium between rate of supply of material, its temporary incorporation into a dynamic depositional landform and its subsequent removal from this form, is being applied increasingly in the study of some bedrock landforms termed surfaces of transportation. The form and inclination of these surfaces appears to be closely related to the form and inclination of the mantle of debris where there would be a quasi-equilibrium between supply and removal. The implication is that the processes of rock breakdown, the transport and deposition of sediment and the ways in which depositional landforms develop, may have some decisive influence on landform development as a whole. However, it is extremely difficult to decide at present whether, over longer periods of centuries or millennia, aggradation, degradation, or balance might become increasingly well-established. There is also the probability of widespread changes during the last few millennia accompanying man's removal of natural vegetation covers and in considering the possibility of balance or grade in streams, the profound implications of the reality, scale and recency of postglacial sea-level rise are too often completely ignored. Even where a surface of transportation exists with a formal resemblance to the overlying or adjacent debris mantle it is very difficult to specify in detail the mechanisms by which this similarity arose. To explain their development it seems that the most accurate statement which can be made as yet is simply that some balance between processes may be such that any reduction in thickness of the debris mantle leads to accelerated processes of rock breakdown on the exposed bedrock, rather than to self-enhancing mechanisms which lends progressively greater immunity to the area of exposed bedrock. Where there is an even solifluvial sheet flatter slopes tend to be associated with finer mantling material. In the eastern parts of the Paris Basin, slopes on chalk, where debris is less than 7 cm in size, are only 2–3 degrees, whereas on more resistant limestones furnishing fragments 10–20 cm in size, the slopes are about 15 degrees. In areas where debris is fine or at the base of slopes, if diminution of size takes place during downslope movement, the excavation of debris requires the maintenance of progressively less steep declivities. Thus current interpretations of pediments regard these geometric rock-cut surfaces as a transportional or equilibrium surface where the prevailing hydrological conditions and that declivity provide just sufficient force to keep material moving downslope. A decrease in sediment yield or in grain size would favour erosion as would a change towards greater efficiency in runoff regime. If any of these three conditions changed in the opposite direction aggradation arrests the erosion of the buried rock surface. A great deal of work on actual pediments and slopes will be required before more specific statements can be made about slopes as transportation surfaces.

In some areas a theoretical quasi-equilibrium between inter-related processes and the development of an area as a whole may be postulated. For instance, if deflation produces a narrow basin in weakly consolidated rocks it might be assumed that this feature cannot grow deeper as the velocity of the airstreams diminishes rapidly close to its floor. On the other hand, the broadening of the basin by sheetwash cannot proceed indefinitely unless excavation of debris by deflation continues. Ultimately, deepening may be self-arresting because, when lowering reaches close to the water-table level, increased dampness on the playa floor increases the resistance of the particles to dislodgement by wind. Similarly, the recession of a coastal cliff depends on fragments being eroded to sufficiently small sizes to be moved down the wave-cut platform.

The same conclusion might apply to the similar problem of adjustments between stream gradient and the entrained sediment load for which theoretical inter-relationships have been discussed for nearly a century. Again, for the transportation of coarse debris a critical gradient would seem necessary to provide the velocity required both to move the material and also to overcome the increased channel roughness created by coarser debris. Thus it has been reasoned that in a given segment of a river there is a tendency for gradient to adjust to a load of given quantity and calibre. However, the controlling factors are likely to change, and stream diversions, abrupt structural movements, climatic change, or man-accelerated erosion might introduce into a channel debris of quantities and calibre beyond the transporting competence of the stream. Deposition then takes place in the upper part of the stream segment until the steepened gradient provides the velocity required to move the coarser material. Downstream, as seen on alluvial fans, any reduction in particle size would require less steep slopes to ensure their removal. These arguments have been used to explain why the longitudinal profiles of many streams tend to be concave upward.

The value of theoretical discussions on the concept of the adjusted or 'graded' stream is seriously impaired by lack of actual measurements, particularly of the effects of heterogeneous sizes of particles, the influence of channel cross-section and the possible role of channel vegetation. If 'load' does influence channel gradient, it has yet to be established which parameter is most significant. It is more likely to be debris of such a size to remain as a channel-lag deposit, unmoved by the extreme flood rather than the total load, which, if made up largely of material no coarser than fine silt could, as is the case in many tropical rivers, be evacuated on the gentlest gradient by comparatively small flows. Some studies have shown that dislodgement and amounts of transportation in stream channels rather than being correlated with the factors commonly assumed to be the dominant influences, such as gradient, velocity, or bank material are closely correlated with the inflow or outflow of ground water. Substantial if not total inflow may characterize a flood in a semi-arid area; meander bank caving appears to some investigators to be related in part to ground-water outflow after a flood. In other areas, particularly in some limestone terrains, deep chemical weathering of the sub-channel zone might reduce the significance of purely mechanical processes. Apart from neglect of the influence of ground-water disposition, theoretical discussions on grade are also made less realistic by inattention to the actual events of the recent geological past, particularly the postglacial rise of sea-level, and to actual examples which, apart from showing some approximate concavity, inevitably the result of flowing from mountainous source regions on to surrounding plains, show little resemblance to the smooth curve of the theoretical equilibrium profile. For instance, in contrast to some rivers in central Africa which have pronounced falls comparatively close to the sea, the Orinoco flows through an alluvial plain as a broad braided stream for more than 2000 km of its 2500 km length at gradients generally less than 1 : 15 000. The gradient of the lower Amazon is 1 : 50 000 and even in headstream areas in north-east Bolivia, gradients are similarly very low and aggradation by the silt-laden waters has forced streams like the Beni and Rio Grande into major diversions. On a smaller scale, rivers in western Malaya have gently sloping often swampy valley floors penetrating far inland.

It seems that in view of the many sources of variation and the relative paucity of actual observations any attempt to generalize here about the inter-relationship between hydraulic and sedimentological parameters of a stream and its gradient would be premature speculation.

E. Inter-relationships between Form and Process

The Lake District mountains, '. . . were it not that the destructive agency must abate as the heights diminish would, in time to come, be levelled with the plains.' William Wordsworth (1822)

In many situations one of the major influences conditioning the course of erosional and depositional processes is the characteristic of the existing land-forms themselves, posing a fundamental problem in the construction of con-cepts of landform development. On the largest scale, and within the pattern of the general atmospheric circulation, it is the major, structurally generated relief forms which largely control the distribution of glacial, humid, and arid environments. Most arid regions lie to the lee of mountain ranges which bar the passage of moisture-laden air currents and the detail of the course of climatic change may reflect similar influences. R. F. Flint (1943) believes that glacial advances in North America terminated farther north in the Dakota area than in the Mississippi lowland, not only because increasing altitudes hindered advance into this western area but also because of the rain-shadow effect of the Rocky Mountains. Even on the smallest scales, forms and features may show a certain inertia to changes and persist to influence the effectiveness of processes established under the new conditions.

The greater the available relief the greater the hydraulic potential and kinetic energy. Also, where horizontal distance is limited the propensity for streams to deepen their channels is also greater. Conversely, 'the degradation of the last few inches of a broad area of land above the level of the sea would require a longer time than all the 1000's of feet that might have been above it'

(J. W. Powell, 1875). It is therefore widely assumed that a rate of erosion tends to be proportional to the average elevation above sea-level. It is, however, often extremely difficult to identify, where some correlation between altitude and a geomorphological phenomenon exists, the exact nature of the mechanisms which the factor of altitude changes. It is even difficult to generalize about the well-understood effect of altitude in producing orographic rainfall on the uplands and in leading to greater evapotranspiration losses from lower areas. One complicating factor is that precipitation amounts depend very much on the closeness of the mountainous areas to the sea, like the situation producing the intense downpours in Central America and Java. In fact, one of the most obvious controls of landform distribution reflects the progressive eastward rise of snow-lines in mountainous areas on western maritime fringes in higher latitudes. In many such areas the elevation of cirques show a marked and consistent eastward rise. At 41°N cirque elevations in northern California rise eastward at about 3·4 m/km. At 39–41° in the Argentine Andes similar gradients range from 9 to 16 m/km. In the Ruby Range in Alaska the inland rise is about 7·3 m/km (fig. IV.43).

There are some actual measurements which suggest that greater mechanical erosion does take place with higher elevations. A. Cailleux (1948) suggested that denudation in mountainous areas might be up to ten times greater than that from adjacent plains. It is never easy, however, to decide whether it is primarily due to greater kinetic energy of available relief, that higher areas tend to be steeper so that altitude is merely an indirect expression of steepness, whether disintegration processes are more intense at greater altitudes, or whether the significance of chemical weathering might be greater in warmer lowland environments. A classic problem is that of an area of valley glaciation and whether erosional activity is greater on the ice-free peaks exposed to freeze–thaw than on the valley floors beneath ice tongues. On the other hand, water equivalents on watershed divides may be minimal in late spring thaws because snow depths are minimal on such sites. Ridges therefore might be ice-free because they are unsuitable for ice-cap accumulation not because they are the product of ice erosion. One example of a measured relationship between altitude and denudation comes from south-west United States, where some of the thickest alluvial fans occur at the outlets of basins which in parts exceed 2250 m in maximum elevation. Rivers on the flat swampy lowlands on the left bank of the Dneiper have sediment concentrations of less than 20 gm/m^3, whereas in headwaters on the north-east slope of the Carpathians turbidities increase to 250–500 gm/m^3. A further point is that much of the material eroded from upland areas is evacuated whereas the amount of material that is subsequently removed from flat terrain may be only about half of that dislodged. One effect of rugged mountainous relief is to make chemical weathering relatively negligible, partly because debris is removed too rapidly by

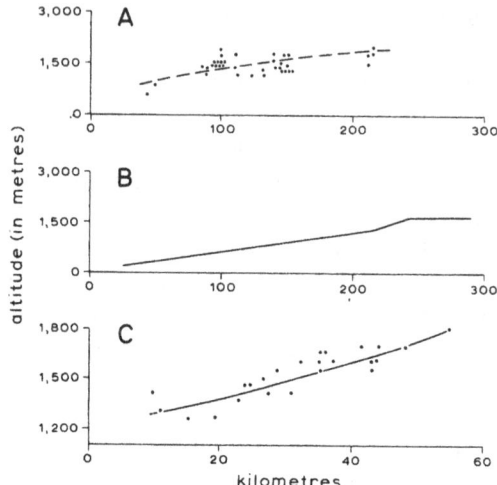

Figure IV.43 Altitudinal relationships of cirques.

A. Scotland, south of National Grid line 700 (*from Linton, 1959*); distance measured eastward from the National Grid line 100 east.

B. St Elias Mountains, from the Malaspina Glacier to Ruby Range (*from Denton and Stuiver, 1966*); distance measured eastward from the Gulf of Alaska.

C. Chile–Argentine Andes, in the latitude belt 41° 00′–41° 20′ from Llao-Lao to Bariloche (*from Flint and Fidalgo, 1964*); distance measured eastward from long. 72° 00′ W.

mechanical processes, even where there may be a combination of climatic factors favourable to chemical weathering. Springs in the highland interior of Hawaii may have only 2·5 ppm silica in solution, whereas in shallow wells on the coastal plain values may rise to 30 ppm. In steeper areas water runs off quickly and in cooler environments snow cover provides an increasingly large proportion of the runoff, adding to the reasons why the relative importance of solution should decline as gradient and altitude increase (fig. IV.44). Also, reduced vegetation growth at higher altitudes leads to slower rates and amounts of biological activity and leaves the ground more exposed. On exposed summits above 500 m in the Southern Uplands of Scotland, the vegetation consists of *Rhacomitrium* heath and contains stunted grass species such as *Festuca vivipara*, offering little insulating cover. In less severe conditions below 500 m ericaceous species like *Calluna vulgaris* and grass species like *Nardus stricta* form a much tougher surface mat.

There have been many speculations on how steepness of slopes might affect erosional processes. W. Penck supposed that the loss of soil mass on a slope was exclusively a function of gradient, and L. D. Baver observed the relation $E = 0.065\ S^{1.49}$, where E is total soil loss in tons per acre and S is slope

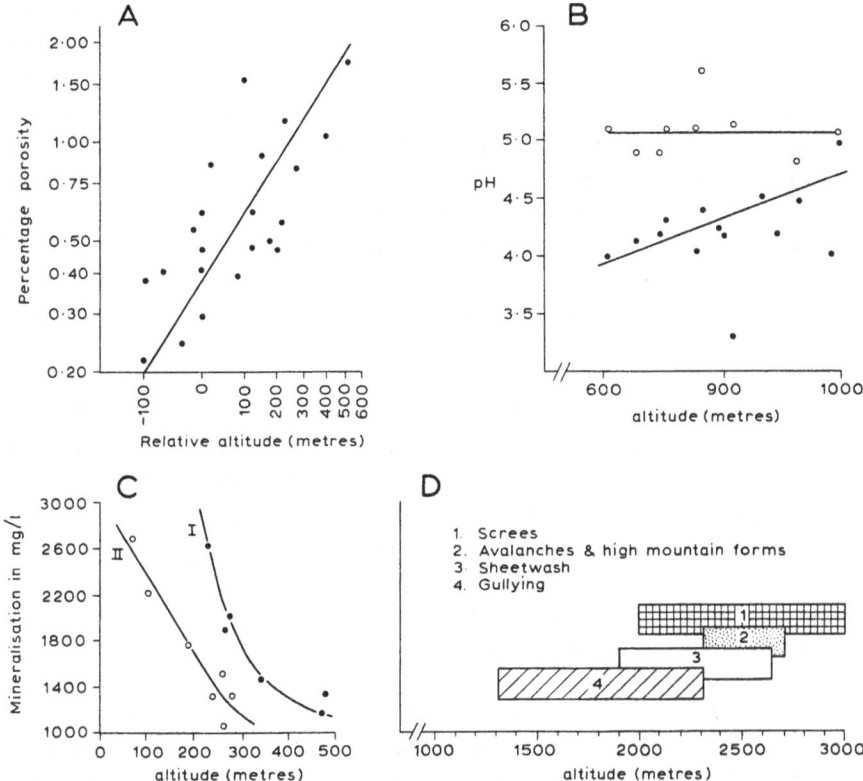

Figure IV.44 Changes in bedrock, soil, solution, and surface form characteristics with increasing altitude.

A. Increases in porosity of granites in central North Scotland (*after Godard, 1965*).

B. Increase in pH at the mineral surface (solid dots) in alpine soils in north-east Scotland (*from J. C. C. Romans et al., 1966*, Jour. Soil Sci., *Vol. 17*). There is no progressive change in pH 60 cm below the mineral surface (open circles).

C. Decrease in mineralisation of stream water in eastern Georgia, U.S.S.R (*from Z. I. Chantladze, 1965*, Soviet Hydrology). Altitudes are for the average elevation of watersheds, I, for the rivers Khadis-Khevi, Black Aragvi, Tama, and Lekhura, and II, for the Ktsiya-Khrami, Alazani, and Aragvi.

D. Distribution of certain processes and associated forms in the Barcelonnette area (*after L. Ottman, 1961*, Mém et Doc. Centre Réch. Doc. Cartog. Géog., *Ed. C. N. R. S., Vol. 8*).

gradient in per cent. One difficulty in generalizing about the effect of slope steepness on erosional processes is that beyond certain declivities certain factors diminish in significance or cease to operate. Snow, for instance, will not remain on very steep slopes exceeding perhaps 40 degrees and in consequence nivation processes are less effective in these comparatively dry environments. On steeper slopes the effectiveness of precipitation may tend to be greater because more circulation of surface and underground water reduces water losses by evapotranspiration. A more pronounced tendency, measured on slopes in Japan with declivities between 15 and 45 degrees, is for infiltration to decrease with increasing slope gradient. Landslips inevitably occur only where gradients are steeper than a critical value. During late autumn storms in 1965, landslips occurred in the San Gabriel Mountains in California only on slopes steeper than 38 degrees. In coastal areas the offshore gradient can be one of the most significant factors influencing erosional processes. When Hurricane Audrey flooded west Louisiana coasts in 1957, the width of inundated land was as much as 50 km because the continental shelf here is 160 km wide, whereas flooding along a coast flanked by deep water would have been insignificant. Another way in which the shape of the water body being filled is significant is in its influence on the form of a delta.

Apart from ways in which steepness and elevation of relief form can influence their own erosion, the pattern of an existing landform may have a profound influence on depositional processes due to perturbations introduced into flow-lines of transportational fluids. Flow-lines are greatly influenced by slight variations of ground surface and therefore become erratic and unpredictable. An obstacle decreases the cross-section through which the lower layers of the atmosphere or water-flow must pass. When flow-lines converge in by-passing an obstacle the fluid accelerates and the drag velocity is greater close to the ground surface than higher up. Conversely, on the lee-side of the obstacle, flow-lines diverge, the fluid motion is retarded, standing waves of air or water may be set up, and deposition may result. When transportation is particularly susceptible to deceleration in reduced fluid velocities, deposits elongate in shape often taper downcurrent from obstacles. Linear hills in the lee of pre-existing hills, termed *zavieja* occur in loessial areas of Czechoslovakia. In some areas, a systematic linear pattern may cover 100 000 km² and influence drainage patterns, as in the Palouse region of the Columbia basin.

The same three characteristic features often occur where an obstacle interferes with flow-lines, even down to the smallest-scale features generated by a pebble on the beach, desert floor, or in a stream bed. There is, first, a resistant object standing above the surrounding surface. Secondly, a crescent-shaped furrow is cut in front, or to one or both sides, of the obstacle by accelerating currents. Thirdly, there is a tapering ridge of sediment deposited or left as a protected remnant in the lee of the obstacle. Material may start to accumulate on the side exposed to the current if the obstacle is low and squat, as the horizontal divergence of airstreams is stronger than the vertical convergence. A large number of depositional landforms appear to string out behind a rocky protuberance. All the complex esker systems near the Casement glacier are on the lee-side of a series of solid rock ridges 60–90 m high, where ice might have been stagnant for a considerable period. In the vicinity of Sebhah, in the Libyan desert, seif dunes appear to originate where either vegetation or small rock masses, or both, form sand-traps rising above the general level of the sand sea. Arrow spits may develop in the lee of an island, like the 'comet-tails' off the Brittany coast. In areas of lowland glaciation 'crag and tail' features form one of the best examples of the lee-effect depositional landforms and small morainic ridges may even form in the shelter of individual boulders. In Kluane drift in Alaska, till ridges to the leeward of moulded bedrock knobs and ridges may be up to 10 km in length. Such distal accumulations are sometimes termed lee-moraines. Relief prominences on the sea floor can, in some instances, form a nucleus for barrier island development during emergence or sediments may accumulate in intertidal areas protected by offshore islands.

In some instances an obstruction may be sufficiently extensive to be a barrier rather than an isolated nucleus precipitating a lee-side effect. In Hardanger fiord, the position of distinct terminal moraines of late-glacial readvance appears to reflect the position of the rock threshold acting as a barrier.

Some initial relief irregularities often influence the track of transportational movements without, particularly on larger scales, producing a clearly defined depositional form. Such irregularities often influence the splitting and closing of channels and the formation of central and marginal delta basins. If the position of an obstacle in a stream is asymmetrical, an obliquely orientated line of scour starts upstream due to deflection of floodwaters. Similar diversions of ice-flow may lead to fluting and grooving of adjacent surfaces by the deflected ice. On the smallest scale the processes of ice-flow round an obstacle were observed in relation to a 2·5-cm cube bolted on to the bedrock at the base of the Casement glacier. Ice flowed over and around the cube, revealing a thin layer of clear ice and a cavity on the down-glacier side 1 cm narrower than the block itself. On a larger scale, comparatively high ground has often provided an obstacle in East Yorkshire, forcing ice-flow mainly to the east of Flamborough Head. In the lee of the headland the ice was able to fan out westward over the low-lying ground of Holderness, though in southern Holderness the south-south-west regional direction of movement was maintained (fig. IV.45). On a larger scale still, many of the lobes of ice that advanced in eastern North America were influenced, in position and direction

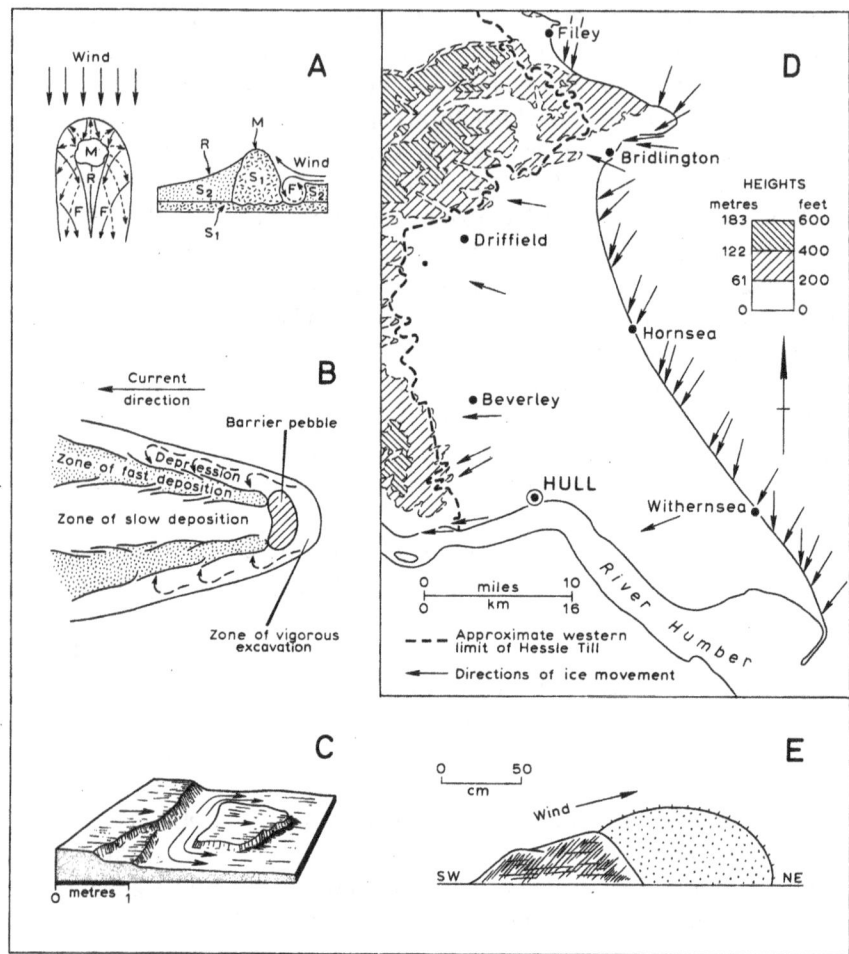

Figure IV.45 Features produced by erosional reworking and depositional processes in the proximity to obstructions and in their lee.

A. Scour marks in snow (*from J. R. L. Allen, 1965, Jour. Sed. Petrol., Vol. 35*). (*Left*) View from above of flow pattern in symmetrical current crescent of snow; M = resistant snow mass, R = ridge of deposited snow, F = furrow eroded round resistant snow mass. (*Right*) Flow pattern in longitudinal section; S_1 = first fall snow, S_2 = second fall snow.

B. Creation of current crescents of sand by a pebble deflecting the direction of stream-flow (*after S. Sengupta, 1966, Jour. Sed. Petrol., Vol. 36*).

C. Secondary ice–scour directions due to deflection by unevenness in the bedrock surface (*from M. Demorest, 1938, Jour. Geol., Vol. 46*).

D. The effect of the Chalk escarpment on directions of ice movement in East Yorkshire (*after Penny and Catt, 1967*). This effect was sufficiently pronounced for successive ice-sheets invading the area to follow the same pattern of movement.

of movement, by the major bedrock relief of the area. Many submarine canyons occur on the upcurrent side of points of land, the decrease in transportability of sediments by longshore currents leading to accumulations which eventually become unstable and move seaward. Some lagoon entrances are 'rock defended', being close to offshore reefs or foreshore rock outcrops which break-up constructive waves and prevent the completion of a barrier to the lee of this point. E. C. F. Bird (1968) cites the example of Lake Illawarra, protected by Windang Island, immediately offshore from the New South Wales coast.

A gap in a barrier produces an analogous effect on flow-lines to that of the local acceleration around an obstacle followed by deceleration in its lee, with acceleration and increased competence through the constriction and subsequent deposition in the diverging flow downcurrent from the constriction (fig. IV.46). In many rocky areas partially covered by moving sand, like the Cronese mountain area in California, the sand may assume unusual shapes, mainly due to the funnelling influence of local relief features. Sandy areas themselves may in turn be due to the funnelling of sand on to a small area by a local constriction in the relief barrier, like the Salton dunes in the Imperial valley. The frictional forces of valley walls on moving ice is often noted, and A. P. Crary's (1966) measurements in Antarctica suggest that the rate of thickening inland is a function of the width of the downvalley end, thickening inland from a 10-km-wide fiord being at a rate of about 20 m/km. An initial depression in the ground surface may tend to be the site of progressive enlargement of a hollow form as appears to happen in the enlargement of dolines in karst areas. Instances of the opposite effect have also been recorded. A striking feature of parts of the Columbia lava plateau, stripped of a former loess cover, is the small mounds 3–10 m across and approximately 1 m high. They occur above depressions in the basalt surface which appear to have trapped dust, favouring in turn the growth of vegetation which continued to gather and hold more dust. After the same process had operated for thousands of years the mounds gradually rose above the rims of the original depression.

A final influence of landform on land-forming processes is seen where certain forms develop only in specific geomorphological sites, although in

---

They were all forced to flow mainly to the east of Flamborough Head, north-east of Bridlington.

E. A gorse tussock building up in the lee of a granite block in the Scilly Isles (*from E. Lenze, 1966*, Erdkunde, *Vol. 20*). Similar features may also be produced on slopes (see fig. IV.14), there is classic 'crag-and-tail' (fig. IV.23) and in contrast to (E), above, fig. III.14 shows how vegetation can obstruct as well as flourish in shelter.

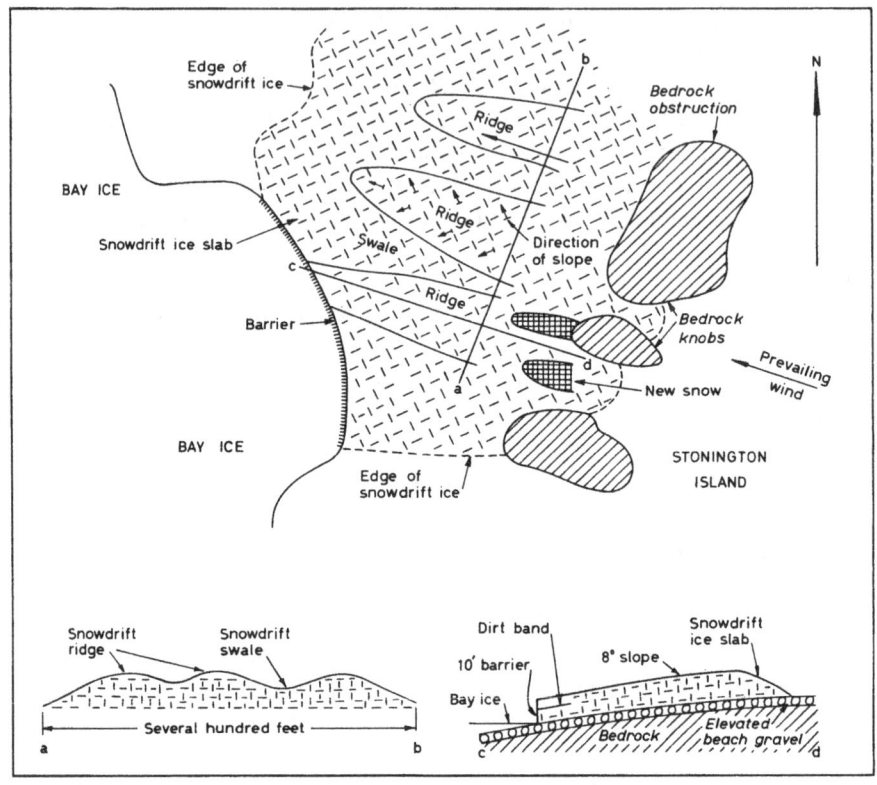

Figure IV.46 Snowdrift ice slabs to the lee of a discontinuous bedrock obstruction (*after Nichols, 1960*). The slabs, observed on the east side of Stonington
Island, Antarctica, form by the accumulation of wind-blow snow like sand
shadows.

many instances such associations may pass unnoticed. Such restrictions apply
to many glacial features. In the north European plain, ice-pushed ridges show
a marked correlation with river valleys, having been pushed from the valley
axis outwards. In the Illinois drift area, bulky crevasse deposits and narrow
segmented aligned ridges occur on plateau tops but not where the ice over-
rode a rolling relief like that of the Mount Vernon hill-country. Conversely,
because the length of sand-grain trajectories increases over a rough surface,
sand-dunes may fail to develop in such places, although much sand may move
across the area. Cuspate spits, dependent on currents from two directions,
occur only in narrow bodies of water where the width is about half or less of
their length. At the most obvious level, a hillslope facing a plain is more likely
to have a pronounced concave footslope than one facing the opposite side of a
valley.

F. The Significance of Changes in Weather and in Climate

1. Local changes due to differing aspects

As much a continuation of the discussion of the influences of landform on process, as a starting-point for considering changes in climate, is the changes in land-forming processes accompanying shifts in the compass direction towards which land surfaces face. A major influence is the orientation in relation to the sun's position. A secondary factor, frequently superimposed on the effects of the first is orientation with regard to the direction of prevailing moisture-bearing winds. Thirdly, wind direction may be important where conditions are dry enough and winds sufficiently strong for transportation of material by wind to be significant on land, and on all shores the length and direction of fetch are always major considerations.

On the 10-degree slopes of a valley in the lowlands of southern England, summer temperatures tend to increase up both north- and south-facing slopes. However, while the north-facing slope may increase by 4°–8°C, upper parts of south-facing slopes may be 5°–20°C higher at the ground surface. In winter temperature contrasts between valley floor and sides or between opposed aspects may be small, although on brighter days temperatures may be as much as 6°C higher on the south-facing slope. For inclined surfaces in higher latitudes it is variations in intensity of cold that might be more significant, but it is difficult to specify how mechanical weathering might be affected. At Valday in the Russian plains, with a December–March average temperature of − 7·7°C, soil froze to a depth of 31 cm according to measurements on the north-facing slope of a ravine in contrast to a depth of 22 cm on the south-facing slope, and the water stored in drifting snow was 30 per cent greater. The consequent greater spring runoff amounted to an extra 50 mm on the north-facing slope. In the Mackenzie delta area, disintegration by frost in east–west gullies appears to be more intense on north-facing slopes. In the Jura, by contrast, greater accumulation of debris at the foot of south-facing slopes might be due to greater amplitude of temperature changes and more diurnal freeze–thaw cycles. Yet in the Kenai peninsula in Alaska, south-facing slopes are steeper than longer, highly dissected north-facing valley-sides where winter snow lasts into June. At Resolute in North West Territories, asymmetry is reversed, perhaps because the maximum thawing of permafrost on south-facing slopes favours solifluction but without conditions becoming sufficiently mild to encourage vegetation growth.

Although the relationship of land surface to the direction of sun's rays is more rigorously invariable over spans of geomorphological time, situation in relation to prevailing winds is an equally important inter-related factor. This is mainly as a control on the supply of moisture on a regional scale. The rainfall range in Puerto Rico is from 750 mm on the south coast to 2500 mm in the north-east, due to the prevalence of the north-east trades, and appears to correspond with the transition from knife-edge ridges with 30–45-degree slopes, clayey soils, and the sheetwash and gullying processes which accompany torrential rains, to the rolling hills and more rocky soils of the drier south. Although it is possible to point to many examples where a leeward position is arid, like north-east Iceland, eastern southern Italy, or east South Island, New Zealand, it is easier to indicate instances in cool environments where the influence of snow drifting into a leeward position appears to produce the opposite effect, with clear morphogenetic results.

Within a given small area aspect is a crucial factor in its control on soil moisture as exposure to sunshine accelerates evapotranspiration processes. In the Salskaya steppe, the contrast in an undisturbed area was 97 per cent average soil moisture for a north-facing slope and 80 per cent for a south-facing one. Under a 75 per cent afforested cover soil moisture was slightly higher but the contrast was reduced, comparable figures being 102 and 90 per cent. In sub-humid areas there is sometimes a correlation between steeper north-facing slopes, their soil moisture conditions and denser vegetation. In San Diego county, vegetation reduces the amount of effective precipitation, runoff, and the erosional impact of falling rain, and exerts a binding force on the soil. Generally, however, the effect of shade on a northern exposure is to favour higher soil moisture contents despite the concomitant increase in vegetation growth. In meadow chernozem soils the spring moisture deficit on a north-facing slope may be less than a third that on south-facing slopes. In cooler environments where increased exposure to sun's rays does not lead to critical losses of soil moisture, biochemical processes, at least, are more active on the south-facing slope. In the Medicine Bow Mountains, Wyoming, at 40° N and at an altitude of more than 3300 m, productivity on a south-facing slope with 73 per cent plant coverage was measured at 585 kg/ha, compared with 224 kg/ha on the north-facing slopes with 54 per cent coverage. In parts of Sweden under natural pasture, south-facing slopes, whether 20 degrees or 4 degrees, are subject to less leaching, have higher pH values, greater pore space, water capacity, and infiltration rates, and brown earths may form, whereas on the north-facing slope, where evapotranspiration is less, podzolization prevails.

Added to the complicated contrasts between processes operating on opposed slopes is the possibility that greater sediment production on one side might force the valley floor stream against the opposite bank and the undercutting leading to its steepening. In other instances it is difficult to decide whether present-day asymmetry of form is attributable to contemporary contrasts in processes or whether the greater steepness of one side is a relict feature of past conditions. For instance, in the Bitterroot Range in western

Montana, the north-facing wall is much dissected by cirques and hanging valleys, whereas there is evidence of only gullying on south-facing slopes. C. B. Beaty (1962), by counting the mass-movement phenomena, active rill-and-gully systems, and nivation depressions estimates that contemporary gradational processes are perhaps two to three times as active on the shaded north-facing slopes. Table IV.11 lists the contrast in mean slope angles.

Table IV.11 Slope inclinations of selected trunk canyons in the Central Bitterroot Range (*from Beaty, 1963*).

DRAINAGE	SOUTH WALL (facing north)	NORTH WALL (facing south)
Bear Creek	17	23
Fred Burr Creek	22	27
Mill Creek	21	27
Blodgett Creek	18	25
Sawtooth Creek	17	24
Roaring Lion Creek	21	35

In cold environments aspect may be critical in determining where ice accumulates. As well as exposure to sun's rays, position in relation to prevailing winds is again significant because of the lee-side accumulation of drifting snow. For example, on the Decade glacier, Baffin Island, there is a mountain ridge running almost parallel to the south-west margin of the glacier which probably leads to the snow accumulation maximum in its lee (fig. IV.47). In the Alaskan Range the last remnants of the dissipating glaciers are usually preserved only on slopes descending north and north-westwards, although in terms of climatic statistics the southern side is wetter and cloudier. Between 46° and 51°S the Patagonian ice-fields show a marked contrast between the western slopes with a maritime climate and the abrupt transition eastward to continental conditions. About 1945 the contrast was particularly marked with western glacier fronts close to the forest edges, whereas virtually all eastern glaciers had retreated markedly from recent end moraines. In east-central Africa, former glaciers descended lower on the eastern than on the western flanks of Mount Elgon and Mount Ruwenzori. A similar relationship appears to have existed on Kilimanjaro during the last major glaciation, but altitudes reached on the southern flank were even lower than those on the east.

There are many landform contrasts which reflect the cumulative effects of contrasting processes on land surfaces of different orientation. Some are perhaps too readily overlooked, some studied only in one or two instances,

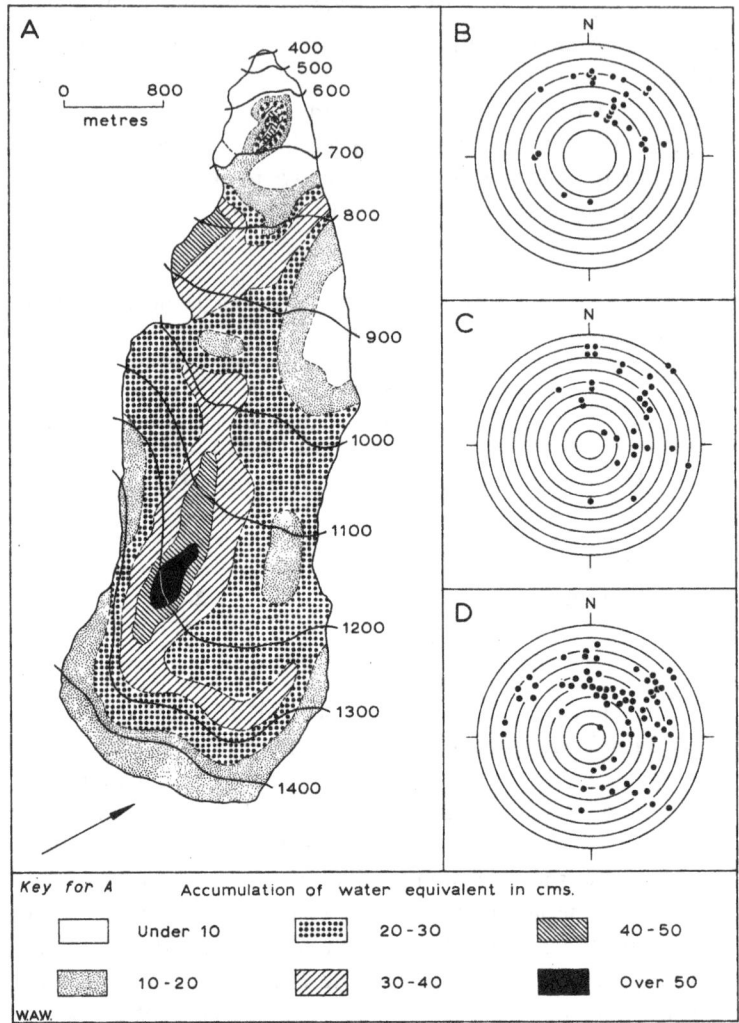

Figure IV.47 Influence of aspect on glacier accumulation and disposition.

A. Ice accumulation map 1964–5 of the Decade glacier, Baffin Island (*from G. Østrem* et al., *1967*, Geografiska Annaler, *Vol. 49A*) showing the effect of a mountain ridge, not shown on the map, running almost parallel to the south-western outline of the glacier in causing the snow accumulation maximum in the leeward position along the left-hand part of the map.

B–D. Polar diagrams showing the relation between cirque floor altitude (concentric circles) and the preferred north-easterly aspect (radii).

B. Northern Nain–Okak part of Labrador (*from J. T. Andrews, 1965*, Geog. Bull., *Vol. 7*) circles 0–915 m.

C. North Caernarvonshire, North Wales (*from Seddon, 1957*) circles 270–750 m.

D. West-central Lake District, North-west England (*from Temple, 1965*) circles 300–780 m.

but for others, like the glacial cirque, the influence of aspect has been widely studied and its striking degree of control clearly established. One of the first observations was made nearly a century ago when approximately 70 per cent of the cirques in part of the Jotunheim, in Norway, were seen to lie on the north side of the massif. For large parts of the mountains in northern Scandinavia the orientation of glaciers and snow patches, as well as cirques, is generally eastward. Near Nain, Labrador, the comparable figure is 89 per cent, with 72 per cent of the cirques orientated between north and east. In the Trinity Alps, California, 45 per cent face north, 20 per cent north-east, and 15 per cent north-west. Comparable figures for the north-west Highlands of Scotland are 24, 31, and 5, with 16 per cent facing east. Generally, cirque development in the northern hemisphere tends to be on north-facing slopes with few facing south. In addition to examples mentioned already, other instances include the Uludağ Massif at 40° N in Turkey, the high land above the South Fork of the Cœur d'Alene river, similarly in the Bitterroot Range farther to the east in Montana and in many valleys in the Scottish Highlands like the Great Glen and Glen Coe, or in the Vosges. There are local variations in the preferred orientation as in Newfoundland where eastern cirques are neither as well developed nor as large as those in the west, but nonetheless fig. IV.47.B–D illustrates typical patterns.

The influence of aspect on the pattern of glacier alimentation is sufficient for its possible influence on the direction of ice-orientated features to be considered. Leverett (1929), for instance, suggested that the ice-sheets of the North American glaciations tended to advance westward because storms came from the south-west. Other features associated with ice may show a preferred orientation, particularly towards the north-east. Nearly all the end moraines in the north-east part of the Jotunheimen are ice-cored, whereas all end moraines on the south-west side are ice-free. Another example, again due to snow drifting by westerly winds, is the strong predominance of avalanches on lee-side slopes. These are mainly east-facing in northern Lapland. In south-central New Mexico, rock glaciers in the San Mateo mountains occur on slopes with north or north-west exposures, protected from the sun's rays. On Niwot Ridge, 24 km west of Boulder, Colorado, the restriction of stone-banked terraces to lee-slopes suggests that wind-drifted snow supplied the moisture which facilitated movement. However, in highland areas in eastern Australia, similar features are restricted to the upper portions of windswept south-west to north-west exposures, and the steps may even have an off-contour dip towards the wind. Landslides are often characteristic of north-facing slopes in the northern hemisphere. In some valleys in eastern Montana, nearly all the landslide blocks are on escarpments facing north-east. Farther south and west the thickest alluvial fans tend to be associated with basins having a north or north-east aspect. As these last few examples show, milder

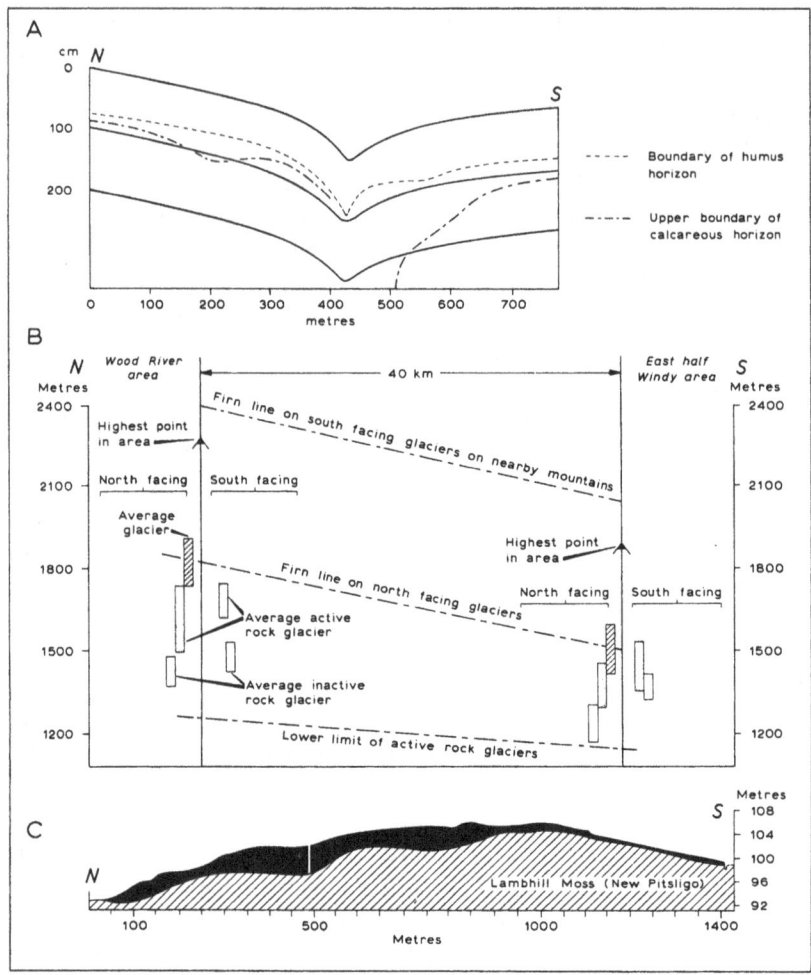

Figure IV.48

A. Profile across a ravine in a chernozem area, Central Russian Highland (*after V. V. Gertsyk, 1966,* Soviet Soil Science).

B. Altitudes of glaciers and rock glaciers in two areas in the Alaskan range, showing contrasts according to aspect (*from Wahrhaftig and Cox, 1959*).

C. Contrast in depth of blanket peat according to aspect, north-east Scotland (*from R. Glentworth and J. W. Muir, 1963,* The soils of the country around Aberdeen, Inverurie and Fraserburgh).

ice-free environments still show strong correlations between landform and aspect, although the processes involved are usually inferred rather than separately and specifically studied.

Exposure to wind may in itself be a further factor where wind action causes erosion, transportation, and deposition of weathered materials. After 6 years of exposure to natural sandblast, the exposed face of a brick in an experimental site set up by R. P. Sharp (1964), was cut back a maximum of 4 cm on its westerly face, whereas the north and east faces showed no wear. In wooded areas certain exposed points and ridges can be particularly susceptible to root-disruption by wind-thrown trees. Perhaps fetch in relation to sea and lake shores provides the clearest controlled effect of exposure to wind direction. On coasts with rocky headlands, the more extensive abrasional platforms may develop where there is some shelter from the direction of dominant wave activity. In intertropical atolls the importance of aspect is seen most clearly, with the more irregular, open side on the protected leeward shores. J. A. Steers (1937) observed that the Australian sand cays are orientated to the east, at about 45 degrees to the prevailing south-east wind. Steep ridges of coral rubble occur on the exposed sides of sand cays facing the heavier waves, with finer material and depositional forms on the leeward side. Wind direction may control lake shape in certain circumstances. In the coastal plains of Alaska, the remarkably systematic elongation of lakes at right-angles to prevailing wind directions appears to be due to differential thawing of permafrost by wind-driven waves and eddies.

Many depositional landforms influenced by wind direction have already been discussed. If there is any geomorphic significance in wind direction and the transportation of airborne salts, such as sodium and chloride from windblown sea spray in coastal areas, this awaits investigation. But with reference to sources of moisture and heat, the factors of aspect and exposure in landform studies cannot be underestimated nor too carefully investigated.

2. Changes in the effectiveness of processes with time

Study of short-term hourly, daily, or seasonal changes in the intensity of processes is an integral part of landform study. Whether the time interval is minutes or millions of years the relative intensity of processes changes, if not the nature of a process itself. Some patterns of change tend to be self-regulating, like the summer replacement of beach sand removed in winter storms. In other instances the very effectiveness of a process depends on the complementary effect of variations in processes. Pipkrake development depends on frost at night and thaw in the morning, and is most effective when similar cycles last a few days. The effectiveness of erosion may well be

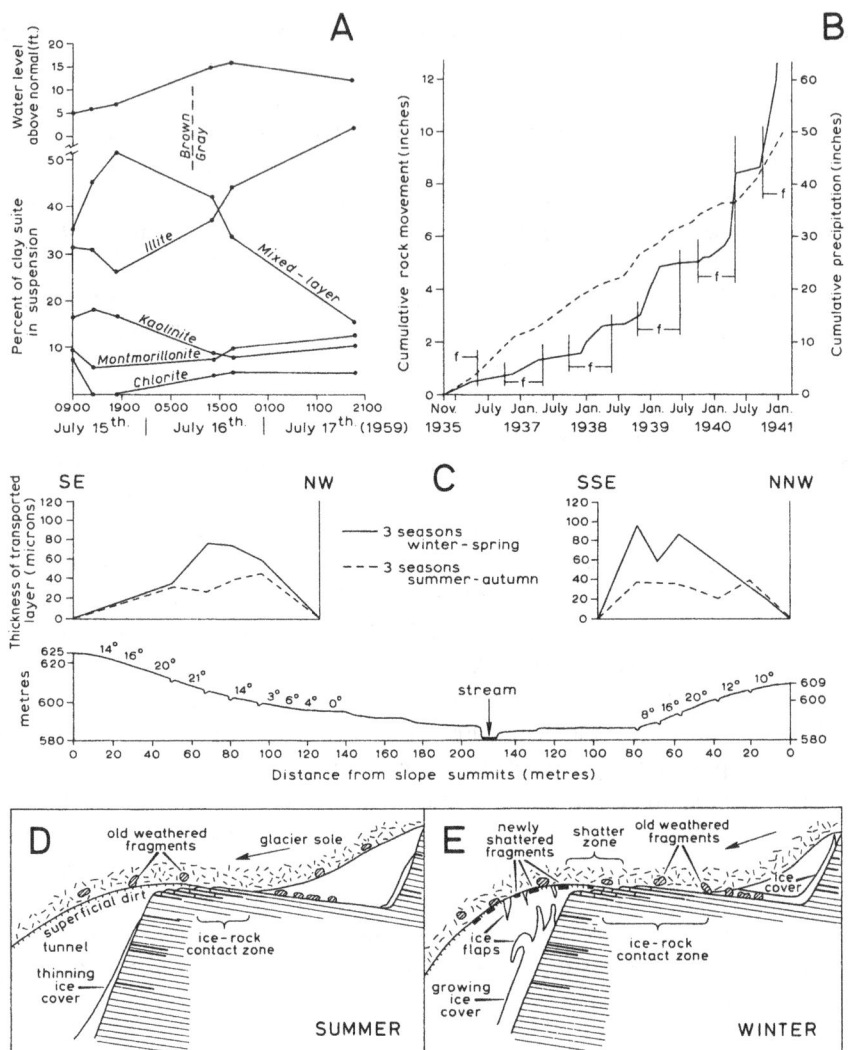

Figure IV.49 Variations in the intensity of transportational and erosional processes.

A. Variations in suspended clay suites of the Arkansas river at Ponca City, Oklahoma, during a major storm (*from C. E. Weaver, 1967*, Jour. Sed. Petrol., *Vol. 37*).

B. Cumulative movement of Threatening Rock (solid line) and cumulative precipitation (dashed line), Chaco Canyon (*from S. A. Schumm and R. J. Chorley, 1964*, Am. Jour. Sci., *Vol. 262*). Vertical lines indicate movement during periods of freeze–thaw, f.

C. Seasonal contrast in amounts of material transported off grassland slopes by rillwash (*from T. Gerlach in Macar, 1966*).

increased by the dominance of mechanical weathering in cooler parts of the year or in a glacial period, and chemical weathering in warmer parts of the year or during an inter-glacial.

The study of short-term changes in the intensity of processes is necessary, in the first instance, to gauge the representativeness of observations contributing to an annual mean. Secondly, a clarification of the nature of short-term changes may be the most reliable approach from which to comment on more scanty data related to changes over much longer spans of time. Despite some obvious differences, the changes in a 24-hour day have some similarity to those within a year; the year might be a useful model for a part of Pleistocene time, as might year-to-year changes in considering the order of change from one millennium to the next.

Small-scale features in unconsolidated material may be changed in a matter of minutes. A wind of 30 km/hr or more is sufficient to reverse the asymmetry of a sand ripple in a minute or two. Certain parts of the day may be more critical than others, particularly where biochemical processes are involved. Precipitation of calcium carbonate is greater during the day than at night. Flow of glaciers in northern Greenland is 5–10 per cent more rapid during the day. For less obvious reasons landslide movements in the Upper Columbia river basin appear to be most frequent about 3 a.m. J. Tricart (1965), observing that in intertropical latitudes even small glaciers have huge terminal moraines, but that fluvioglacial forms are very poorly developed, attributed this contrast to the reduced effectiveness of melt water produced by diurnal thaw compared with that of the seasonal thaw in mid-latitudes.

Day-to-day changes within a given season depend on short-term extremes of temperature or precipitation. In stream transport changes may even depend on where, in a large catchment, rain falls. Studies of the Arkansas river at Ponca City showed that storms in northern Oklahoma and central Kansas increased the amount of mixed-layer illite–montmorillonite in the river, whereas storms farther to the west, in the Rocky Mountain foothills, eventually cause a major increase in the illite in the suspended load (fig. IV.49.A).

Seasonal factors can be among the most significant events of geomorphological processes. In northern India the silt load other than in the monsoon season is a trivial fraction of the whole, an important factor being the abrupt increase in discharge which may still be relatively feeble in June. It is impossible to generalize about the time of year when sediment removal is greatest

---

D and E. Differences in frost-shatter between summer and winter conditions in cavities beneath a Norwegian glacier (*from R. G. Bennett, 1968*, Norsk geog. Tidssk., *Vol. 22*), as observed under the margins of Østerdalsisen, Svartisen, Norway.

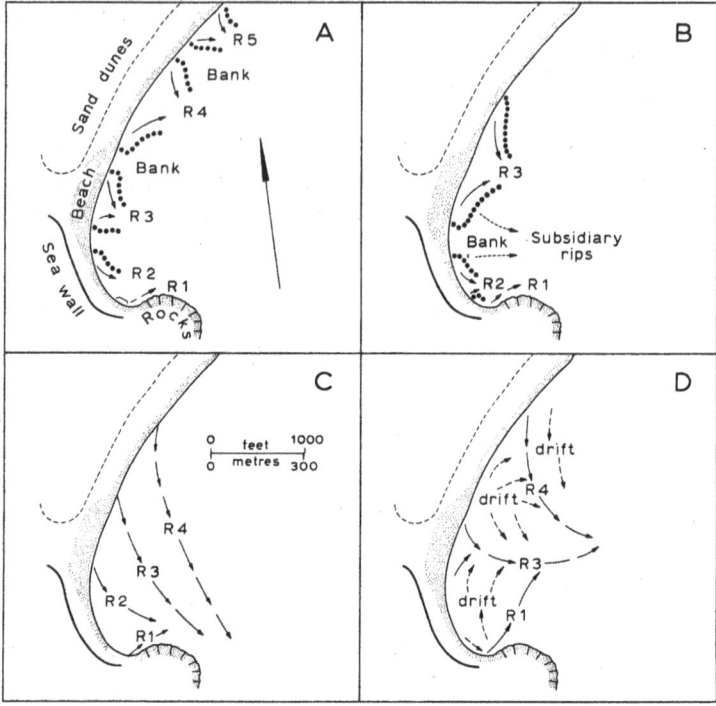

Figure IV.50 Changes in position, size, and number of rip currents due to changes in sea condition (*after P. McKenzie, 1958*, Jour. Geol., *Vol. 66*). Dee Why beach, New South Wales. (R) indicates positions of rips.

A. Normal positions under steady northeasterly winds.

B. Changes in position as tide falls, before the currents become consolidated in their channels.

C. Enlargement, joining together, and moving south of rips under heavy storm from the north east.

D. Joining together and moving north as moderate swell from the south increases to storm.

because of the differing times at which controlling factors are most effective. A study of bank cutting near Rockville, Maryland, revealed that although the largest discharge occurred in summer, soil wetting was more thorough in winter, which, in combination with freeze–thaw, led to maximum bank erosion. In Baja California, typical summer winds are unidirectional and onshore due to the development of low pressure over the inland desert. The winter wind pattern is bi-directional, with the strongest sea-breeze of about 9 m/sec compared with a summer maximum of 12 m/sec in mid-afternoon.

Associated movements of sand are about 8 cm/day in summer, but only 2 cm/day in winter. On some coasts the pattern of drifting varies seasonally. On Boomer Beach, California, waves from the south-west move sand northward in summer and waves from the north-west drive it back in winter. A summer beach in many parts of the world is higher, wider, and composed of finer sediments. In stormier winter conditions high, steep-fronted waves tend to remove much of the beach and leave low-level flat areas covered with a lag of coarse material. On Hawaiian beaches the average increase in grain size is about 0·5 *phi* units. In general, drift rates in periods of heavy surf in winter-spring months may be three to ten times greater than those prevailing during the major portion of the year. In higher latitudes, although seasonal contrasts remain, the effect of freezing spray and frozen sand is to protect a beach. However, with little infiltration or surface roughness, uprush extends further than under summer conditions. Deposits may reflect seasonal changes too, particularly in cooler environments with a rhythm from summer silt to clay in winter. The clay settles out when flows are low and lakes ice-covered. Compactness of surface materials may vary seasonally. The greater activity of benthic animals in summer probably explains why a sandy beach may be significantly less compact in summer than in winter. In contrast, the badland slopes studied by S. A. Schumm (1956) are compacted in summer by rain-beat, become less permeable and the proportion of surface runoff increases. On a larger scale, while washboard or de Geer moraines were assigned too hastily the genetic name, annual moraines, the possibility of their deposition annually remains. J. T. Andrews and B. B. Smithson (1966) describe some 670–790 moraines in north-central Baffin Island where deglaciation lasted 700 ± 180 years.

The disappearance of seasonal snow cover marks a drastic change in the physical and biochemical characteristics of the earth-atmosphere interface and the way in which interactions operate. Spring thaw is a time of considerable transportational activity, even if the debris entrained was loosened at some earlier period. In seasonally frozen streams, a distinctive pattern of sediment transport may follow the spring thaw. Typical of many streams of this type is the Colville river, Alaska, which is frozen for 7–8 months in a year. A distinctive characteristic of the first flood is the very low content of solids per unit volume of snow melt water, even at a high stage (fig. IV.51). During summer the fluctuations in suspended silt correspond fairly closely with discharge and thus with precipitation, but with the maximum load per unit volume tending to occur before the peak discharge. Turbidity is greatest when all the catchment thaws. However, later in the summer, the silt content is virtually the same whether the stage is rising or falling. Turbidity is reduced as frost begins to affect alpine heaths and because the first floods, once the soil melted, had already washed out much of the loose material.

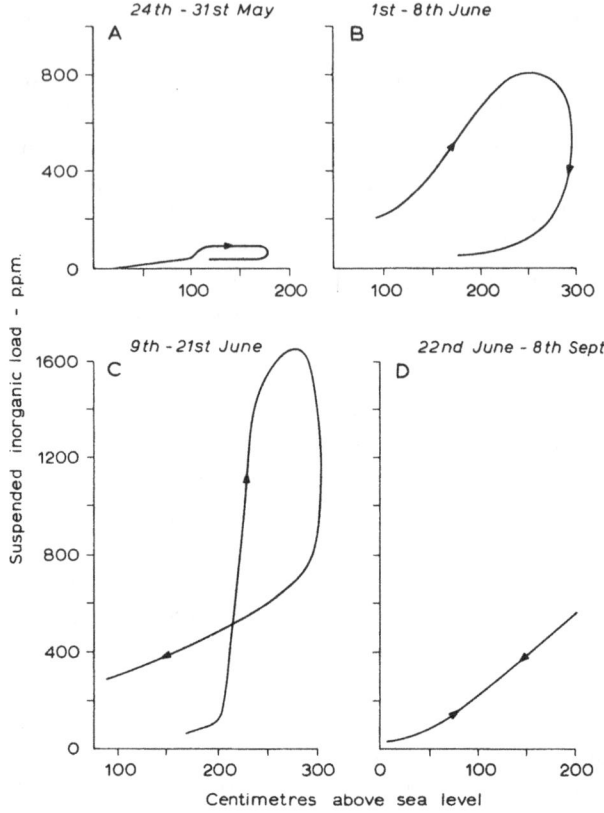

Figure IV.51 Changes in suspended inorganic load in relation to water stage at different seasons (*from Arnborg* et al., *1967*), illustrating the effect of the 1962 thaw on the Colville river, Alaska.

Dissolved loads of streams in cooler environments may also show sharp changes as snow melts. Fig. IV.52 illustrates the case of the Pyalitsa, a lowland river in the U.S.S.R. With snow melt there is an abrupt flood with associated reduction in mineralization of the water (line 2, fig. IV.52) and rapid variation in the proportion of the main ions. Sodium follows the chloride curve (line 3) as both are typically present in water from a snow cover whereas the bicarbonate ions show corresponding falls. As soil and ground water begin to enter channel systems in significant amounts as the high water recedes, mineralization increases and the relative concentration of bicarbonate ions increases at the expense of chloride ions. In middle and higher latitudes there may also be a relative surge of biochemical activity in early spring when plant activity recommences. On the steeper slopes and cliffs, both inland and coastal, mass movements of smaller material loosened by freeze–thaw and of larger blocks due to the saturation of the ground with melt

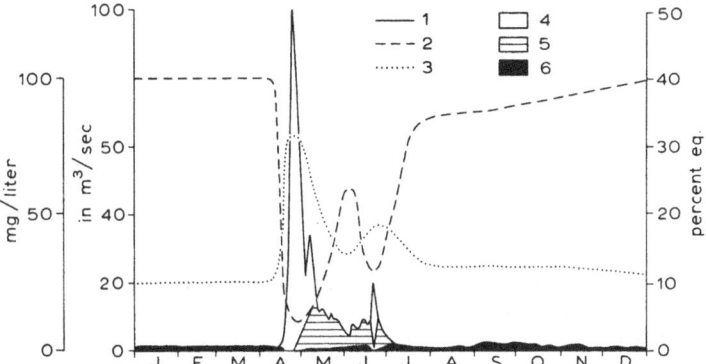

Figure IV.52 Hydrochemical regime of the Pyalitsa river, European U.S.S.R., reflecting changes in origin of drainage water *(from B. G. Skakalskiy, 1966, Soviet Hydrology)*: (1) discharge in m³/sec., (2) total ions, mg/l., (3) chloride, mg/l., (4) water of slope origin [surface-slope and soil-surface], (5) Water of soil–ground origin, (6) ground water.

water follow the spring melt. The latter process was probably involved in the accelerated winter movements of the 30 000-ton Threatening Rock (fig. IV.49.B). The frequency of rockfalls on three railways near Bergen, 1920–58, reaches a marked maximum when air temperatures have passed above zero, and frequency of slides in quick clay in the same area shows a similar pattern (Table IV.9). Amounts of slope erosion may also vary according to season (fig. IV.49.C).

In savanna areas a profoundly significant effect is the seasonal contrast between wet and dry conditions. The end of a savanna dry season is most critical, as the screening effect of parched vegetation over a desiccated soil is at a minimum. The first winter storms which are generally abrupt and substantial, lead to rillwash on the unprotected ground. Chemical changes are equally significant with evaporation leading to the precipitation of minerals hydrolysed in the preceding wet season.

Compared with seasonal changes within a year, fluctuations of their net average annual effect usually involves changes in degree only from year to year. The main interest lies in observing the degree to which an individual set of observations might bias short-term observations. There is also the possibility of observing contemporary forms, mainly in unconsolidated materials, which are relics of some extreme activity in a previous year rather than a form inter-related with the intensity of processes usually observed. C. G. Tuckfield, studying gullies in the New Forest, Hampshire, noted that although the period April 1960 to April 1961 was unusually wet, with exceptionally heavy rain-storms in the summer, there was not enough heavy rain in the subsequent two years to remove the fallen material completely. With the additional factor of relatively very cold winters in 1962 and 1963, widening

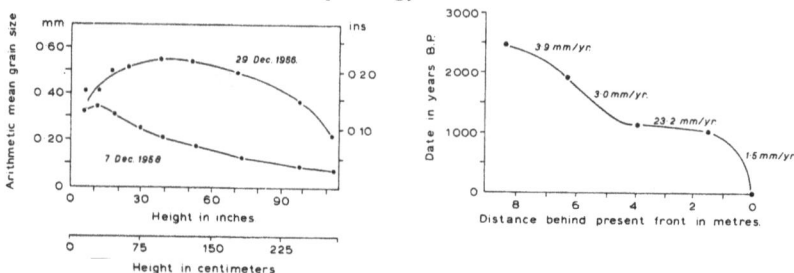

Figure IV.53 Variations in movement rates of material. (*Left*) Arithmetic mean size of wind-transported sand plotted against height above the ground (*from Sharp, 1964*); 29/12/1956 collected for 83 days of strong wind regime; 7/12/1958 collection followed 170 days of gentle wind regime. (*Right*) Rate of downslope movement of the front of a stone-banked terrace, Niwot Ridge, Boulder County, Colorado, based on radiocarbon dates (*from Benedict, 1966*).

was more effective than deepening. For larger streams year-to-year variability may be less marked. Annual runoff in the United States over a 25–30-year period is usually within 10–20 per cent of the median. Sediment loads, however, show greater contrasts. For example the 560 000 m³ of sediment deposited in Lake Constance in 1949 was only one-tenth that of the

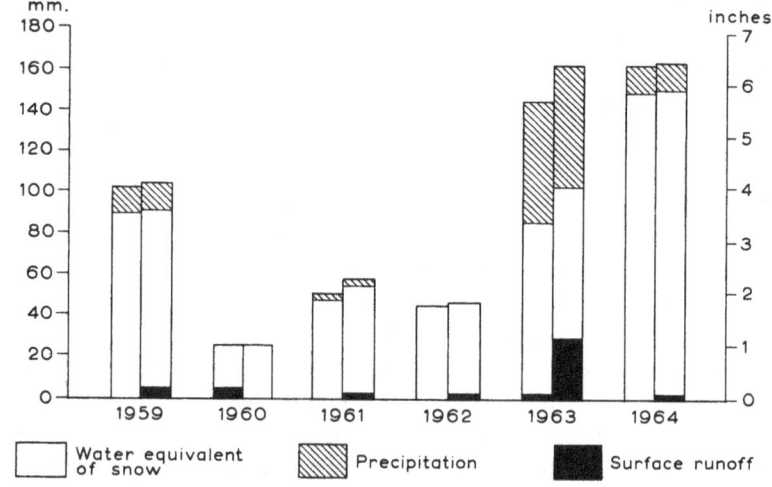

Figure IV.54 Year-to-year fluctuations in surface runoff in spring following snow-melt, Donets River area, Ukraine (*from A. V. Plashchev and L. A. Uvarov, 1966, Soviet Soil Sci.*). For each year the column on the left is for a dense oak plantation, that on the right for a maple plantation which stands on the lower part of the same slope as the oak plantation. Some of the factors causing the fluctuations is protection of soil from freezing by a thick snow cover with infiltration of snowmelt water consequently unimpeded, e.g. 1964; small amounts of snowfall, e.g. 1961, also reduce runoff. In 1960 deeply frozen sub-soil caused runoff under the oak forest; measurements were not made in the maple plantation in 1960.

maximum during a 30-year period. This maximum, which amounted to 5 million m³, was observed in 1935. Annual changes in dissolved loads are much more closely a function of discharge, overriding the effect of retardation of rates of biochemical processes in drier or cooler years. The direction of dominant winds may vary from year to year. For instance, records of onshore winds at Blakeney Point indicate that north-west winds were dominant in 1958, and north-west winds in 1960. There may be a slow shift over several years as appears to have happened on small reef patches in Djakarta Bay where coral ramparts have changed position over at least a 75-year period. The mean annual velocity of movements in dynamic depositional landforms varies also; for instance, dunes in Peru travelled 41–226 m/year faster in 1958–64 than in 1955–8. An example of the variability of beach sand transport is the accretion behind the Santa Barbara breakwater, California, ranging from 225 m³/day to 1095 m³/day during the 1938–50 period. Ice accumulation varies, different combinations of atmospheric conditions producing contrasted or similar results. The average ablation gradient for the Storglaciären, Kebnekaise, Swedish Lapland, is 55 cm/100 m, but over a 20-year period it ranged from 40 to 70 cm. On a larger scale and of profound significance is the change in the ice boundary in the North Atlantic. This is one of the most variable physical features of the earth's surface, varying from year to year by several hundreds of kilometres. Seasonal changes may also influence glacial erosion and transportation (fig. IV.49.D–E). The degree of frost penetration into the ground varies. In Alsace the average maximum depth reached annually is 0·2–0·3 m, increasing to half a metre in more severe winters such as the one of 1955–6, and the exceptionally persistent frost with a recurrence interval of about 100 years may reach a metre in places.

3. Examples of past climates

Glancing over longer spans of time consistent trends are evident in the last century in the levels of enclosed lakes. Those in Oregon and Washington were at high levels in the 1870s and 1880s, and again in the early 1900s, followed by extreme desiccation until the middle and late 1930s. In many areas the results of human activity, such as pumping and the indirect effects of accelerated erosion are difficult to dissociate from natural changes. It is also difficult to decide whether trends in levels are global or local, for example, whether the pattern of years when water levels in small lakes in the south Ukraine were high – 1882, 1906, 1911, 1912, 1924, 1927, 1928, 1931, 1932, 1936, 1939, and 1947 – is significantly similar to conditions in Oregon and Washington This instance illustrates a general problem of interpreting climatic changes and one which becomes increasingly difficult to resolve the farther back in time that investigations lead.

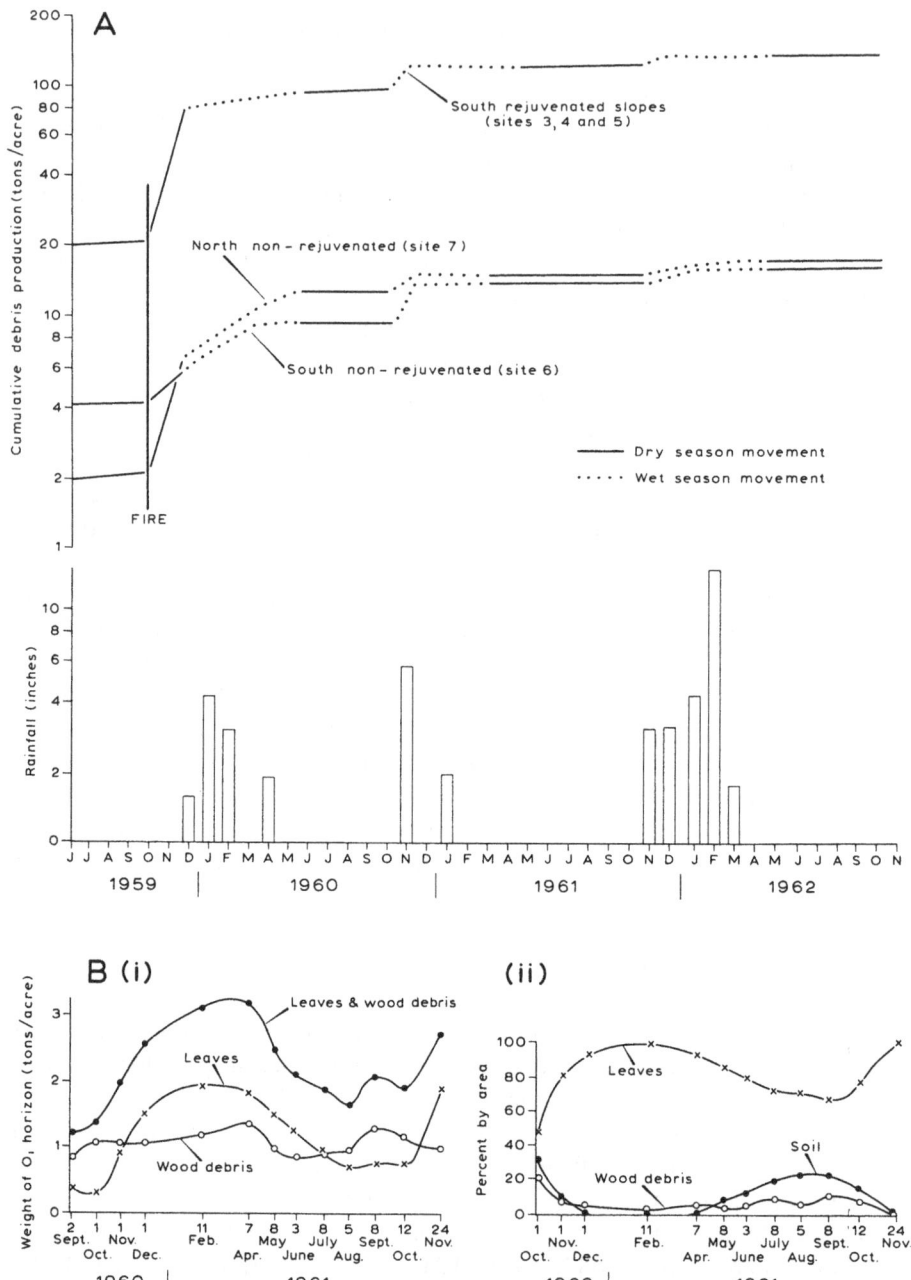

Figure IV.55 Artificial and natural changes in the litter cover on forest floors.

A. Erosion accelerated by four to sixteen times in the San Gabriel Mountains, southern California following fire *(from J. S. Krammes, 1963*, U.S. Dept. Agric.

Table IV.12 Chronology of Late Glacial and Holocene time *(from Fairbridge, 1968).*

YEARS BP	POLLEN ZONES	VEGETATION	SUB-STAGE NAMES	NORTH EUROPE ARCHEOLOGICAL CULTURES
0– 1 000	IXb	Beech and heath	Late Subatlantic	
1 000– 2 300	IXa		Early Subatlantic	Roman Iron Age
2 300– 3 700	VIIIb	Oak and ash	Late Subboreal	Bronze Age
3 700– 5 300			Early Subboreal	Neolithic
5 300– 6 600	VIIb	Oak–elm and mixed forest	Main Atlantic	Mesolithic
6 600– 7 500	VIIa		Early Atlantic	
7 500– 8 700	VI	Hazel–oak	Late Boreal	
8 700– 9 800	V	Hazel–pine	Early Boreal	
9 800–10 300	IV	Birch–pine	Preboreal	
10 300–10 900	III		Younger Dryas (Arctic tundra climate)	
10 900–11 800	II		Alleröd (Subarctic climate)	Late Paleolithic
11 800–c. 15 000	I		Older Dryas (Arctic tundra climate)	

From A.D. 980 to about A.D. 1540 Norse colonies existed in south-west Greenland under generally warmer conditions than prevail today. Climate deteriorated markedly about the beginning of the thirteenth century. From A.D. 1000 to A.D. 1450, the number of major floods on British coasts increased from seven (A.D. 1000–1200) to nineteen in the latter part of this period. In South America, ancient Indian sites at points along the Pacific coast in areas now completely devoid of any permanent fresh water suggests that desiccation has been progressive even within the last millennium in this region.

The marked decrease in rate of sea-level rise about 5000–3500 BP may be due to retardation of ice-cap melting and therefore probably indicates global

Misc. Publ. No. 970). The effectiveness of subsequent precipitation on the burnt slopes is also shown.

B. Seasonal changes in soil exposure and volume of litter on a forest floor, Madison, Wisconsin *(from G. A. Nielsen and F. D. Hole, 1964,* Soil Sci. Soc. Am. Proc., *Vol. 28)* showing: (i) that the weight of leaves and wood debris is greatest in early spring, and (ii) that soil exposure is at a maximum in late summer.

cooling. This period marks the end of the warm Hypsithermal interval and the start of the following cold phase or Neoglacial. Palaeobotanical investigations indicate that during postglacial time amelioration of climate reached an optimum in the warm and dry sub-Boreal, about 4500–2500 years BP. At that time the timber-line in southern Norway and presumably even the firn-line was 300 m higher than at present. Previously, a large portion of the Atlantic Ocean underwent a temperature increase of 6°–10°C during a period of less than 2000 years, with the midpoint of this change within 300 years of 11 000 years ago. At this time about half the water had returned to the oceans from the ice-sheets of the last glaciation. This abrupt world-wide climatic change was first recognized from deep-sea cores by the study of foraminifera (Ewing and Donn, 1956) and of the temperature-dependent oxygen-isotope fractionation that occurs during the formation of carbonate by C. Emiliani (1955). Supporting evidence includes glacial retreats and pollen profiles in Europe and North America and pluvial histories of lake basins. Climates rapidly lost their full glacial character and became more like climates of today. In some areas, like Alaska, the climate might have been warmer than at present by about 8000 BP, the evidence including that of fossil dams built by beavers in situations beyond their present range. For many parts of the world, however, the highest postglacial temperatures occurred several thousands of years after the end of this early Holocene warm interval.

Many theories have been advanced to explain the climatic variations associated with the Pleistocene, but none has been widely accepted for long. Suggestions include variations in the amount of carbon dioxide or of dust in the atmosphere, continental drift introducing critical changes in the distribution of land and sea, and the astronomical hypotheses of J. Croll (1875) and M. Milankovitch (1938) invoking variations in the earth's orbit. To all these suggestions there are cogent objections. Neither is it clear whether ice retreats were an inevitable result of self-arresting mechanisms once a glaciation had advanced beyond certain limits or whether a separate set of factors operated to induce ice shrinkage. However, the essential features of atmospheric circulation imply that there is always a tropical moist zone between the north and south arid belts. The equator is therefore always between two arid belts. Polewards of the latter lie temperate climatic zones. In this sense the specific features of Quaternary climatology incorporated no radical change from the general plan of the present climatic zones. Whatever may have been the causes of the ice advances or of their subsequent retreats some facts have been reasonably clearly established. For the Würmian glaciation J. Büdel (1963) suggests that latitudinal variations in the longitudinal strip between 0° E and 15° E included a glacial zone from 90° N to 55° N compared with a present southerly line at 77° N. The periglacial environments today commonly encountered between 69° N and 77° N spanned 45°–55° N in the Würm,

Table IV.13 Tentative correlation of Quaternary glacial and interglacial stages (*from Fairbridge, 1968*).

Glacials are in capitals and interglacials in lower case. Datings and correlations such as these are under continuous revision; more detailed local sequences will be found in the appropriate references in the bibliography.

ABSOLUTE DATING	N. AMERICA	ALPS	NORTHERN EUROPE	POLAND–U.S.S.R.
0– 67 000	WISCONSIN (early W. = Iowan)	WÜRM	WEICHSEL (VISTULA) (early W. = Warthe)	VARSOVIAN (VALDAI)
67 000–128 000	Sangamon	Uznach	Eem (Hoxnian)	Masovian I
128 000–180 000	ILLINOIAN	RISS II RISS I	WARTHE (DRENTHE) SAALE	CRACOVIAN (DNIEPER)
180 000–230 000	Yarmouth	Hötting	Holstein	Sandomirian
230 000–300 000	KANSAN	MINDEL I and II	ELSTER	JAROSLAVIAN (Likhvin)
300 000–330 000	Aftonian	—	Cromer	
330 000–470 000	NEBRASKAN	GÜNZ I and II		MENAPIAN
470 700–538 000			Waalian	
538 000–548 000		DONAU II	WEYBOURNE	
548 000–585 000			Tiglian (Tegelen)	
585 000–600 000		DONAU I	RED CRAG (BUTLEY)	
c. 600 000–c. 2 m.		Villafranchian		

thus reaching to the margin of the present-day Mediterranean climate. In the Worcestershire Avon valley 38 000 years ago the summer temperature range was approximately that found today in the high birch scrub of the Scandinavian mountains. In this environment in north-west Sweden present-day summers are very short with only 45 days when the mean daily temperature rises above 10°C. July is the warmest month with an average of 11°C, suggesting that the Avon valley was on average 5·5°C cooler than at the present day. In central Europe the mean annual temperature in former periglacial areas was 6°–10°C lower than at present, and the global cooling might have been, on average, as much as 5°C.

On three or more occasions in most continents an apparent increase in dampness accompanied a general cooling. C. C. Reeves (1965) lists the results of several workers in addition to his own from the southern High Plains which indicate that the drop in summer temperatures during pluvial periods was often very close to twice the mean annual drop. These observations might imply that the characteristic of pluvial periods in the Pleistocene was perhaps the influence on the hydrological cycle of the relatively lower summer temperatures rather than an increase in rainfall amounts. Some authors still consider that the comparison of Quaternary pluvials with glaciations is full of uncertainties. Particularly for low latitudes speculations concerning glacial and interglacial climates are controversial and sometimes internally inconsistent. Inevitably, away from the highest mountaintops the changes were essentially those involving amounts of water, lake fluctuations being again one of the more reliable records. Farther back in time the climatic history of even the early Pleistocene is still largely obscure because very few continuous marine or continental sections of deposits have survived to be discovered and described in adequate detail. However, it is hoped that increased sampling of the bottom deposits of some larger present-day lakes may yield some information. In the Cenozoic as a whole the constancy in position of climatic belts does not signify that the climates remained uniform. Apart from the moisture increases and temperature decreases in the Quaternary, there were marked temperature fluctuations even in the Pliocene. In the Neogene there was a wave of progressive cooling leading to the Quaternary glaciations and the evidence of fossil plants and animals clearly indicates that temperatures have decreased more or less steadily throughout the Tertiary. In the Eocene there was a sharp warming trend extending the boundaries of tropical flora and bauxite formation to present-day temperate areas beyond the arid belt. At this time a deep-weathering crust developed in Ireland at 55°N, and in the Oligocene there were pines and firs in Greenland. Flora indicate warm and moist climates in the Canadian Shield in Upper Cretaceous times with temperatures averaging more than 20°C where today mean temperatures are $-3°$C, and about this time there were dinosaurs in Spitzbergen at 78°N. During the late Mesozoic and Tertiary savanna conditions prevailed in the Sahara and seasonal floods transported huge quantities of alluvium to the sea on the North African shore. It was only with increasing aridity at the beginning of the Quaternary that the desert plains of the southern Sahara started to form. However, in most instances any link between present-day landforms and palaeoclimates different from the present remains obscure.

Landforms and time

A. Dating

Although there is available an increasingly large number of precise, scientific measurements of recent geological time, great expertise is essential for sampling, analysis, and interpretation. The sample must be virtually unaffected by weathering or post-depositional chemical alteration or from contamination in a borehole by younger material falling from above. The interval between the deposition of a sample and the geological event to be dated may be wide and uncertain, partly because the geological history may be incompletely known, and partly because deposition or growth in some places began only thousands of years after the drainage of a lake, withdrawal of the sea, departure of the ice, or some similar event. Usually some independent method of ranking events in time is sought to confirm the accuracy of an estimate of absolute age. Conversely the correspondence between the ranks of a time sequence and a series of absolute ages confirms the efficiency of the ranking method. A distinction between absolute and relative methods of dating becomes less useful as radiometric determinations are increasingly used stratigraphically to infer simultaneity of events in separate areas. Like the estimation of adjustments to absolute ages, methods establishing relative ages also usually depend on a high degree of sophisticated understanding and the methods are perhaps best reviewed according to the specialist expertise and knowledge required rather from whether the aim is to establish an absolute or a relative age. Although most dating techniques are specialized skills in themselves, it is evident to the geomorphologist that he cannot use dates without circumspection and that therefore some awareness of the nature of these methods and their limitations is necessary.

The discovery that so-called radioactive atoms have unstable nuclei which decay at an exponential rate to lower energy states has thrown the entire geological record into sharper focus. The unit of measurement of the decay rate is the half-life, the time taken for half of the radioactive atoms in a system to decay. For instance when potassium-40 decays to argon-40, the crystal lattice of the mineral traps the argon, an inert gas, and the ratio of radiogenic argon-40 to potassium-40 is directly related to the time at which the mineral crystallized. The time-range limits are 300 000 and 100 million years old; the amount of radiogenic argon is very small due to the long half-life of potassium-40 (1310 million years) and dates less than 0·5 m years are seldom precise to within 10–20 per cent. Potassium-bearing minerals are usually found in igneous rocks, like bioxite, muscovite, sanidine, plagioclase, and hornblende, but glauconite by being a sedimentary mineral greatly extends the possible usefulness of this method, and there are optimistic hopes for establishing a time-scale of Tertiary sediments. So far, however, it is Tertiary volcanic rocks, particularly those in north-west United States, which have been dated most systematically, resulting in an invaluable control on paleo-floras and age rankings based on floristic criteria. These ages have also defined and illuminated time spans for landform development involving the dissection of the lava flows.

There seems little doubt that extension of results from the expensive and highly skilled operation of potassium argon dating to the widely scattered areas of Tertiary volcanism will lend unparalleled clarity to concepts of geomorphological time. Partly because the decay-rates are too slow, the uranium–lead and rubidium–strontium methods are usually employed on rocks too old to provide a significant datum in landform studies.

Thorium-230, formerly known as ionium, is produced in the uranium-238 decay series, and being readily precipitated in sea-water may be incorporated into ocean-floor sediments. The method gives an estimate of ages up to 300 000 years ago. As Thorium-230 has a half-life of only 75 000 years it is assumed that when, at a particular depth in a deep-sea core the Thorium-230 concentration is only half that at the top of the core, the sediment is 75 000 years old. A grave restriction on this assumption is the necessity for the sediments to have remained undisturbed. A related method, which may date ocean-floor sediments up to 150 000 years in age, is the use of the ratio Thorium-230/Protactinium-231. Pa^{231}, produced by uranium-235 decay, also precipitates quickly in sea-water, but with a half-life of 34 300 years, it decays twice as quickly as Thorium-230, thus providing the basis for the ratio. Although the methods based on progressive decay of radioactive elements incorporated into deep-sea sediments have provided valuable sequences of events for late Pleistocene times, the usefulness of estimates of absolute dates remains uncertain.

C. Emiliani (1955) has used the ratio of the oxygen isotopes O^{18}/O^{16} in fossil foraminiferal tests as an indication of the temperatures of the environment when they formed. However, reconstructions of past sea temperatures based on assemblages of foraminiferal species give different results from those obtained from the oxygen isotopic analyses of the tests. As the basis of these working hypotheses, the nearly instantaneous heat exchange between ice-caps and ocean water accompanying oscillations in continental ice, appears sound. Further work and the devising of further stringent criteria might help to resolve the discrepancies in interpretation.

A technique currently under investigation which might, if accepted as a viable dating method provide invaluable dates for landform interpretation, is the study of thermoluminescence. Thermoluminescence is the light emitted by many minerals when heated below the temperature of incandescence. If the temperature gets high enough the thermoluminescence may be completely annealed away. The amount of thermoluminescence produced by a crystal is proportional to the number of electrons trapped in some imperfections in the crystal lattice structure. If temperature is constant the number of escaping electrons tends to equal those escaping from the traps, thus approaching a dynamic equilibrium position. To establish such an equilibrium under present Antarctic temperatures would take at least 2·6 million years. The leakage rate is a function of a given constant temperature, but any rise in *maximum* temperatures leads to the rapid release of trapped electrons. If an equilibrium value imposed by a former environment of constant temperature can be estimated, the length of time since the change in temperature in the rock's environment can be calculated. For instance, in Spring Cave, near El Dorado, Kansas, a perennial stream keeps summer temperatures close to the average annual surface temperature. In the gully outside the cave the temperatures are much higher in summer and amounts of thermoluminescence are reduced towards a new and lower equilibrium level in proportion to the length of exposure to conditions outside the cave (fig. V.1). It has been suggested that high rates of erosion might expose deeply buried material in a time substantially shorter than that required for the sample to approach an equilibrium with the climatic conditions at the ground surface. If the difference between the sub-surface temperature and the prevailing climatic temperatures can be estimated closely it might then be possible to estimate the rate of erosion. Although this argument is perhaps as yet somewhat hopeful it does demonstrate that there are reasonable chances of techniques being developed which would be focused on some of geomorphology's most imponderable questions.

Despite recent formulation and use radiocarbon dating has provided the basis of a rapidly accumulating stock of precise knowledge for stratigraphers studying late Pleistocene chronologies, from which landform study has also benefited considerably. There are, however, still only a few localities for which there are a series of dates. It is bombardment of nitrogen atoms by the neutrons of cosmic rays in the upper atmosphere which produces a radioactive isotope of carbon, by knocking a proton out of the nitrogen atom nucleus. Because of a delicately poised dynamic equilibrium, diffusion of new C_{14} to the lower atmosphere balances its loss by radioactive decay under natural environmental conditions. Both radiocarbon, with a half-life of 5730 years, and normal carbon, combine with oxygen to form CO_2 which is incorporated into the tissues of living organisms. After death the amount present declines steadily due to radioactive decay. Sources of organic material are usually wood fragments in tills and in beach deposits, but may be varied. For example, small organic particles originally transported on to snowbanks by wind provided the basis for a radiocarbon chronology of ice-cored moraines in the Jotunheimen, Norway. Determinations are possible of some bulk shell samples by a comparison of the C_{14}/C_{12} ratio in the carbonate material. With great expertise C_{14} measurements may be valid for 70 000 years, but very few results have been obtained beyond 45 000 years. No reliable dates are possible since 1850 because the combustion of industrial fuel has caused a decrease in C_{14} activity in the biosphere. In addition to providing absolute dates, the C_{14} method, like other radiometric techniques, also provides a means of inter-continental correlation by identifying contemporaneous events in distant geographical or ecological areas. There are some difficulties here, however, like the C_{14} date of the Valders glacial maximum being 10 700–11 000 years at Milwaukee, correlating in time with the relatively warm Alleröd in northern Europe. In this situation a few workers prefer to believe that broad intercontinental climatic conformity is probable, even to within a

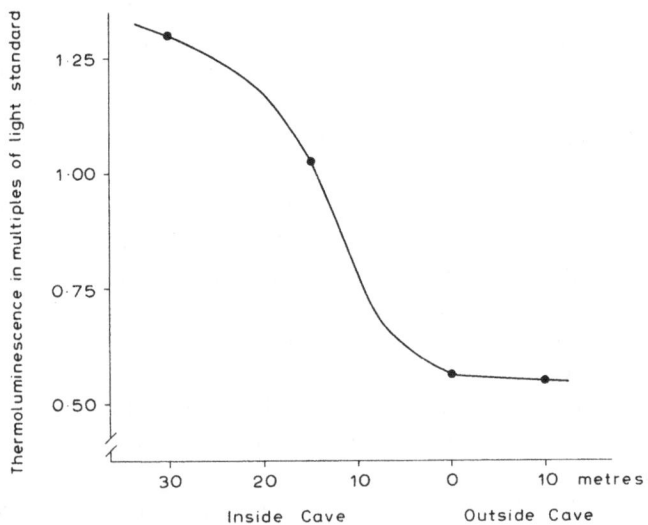

Figure V.1. Amount of thermoluminescence at points inside and outside a cave (*after Ronca and Zeller, 1964*). The scale of contrast could offer an indication of the length of time during which the gully wall outside the cave has been exposed to subaerial conditions. The example used is Spring Cave, south-central Kansas.

millennia or two, and to emphasize inaccuracies and untrustworthiness of C_{14} dating.

In all but historical times the traditional cornerstone of time sequences is the evolutionary changes in organisms, particularly vertebrates, based on composite aggregates of genera and species, with no genus or species necessarily restricted to the age that it helps to characterize and not by itself providing a guide or index for that age. Within the range of time which concerns the geomorphologist the rapid rate of morphological evolution of land mammals and in particular their extremely rapid dispersal has provided the basis for a widely applicable chronology of Cenozoic times. A major recent achievement has been the closeness of correlations achieved in intermeshing the North American Land-Mammal Ages with potassium-argon dates. In their studies of fossils, Quaternary palaeontologists, who include geomorphologists like B. W. Sparks in their ranks, can distinguish 'cold' from 'warm' floras. For example, in association with possibly the most definite of the revised Pleistocene transgressions of the Mediterranean sea, the Tyrrhenian II level found at 5–10 m on stable coasts, is a varied thermophile fauna, distinctive because it includes Senegalese species now extinct in the Mediterranean.

There are several methods not dependent on radioactive decay or evolutionary changes in organisms, which may provide some indication of age over the past few centuries. The counting of tree rings, dendrochronology, is particularly useful for the study of certain depositional landforms, but may depend on an estimate of the time taken for the trees to become established on initially barren ground. For the Donjek moraines in the Icefield Ranges, Alaska, the establishment of spruce trees appears to take at least 30 years and poplars at least 23 years. Fig. III.17 illustrates the flood-training of cottonwood saplings and the ages of various portions of a floodplain based on tree-ring counts which differed by no more than 10 per cent for each grove.

Tree damage as well as tree growth may assist in the problem of chronology. In higher latitudes where ice might damage the bark, cambium, and outer part of the wood, a dead area of wood knobs forms at the edges due to the growth of the living cambium cells and subsequently the annual ring here is dark compared with other layers. This feature may offer some approximate dating of extreme floods, their height and their frequency. Floods may damage trees to a far greater extent than the notching of bark by ice, and some well-rooted trees may contain a tangled record of past geomorphological events in their branches (fig. V.2).

Like tree-trunks lichens increase in radius with age expressed by their diameter since growth rates are often constant. J. T. Andrews and P. J. Webber (1964) calculated lichen growth on the north-west margin of the Barnes ice-cap, *Rhizocarpon geographicum* being 0·057–0·067 mm/year and

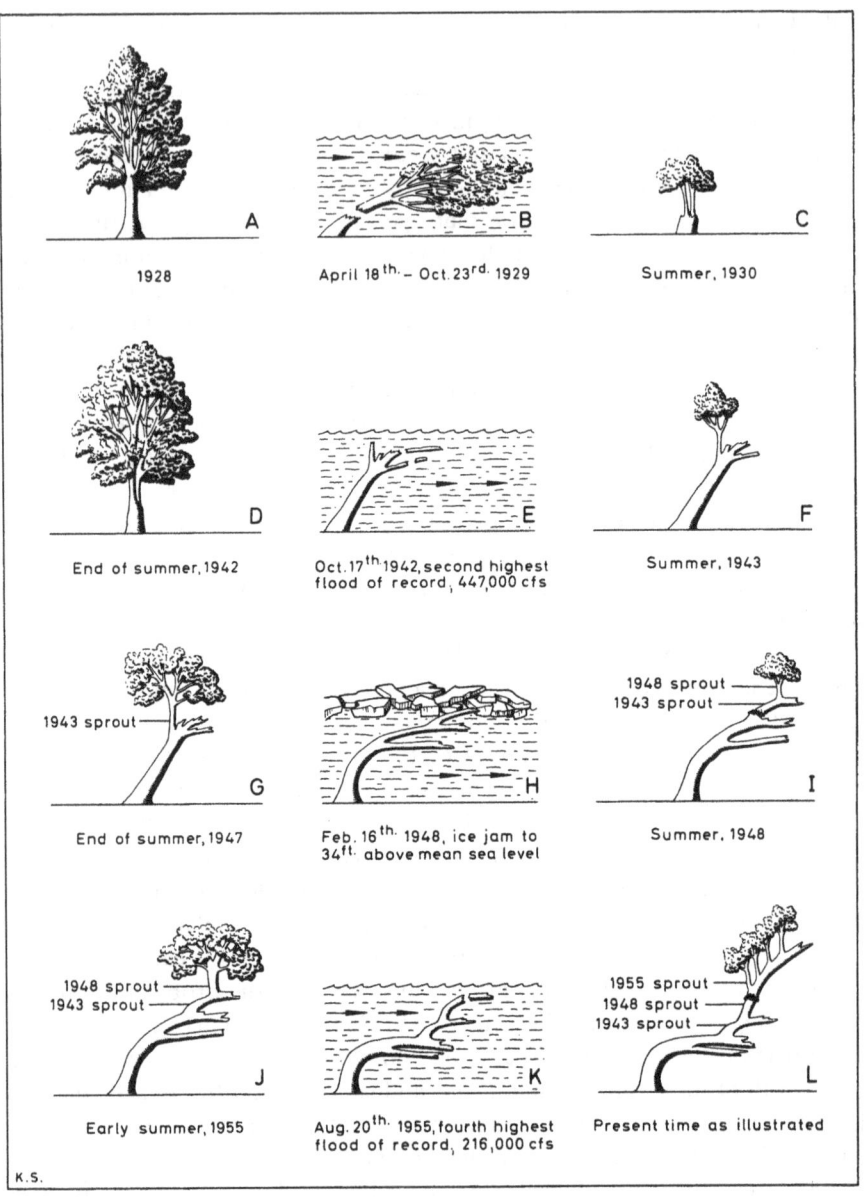

Figure V.2 Sprouts of four ages on a flood-damaged ash tree, Potomac river floodplain, near Chain Bridge (*from Sigafoos, 1964*).

Alectoria minuscula 0·40 mm/year. The rate of growth of *A. minuscula* on a delta in McBeth fiord, Baffin Island, was 0·41/0·44 mm/year. Despite great difficulties in allowing for local ecological conditions like varying water and nutrient supply, differences in exposure and surface roughness of the rock surface, and problems of understanding the physiology of lichens, some progress in dating by lichenometric methods has been made since 1950, particularly by R. Beschel. He has shown that for certain species of lichen the rate of growth increases ten-fold for a four-fold increase in the ratio of precipitation to altitude. Reliable dates for individual localities are a prerequisite for establishing a time-scale, sometimes provided by old photographs or old dated stones. In uninhabited areas C_{14} dating of depositional landforms may provide a basis for a lichen chronology. Suitably calibrated, any crustaceous or foliaceous lichen may yield important data, as it spans the last 1000 years under alpine conditions, and in polar regions lichens might be useful for twice that time-span. In many situations lichenometry may make more precise statements about events in recent decades or centuries possible. The amount of lichen cover tends to indicate the degree to which loose rock fragments in a rock glacier, scree, or similar accumulation are stable. Similarly, it is well known that shore gravel not reached by breakers may be grey because of a cover of grey crustace lichens.

There are many situations where a uni-directional sequence of geomorphological events has taken place and where the vegetation cover gives a clear indication of relative age. Many abrupt geomorphological events, like glacier retreat, sand-storms, drainage of a lake, landslipping, or volcanic eruption expose initially sterile habitats on which the subsequent plant succession is conspicuous and easily observed, like the moss successions in deglaciated parts of Scandinavia. Within a few years of ice recession, lichens, ordinarily crustose types, are the first colonizers of the bare ground, with umbilical species (rock tripes) appearing later, followed in turn by other foliose lichens, the so-called reindeer-mosses (a fructicose lichen) and mosses. Cryptogram communities lead to ground stabilization and humus layer formation. Similarly, successions are well marked in areas of recent volcanic activity where lichens may appear on a flow 4 years after the congealing of the lava. Following the steadying in postglacial sea-level rise many submerged coral reefs have enlarged into areas of dry land, to be colonized in a fairly regular manner by creepers, shrubs, grasses, and finally trees. Alluvial islands in some temperate-latitude streams similarly starting with barren sediment acquire first a willow cover and finally a hardwood forest. There are many other environments in which the vegetation could be studied to elucidate changes in landform or process intensity within recent historic times. For example, the vegetation cover around perennial snow patches may reflect long-term variations in their extent. With the Chaos Jumbles, a 5-km-square

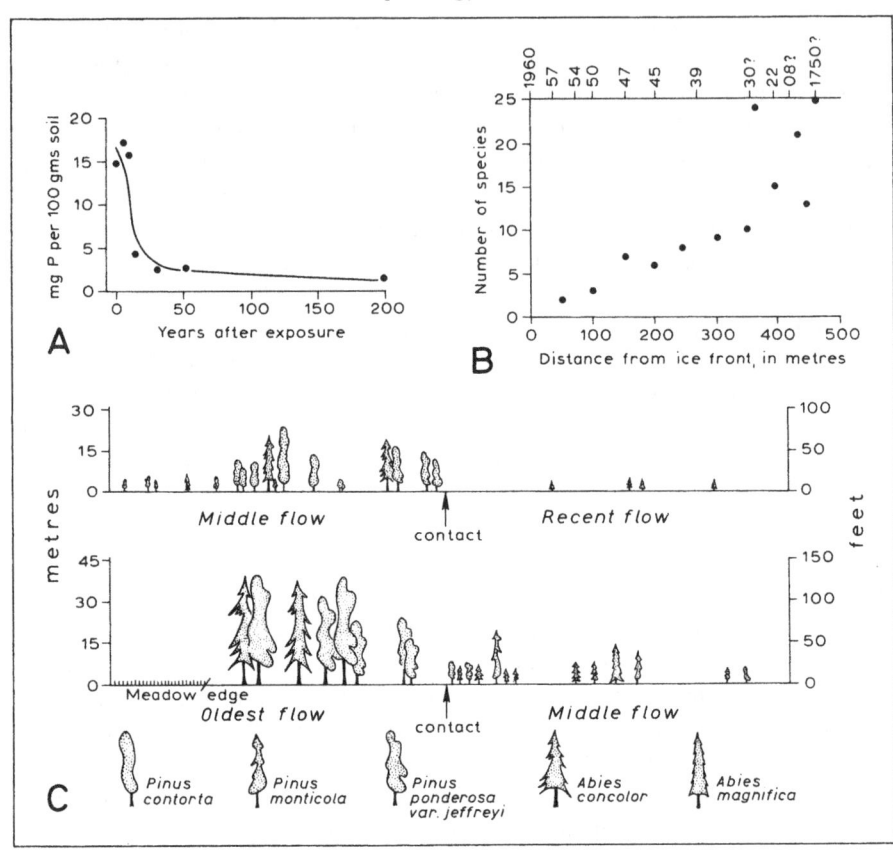

Figure V.3 Soil and plant evidence in dating depositional landforms.

A. Decline of phosphate content in the surface layer of till with increasing age of deposit *(from Stork, 1963)*. The morainic debris was left by the retreating Storglaciär, in the Kebnekaise area, northern Sweden.

B. Progressive increase in number of plant species in front of Storglaciär following deglaciation *(from Stork, 1963)*.

C. Changes in species and progressive increases in growth and density of the tree cover with increasing age of rock avalanche deposits *(from Heath, 1960)*. The two transects were laid out in the Chaos Jumbles, Lassen Volcanic National Park.

avalanche deposit of dacite blocks on the north-west part of Lassen Volcanic National Park, the evidence of the vegetation points to three separate avalanches (fig V.3). On the youngest deposit, a range of colonizing pines is usually dwarfed and distorted due to reduced root space in the rocky soil. On the second deposit, the Yellow pines are often normal in pattern, but have sharply conical trunks that indicate exceedingly slow growth. On the oldest deposit where there is a soil mantle, the Yellow pine cover is nearly mature or sub-climax forest. Some of the species which invade the barren areas, like the Western White pine (*P. monticola*), surviving as long as the primary competition is with the elements, are absent on the oldest deposit, where the Yellow pine becomes well established.

By far the most powerful botanical technique for estimating relative ages, spanning times from the recent historical past to well back into Tertiary times is the study of fossil pollen-rains. These were first analysed systematically by L. von Post in 1916 and have since flourished to provide one of the most reliable aids in establishing chronologies of recent geological times. Pollen shed by trees and by other plants may be carried by air currents and be washed down into lake sediments, peat bogs, or a similar environment where the very tough pollen seeds resist decomposition. G. W. Dimbleby (1961) has shown that pollen may also be preserved in acid soils for several millennia. If the kinds of pollen are identified and their numbers counted for a succession of points in a vertical series, the pollen in a given layer may give the skilled investigator some indication of the type of vegetation growing in the area at that time. The vegetation gives some indication of the climatic conditions prevailing at that time, and these characteristics indicate a specific time zone now usually calibrated by radiometric determinations. A break in the pollen charts usually indicates that a former ground surface was abruptly buried by more recently deposited material. The interpretation of pollen diagrams, however, depends on the intuitive allowances made by skilled ecologists for several potential imbalances in pollen diagrams for which no absolute correction exists as yet. These include over-representation by a species dominant in a restricted habitat close to the sample site, the rapid change of plant species with micro-environmental changes, such as hydrological conditions over uneven or sloping ground, the probability that some species contribute proportionately more pollen than others, and that drier, windier conditions in the past could cause changes in the ranges of pollen dispersal.

Of man's activities useful in dating events in landform history, the value of old photographs increases annually and now includes older air photographs. A photographic survey carried out in 1910 by F. Enquist in the Kebnekaise massif, northern Sweden, has proved of greatest value to present-day glaciologists. C. A. Kaye (1964) searched for old photographs of upper barnacle limits for localities on the New England coast and was eventually able to demonstrate that while sea-level just over a century ago was comparable to that of today, it was about 15 cm lower at the turn of the century. These fluctuations matched tide-gauge records at nearby harbours. The value of photographs is not limited to the comparisons that they themselves may provide, as they may also provide a check on the reliability of a range of dating methods that might be projected further back in time. Documents can provide such a wealth of information about past geomorphological events that it is difficult to illustrate the range of possibilities. One example is the long-term records of ice conditions on Churchill river and the Hayes river entered in the Hudson Bay Company archives from the early eighteenth century onwards. For early maps and diaries independent corroborating evidence usually has to be sought, but again useful information may be gleaned. For instance, it appears from the Rev. F. Consag's records that the last eruption of Tres Virgines volcano in the Gulf of California was in 1746 (R. L. Ives, 1962). For human records, remains and artifacts farther back in time the study of archaeology is a long-established specialism. An observation perhaps worth making in the present context, however, is that the implications for landform study of such evidence may not be directly related to their intrinsic archaeological interest or significance. For instance in the 1930s several arrows, with wooden shafts intact, were found in Norwegian mountains at sites from which snow banks had recently disappeared. The oldest date from A.D. 400–600 and the most recent from post-Reformation times. As specimens these finds were invaluable, but equally striking is the fact that their preservation suggests that the associated snowbanks had similar volumes between A.D. 400 and 600 and again about 1930–40, and that during this interval they were larger. Another noteworthy point about human artifacts is that those of very recent origin have immense potential value in the dating of events and in determining erosion rates over the span of the last few decades. For example, the incorporation of crown-shaped bottle-tops into tropical beachrock formations provides one of the best confirmations of the rapidity of concretion formation in warm environments.

In many ways soil and sediments may help in establishing chronologies either by providing stratigraphic horizons or by reflecting a time-span in the degree to which certain characteristics have changed. The possibilities for absolute dating are normally confined to those glacial clays where fine laminations of varves can be regarded as annual, C_{14} datings confirming the principle of De Geer's classic method of geochronometry that there might be a close connection between the periodic laminae of the clays and the annual ablation of the land ice. Currently there is an extension of the search for laminations due to the seasonal factors influencing settling of particles on to the floor of sedimentary basins in non-glacial areas.

Of the other soil and sediment evidence, all of either stratigraphic significance or indicative of relative ages only, distinctive deposits include wind-blown silts. On moraines and other glacial features the thickness of wind-blown silts may indicate the relative age of the forms. Volcanic ash layers in particular can provide an excellent datum. The Pearlette Ash is a unique means of correlating late Kansan sediments throughout the Great Plains and into the glaciated region of the Missouri river valley. The specialist study of distinctive pyroclastic layers of known age, termed tephrochronology, is employed in areas of recent volcanic activity like Iceland and Alaska. Other accumulations occur within soils due to distinctive pedogenic processes. Caliche accumulations whitening and indurating a soil zone in sub-humid areas may persist because the nodules are resistant to later elimination by leaching. These and similar secondary carbonates appear to offer promising means of identifying relatively dry climates in the Pleistocene, provided that other influences can be eliminated. However, yet again disagreement in interpretation is well known, due to problems reminiscent of those which complicate the interpretation of pollen diagrams. Some workers like K. Butzer have concluded that eolianite, for instance, indicates earlier phases of continental glacial advances, whereas D. Yaalon stresses the importance of local relief, microclimate, and water-table conditions in their formation. In contrast the evidence of salt accumulations in arid areas is independent of geological and biological factors.

Truncated as well as depositional horizons may exist in soils, the former recording phases of instability and erosion. If not actually visible in the field section, soil analyses at a succession of depths may, like pollen diagrams, show breaks as composition changes abruptly at a truncated horizon (fig. V.4). In Australia, such soils are widely used as records of minor cycles of erosion separating intervening periods of stability when soil formation took place.

The study of the degree of soil-weathering is, like the study of varves, a classic method in establishing chronologies. The depth of penetration of

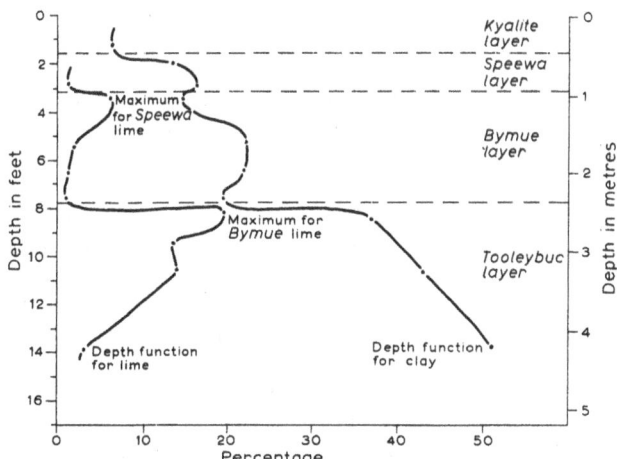

Figure V.4 Abrupt changes in soil properties, indicating the separateness of buried soil layers (*H. M. Churchward, 1961*, Jour. Soil Sci., *Vol. 12*), at Swan Hill, Victoria, Australia.

pedogenetic processes is often a vital indicator and has helped materially in differentiating many major drift sheets since A. Penck first studied on a large scale in 1899 soil formation on moraines in the Alps. In the Puget lowlands of western Washington, the oldest till, the Shakwak, is distinct in being oxidized to a minimum depth of 4·5 m, and in the same locality the highly oxidized Icefield outwash III contrasts sharply with the younger unoxidized Kluane outwash I which is medium grey in colour. Dune sands may show similar changes, appearing to acquire a coat of ferric oxide which affects an increasingly large number of grains and perhaps increases in thickness with the passage of time, changing the grain colour from that of unoxidized sand through light yellowish colours to dark reddish-brown or even brownish-black in very old dunes. Apart from the translocation of iron, weathering may produce progressive changes in particle size and also in the composition of suites of minerals in sand grains. Blackwelder was the first to use boulder-weathering ratios as an indication of age. Yet again micro-environmental conditions like dampness, aspect, and position have a marked significance. There is also the problem of defining when a boulder is weathered; F. Birman recognizes a boulder as weathered when more than half of its surface contains grains loosened to the depth of the average grain diameter. R. P. Sharp (1960) found several differences between boulders on Early Middle and Late substage moraines in the Trinity Alps, California. Table V.1 lists the boulder-weathering ratio. Remnants of mechanically worn surfaces were found on Late boulders and then only rarely. Spalling and fracturing along joints are marked on Late boulders, seen to a small degree on Middle ones, and almost entirely lacking on Early boulders. Loss of rounding due to differential weathering is most marked on Early boulders, on which a

Table V.1 The ranking in age of Wisconsin moraines by the degree of weathering of boulders (*from R. P. Sharp and J. H. Birman, 1963*, Bull. Geol. Soc. Am., *Vol. 74*). The Mono basin glaciation may be Illinoian.

	PER CENT WEATHERED/PER CENT UNWEATHERED GRANITIC BOULDERS			NUMBER OF BOULDERS OVER 1 FOOT DIAMETER IN 100 × 20-FOOT STRIP
Glaciation	1	2	3	4
Tioga	30/70	30/70	10/90	300
Tenaya	51/49	49/51	50/50	180
Tahoe	73/27	67/33	80/20	115
Mono	—	—	95/ 5	60

1 = Sequoia–King's Canyon National Parks. 2 = San Joaquin river drainage. Yosemite and eastern slope of the Sierra Nevada. 3 and 4 = Walker Creek (Bloody Canyon) moraines, Mono basin.

Table V.2 The mean–maximum protuberance of quartz veins and inclusions on carbonate cobbles as a means for ranking alluvial fan deposits by age. Gorak Shep fan, Eureka valley, California (*from Hooke, 1967*).

	MEAN — MAXIMUM DEPTH OF WEATHERING IN MILLIMETRES	
	Upper segment	Lower segment
Recent channel deposit	1·8	2·5
Oldest channel deposit	2·5	4·3
Youngest fan surface	4·0	5·8
Oldest fan surface	12·5	16·3

ground level weathering platform a few inches wide is also relatively common and where boulder burial by grus is significant. Cracks are etched 2·5 cms deep on Middle boulders and 10 cms deep on Early boulders. Conversely inclusions etched into relief project by 2·5–5 cm on Middle boulders and by more than 5 cm on Early ones. A similar property has been measured on an alluvial fan where quartz inclusions stand out in varying degrees from carbonate-rock pebbles, according to the degree of weathering of the fragments (Table V.2). Beaches may be similar to moraines in that older material may have small weathering cavities, due to the boring activity of algae. Considering smaller particle sizes, A. R. van Wambeke (1962) believes that the silt content (2–20μ) of tropical soils derived from identical parent rocks is closely related to the actual amount of weatherable minerals, and on the basis of more than 1000 samples, mainly from the Congo, finds that a silt–clay ratio of 0–10 appears to characterize soils older than end-Tertiary (Table V.3). Similarly, the average measured median grain size of heavy minerals may with the increasing age of sediment, decrease in relation to the average theoretical median diameter of the heavy mineral suite, due to chemical decay of less stable heavy mineral species by interstratal solutions. With the heavy minerals composition changes. In the Jekyll island area, a general impoverishment in hornblende and epidote and enrichment in

Table V.3 Silt–clay ratios of parent materials in some soils from Central Africa (*from Van Wambeke, 1962*).

AGE OF PARENT MATERIAL	GRANITE		GRANO-DIORITE	MICA SCHIST	BASALT
	1	2			
Recent	0·39	0·36	0·74	0·96	1·03
Pleistocene	0·27	0·17	0·32	0·18	0·32
Pre-Pleistocene	0·11	0·08	0·12	0·11	0·09

sillimanite characterizes the older Pleistocene terraces. Hornblende averages 19·3 per cent of the heavy mineral suite in samples from adjacent rivers, about 7 per cent along the coast and less than 0·5 per cent in the Pleistocene terraces. One aspect of pedogenetic processes, that of clay-mineral alteration, appears to offer great potential because the changes take ten thousands of years or more to become clearly established. The time factor is highly significant in the formation of kaolinite. Study of a kaolin-rich residual soil near Thornaston, Maine, suggests a pre-Wisconsin and probably a pre-Pleistocene age. Clay-mineral alteration data reveal that the post-Wisconsin weathering profile is distinctly different from those of the Sangamonian and Yarmouthian. The Yarmouth weathering of the Kansas till, for instance, went on for much longer than examples of the other two tills and led to a complete loss of feldspars and chlorite from certain zones and the substantial alteration of illite structures. X-ray analysis by A. Godard of ancient red soils from fissures in Cambrian limestone found at many points in north-west Scotland suggested a similarity with Mediterranean terra-rossa soils, a relic of an environment slightly wetter and warmer than the present-day Mediterranean climate.

In addition to some specific characteristics which promise the possibilities of quantitative tests, some soils show general structures which may reflect time of weathering or may add to their usefulness as a stratigraphic horizon. In the first case, for example, latosols in Natal believed to be associated with older land surfaces are typically platy vesicular structures with predominantly red and purple colours, whereas those believed to be developed on younger late-Cenozoic surfaces are more nodular and pisolithic in structure with a predominance of colours between red and brown. In the second instance perhaps the best examples, well-known for their chronological significance, are soil structures due to permafrost activity. During the past two decades, German workers in particular, like J. Büdel and H. Poser, have made considerable use of periglacial patterns in soil and sediment structures. Despite the general usefulness of soil structures it should be added that their appearance may sometimes be misleading. For instance, before analytical techniques were available to demonstrate the length of the Yarmouth interglacial, it was estimated to be the same as that of the Sangamon because the two soils appear to have developed to about the same degree.

Another general point is that while several distinct characteristics have been discussed separately, invariably the usefulness of soil and sediment studies in the search for chronologies depends on the consideration of more than one characteristic. For instance, three drifts bordering the Reedy glacier, Antarctica, are readily correlated by their degree of cavernous weathering, the amount of staining by ferrous compounds, and by their altitude. Boulders of the oldest and highest Reedy I drift are stained by ferric oxide and have deep cavernous weathering pits. Reedy II boulders are similarly stained, but cavernous weathering is absent or only incipient. On Reedy III boulders the surface is fresh, with little or no oxidation.

While the range of methods for establishing chronologies has expanded, the traditional reliance on altitude of land surfaces as a single criterion has been largely discarded. Difficulties of interpretation, problems posed by the variable amounts of movement of land and sea, the indistinguishability of eustatic and crustal movements, and the realization that altitudes of former sea-levels are not necessarily related consistently to their relative ages and the difficulty of tracing continuous features in the field are some of the reasons for this change. Even where depositional terrace-like features have clearly defined altitudes, careful levelling has cast doubt on the reliability of extensive correlations and generalizations. For instance, J. B. Sissons (1963) has levelled altitudes on a delta remnant in the Forth valley at 35·4–35·9 m OD, at 32·8–33·1 m OD, and at 25·5–26·8 m OD, none of which appears identifiable with the former widely recognized 30 m raised beach in Scotland.

Despite the wide range of methods that might provide information for dating or the ranking in a time-sequence of significant geomorphological events, some generalizations apply to most techniques. The most important consideration is that the sampling and interpretation is usually in the hands of a small number of highly specialized experts utilizing sophisticated and expensive analytical procedures and whose primary interest is often in dating as an end in itself, and whose secondary interests do not usually include the study of landform development. The geomorphologist therefore is usually dependent on the findings and progress of these specialists rather than being in a position to tackle most dating problems himself. On the other hand there are some relatively straightforward procedures which have been almost totally overlooked until the last decade. Their disadvantage is that they demand a great deal of work without necessarily providing a guarantee of a meaningful reward in return. Another common problem affecting a wide range of phenomena susceptible to environmental change is the degree to which a given pattern might reflect local micro-environment characteristics rather than giving a broader general picture.

Few would now agree with W. M. Davis's firm belief that landforms themselves are the best guide to their age. There is the growing realization that one cannot tell whether a landform is young or old simply by looking at its shape, and that there is a need for some independent indication of how forms and processes have changed over decades, centuries, or millennia. Intensified interest in the testimony of sediments, soils, and plants is a current trend and likely to lead to a greater precision in the study of sequences in landform changes than the scant factual information in the following six sections might suggest is possible.

B. The Initial Form

> 'Many familiar geomorphological features of the emerged lands may be recognised below the surface of the sea.' C. Emiliani, 1958

It was only in the 1930s that soundings of continental shelves and ocean basins became detailed, and it is only within the last decade that surveys resulting from continuous profiling have appeared in large numbers. As a result geomorphologists need no longer accept that the constitution of an initial land surface on which landforms would develop subsequent to uplift need necessarily be a matter of speculation. The relief of the surface that would, with a 100-m drop in sea-level, increase the earth's land area by 2·7 per cent is now better known. In many areas this marginal shelf has the same degree of gross irregularity as the adjacent land areas. Off the north-west coast of the United States a rough part of the continental rise is called the Ridge and Trough province and, off southern California, basin and range relief is very marked and resembles that of the Basin and Range province. The continental shelf off the east coast of the United States is much broader than the narrow Californian shelf, averaging 135 km, but is as little as 8 km off Palm Beach, Florida. Off Maine it is 420 km wide, but like the shallow shelves off Quebec, New Brunswick, and west Newfoundland, the surface is relatively rough. There is also a series of seamounts extending 1600 km from Bermuda to points off New England, believed to be late-Cretaceous volcanoes. Off Vladivostok the shelf is 60 km at its widest, but is mostly less than 20 km wide and very narrow north of 67° N. Several submarine features differentiate its surface. Tectonic and structural lineaments as well as submarine canyons are significant features off many coasts. Relatively long and narrow depressions are found parallel to the main trend of the Norwegian coastline and similarly on other shelf areas off glaciated coasts such as flank Spitzbergen, Greenland, Labrador, and Scotland. One on the Labrador shelf is 400 km long, with an undulating bottom relief including depressions exceeding 730 m in certain places. H. Holtedahl (1958) considers that the fracturing may be related to late-Cenozoic upwarping of the crust. Some irregularities on shelves show some alignment like the south-west trend of 100–400-m scale of relief east of Madagascar. Similarly in the western approaches to the English Channel, there appears to be a well-defined WSW–ENE elongation of structures in the metamorphic and igneous basement which are traceable to the continental shelf edge. Nearer the mainland there are isolated trenches along the strike of Jurassic clays and soft sandstones off the Dorset coast and Bristol Channel. Near Weymouth one depression is 22 km long and 1·5 km wide, and 45 m deeper than the surrounding ground. On the floor of the Baltic a Silurian and Ordovician cuesta system is well developed. On many other shelves there is irregularity but with little systematic trend.

Another reason for the irregularity of the sea-floor surface offshore is that most sediment delivered from many continental areas has been funnelled off the shelf along submarine canyons. Even in areas where a sediment cover exists a considerable amount of buried relief exists, as in the North Sea between Scotland and Norway. Similarly sediments on the submerged coastal plain off Buenos Aires province, although averaging 0·5 km in thickness, are merely fills in two depressions in a pre-Cambrian basement complex. These depressions run seaward, are at least 700 m long and 150 km wide, and one includes the Rio Negro basin. A further point about sediment accumulation on the continental shelves, apart from its general sparsity, is its sporadic distribution which may assume distinctive trends. In the English Channel the similarity in orientation of ribbons of sediment and tidal currents is most striking. Sand waves may be up to 20 m in height with a wave length of 900 m. In the Irish and Celtic Seas, lense-shaped masses are up to 30 and 110 m thick respectively and reflect how sediments pass off the shelf along relatively restricted paths. In some shelf areas off the eastern coast of the United States rising sea-level left several terraces of both erosional and depositional origin and many submerged bars which once may have separated long lagoons from the open sea. Within the Gulf of Maine the surface relief reflects the activity of Pleistocene glaciers. Thus where it exists, a sediment cover may not provide a smooth-surfaced veneer but may have well-developed trends. Also, far from blanketing the underlying bedrock surface, irregularities in the latter are often largely the controlling factor in the dispersal patterns of sand moving off the shelf.

In terms of their function, age, and size the most striking feature of the continental shelves is the submarine canyons. Their dominant role is to funnel sediment from the continental shelf to the deep-sea floor. The age of deposits partially filling some canyons, indicates their formation in pre-Middle Tertiary time. Many are huge in size. One canyon between Norway and Iceland has a depth of 1000 m. Off south-east Alaska, there are three marked submarine canyons, that off Cross Sound being 22 km wide and 440 m deep. Oil wells drilled into a valley fill inland and aligned with the Monterey submarine canyon failed to reach basement at a depth of 2380 m. The fill below 900 m may be middle to early Miocene in age. Apart from the remnants of old aggraded surfaces forming matching terraces high above the present channel, as in the Newport canyon, many erosional features of submarine canyons resemble those on land as off the west coast of Corsica or off the Azores. On a smaller scale than submarine canyons are sea gullies generally considered to have a relief of less than 60 fathoms which are common in a wide variety of environments. Those in the San Clemente area have average wall declivities of 15 degrees with local maxima up to 33 degrees. The long profiles are concave, $6\frac{1}{2}$ degrees in the upper portion and $2\frac{3}{4}$ in the lower part. Even on the Bering–Chukchi platform which has been described as one of the flattest erosional surfaces on earth there is a shallow but well-defined sea valley about 3 km wide.

Recently acquired evidence demonstrates that it is probably unrealistic to assume an idealized flat surface as an initial stage in landform evolution. It seems that any relative lowering of sea-level would tend merely to lead the outflow of a stream, already established inland, into the pre-existing pattern of irregularities on the exposed part of the shelf. Also the evidence of unconformities, now known in the detail of three-dimensions rather than in the two of the cross-section, show that many new phases of landform evolution started in the geological past on irregular surfaces. In the Smith river basin, Montana, the volcanic ash was deposited on a much dissected pre-Oligocene relief. Pliocene–Pleistocene erosion both followed the dendritic drainage pattern of the ancient land surface and also initiated several changes in the new pattern. Similarly it has been suggested that any belief that the parallel ranges of hills of south-west Ireland is due to the carving out of synclinal valleys during the Tertiary from a surface continuous at summit levels has been shown to be no longer tenable. The existence of chalk outliers on valley floors near synclinal axes suggest that the present landscape is essentially an exhumed late-Cretaceous surface.

Perhaps the starting-point for the study of the effects of time in landform evolution could most realistically be a structural surface, and the stages investigated those of the erosional modification of this initial surface, rather than the erosional modification of a hypothetical erosional or depositional surface.

C. Stages in Slope Evolution

'The secret of landscape evolution lies, evidently, in the mode of development of hillslopes.' L. C. King, 1953

One of the most controversial topics in landform study is the manner in which slope-profiles change with time. Disagreement is inevitably the outcome of discussion in which imagination rather than observation serves as a starting-point. For instance, soil scientists who have remarked on the considerable conflict of opinion on both the character and the genesis of steep-slope soils note that detailed studies are few. Similar remarks about the evolution of pediments are numerous. Amid the uncertainty of the theoretical assumptions about time sequences due to their limited or non-existent observational base and the sheer variety of ingenious but unrealistic idealizations of the problem, it is difficult to isolate some facts which are known with some certainty. One of the main practical difficulties is to decide when and under what circumstances slope angles tend to decline with passage of time and when periods of parallel slope retreat, as suggested by W. Penck and K. Bryan (1922), operate on certain parts of the slope. A related problem is to decide whether a caprock which will inevitably ultimately disappear, may supply a comparatively steady excess of material to the lower slopes for an open system to be recognized, in which the lower slope for a significant length of time, might remain essentially the same in form. Another problem is the lack of understanding of how, in contrast, changes in the proportions of bedrock exposure to colluvial mantle may affect the total shape.

Three main factors may influence the course and nature of stages in slope development. First, changes may be either an adaptation or re-adaptation to new conditions created by random changes in the morphogenetic environment, such as a cooler phase of climate in the Pleistocene, or due to some systematic change related to the erosion of the slope itself like the progressive removal of a critically strong or weaker stratum or the increasing influence of some local controlling base-level. A second factor is the different rock types which introduce fundamental distinctions into the modes of slope form change. In Central Australia, gradients on granite or gneiss pediments do not generally exceed 3 degrees, whereas on schists gradients commonly attain 5–8 degrees near the piedmont junction. A marked lithological control on pediment gradients had also caught J. Gilluly's attention in 1937, and more recently C. R. Twidale (1967) concluded that it cannot be too strongly emphasized that structure is the most important factor determining the morphology of slopes and of the piedmont angle or zone. Slopes in alluvial material are also distinctive and are therefore unlikely to provide models of erosion on bedrock slopes. Thirdly, it is necessary to consider carefully some realistic reconstruction of an 'initial' surface. This might be the result of differential erosion of adjacent bodies of rock of varying resistance as in the Appalachians, tectonic movement as along the sides of the central African rift-valley, or the incision by the diversion of a stream into a shorter route to lower ground. According to S. A. Schumm (1966), most hillslopes owe their existence to the incision of a terrain by streams.

For most slopes in sedimentary rocks and in many areas of extrusive or shallow-depth volcanic activity, the caprock, its disposition and the degree to which it has become dissected into disconnected outliers is the crucial factor in the stages of slope evolution. Slopes steeper than $26\frac{1}{2}$ degrees can be maintained in the Snake river plain only by a protective rim of rock. Even in tropical areas where differential erosion of bedrock is often obscured by deep weathering mantles, the development of weathering crusts plays a very important role in controlling slope evolution. In fact there is the possibility that as they form on valley floors, at the foot of slopes and over piedmont slopes where dissolved products concentrate, they eventually become sufficiently indurated to produce frequent inversion of relief as, with the passage of time, they become the caprock of residual plateaux. This, however, has not happened in central Australia since the Tertiary, as relics of low platforms capped by weathering crusts occur close enough to adjacent hills for the reconstruction of former piedmont profiles. It seems that the present plains were already the lowlands on the Tertiary surface, topped by the hills of today and that subsequent slope evolution involved mainly the dissection of the crust, the etching away of less than 10 m of subjacent soft weathered rock, and the concomitant shaping of the piedmont profile from broadly concave to angular. In most sites the hill base was set back to a structural boundary for which there appears to have been little subsequent retreat, possibly in part due to a change to a drier climate. Farther west in Western Australia the old plateau is reduced to smaller remnants. In other sub-humid environments a caliche cap resisting erosion gives rise to tablelands.

A. Rapp (1960) was one of the first to observe caprock-face processes in detail. Due to the greater kinetic energy of the greater drop, it is possible that high widespread caprock-face falls tend to create more stable, concave talus slopes with a fringe of large boulders at the base, while falls with smaller vertical drops tend to form steep unstable talus. Many theorists have suggested that during the growth of talus slopes, the higher bedrock walls diminish in height until they are totally covered. Rapp's observations extend this view by suggesting that as falls become progressively shorter the sorting of boulders becomes increasingly that of the unstable talus. The scars from which the material dropped were scattered all over the steep walls suggesting that the mode of evolution of the rock wall was by parallel retreat. The next stage is essentially a development in plan when chutes and funnels begin to break up the continuity of the simple wall, but it also involves the sections of the slope affected in downwearing by linear dissection. It might be supposed that if these notches then intersect, the isolated portions of caprock become ruiniform in appearance and become progressively reduced.

In situations where there is no caprock or mountain wall, low domes form on the interfluves between pediments or similar lower slopes. Average slopes are usually noticeably less. In this way parallel retreat becomes ultimately a self-extinguishing process by removing the upper slope which supplied the material for transport across the lower slope and the former lower slope ceases to be a debris-controlled slope once the supply of debris is eliminated. S. A. Schumm (1956) observed that even with the disappearance of a miniature badland residual, the broadly concave pediments coalesce but with a convexity developing at their junction on the divide.

A quantity of major importance needed in establishing stages of slope evolution is the rate at which caprocks retreat. The maximum distance of cliff recession caused by several rockfalls from Sawtooth Ridge with the Pleistocene was probably about 400 m. P. Birot (1965) guessed that the 60-degree flanks of limestone domes in the tropics might retreat 5 m in 100 000 years. M. A. Melton (1965) supposed that pediments might extend laterally at about 800 m/1 million years, with a lowering of the surface at about one-tenth this rate. The ratio of these two rates, and an assumed differential lowering due to progressive exposure of the pediment bedrock, suggests that the latter would theoretically slant upward to the base of the retreating wall at the angle of 5·6 degrees.

In contrast to stages in slope development accompanying the reduction and removal of a caprock, dome shapes in certain massive rocks appear to be self-perpetuating by favouring the development of expansion joints and their concentric patterns. B. P. Ruxton and L. Berry (1961) describe how on the stripping of rock waste off granite batholiths in east-central Sudan the bare rock hills at first retain their simple slope profiles, but later large pits are weathered out at the intersection of joints, cavities are formed beneath some sheets and depressions are excavated just below the angles at the top and bottom of the encircling rock fans. However, it appears that as a stripped residual hill gets smaller the dome form extends downward, eliminating the angular junctions bounding the dome and rock fans until a steep rockface meets the flat clay plain at an abrupt angle. Usually this change affects only one side of the hill producing a half dome. If dome-shaped hills in massive rocks tend to be self-perpetuating the same control of expansion joints in bowl-shaped or concave slopes might operate to maintain their shapes.

Some of the most incontrovertible stages of slope evolution in solid rock occur in areas where mass movement is dominant. Three main types might be recognized. First, there are landslides with negligible rotation, like the Chuska Mountains landslides where eastward from the escarpment crest there are up to seven successively lower ridges. In the highest of three main landslide areas there are huge caprock blocks ranging from 9 to 300 m in length, 9 to 30 m wide and high, broken apart slightly along the vertical planes of intersecting joints but still essentially horizontal. Troughs have flat sand-covered floors but contain little coarse debris. In the second area, caprock blocks are broken into a jumbled mass of individual blocks and erosional debris nearly fills the troughs so that the ridge and trough form is obscured. Along the base of the steep slope separating the highest from the second area of landslide debris a series of springs emerge which have helped to erode a wide flat area. In the third and lowest area of landslide debris, the main continuous remnant is about 0·7 km wide, but debris is scattered over the Cretaceous shale eastward for a further 8 km. Canyons of intermittent streams exceeding 30 m in depth are the main relief form, the blocks of

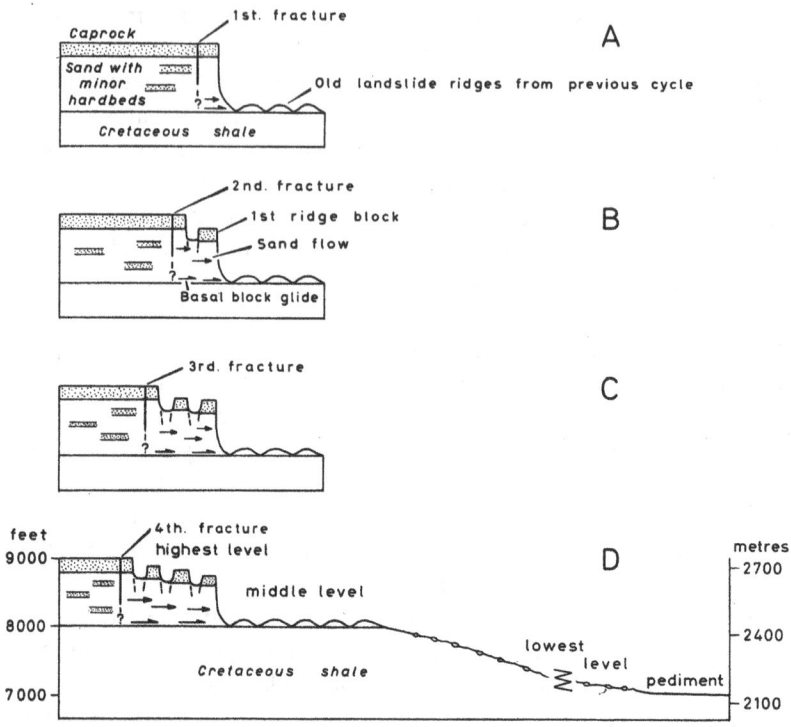

Figure V.5 Evolution of escarpment slopes with progressive block glide of ridge-blocks on sand (*from Watson and Wright, 1963*). This example, observed on the east flank of the Chuska mountains, north-western New Mexico, also shows lateral spreading on shale of loose sand during block glide of the entire mass and the progressive diminution in scale of the ridge and trough forms.

sandstone, now less than 1·5 m forming the major relic of the landslide topography, protecting ridges up to 6 m in height. There is a reddish soil mantle 0·6–0·9 m thick, a horizon of secondary carbonate and sandstone blocks blackened by desert varnish. The entire area has been lowered by erosion and downwasting of the landslide mass and the underlying Cretaceous shale and, as in numerous similar situations, it is difficult to decide on the degree to which detached masses have slid laterally and the degree to which caprock remnants have been essentially lowered by the removal of the adjacent or underlying less resistant material.

A second type of movement in which stages of slope evolution can be clearly deciphered is where rotational landslipping occurs. Stages may involve the pre-slip surface, its abrupt disorganization by slippage then either its gradual regularizing by slopewash or the enlargement of the scar. The relatively undissected nature of the slide material of the many postglacial landslides in the Central Pennines gives some indication of how long stages of

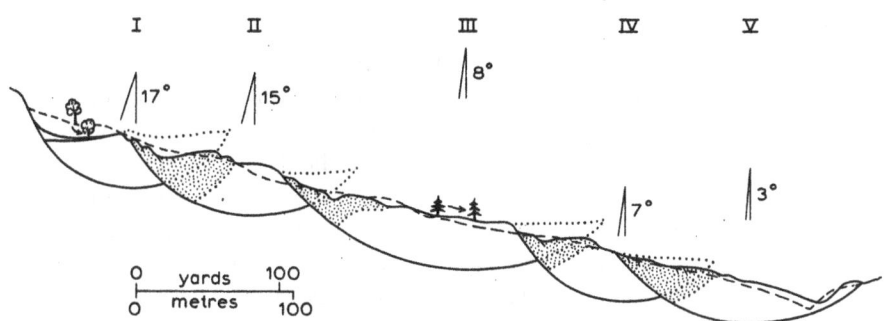

Figure V.6 A downslope series of rotational slips (*from G. Reichelt, 1967, Erdkunde, Vol. 21*), on the western slope of the Eichberg, north of Achdorf.

slope evolution in such circumstances might be. There may also be stages within the slipping phase. In the Eichberg slip near Achdorf, it seems that the slumped material of an initial rotational slip near the slope summit loaded a downslope section of the hillside, inducing a second rotational movement and that this initial disturbance perpetuated a sequence of five rotational slips in all, with the amount of rotation decreasing downslope. Similarly, the over-steepening of the toe of a slope could, by removing the lateral support from segments of a hillside, propagate upslope a chain reaction of slips analogous to that produced from above by overloading. A third type of mass-movement involves the free fall of individual rock fragments which may form definable stages of depositional forms in cool environments where there are seasonal snow banks beneath rock walls. There are two clearly distinguishable stages of slope platforms with protalus ramparts. Younger platforms have well-defined ramparts and distinctive ditches made up of little weathered material with a scant vegetation cover. Older talus platforms have only moderately clear ramparts and poorly defined ditches, made up of moderately to slightly weathered material largely buried beneath a thin soil veneer.

This would not apply to features as small as many dolines, the funnel-

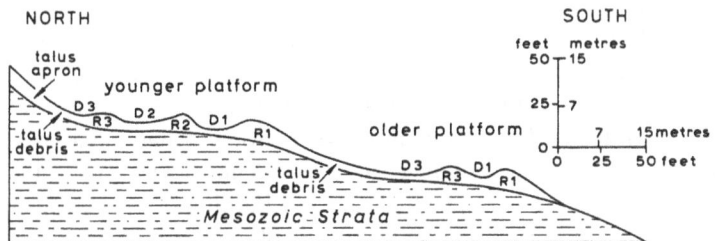

Figure V.7 A slope on the south flank of Navajo mountain showing platforms of two ages each with a sequence of protalus ramparts, R, and associated ditches, D (*after Blagbrough and Breed, 1967*).

shaped hollows which are one of the most widespread characteristics of limestone relief. In this case close interdependence with an underground drainage system is an important factor in slope development, and a complicating factor is the different modes of initiation of depressions. Nonetheless, at least some stages in the evolution of slopes on the flanks of enclosed depressions are observable. These slopes therefore hold general implications for landform study as well as being of particular interest to karst specialists. In a bare karst, where an appreciable area of limestone is exposed at the surface, there is often no reason why a depression should develop at a particular point, and solution tends to lower the limestone surface as a whole. One of the most effective ways in which solution concentrates at a particular point in limestone depends on an initial stage when the limestone, as a covered karst, is still overlain by a largely non-soluble stratum; thin or more permeable points in the cover rock then localize solutional activity at particular points on the underlying limestone, perhaps for substantial periods of geomorphological time. A pocket of deeply weathered limestone may therefore already be in existence before the first stage in the surface slope development appears as the cover rock founders. In part the initial hollow may be the result of removal of some of the cover rock by piping. The Mendip area, therefore, is typical by having a very large number of its enclosed depressions on the Mesozoic rocks which flank the outcrop of massive Carboniferous limestone. If a connection with an underground drainage system is maintained, the depression enlarges. In some situations the greater biological activity in the deeper soils of the depression favours more rapid solutional activity, but it seems common for rillwash to develop on the slopes of depressions enlarged beyond a certain size. This induces progressive infilling of the floors which reduces permeability, arrests deepening at the base of the slopes and favours relatively accelerated solution on the upper slope. As in the Indiana karst, ponds may form, sometimes as rapidly as collapse features appear. Their infilling with silt and organic debris is particularly rapid. Similar developmental stages have been envisaged for the slopes of depressions created in calcareous rocks by water draining off an adjacent insoluble or impermeable rock. An example is the pockets developed in the chalk of northern France at the contact with the overlying Tertiary strata. Occasionally surface depressions appear above an underground stream which may evacuate much of the debris, leaving the walls of a cylindrical shaft as the initial form. Some collapse features, like the shaft doline of Modro Jezero, near Imotski in Yugoslavia, have scree banked up on their lower slopes. From this stage the infilling of finer material would then lead to slope forms similar to those of covered karst origin, which become increasingly concave as the debris fill grows higher. In the initiation or reactivation of slope development in enclosed depressions seismic shocks are sometimes significant.

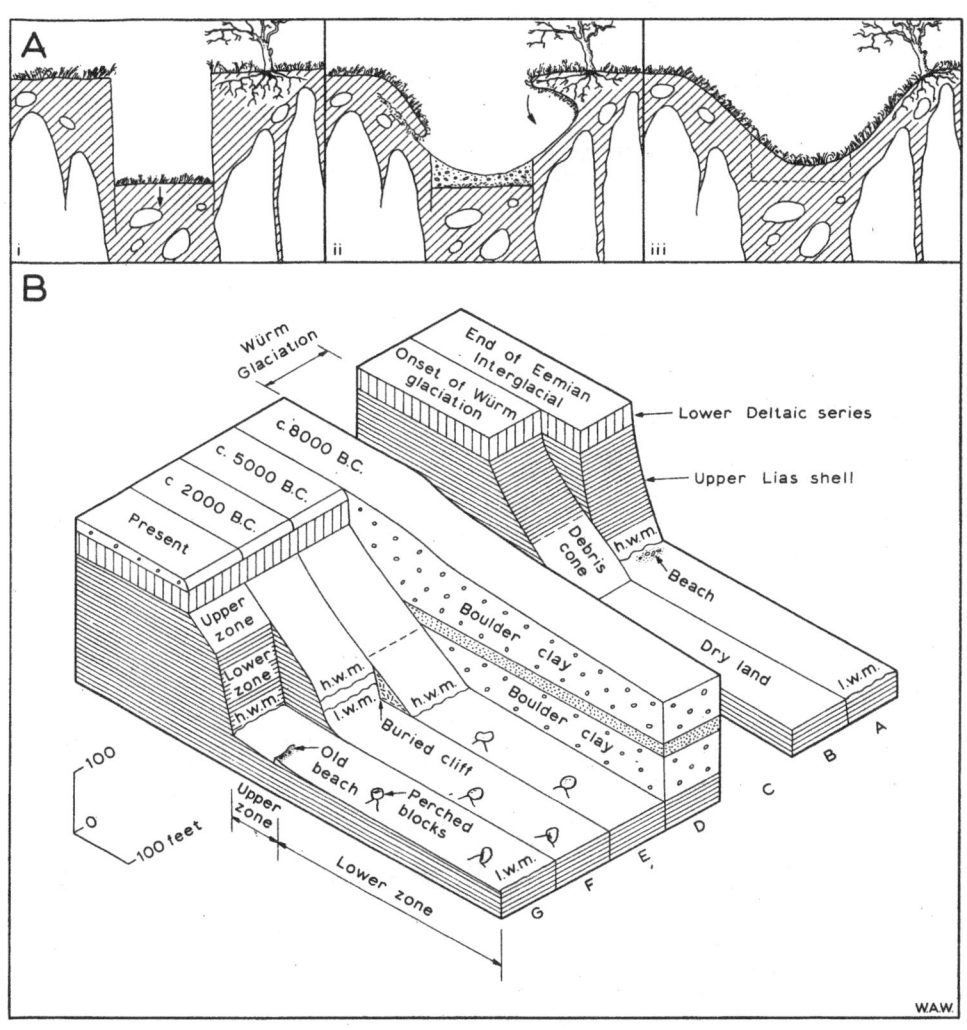

Figures V.8 and V.10 Observed changes in slope form. (*Above*) Evolution of sides of a doline (*H. Alojzij, 1953,* Kraška Ilovica). (*Below*) Stages of coastal cliff development in north-east Yorkshire (*Agar, 1960*).

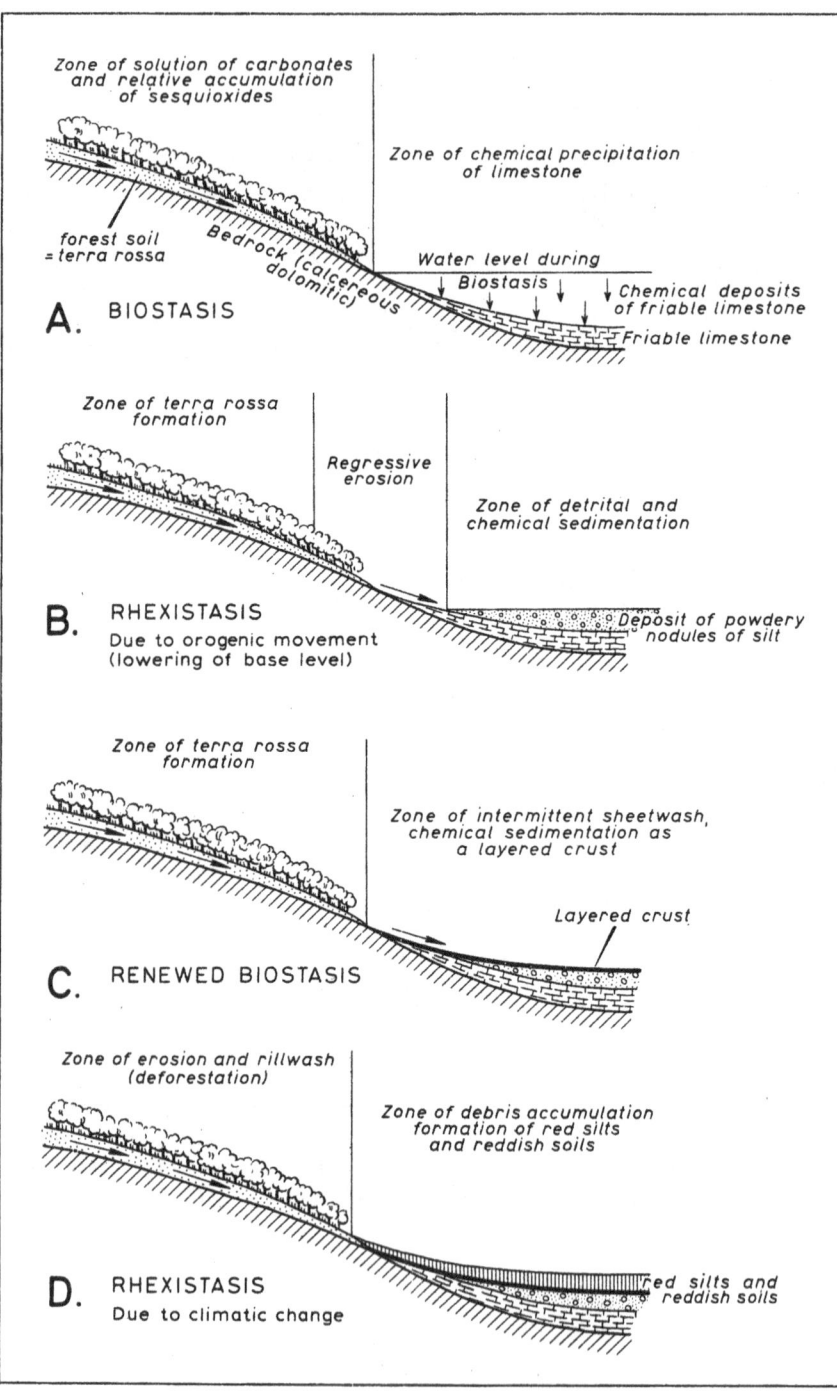

Zone of solution of carbonates
and relative accumulation
of sesquioxides

Zone of chemical precipitation
of limestone

forest soil
= terra rossa

Bedrock (calcereous
dolomitic)

Water level during
Biostasis

Chemical deposits
of friable limestone

Friable limestone

A. BIOSTASIS

Zone of terra rossa
formation

Regressive
erosion

Zone of detrital and
chemical sedimentation

Deposit of powdery
nodules of silt

B. RHEXISTASIS
Due to orogenic movement
(lowering of base level)

Zone of terra rossa
formation

Zone of intermittent sheetwash,
chemical sedimentation as
a layered crust

Layered crust

C. RENEWED BIOSTASIS

Zone of erosion and rillwash
(deforestation)

Zone of debris accumulation
formation of red silts
and reddish soils

Red silts and
reddish soils

D. RHEXISTASIS
Due to climatic change

One of the main limitations on the study of slope evolution is the typical absence of sequences of sediments from which the phases of erosion on the upper slopes might be inferred. However, in many areas the scope of the study of lake sediments in this context awaits investigation and in many semi-arid or arid environments such sedimentary sequences may remain on lower slopes in basins of internal drainage. J. H. Durand (1959) describes such a situation in Algeria and sketches some typical changes in soil and slope development (fig. V.9). Here terra rossa formed on upper slopes under a damp and warm climate due to the relative enrichment in sesquioxides of the soil in relation to the parent material of limestone or dolomite, and also due to the addition of soil material from an external source. Water draining from the slopes, rich in dissolved calcium carbonate, led to the precipitation of friable limestones on to lake floors at the foot of the slopes. The absence of clastic material in these limestones indicates a period of biostasis with no erosion at this stage (fig. V.9.A). In a subsequent phase lowering of local base-levels and the drying-up of lakes encouraged erosional activity which extended some way up the slopes. At this stage the rhexistasis, indicated by clayey silts and by the powdery limestone nodules formed in contraction fissures by the segregation of calcareous material from other sediments, was due to tectonic movements rather than to climatic change as the moisture regime conditions had not changed (fig. V.9.B). As a new equilibrium was established erosion ceased and the drainage water again became sediment-free. The calcium carbonate was then deposited on the lower slope as a layered crust due to its precipitation from sheetwash, each layer hardening between successive sheetwash episodes (fig. V.9.C). In a fourth phase, characterized by desiccation, the reduced effectiveness of the sparser vegetation in filtering sediment out of surface water led to some erosional downwearing of the upper slope and to the transportation of red silts and reddish soils to add to the depositional layers on the lower slope (fig. V.9.D). The red silts are the product of the more intense erosion.

Of rapid changes, forms in unconsolidated materials provide the clearest examples. In Rhodesia, sand ridges up to 120 m apart are now only 3–4 m high, sheet erosion probably being the main agent in the destruction of inferred former dunes. It is likely that erosion and trough-filling would be most rapid initially when steep slopes still prevailed. Similar in the degree to which unconsolidated material undergoes rapid change is the instructive sequence of forms observed, in the mollisol banks in the Mackenzie delta area, by J. R. MacKay (1966). In early summer, the lower scarp face and

Figure V.9 Phases in slope development in semi-arid Algeria linked with stability of the vegetation cover, erosion on the upper slope and the accumulation of stratified slope deposits at the base (*after Durand, 1959*).

frequently also the middle is covered with slumped debris which accumulated in the previous autumn. In general, the scarp face is cleaned off by mid-July. When first exposed, the surface is usually smooth from top to bottom, but it may roughen as the season progresses. As thawing proceeds, 'caprock' hummocks, often bordered by roots are undermined, overlying and frequently falling to the base of the scarp. A ridge and gully system encroaches on the scarp face until it becomes partially or totally ridged. Towards the end of summer the ribbing tends to disappear, perhaps reflecting the decreasing effect of stream erosion towards the end of the summer. As it appears to be a universal characteristic of slope development, the transportation of material from the toe of the slump as quickly as it accumulates favours active slumping and the maintenance of a steep profile, and other features of this small scale example might be relevant to the study of slope development on larger scales in space and time. S. A. Schumm (1956) observed seasonal changes on micropediments showing comparable contrasts in degree. During the spring much of the runoff from these slopes flows beneath the layer of aggregates to reappear on the pediment at the base of the slope. However, the proportion of surface runoff increases during the summer as the slopes are compacted by rainbeat and become less permeable.

The erosion of marine cliffs, where rapid, may also point to some more broadly applicable generalizations about slope development. For instance, in 1956, January–March storm-wave erosion of the cliff base at Highland, on Cape Cod, left a basal scarp 1·5–4·2 m high. Although this scarp was slowly buried by colluvial material from the cliff face and was gone as a cliff feature by May 1956, the line marking the top of the scarp was still distinguishable in September 1956. This example lends support to the belief that important stages in the history of a slope's evolution might be no more than mere subtleties in the present-day form. It also points to the usefulness of the study of coastal slopes or cliffs as models of situations where basal undercutting is undoubtedly effective.

Equally significant is the cessation of undercutting, and particularly instructive is the contrast between abandonment due either to base-level fall or to accretion at the foot of the coastal slope, transportation of debris tending to be more rapid in the former case. Even the undercutting of solid-rock coastal cliffs and slopes may lead to well-defined evolutionary morphological stages. Examples occur along the north-east Yorkshire coast, as described by R. Agar (1960), due to the undercutting of cliffs made up of Jurassic shales, sandstones, and shales with thin hard ironstone bands. In the last inter-glacial the cliff-line receded from about the present low-water mark to within 20 m of the present cliff. Large fallen blocks remained on the foreshore (fig. V.10.A). With the onset of the Würm glaciation, undercutting ceased with the lowering in sea-level, and the abandoned foreshore underwent weathering, colonization by vegetation and soils formed. The shale under the sandstone boulders may have been protected from these changes. The cliff-top weathered back, and debris cones and landslide debris filled the cliff-foot notch (fig. V.10.B). In late-glacial times narrow stream gorges developed in the drift but it was rapid postglacial sea-level rise reaching to within 5 m of present sea-level which led to the clearing away of much of the drift by undercutting; there was, however, no appreciable erosion of solid rock (fig. V.10.E). Renewed rise of sea-level from the pause at −5 to −3 m OD level, passing the present level and rising to a maximum of 9 m OD, led to undercutting and removal of debris cones from most of the old interglacial cliff foot. Undercutting of the solid cliff also started but was apparently insufficient to undermine the cliff-top. The slow undercutting of the solid cliff and slow down-wearing of the foreshore started about A.D. 1000 and continues today (fig. V.10.G). The result of undercutting in postglacial times is a cliff on average about 30 m high. At points along the coast where the lower zone of active undercutting has not yet affected the full height of the cliff, recession of the cliff-top is negligible (fig. V.10.E–G). Conversely, with the full height of the cliff affected, parallel retreat takes place with measured rates of recession being similar for the basal notch and for the cliff-top. At Huntcliff, the portion of a Roman Signal Station that has fallen away indicates a cliff-top recession of 3·6 m/100 years at a point where undercutting affects the full height of the cliff.

Fig. V.10.B illustrates a common pattern in the interglacial evolution of coastal slopes and cliffs with the lowest sections of an old solid-rock cliff being present early in the phase of sea-level fall and retaining its original steepness, but with the cliff above the level of protection tending to wear back. In south Devon such upper slopes are irregular and have declined to angles of 20–35 degrees (Orme, 1962). This formation of an upper convexity contrasts with the essentially static nature of the north-east Yorkshire cliff-tops during the shorter time span of postglacial times.

D. Stages in Drainage System Evolution

'The principle of first importance in the writer's view of geomorphology is the concept of essentially universal superposition (or superimposition) of drainage in flatland regions.' F. A. Melton, 1965

'... superimposition of drainage on the basement may not be a necessary consequence of the removal of coastal-plain sediments.' R. F. Flint, 1963

One of the basic considerations in studying the history of drainage pattern evolution is the consequences of an actively downcutting stream encountering different subjacent rocks. If a series of strata are essentially conformable the

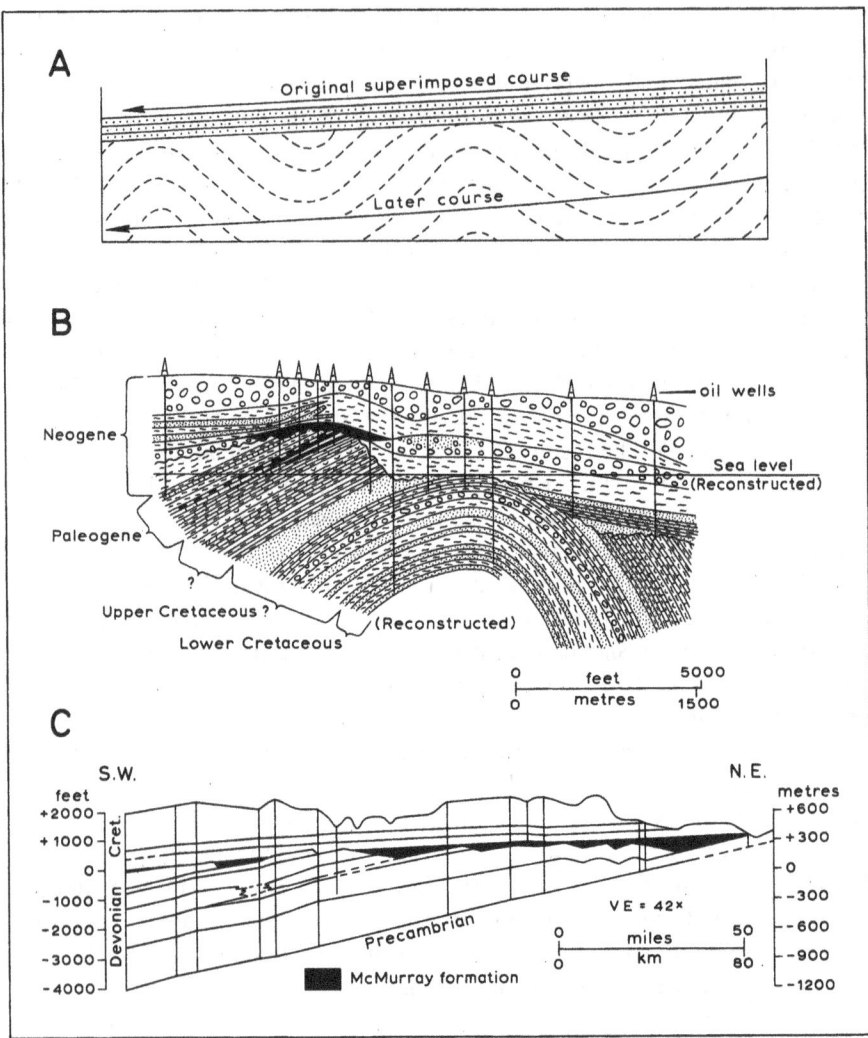

Figure V.11 Buried land surfaces showing the contrast between the hypothetical peneplain and actual reconstructions based on the boring logs from intensively drilled oil-fields.

A. The buried peneplain, assumed to have been bevelled by erosion, on which superimposed drainage could develop with the downcutting of drainage established on the unconformable cover of younger strata (*from Sparks, 1960*).

B. Cross-section of the south Alamyshik oil-field, Uzbek S.S.R., Turkestan (*from Martin, 1966*), showing an escarpment on the flank of an anticline and its economic significance as an oil trap. The position of wells is also indicated.

lowering of a stream-bed from one stratum on to the next is accepted simply as valley downcutting in which any attendant changes might be limited to those of slope form, with the pattern of drainage remaining unchanged. However, since many lowlands are made up of unconsolidated sediments lying unconformably on much older and harder rock, the possibility is frequently considered of a stream actively downcutting through a cover rock and ingraining its pattern on to subjacent rocks with disregard for their different structural attitude. The likelihood of this development was enhanced in W. M. Davis's scheme of drainage superimposition by the assumption that the unconformity had the virtual flatness of the hypothetical peneplain with the differing resistances of underlying structures planed off.

Although the lowering of a drainage system from a high rock on to one beneath is inevitable, one of the difficulties in accepting the idealized scheme as a starting-point in drainage development is that the assumption of a relatively subdued erosional surface at the unconformity beneath the cover rock is increasingly found to be mistaken. Borings now indicate that there were waterfalls at least 25 m high on the Paleozoic bedrock relief now buried in the Lower Mississippi valley, and that all the way from Cairo, Illinois, to the Gulf of Mexico, the buried pre-Recent surface is broadly rolling and markedly incised by an intricate valley system.

Another difficulty, posed by the hypothetical initial conditions, is that of visualizing how the initial surface, unless it be a volcanic cone produced overnight, comes into existence without being subjected to the existing pattern of drainage in the process. For instance, the emergent marine areas frequently supposed to provide an ideal gently inclined smooth surface on which 'consequent' streams develop are in reality more likely to be mere seaward strips in the path of outlets from existing drainage systems with an existing pattern perhaps closely tied to structures continuing seaward beneath the covering fringe of marine sediments.

A problem of applying this concept in the field is the absence in many areas of any geological evidence to demonstrate that a cover rock once spanned the area. In such instances the base of a possible cover rock is projected geometrically from its outcrop in an adjacent area. However, this could be a misleading reconstruction as in many areas Tertiary deformation was substantial and widespread. G. H. Dury (1959) considered that a Carboniferous cover as an

---

C. Cross-section of Athabasca 'Bituminous Sands' area, north-eastern Alberta (*from Martin, 1966*). The basal Cretaceous sandstone units of the McMurray formation were laid down in a connected valley system where erosion-resistant layers in the underlying Devonian–Mississippian sequence stood out as north-west to south-east trending escarpments. Note the thinning north-eastward of the underlying Devonian strata, due to interstratal solution in salt-bearing formations, and the resultant subsidence in post-Devonian strata.

initial stage in superimposition of drainage in Donegal could be ruled out, 'especially since the projected base of the Carboniferous passes well above summits in west Donegal'. This hypothetical geometry, however, is mistaken as there are small Carboniferous outliers actually in existence below the present summit levels. Conversely, some reconstructions which are geometrically possible for one area are irrelevant or geologically impossible in others. The extension north and west of the Cretaceous outcrops appears feasible in lowland England; yet it is pointless to extricate the same strata from beneath the Tertiary lavas in Northern Ireland as the latter have the discordances between structure and drainage which the Chalk cover might be invoked to explain elsewhere, and impossible to extrapolate from its small outlier enclosed at the base of exhumed late-Cretaceous valleys in Co. Kerry in southwest Ireland. In other instances the circular argument, that observed discordances between drainage and structure point to the existence of a former cover rock, is sometimes used where independent evidence of the existence of a former cover is lacking.

A second basic consideration is the mode in which a river enlarges its valley. After W. M. Davis's imaginary, initial stage of drainage development, the 'consequent' streams became incised across a structural grain of the buried peneplain assumed to strike at right-angles to their course. The remainder of the cover rock was gradually removed as 'subsequent' streams started to hollow out strike vales by a mechanism termed headward erosion. This is assumed to be a self-accentuating process which gets more powerful the larger a catchment area it cuts. Thus the most powerful stream in an area with no self-regulating mechanism envisaged to check it, breaks through divides with Davisian zeal and cuts off the line of descent of less powerful neighbours by decapitating them. Amid all this excitement and the large number of interpretations tacitly depending on certain assumptions about the efficacy of headward erosion it is difficult to realize that the mechanism itself has scarcely ever been studied or described. Almost invariably it is the form of the ground that provides the evidence.

A search for the source of the boundless energy of headward erosion is hampered by features due to artificially accelerated erosion in many headstream gullies. However, it is clear that in many instances such gullies are not cut into an unmodified surface but in some cases represent the transformation of lines of concentrated flow of sub-surface seepage into surface gullies. This change seemed to explain the headward extension of gullies in the New Forest, Hampshire, where longitudinal growth appeared to result from the coalescence of discontinuous hollows rather than by upslope migration of a channel head. In fact, R. E. Horton, defined the summit of a divide as a 'belt of no erosion'. M. A. Melton (1958) considers that the maintenance of a length of stream channel depends on a certain catchment area needed to supply runoff sufficient to remove the products of mass wasting tending to obliterate the channel. As B. T. Bunting (1961) also argued, the extension of incipient drainage or seepage lines into crestal areas must gradually cease when the crestal area, reaching some minimum width, can no longer yield sufficient moisture to the adjacent seepage hollows or cusps for the processes of corrosion to continue.

An appreciable exaggeration of the erosional potential of first order head-streams results if headward erosion is applied loosely to include knickpoint or waterfall recession and spring-sapping in this term. The last two are fundamentally distinct due to the organized drainage system that exists above the rapids, fall or spring. From a study of laboratory streams L. M. Brush and M. G. Wolman (1960) concluded that in actual streams a knickpoint probably would not be recognizable more than several miles upstream from its initial position, unless the original fall were extremely high, and that it would be most unlikely to travel the entire length of a natural stream. Added to this is the field evidence that most falls of great size are not situated in high mountains but on plateaux often of great age and are sometimes relatively close to the sea.

It seems that if a history of drainage evolution emphasizes the formation and headward growth of new channels this view is largely a logical deduction from the basic assumption of an initial, ideal undissected surface, following inevitably as an artifact of the initial premise rather than as the result of observation of actual landforms.

In some areas where structurally controlled drainage appears to have flowed for geological spans of time, the concept of superimposition is superfluous. R. F. Flint (1963) considered that superimposition of drainage on a basement may not be a necessary consequence of the removal of coastal plain sediments, as the general drainage pattern north of The Fall Zone seems to be much the same as it was before Cretaceous sedimentation began. J. W. Ambrose (1964) concluded that for a 5 million-square-km tract of the Canadian Shield a drainage pattern, once adjusted to structure, will persist indefinitely, and if buried and exhumed would be reactivated virtually intact. The hills and valleys, of the order of 2000 million years old, are closely related with bedrock strike and type. The inter-stream areas are more resistant granites and the streams and lakes lie in softer gneisses and schists. In northern Scotland, the valleys of the River Lossie and the Black Burn, southwest of Elgin, correspond in part with pre-Devonian valleys.

A factor similar to that of the persistence of rigid geological controls is that many contemporary major rivers had precursors following a similar approximate trend to their own at times in the geological past. The St Lawrence drainage has probably emptied near where it now does since well back in the Tertiary. Historically the area that is now the Amazon basin existed as a

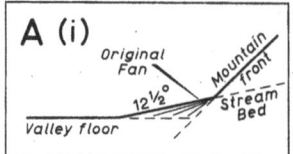

A (i) Original Fan
Mountain front
12½°
Valley floor
Stream Bed

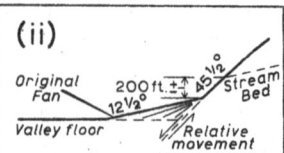

(ii) Original Fan
45½°
200 ft. ±
Stream Bed
12½°
Valley floor
Relative movement

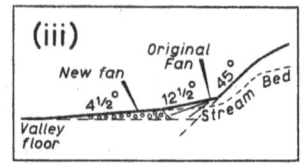

(iii) New fan
Original Fan
45°
4½°
12½°
Valley floor
Stream Bed

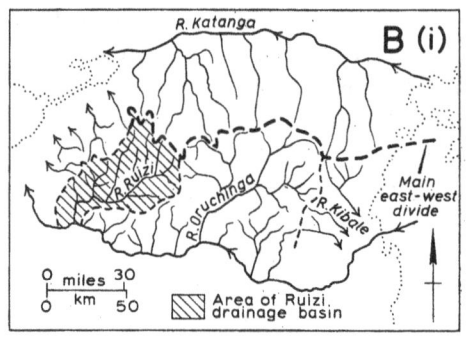

B (i)
R. Katanga
R. Ruizi
R. Oruchinga
R. Kibale
Main east-west divide

0 miles 30
0 km 50
Area of Ruizi drainage basin

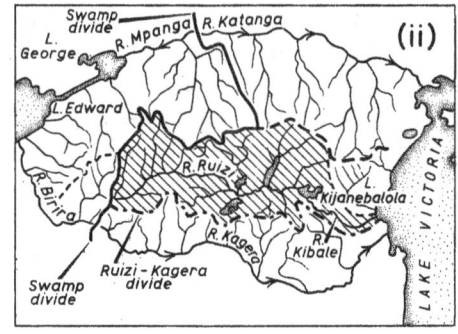

(ii)
Swamp divide
L. George
R. Mpanga
R. Katanga
L. Edward
R. Ruizi
R. Birira
Kijanebalola
R. Kagera
R. Kibale
Swamp divide
Ruizi – Kagera divide
LAKE VICTORIA

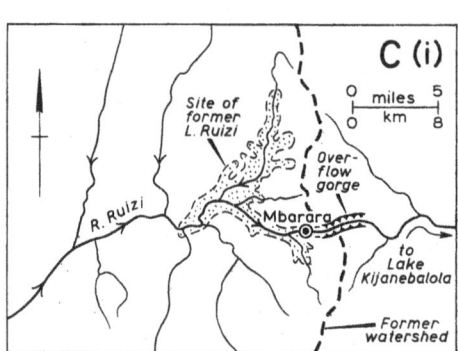

C (i)
Site of former L. Ruizi
Overflow gorge
R. Ruizi
Mbarara
to Lake Kijanebalola
Former watershed

0 miles 5
0 km 8

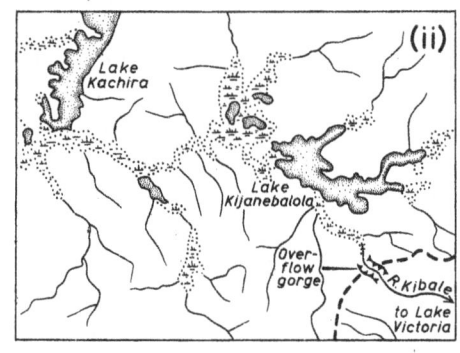

(ii)
Lake Kachira
Lake Kijanebalola
Overflow gorge
R. Kibale
to Lake Victoria

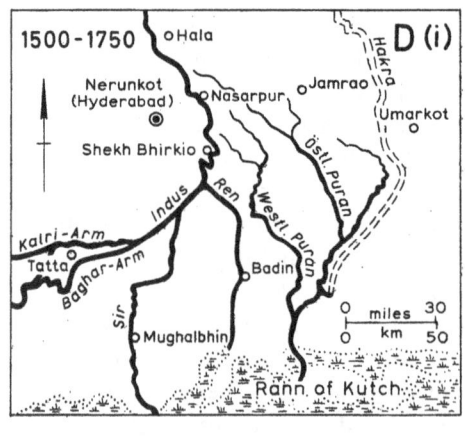

D (i)
1500-1750
Hala
Nerunkot (Hyderabad)
Nasarpur
Jamrao
Umarkot
Shekh Bhirkio
Indus
Ren
Westl. Puran
Östl. Puran
Katri-Arm
Tatta
Baghar-Arm
Badin
Sir
Mughalbhin
Rann of Kutch

0 miles 30
0 km 50

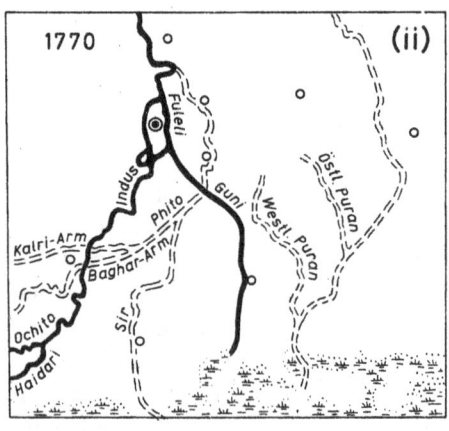

(ii)
1770
Fuleli
Indus
Guni
Östl. Puran
Phito
Westl. Puran
Katri-Arm
Baghar-Arm
Ochito
Sir
Haidari

trough throughout most of the Mesozoic and Paleozoic eras and possibly into the pre-Cambrian era. The course of the Lower Senegal river has probably not changed significantly since Miocene time. The Congo has been evacuating its cuvette along the same course since Miocene or even Oligocene times. F. Dixey (1938) considers that the upper parts of the main rivers in Madagascar and East Africa are of ancient origin and might date back as far as Jurassic times. In the more immediate past and affecting innumerable rivers of much smaller size is the deepening of channels by the order of 30 m or more beneath their present levels by downcutting during Pleistocene low stands of sea-level. The buried channel beneath the Quinnipac river as it enters Long Island sound is at least 50 m below sea-level and possibly more than 60 m. In northern Connecticut, pre-Wisconsin entrenchment is more than 30 m below present sea-level as is that of the Pleistocene Elizabeth river; the entrench-ment beneath the Pleistocene James river is 50 m. Drowned valleys filled to varying degrees by alluvial material are common around Australia like the River Hunter where depths of 75 m have long been known. The preglacial Teifi is typical of many streams in Wales by having a maximum depth at its mouth of 22 m below present sea-level, although a depth of 40 m is known in the River Tawe valley 10 km from its mouth. Depths of at least 20 m occur in the Avon, 35 m in the Severn, and 25–30 m in the Solent. The broad implica-tion of the existence of these buried valleys for stages in the development of the lower courses of rivers seems clear. A relative downward movement of sea-level by several dozen metres would not lead to any substantially new river valley development but merely to the re-excavation of existing buried channels. For the Bristol Avon, as with the River Tawe, it has been suggested that the buried channel was already eroded by the First or Second Phase of the Last Glaciation. With ice recession and subsequent marine transgressions the channel was probably infilled, subsequently to be re-excavated during

---

Figure V.12 Some examples of stages in drainage system development.

A. Development of a dissected alluvial fan in south-west United States (*from Beaty, 1961*): (i) initial conditions, (ii) immediately after faulting, and (iii) present situation.

B. General changes in the drainage of southern Uganda due to tilting (*from J. C. Doornkamp and P. H. Temple, 1966*, Geog. Jour., *Vol. 132*): (i) before the middle Pleistocene reversal, and (ii) the present lakes and rivers.

C. Effects of back tilting on tributaries in part of southern Uganda (*from Doornkamp and Temple, 1966*): (i) lacustrine deposits of a former lake west of Mbarara, and (ii) the system of lakes and marshes (stippled) to the east and their overflow into Lake Victoria.

D. Drainage system in the Lower Sind before and after the Indus diversion in 1758–9 (*from H. Wilhelmy, 1966*, Erdkunde, *Vol. 20*), based on information from M. R. Haig, H. T. Lambrick, and others.

marine recessions of the Third and possibly the Second Phase of the Last Glaciation and finally to be refilled with post-glacial deposits before and during the Flandrian Transgression.

Perhaps because of the artificial separation of uplift from erosion in W. M. Davis's imaginative models, the degree to which the history of many drainage networks reflects tectonic changes has perhaps been underestimated. Although identifiable alterations due to strike-slip faulting are very localized, modifications due to other crustal disturbances are more widespread. In northern Utah, when the Bear Lake was 2 m above its present height an outlet developed at the west side of the valley. The most logical explanation for a subsequent easterly migration of the outlet is movement along the Bear Lake fault. Similar movements might explain why all fans on the western flank of the Black Mountains in Death valley have deeply incised channels cutting far into their apex region, leading down to a newer part of the fan (fig. V.12.A). In another tectonically active area, westerly uplift resulted in an eventual reversal of the rivers Kagera, Katonga, and Kafu from their former courses to the Albert and Edward into the present Kyoga–L. Victoria system (fig. V.12.D). If the rivers like the Kagera, Katonga, and Kafu were originally Congo headstreams flowing to the Atlantic, this connection was severed during or before early Miocene times and subsequently were controlled by successive rejuvenations due to tectonic movements in the rift valley. In the Murray basin in Australia, a displacement of 15 m on the Cadell Fault near Mathowa diverted the ancestral Murray to the north. The abandoned section of its former course, Green Gully, is now a linear depression on the elevated tilt-block. Multiple branching indicates that more than one phase of ancestral river activity was responsible for the pattern of ancestral rivers. The example in the best traditions of classical Uniformitarianism is the Casiquiare Channel curiosity which connects the Rio Negro, a principal tributary of the Amazon, with the Orinoco. The direction of flow in this channel depends on the flood stages of the two river systems and reverses seasonally. Here only a minor regional tilting would cause the diversion of a sizeable portion of the Orinoco drainage into the Rio Negro.

Although tectonic movements may initiate diversions or even reversals of drainage, some rivers maintain their courses during such movements to the extent that this applies to most drainage systems with the term antecedent describing the case where the maintained course becomes a valley discordant to the new structure. In the Lower Truckee Canyon area the surface of Pleistocene volcanic lavas was initially near-horizontal and the river cut down concurrently with a westward tilting so that the present canyon is antecedent. Because no sediments have been found upstream to indicate ponding during uplift, the rate of downcutting in the Lower Truckee Canyon appears to have equalled or exceeded the rate of uplift. The nearby upper Missouri river is also to some extent antecedent. In India, slow or periodic tilting in the Pleistocene caused the old Brahmaputra river to become antecedent in places by necessitating river scour into comparatively resistant Pleistocene sediments.

In addition to active tectonic movements initiating diversions this significant stage in the history of a drainage system may be due to a combination of the erosional and depositional activities of a river itself, in some situations operating in conjunction with static geological factors like lithology. At the downstream end of rivers the very low gradients on some lowland plains mean that either heavy sedimentation in a channel, or only a slight regional tilt or subsidence, may favour diversions of river courses involving huge volumes of water. For instance, continued subsidence on the western flank of the Mississippi modern sub-delta is causing diversion of the river to the coast 160 km west of the present bird's-foot delta. A feature of such situations is the possibility of a chain reaction as the excess flow in turn causes diversions in a receiving river. The sudden change of course of the Tista river with resulting addition of its waters to the Brahmaputra river may well have been a contributing factor towards the diversion of the latter (fig. V.12.D).

In middle courses of streams, stages in drainage pattern development may accompany and follow lateral stream cutting. In areas of horizontal strata or homogeneous rock, a river deepens its course with little lateral erosion, but where there is even a slight regional dip or the systematic development of less resistant strata on one valley wall compared with the other there is a greater or lesser element of lateral shift in the position of a downcutting stream. F. L. Stricklin (1961) describes slopes of the Seymour valley, Texas, which suggest that the valley floor was cut primarily by two major streams. The bedrock floor beneath the Seymour has a dominant northerly slope on the south side of the Brazos and a dominant easterly slope on the north side ranging from 1–4 m/km. The bedrock surfaces are smoothly bevelled and overlain by coarse, cross-bedded alluvium, indicating their origin by lateral stream planation, and appear to reflect a northerly shift of the Brazos and an easterly shift of one of its large tributaries. The position of the Pearlette ash indicates that the shifting in channel position has probably been in operation for a few million years. In an area farther north, in the Three Forks basin, Montana, the Bridger Range began to rise in late Pliocene or early Pleistocene times and because of the northerly tilt the main streams like the ancestral Madison, Jefferson, and Gallatin tended to shift northward with time. On the eastern side of North America in Tertiary times, the attitude of the consolidated rocks in the Teays river basin apparently exercised some regional control on the river and its tributaries, causing the former River Teays to migrate laterally by down-dip shift. This 1500-km-long river flowed from its headwater region in the Piedmont of Virginia and North Carolina, north-westward and west and finally into the present Illinois. A pre-Pleistocene stage in the dismemberment of this former major drainage feature was its diversion at Scary. E. C. Rhodehamel and C. W. Carlston (1963) believe that down-dip shift of the stream was involved in this diversion, that an uncommon thickness of sandstone formed a local base-level and greatly slowed the rate of downcutting immediately downstream from the point where the diversion eventually occurred, although the final mechanism that accomplished the diversion at the 'elbow of capture' is not known. However,

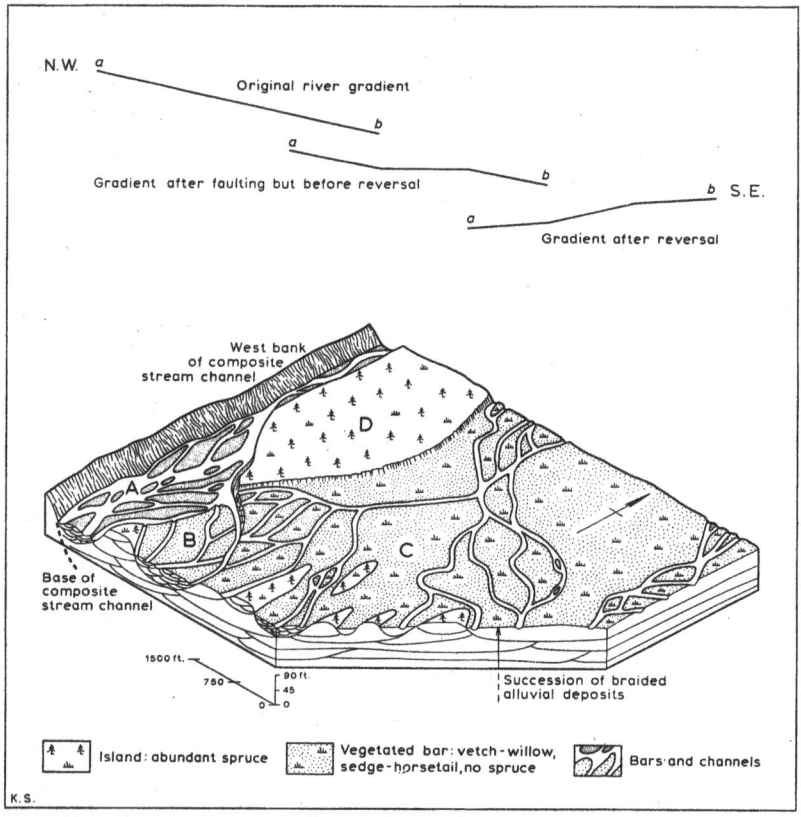

Figure V.13 Further examples of stages in drainage system development. (*Above*) Structural movements causing the reversal of the Lemki river, east central Idaho, in late Pliocene to Middle Pleistocene times, a change typical of much of an area where drainage probably has been structurally controlled since the advent of range-front faulting in Miocene times (*from Ruppel, 1967*). (*Below*) Sequence of alluvial terraces (*from P. F. Williams and B. R. Rust, 1969*, Jour. Sed. Petrol., *Vol. 39*) in the Donjek valley, Alaska. Downcutting may be due to a change in hydrological regime, prolonged glacial wastage in this area increasing melt-water volume and raising total discharge.

as this and some other diversions like that illustrated in fig. V.14 resemble in many details diversions on inclined strata in the southern Pennines, an explanation suggested for the diversion stage in river drainage history in that area might have some broader application. Fig. V.15 illustrates the suggested mechanism, termed down-dip breaching (Pitty, 1965). Stages in drainage diversion may start with indentation along a weaker scarp section by dip-controlled stream erosion, followed by lateral erosion within the indentation, combined with an eventual overtopping of the resultant lowered divide which could involve an accumulation of alluvium on the erosional surface within the embayment. A remnant of this surface may persist as a knoll in the funnel-shaped entrance to the top through the divide cut by the diversion. Alternatively in some parts of the southern Pennine scarplands with a low dip or where the bench formed between two resistant strata is narrow, alluviation in the channel may make diversion possible over the lower stratum forming the up-dip margin of the drainage basin. This diversion mechanism, termed up-dip breaching (Pitty, 1966) leads to a diversion channel cut obliquely to the strike (fig. V.16).

A feature of river diversions is that they may result from a rare or even unique event. One of the clearest examples followed the overflow of Lake Bonneville into the Snake river, perhaps in the early part of the last glacial

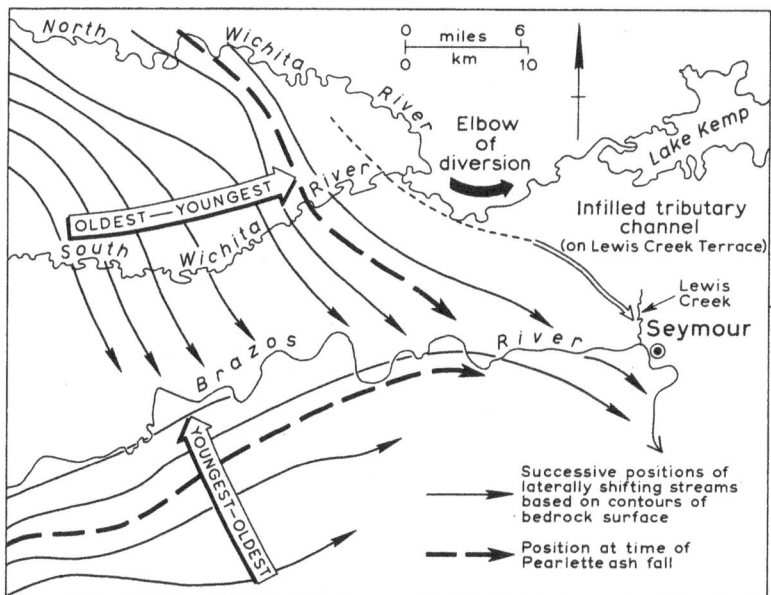

Figure V.14 Successive positions of laterally shifting streams in Texas (*from Stricklin, 1961*), as determined by outcrops of the late Kansan or early Yarmouthian Pearlette Ash.

about 20 000 years ago. The overspill itself was perhaps the result of drainage changes in the Bear river. The flood was of such great volume that at bends and restrictions in the Snake river canyon it was some hundreds of metres deep and it overflowed the canyon walls, took short cuts across the plain and created spectacular bars and plunge pools where the overflow waters returned to the canyon. This diversion followed earlier diversion stages in the history of this river which occurred when the extrusion of the Snake river basalt in the Cedar Butte locality dammed the river, with the water impounded in the ponded lake extending about 60 km upstream.

When a river system exposes limestone on its floor distinctive, although not universally similar, stages in drainage evolution usually proceed until

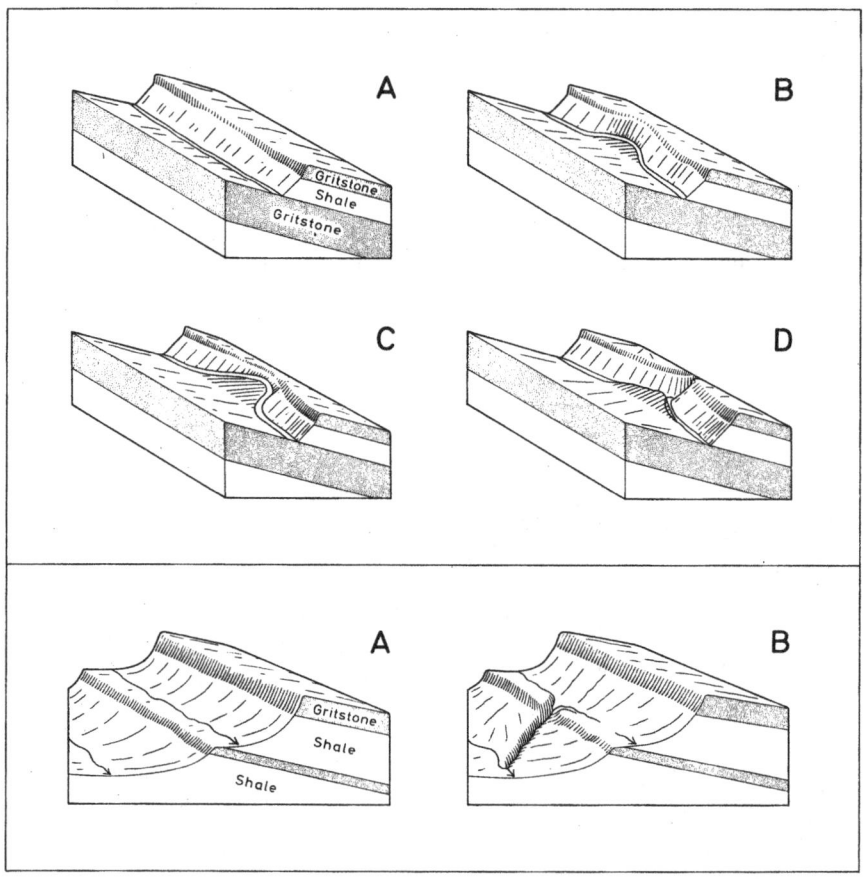

Figure V.15 (*Above*) Stages in the development of a down-dip breach (*Pitty, 1965*).

Figure V.16 (*Below*) Stages in the development of a counter-dip breach.

much of the limestone is removed. The final stage is the re-establishment of normal drainage on the impermeable basement beneath the limestone strata. The gradual diversion of drainage underground depends on more factors than the solubility and jointed nature of limestone. Tectonic movements are particularly significant in the earlier stages. In Yugoslavia, tectonic movements along major lines of weakness have been fundamental in the deranging of stream patterns and in the draining of lakes. Some lakes have actually been observed to drain temporarily after earth tremors. In some areas normal surface drainage persisted before orogenic movements took place as fissures were then few. In the Cracow upland, dry valleys and cave systems developed only after Miocene orogenic movements had occurred. A second factor in the diversion of drainage underground is the necessity for there to be a low point in the limestone outcrop at some distance below the general level of the limestone land surface to serve as a point of hydrographic concentration. In a ponded karst, where impermeable rocks surrounding the limestone form a relatively high rim, development of underground drainage is restricted. An important consideration in the contrasting type of karst, the perched karst, should the limestone extend to the sea, is the degree to which efficient points of hydrographic concentration developed during low stands of sea-level in Pleistocene times. On the Adriatic coast, submarine springs are numerous and water-levels encountered 200 m below the bed of the Riječine seem to have been affected by a base-level below the Adriatic itself. In general, the lower the point of hydrographic concentration the more the karst evolution resembles that of the dinaric type; its periodic lowering, particularly where it is the level of an impermeable rim which controls the unwatering of the limestone mass, may favour the development of a sequence of levels in the abandoned drainage system (fig. V.17.A). The present valley occupied by the Plitvice lakes is an old cave-level, a higher level having been formed by the floor of dry caves which enter the gorge above the present lakes. M. M. Sweeting (1950) has demonstrated how the levels of caves in north-west Yorkshire tend to occur close to certain altitudes. A third factor concerns the degree to which water flowing over a limestone surface is capable of enlarging initial tectonic or structural lines of weakness, and of exploiting the potential depth for cave development where the point of hydrographic concentration is low. In some limestone streams, particularly in chalk areas, the water at its source has as high a concentration of calcium carbonate in solution as at any point on its lower course (fig. V.17.B). In consequence, the most spectacular features associated with stages in the diversion of drainage underground occur where catchments, made up largely of non-calcareous rocks, channel water low in dissolved carbonates on to a limestone outcrop. Quite often such an allogenic stream in fact maintains a surface course over the limestone. However, it seems that a zone of bedrock is sometimes

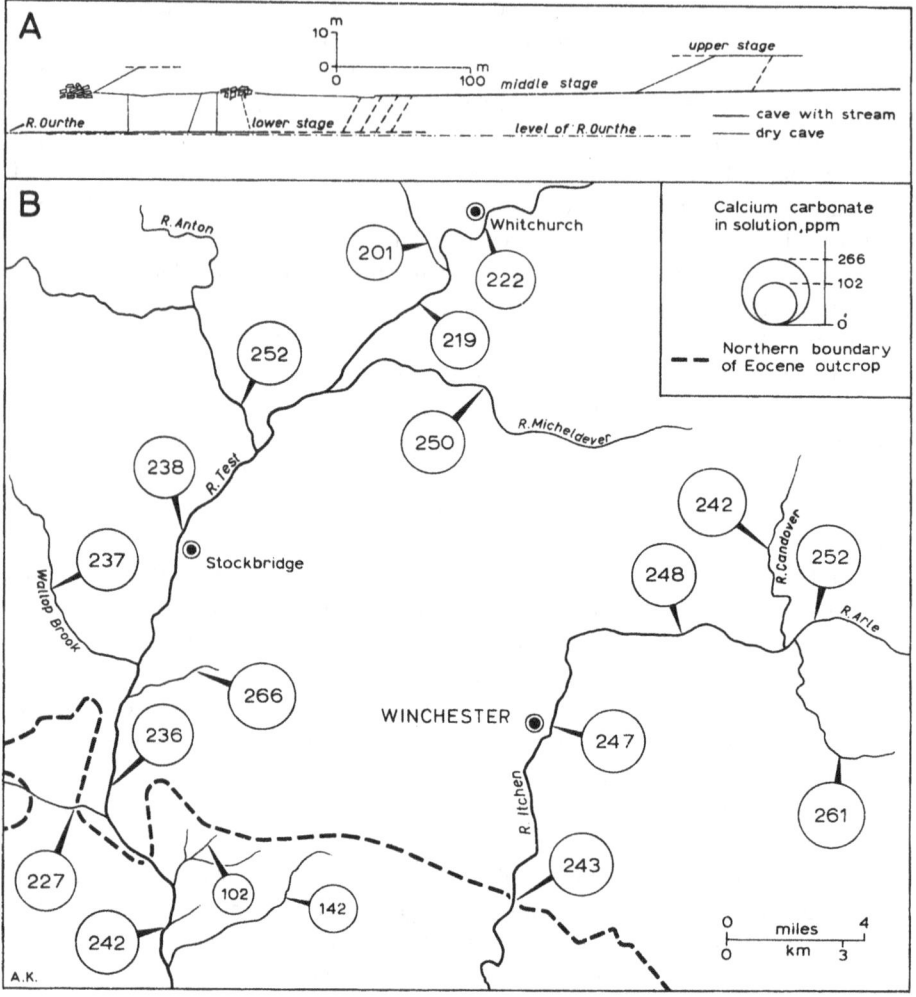

Figure V.17 Aspects of the development of drainage systems in calcareous rocks.

A. Levels in the Sainte Anne cave at Tilff linked with downcutting of the Ourthe river, Belgium *(from C. Ek, 1962, Rassegna, Spel. Italiana, Vol. 14)*.

B. Changes in chemical quality of streams at a geological contact, as observed in the Hampshire basin. Where Eocene sands still cover the chalk the calcium carbonate content of surface streams is well below the consistently higher level of those where chalk is exposed at the surface (the values are means of five observations, 1961–2).

dissolved beneath such a stream. The zone of solutional penetration, with clastic debris filling the cavities, was 70 m deep at the Kentucky dam site in the Tennessee valley. Impermeable beds limit the downward penetration of solutional activity. The larger sub-river openings were 30–300 m long, and joint controlled, but occupied a zone only 20–30 m wide at the top and tapered downwards. From the information provided by the unusually intensive borings in the river-bed of the Tennessee valley it is readily envisaged how, as a general pattern, a fall of local or regional base-level could rapidly transform a limestone valley into the steep-sided gorges which typify many limestone areas. Alternatively an allogenic stream may sink on meeting a limestone outcrop and a cave system develops. If the insoluble material in the headwater area is relatively easily eroded and evacuated through the cave system, the valley upstream from the sinkhole may be deepened by vertical erosion more rapidly than the general level of the limestone surface is lowered by solution. In this case a blind-valley develops near the geological contact, with a horseshoe-shaped cliff or steep slope facing upvalley and the sinkhole near the base of the blind-end. A striking example is the Pazin river at the edge of the shale outcrop in Istria. This drainage system also illustrates another unusual stage in drainage evolution in limestone areas as this river feeds several outlet points (fig. V.18.A). At some distance on to the outcrop of limestone roofing an underground stream the possibility of shaft dolines developing by collapse and subsequently enlarging into karst windows appears to be a distinctive if perhaps rather unusual stage in drainage system evolution. The Rak river in Slovenia is a classic example (fig. V.18.B). The river to which it is tributary, the Ljubljanica, is truncated into six parts with the surface reaches across basin or polje floors each separated by sills. Integration of the surface drainage appears to be progressing towards the headwaters. Like the present Pazin stream the Ljubljanica headwaters are thought formerly to have flowed in several directions.

A final consideration in the study of stages in drainage development is that of paleohydrology, as one of the major uncertainties in drainage histories is how climatic changes, as well as influencing terrace formation, affected the density of drainage. The density of valleys dry and drained during the Pleistocene in Belgium and many other lowland regions were approximately the same for clay, sand, chalk, and shales, yet today's density of drained valleys varied with permeability of rock and a lot of chalk and sand valleys are dry. One possible explanation is that the average groundwater-table level averaged over spans of thousands of years might have been substantially higher perhaps due to greater effective precipitation resulting from lower temperatures and possibly higher precipitation totals. Also, the observed rise of watertables on the banks of reservoirs when first impounded might suggest that some influence of sea-level rise on watertables on a regional scale could

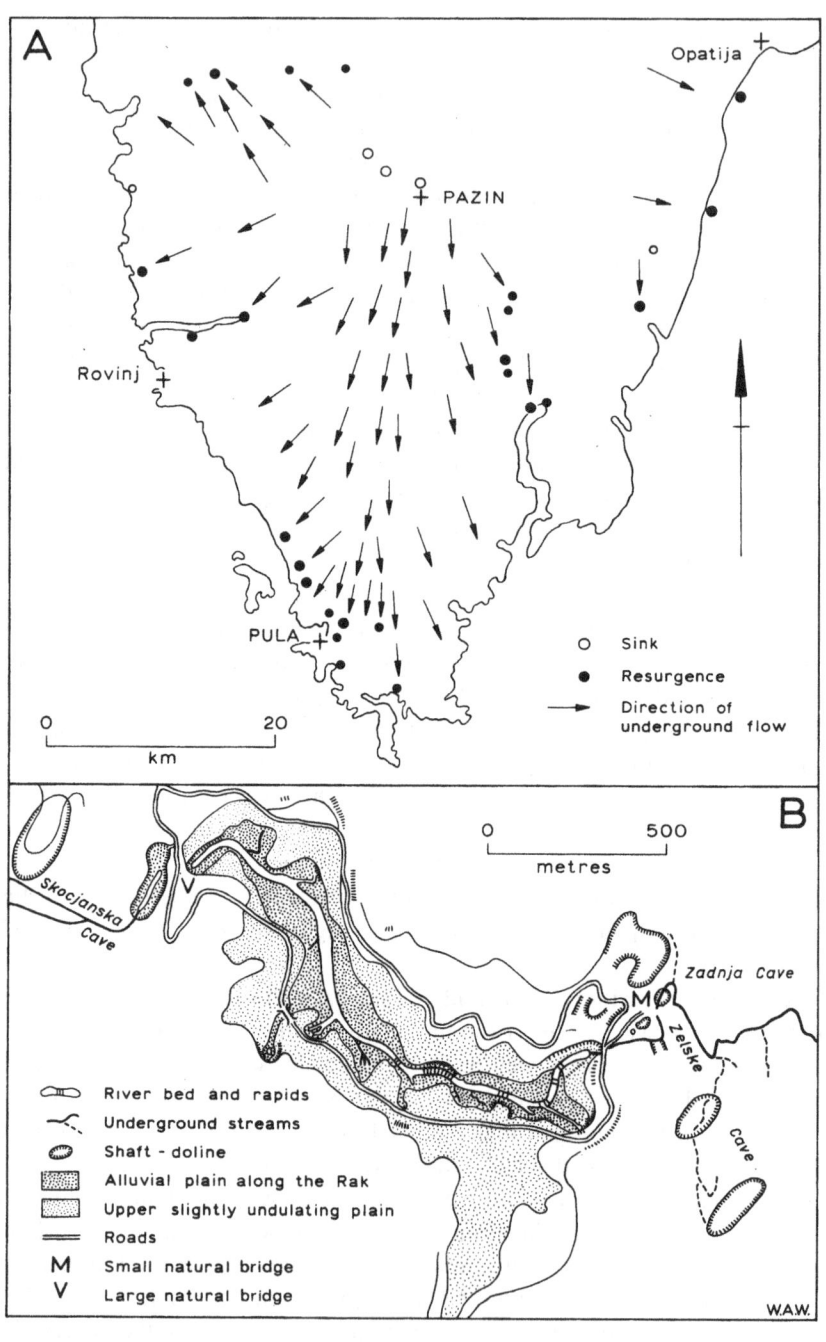

A

Opatija

PAZIN

Rovinj

PULA

○ Sink
● Resurgence
→ Direction of
   underground flow

0   20
        km

B

Skocjanska
Cave

V

0   500
      metres

Zadnja Cave

M
V
Zelske
Cave

⌒⌒⌒ River bed and rapids
↶↶   Underground streams
◯   Shaft - doline
▦   Alluvial plain along the Rak
▨   Upper slightly undulating plain
═══ Roads
M   Small natural bridge
V   Large natural bridge

W.A.W.

be considered. A second speculation that might be worth quantifying is the degree to which the greater viscosity of the colder water would reduce infiltration rates and thereby favour greater surface runoff. The largest changes in drainage density occur in most present-day sub-humid to arid environments. In the Texas High Plains, permanent streams are now virtually non-existent and valleys and associated depressions exist now only as faint but obvious surface features filled with eolian and fluviatile debris, yet are cut in bedrock and may be used during extended periods of wet weather. Similarly remnants of old drainage patterns of once permanent streams exist in the Sahara and the Kalahari. Similar elongate basins appear to be the remnants of dismembered river systems in western Australia. In eastern Australia, ancestral rivers in the Murray basin were of much greater magnitude than those of the present system.

To some extent inevitable was the association of the disintegration of drainage networks due to reduced runoff, with the blocking of channels with dunes of wind-blown sand. Even over short spans of time, with a few seasons of exceptional drought such as is liable to occur in parts of South Africa, wind-blown sand may cause a temporary choking of the watercourse and even reversal of drainage. In some instances the river system of the succeeding phase of greater runoff became re-established in accordance with the dune pattern. Wind-aligned drainage systems have been noted in the northern Great Plains, like the north-west to south-east alignment of tributary streams of east-flowing rivers in South Dakota, in Navajo country in Arizona, and over a large area in the Colorado Plateau where numerous south-west-flowing tributaries incised by 60 m and considerably more are parallel with present wind direction. Most areas of aligned drainage are downwind from large intermittent streams, the beds of which receive periodic replenishments of sand available for wind transport during dry periods. It seems that any deviation away from the direction of prevailing winds by smaller tributaries is counteracted by the accumulation of eolian material in this segment of the channel.

In more humid areas in mid-latitudes there are also indications of environmental changes causing substantial changes in drainage features. One of the

---

Figure V.18 Critical stages in drainage development in limestone areas. (*Above*) A large number of resurgences supplied from a small number of sinkholes (*from M. Malez, 1968*, Geografski Glasnik, *Vol. 30*) showing the significance of the Pazin sink in the underground drainage of central and southern Istria. (*Below*) The Skocjan valley, east-north-east of Postojna, Slovenia (*after A. Serko, 1949*, Geografski Vestnik, *Vols 20–1*) where collapse has revealed the underground drainage system of the Rak River in one large 'karst window' and three much smaller ones and left two natural bridges.

more widespread examples, investigated and discussed in detail by
G. H. Dury (1960; 1964) is the systematically developed large valley channels curving in wide-radius meanders which are now abandoned, partly filled
with sediment, and occupied by much smaller active meander belts.

Changes in load–discharge relationships caused by climatic changes and
associated changes in erodibility of the ground surface were responsible for
the creation of 'climatic terraces'. These terraces do not necessarily give a
very clear history of stages in the vertical changes of a stream because, first,
there may be little correspondence between the gradients of the present and
of the terrace-building stream, and terrace profiles may converge or diverge
downstream. Climatic terraces are most clearly displayed upstream from a
relatively unchanging level, such as might be imposed by a resistant bar
across the stream course. This is the case in the Tapitallee Creek, New South
Wales, where terraces converge downstream on a sandstone barrier.
Secondly, the exact nature of inter-relationships between terraces and
climatic change is complicated by non-linear and even non-monotonic relationships between changes in precipitation amounts, extent and amount of vegetation cover, periodicity of rainfall, and differences in their effects on erosion,
transport, and deposition of sediments. However, in middle latitudes,
aggradation and terrace building seems to be associated with decreases in
precipitation. In many sub-humid areas in Australia it seems that stable
ground surfaces and associated soil formation occurred in relatively humid
conditions when there was greater vegetation cover on the hillslopes,
whereas instability and the removal of hillslope sediment into stream
courses was associated with a relatively dry climate.

Climatic changes also induced significant changes in areas covered by
lakes. For instance, the New Black Sea level 2–3 m above its present height
(2500–500 B.C.) was preceded by the Neoeuxine era when the Black Sea was
merely a small freshwater lake 100 m below its present level.

E. Stages in Glaciated Areas

The discussion and study of stages in glacial activity is of broader interest
than other erosional and depositional agents. However, there are three
main reasons why stages in landform development associated with the stratigraphical reconstructions of Quaternary events are not described in detail.
First, the results of successive phases of glacial erosion appears to be cumulative with a presumed progressive deepening, widening and straightening along
preglacial valleys. In the west-central Lake District many of the best-
developed cirques are located in structurally weak zones where preglacial
erosion had already hollowed out depressions as headwaters to significant
minor streams. Too little is known about the mechanisms of cirque erosion to

support the long-established speculations about stages of cirque expansion progressively encroaching on uplands to form serrated ridges and peaks. It is just conceivable that the absence of divide-lowering processes in glaciated highlands might offer an equally specific explanation for the development of sharp outlines in the relief. The tendency for cirques to occur in one quarter of the compass will probably attract closer interest. Recently, A. P. Crary (1966) suggested tentatively that if glacial erosion in fiords is limited to an area at the junction between grounded ice and floating ice, multiple basins might result from changes in glacial conditions, such as in ice thickness, causing a shift in the junction position from one phase to the next. In general, however, geomorphological studies of the forms of glacial erosion do not often suggest that the recognition of successive stages in development is a significant consideration in interpreting the details of the forms created by progressive erosion of successive phases of glacial activity. Also there are suggestions from many areas that valleys occupied briefly by ice during the Quaternary had attained present-day proportions in preglacial times.

A second consideration relates to a somewhat similar conclusion in relation to deposition forms, for successively younger ice advances most effectively destroy the depositional features of earlier glaciations should they override them. In consequence, if older glacial deposits survive they are isolated fragments high on valley-sides or in the most distal parts of an ice-invaded area, and are usually without enough surface expression to be of direct relevance to landform studies. For instance, Irish Sea ice, perhaps as an early Würm glacier, probably advanced south across the Cheshire Plain to reach its greatest extent at the Wolverhampton line. Here there is a great concentration of large erratics from the Lake District, south-west Scotland, and the Irish Sea. However, since there is no end-moraine feature, this line is a clear instance of a clue crucial to Pleistocene geology but with no direct relevance to the study of landforms as these are largely absent. Similarly in the north-west United States till of the Donner Lake glaciation, commencing 400 000–600 000 years ago has lost all original morainic forms. However, geomorphologists interested in studying Pleistocene stratigraphy investigate phenomena like these, and others benefit indirectly as the chronology of climatic change in the Pleistocene becomes more clearly established.

A third factor is that those areas which have been glaciated longest will tend still to retain an ice cover burying the underlying land surface, as in Antarctica where there may have been intermittent if not continuous polar ice for the last 11 million years.

There are two aspects to stages in the evolution of depositional landforms in glaciated areas. The first is the results directly attributable to advance and retreat of ice. The second is the succession of events during a glacial phase involving the activities of running water, wind, and frost interwoven with the

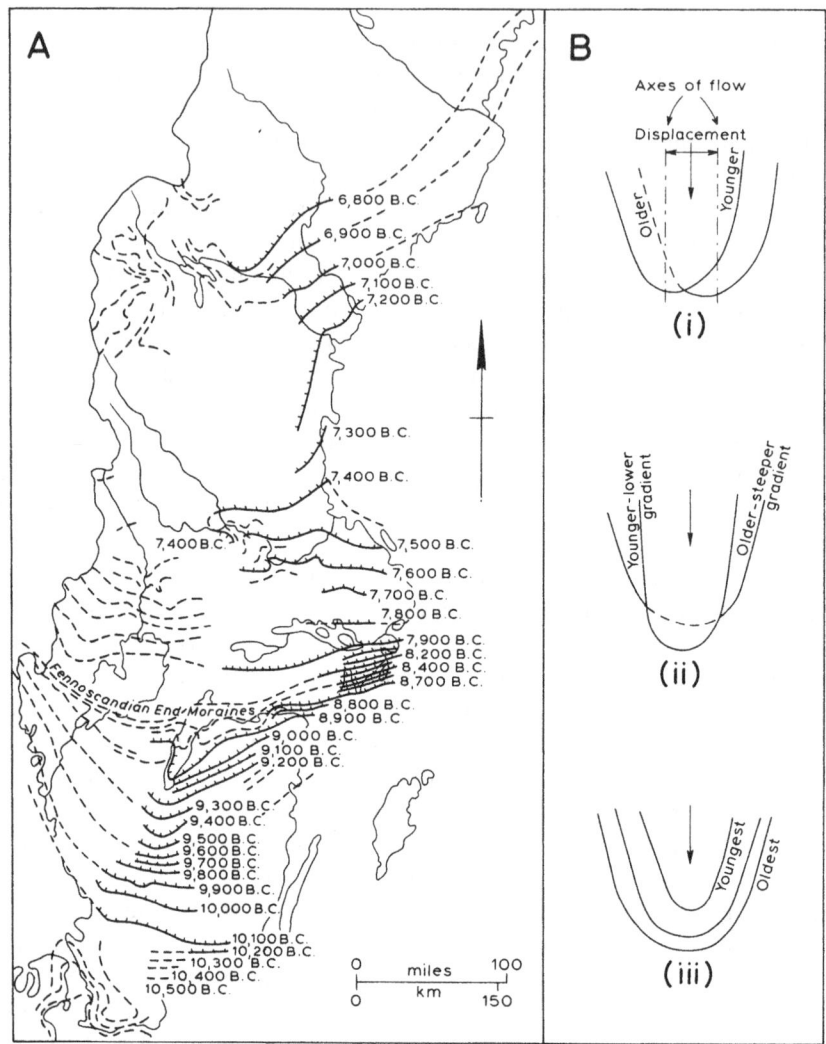

Figure V.19 Stages in glaciated areas.

A. Stages in the retreat of continental ice from Sweden (*from J. Lundqvist in Rankama, 1965*). The ornamented lines provide a geochronological time scale, based on G. de Geer's pioneer work on varves rather than a geomorphological map delineating landforms.

B. Possible relations between successive advances of a valley glacier (*from R. S. Sigafoos and E. L. Hendricks, 1961*, U.S. Geol. Surv. Prof. Paper 387-A). Solid lines show position of moraines that remain after readvance: (i) Similar in down valley extent but with axes of flow displaced laterally, (ii) closely similar in downvalley extent but differing in longitudinal gradient (alternatively the younger advance might have the steeper gradient), and (iii) different downvalley extent but with same axis of flow.

deposition of morainic features, the predominant direct results of stages in ice activity. Although every area of major end-moraine development is a highly distinctive area where more than one moraine remains, it may reflect any of the basic conditions illustrated in fig. V.19. End moraines might epitomize the changing times of landform development, were it not that the identity, age, and continuations of many end-moraines are often problematic. It is easy to underestimate how actual moraines represent prolonged and complicated histories and often may bear little resemblance to textbook models. J. K. Charlesworth's Lammermuir–Stranraer moraine, identified in 1926 along the northern edge of the Southern Uplands of Scotland and interpreted as a major readvance of Highland ice, has subsequently been reinterpreted as essentially stagnant-ice fluvioglacial deposits away from its western end. Also it is probably only in the Lammermuir area that the last ice advance was from the Highlands. The last ice to reach much of the northern edge from the Southern Uplands came from the higher ground to the south. Similarly in the South Wales end-moraine J. K. Charlesworth (1929) saw a feature of the last glaciation, whereas more recently G. F. Mitchell (1960) suggested that it could be Riss in age, with the Würm glacial advance confined to North Wales.

It is on forms produced by phases of ice-retreat and readvance in post-glacial times that stages in ice-deposited landforms are more clearly displayed due to the comparative recency of the events and the clearer documentation, including intensive pollen analysis. Again, however, much of the information available reflects the stratigrapher's interest in problems of chronology and correlations in time rather than in the differences in detail of the landforms.

Selected regional examples of the sorts of pattern of advances encountered are probably more useful illustrations of present knowledge than possible generalizations. The definition of two climatic phases of postglacial time, however, provides a basic starting-point of broad significance. First the Hypsithermal interval was the time 5000–8000 years ago when many glaciers shrank most in postglacial times. The Neoglacial describes the climatic episode covering the last 5 millennia during which, at various times, glaciers grew or were re-created. In arctic Canada, moraines were built about 2400 years ago, with similar features created in western Greenland 3000–4000 years ago. Many glaciers after reaching their Neoglacial maximum, commonly within the latter part of the nineteenth century, began a fluctuating recession until cooler and wetter conditions began in the 1940s. In Alaska the Sherman and Sheridan glaciers, although they advanced several kilometres beyond their present position in Wisconsin times, readvanced in the eighteenth and early nineteenth centuries. There was a readvance again in the late nineteenth century until 1900, when the Sheridan glacier was 700 m beyond its present position and the Sherman 1175 m forward. Recession followed but in 1930 both readvanced to approximately 500 m beyond their present positions. On the opposite side of the continent the Barnes ice-cap re-advanced to the outer moraine in *c.* A.D. 1250, followed by slow retreat until A.D. 1550 when readvance created a major moraine. Subsequent retreat was interrupted by a readvance to another distinct moraine formed about A.D. 1700. Conditions since have been more stable but with a readvance forming a moraine about A.D. 1840. Distinct from the phases quoted so far is the Washington and Oregon area in north-west United States, where the greatest Neoglacial advance was between 1800 and 500 years BP. In Argentina the first postglacial readvance took place 4600 BP and may have reached a maximum about 2000 BP. This is essentially the Sub-Atlantic period when advances occurred in Alaska, Canadian Arctic, Greenland, Iceland, and Europe. Also like many other parts of the world Argentinian glaciers advanced during the last three centuries, although, unlike many areas, this was the greatest position in postglacial times.

Taking a broader view than the minor retreats and re-advances during a major glaciation, certain stages in landform evolution might be linked very tentatively with general intensification to the last maximum glaciation followed by more clearly defined stages in its subsequent decline. The initial phase is an interglacial period with long-continued biochemical weathering of the land surface. It might be supposed that in upland areas, if the onset of glaciation were slow enough, there would be intensive frost spalling on valley walls and the development of screes and rock glaciers downvalley from the glacier front. This would provide significant quantities of debris to the glacial moraines and might be a factor in valley widening. However, in some areas like Central Scotland, material weathered either by biochemical agencies or by intense periglacial action immediately prior to the ice advance was not necessarily picked up by the ice. For stages in deglaciation there is much more tangible evidence, although it has been used to develop a range of contrasting interpretations. It is established, however, that many forms associated with deglaciation do not appear simultaneously but often occur in sequence.

Just as the importance of freezing conditions preceding a glacial advance might have varied widely according to local circumstances, no set of generalizations apply to air and soil temperatures in deglaciated areas. In the area vacated by the retreat of ice of the Late Weichsel stadial in lowland Britain, fossil permafrost features indicate that permafrost conditions developed after this particular retreat. In contrast, in coastal areas, like the fiords of southern Norway, deglaciation probably took place fairly rapidly with floating of ice leading to its sudden break-up. Another variable factor is the amount of melt water released during deglaciation. On the Labrador coast, where the absence of melt-water forms may reflect glacier reduction due to diminished precipitation rather than by temperature change, sharp-crested morainic ridges are in consequence well preserved. Similarly the relatively dry climate in the western Canadian plains meant that features were mainly composed of till rather than of fluvioglacial materials and that they remain well preserved due to only slight postglacial erosion.

The most significant aspect of an understanding of stages in deglaciation has been the fundamental change in basic postulates in the last two decades. The classical concept of deglaciation, elaborated about the turn of the century for upland areas in many countries, was the retreat of a continuous ice margin towards lower ground ponding elongated glacial lakes against the flanks of upland nunatak areas or mountains like those in central Norway and Sweden where the snow-line rose so high during deglaciation that no accumulation area was left. Had more time been available for field surveys these postulated lakes might have been reconstructed on the evidence of shorelines or shoreline deposits, but frequently the lakes were merely inferred to have existed at an altitude determined by the level of the channel floor assumed to have served as a marginal lake outlet. While field workers gradually showed that such shorelines frequently had neither form nor deposits at the postulated altitudes, C. M. Mannerfelt (1945) put forward his ideas on the downwasting of ice-sheets. His interpretations, worked out in southern Sweden, have subsequently been successfully tested in other areas, particularly in Britain. Similarly, in the Susquehanna valley, stages of deglaciation are now thought to have been predominantly those of downwasting and stagnation rather than of frontal retreat. Also stagnation of ice characterized the final stages of the earlier short Illinoian glaciation. At this stage in deglaciation lowland areas are occupied by large stagnant ice bodies with melt water slipping readily between and beneath the decaying ice. In higher areas which were centres of ice dispersal and which were high enough to remain as zones of accumulation while lowland ice melted, the ice-sheet contracted upwards and penultimate stages of deglaciation resembled initial stages of its onset as the ice-sheet separated into progressively smaller units. As the mountain ridges emerged from the ice, melt-water streams formed overflow channels at low points in the ridges. Many glaciers were fed from more than one direction and with some glacier tongues being more active than others, reversals of melt-water flow occurred and cols followed by tongues of diffluent ice as well as by melt water. At this stage in the deglaciation of the Canadian Cordilleran region differentiation of sedimentary products of reworked till took place, with small deposits of lag gravel forming in upland areas and silt and finer material remaining in suspension and being sluiced from the uplands into silt-filled valley bottoms. A study of stages in deglaciation depends on a careful determination of loess and drift relationships. The Farmdale loess, blown from valley trains in the Mississippi valley, for instance, surrounds a drift lobe but does not lie on it, indicating that it was blown chiefly while the glacier was at its maximum stand and subsequently while the till remained ice-covered by

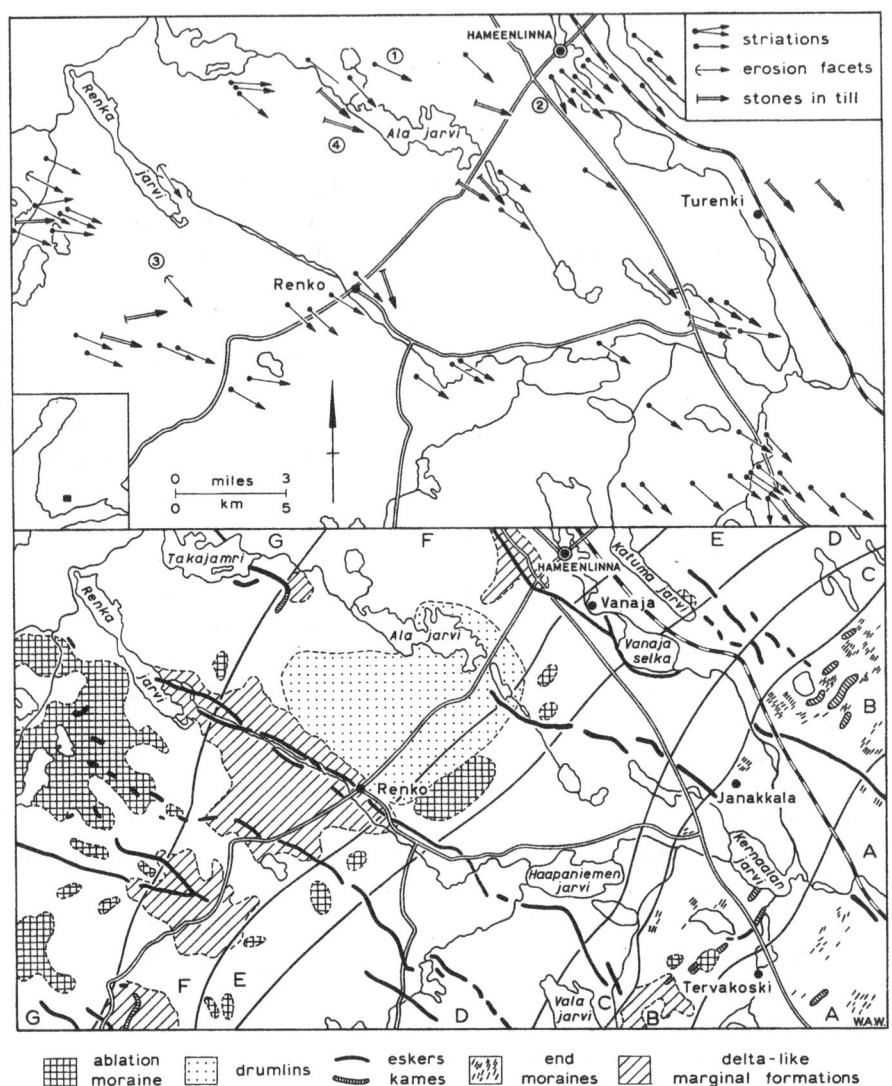

Figure V.20 Indications of continental ice-sheet advance across the Hämeenlinna region, southern Finland (*above*) and depositional forms related to its retreat (*after K. Virkkala, 1961*, Bull. Comm. geol. Finlande, *Vol. 33*). A–G = functional zones of retreat of the ice-sheet.

downwasting ice. By contrast the Iowan loess was blown from the Iowan drift during ice-retreat, and the maximum thickness is at the border of the drift. Loess deposits form an extensive belt along the north European plain, but is only patchily distributed in south-east England and is almost completely absent in Scotland.

Apart from the case of well-defined end-moraines the association of glacial depositional landforms with stages of deglaciation is scarcely practicable nor realistic in most areas. The complex variability of modes and rates of ice-retreat and the associated depositional forms is shown clearly in a study by K. Virkkala (1961) of the Hämeenlinna area in southern Finland. In this area, there are several different zones in the glacial forms, related to stages in deglaciation. Ice retreated first from the south-east part of the area, distinguished as zone A in fig. V.20. It is now occupied by numerous small morainic ridges and kames aligned at right-angles to the direction of ice-retreat. These features probably reflect halts or even slight readvances punctuating the ice-retreat. In zone B large kames, much pitted in the Turenki locality, indicate standstill of the ice margin for considerable periods, an ice condition also favouring delta formation. However, in some parts of this zone the ice behaved differently and washboard moraines are numerous. Zone C is a 2–3-km broad belt in which no ice-marginal formations occur. In the succeeding zone D large eskers reflect a rapid rate of ice-melting and ice-retreat along an even front. In contrast, the fragmentary ablation moraines dominating the 2–4-km-wide zone E indicate that ice disappeared from this zone largely by downwasting. Delta-like marginal formations occupy much of belt F, but between Alajärvi and Renko the behaviour of the ice sheet contrasts with that in the remainder of the Hämeenlinna region; a conspicuous feature is the extensive drumlin field signifying the work of thick moving ice. In zone G there is yet another example of substantial changes in glacial deposition patterns within very short distances. On the northern side of Renkajärvi the ice retreated along an even front, creating imposing eskers which in places spread out on to extensive plateaux. At the southern end of Renkajärvi broad areas of ablation moraine indicate downwasting; the ice here had retreated, for the most part, across dry ground.

Within their span C_{14} dates have challenged so many correlations that, apart from the fact that the problem of chronology is primarily the geologist's speciality anyway, it would be unwise to examine the broader stages of the main Pleistocene glaciations too closely here. However, since the glacial stages are the main support of the skeleton of temporal stages into which all the decipherable stages of landform evolution might be fitted, some brief review is relevant. Three important points might be made. First, the degree to which glacial stages or shorter phases are synchronous over a regional as well as a global range depends entirely on the degree of general correspondence which each expert considers constitutes a correlation. The classification of Wisconsin glacial deposits, for instance, is much criticised or strongly defended by Pleistocene stratigraphers. One reason for difficulties is that an advance in one place need not correspond to an advance in another place, whether these be on separate continents or a mere mountain divide apart. In Scandinavia, the last major ice sheet began to form on the divide of the fjell range, but later moved eastwards, partly due to cyclone tracks moving southeast and southward. This might have caused reduced snow supply to northerly parts and even recession, which might explain why ice began to retreat at quite an early stage from the northern mountain areas in Norway. This reconstruction shows that, due to the significance of local conditions, particularly relief and aspect, the progress of glaciation does not necessarily imply a parallel development with the glaciation of an area as a whole. Local ice culmination may possibly have occurred at different times in different places. From this it follows that broader correlations may be difficult. For instance, although a sequence of three glacial drifts in the Argentine Andes is fairly well established, correlation with other regions remains to be attempted. Similarly, it is only a possibility that four glaciations separated in the Hupei region of China might correspond to the classical four glaciations in the European Alps. Even in Scandinavia it is uncertain whether the Old Baltic ice-stream, which preceded an equivalent of the main flow of Würm ice, developed during final phases of Riss glaciation or during the initial phases of the Würm glaciation.

A second point is that within the 'Ice Age' ideas on the number of major glaciations are increasingly qualified due to growing evidence indicating climatic oscillations which preceded them and which make a fixed notion about four glaciations too rigid a frame.

A third, substantial problem is that of the broad time zone that now defines the beginning of the Pleistocene. A faunal date from East Africa puts the Villafranchian back to 1·75 million years, and ages for an approximate time equivalent in North America, the Late Blancan, are between 1·5 and 3·3 million years. These ages are far older than the period, probably starting approximately 600 000 years ago, covering the classic four glaciations of the Alps. It is now recognized that the Pliocene–Pleistocene boundary cannot be defined by the first incoming of glaciation, but that the earliest evidence of a marked climatic deterioration be taken as the start of the first major phase of glaciation for a given region.

F. Stages in Coastline Evolution

Some studies dealing with changes in coasts and coastlines record some of the most exactly known stages in the evolution of landforms. Coastal stretches where cliff retreat, even in consolidated rocks, is measurable within spans of decades or centuries, provide some of the most precisely definable stages in slope evolution. Therefore, in terms of geomorphology as a whole, and as illustrated in the earlier section on slope development, the significance of such sites is not confined to their own limited strips of coastline. Within the context of coastal development, however, such sites made up of consolidated rocks are unusual. As on rapidly eroding coasts, the stages in the development of depositional landforms in coastal environments may be defined with unusual accuracy, particularly where the records of detailed maps go back several decades or more. However, along all but a few very short stretches of coast information is insufficient and interpretations uncertain. In fact, set against the detail obtainable at one or two sites there is, in current interpretation of stages of coastal development, considerable uncertainty and major recent reconsiderations of basic concepts. Many factors make generalization about stages of coastal evolution difficult if not misleading. In addition to the complications of considering both section and plan, the juxtaposition of erosional and depositional stretches of coastline is difficult to interpret and their separate consideration may be unrealistic. Other sources of variation include changes in the relative level of land and sea due to isostatic adjustments of land levels, postglacial rises in sea-level and the superimposition of both effects on local tectonic movements. In consequence, it is substantially more difficult to support the idealized sequence of forms that might occur in the plan of a coastline than was permissible two decades ago. D. Johnson's (1919) elaboration of early notions explaining a contrast between emergent and submergent coasts is now seen not merely as over-idealized but also as totally unrealistic. Also the stratigraphical evidence and ideas of use in studying stages of heights in land–sea relations in coasts even a decade ago are now obsolete. Although stratigraphical field study is permitting some progress, it is now appreciated that undeformed strand-lines are at present very difficult to identify, and that the eventual complete correlation between rocky and depositional coasts is perhaps intrinsically unobtainable.

Classically sea-level was believed to have been at about + 60 m during the Günz/Mindel interglacial (325 000 BP), at + 30 m during the Mindel/Riss interglacial (195 000 BP), and at + 18 m during the Riss/Würm interglacial (105 000 BP). These three levels were named, respectively, Milazzian, Tyrrhenian, and Monastirian, but their interpretation is now being extensively revised on increasingly stratigraphical evidence. One reason for the revisions is that the Mediterranean region including type localities of the initial scheme was tectonically unstable prior to the last interglacial. There is also independent evidence that sea-level itself might have been at about the same elevations during the three interglacials. Substantial revisions even appear necessary for changes within postglacial times. For instance, in Scotland for the last 80 years or so the raised beaches have been grouped at levels of 30, 15, and 12·5 m, yet after recent careful measurements in the Forth Valley area, J. B. Sissons (1963) is uncertain of the total number of shorelines and how fragments of these might be correlated with other fragments within even this area. One of the weaknesses in D. Johnson's deductive scheme is that, apart from areas like northern Canada and Scandinavia where postglacial isostatic rebound outstripped postglacial sea-level rise, or tectonically active areas where eustatic and crustal movements are scarcely distinguishable, all coastlines have been, in geological terms, very rapidly and emphatically drowned in the 15 000–5 000-BP interval.

Although the difficulties of pointing to stages in the history of erosional planation along coasts are now widely appreciated, there remain a few favoured and restricted localities, particularly on small islands with rock benches for which a wave-cut origin is highly likely. For instance, on the San Benito Islands, a group of three off the Pacific coast of Baja California, the most pronounced features are three wave-cut marine terraces. The uppermost, occurring only on West Island, dips north-west at approximately 5 degrees and ranges in elevation from about 90 to 150 m. The intermediate level, best developed between 30 and 75 m on East Island may also be represented on West Island at the same approximate range in elevations. All three islands have terraces near present sea-level between 6 and 30 m. Some terraces dip symmetrically outward from the centre of the island. Others are tilted, due to differential tectonic uplift subsequent to wave erosion. Similarly, the most striking feature of Middleton Island in the Gulf of Alaska, 110 km from the mainland, is the step-like terraces at 3, 14, 20, 25, and 32 m, cut in Pliocene or Pleistocene bedrock.

The clearest examples of staircases of relict erosional benches notched in higher areas inland from coasts are found in mobile, tectonically unstable areas where abrupt uplift and periods of stillstand occurred during Pleistocene times. It is in these very areas, however, that vertical movements vary widely in amount and even in direction of movement and where, in consequence, present altitudes of benches and terraces have little significance. Their interpretation requires the additional information provided by sea-cliffs, buried channels, alluviated river mouths, and the stratigraphy of off-shore sediments as well as well-authenticated and independently established data on eustatic movements of sea-level during the Pleistocene. The work of E. C. Buffington and D. G. Moore (1963) relating to the San Clemente area, southern California, exemplifies the intricacy of reconstructions in the tectonically active areas where the morphological evidence is best preserved. At least six terrace levels were developed during the Pleistocene in the upper Miocene–lower Pliocene Capistrano sedimentary rocks. These terraced areas are separated by relatively depressed and predominantly depositional areas such as stream mouths graded to a level about 90 m below the present sea-level. Buffington and Moore adopt a figure of 60 m for the greater stand of sea-level above the present during preglacial and interglacial times. Combined with a tectonic depression of land surface 80 m below its present elevation in the San Clemente locality, this early high sea-level would have created conditions for the formation of the fifth terrace in post-Capistrano times. This could have been in the late Pliocene or during one of the early Pleistocene interglacial phases (fig. V.18.A). The formation of the lower four terraces may have occurred subsequently during relative stillstands of the sea during the intermittent re-elevation of the coast (fig. V.21.B). With a continued elevation of land of perhaps 195 m and a eustatic lowering of sea-level by approximately 100 m during the last stage of the Pleistocene, land that is now 250 m below present sea-level would have been exposed to sub-aerial erosion, a new base-level formed (fig. V.21.C), and vigorous erosion cut gullies down the coastal slope. These were then rapidly drowned by a rise of sea-level approximately to its present position (fig. V.21.D), a trend continued by a slower eustatic rise of sea level (fig. V.21.E). The latter phase of sea-level rise was slow enough for the planation of a rock platform and the deposition of 12 m of sediment on it. Since drowning, deposition appears to have been the dominant process.

Generally speaking, evolution of forms on rocky coasts is too slow for changes to be discernible. A. Guilcher (1949) has suggested that some high cliffs in hard rocks might be fossilised features formed a hundred thousand years ago. In consequence stages in coastline changes are usually best seen in glacial or alluvial deposits. However in the case of erosion of soft rocks former stages are only reconstructions of a past that has disappeared beneath the waves and are not part of the present-day forms like those representing stages in coastal deposition. In the latter case several stages are sometimes discernible. This is particularly so in shallow shelf areas where the post-glacial rise in sea-level involved substantial lateral displacements of shorelines and associated reworked shelf deposits at rates of tens of metres per annum. For instance, much of the 600–700-km-wide shallow platform between Timor and the north-west Australian coast, the Sahul Shelf, was exposed 18 000 years ago, as was the entire Orinoco shelf. As the initial rise was rapid even delta shorelines, like the Mississippi, were moved landward. As long as the last major rise of sea-level was taking place, new areas of coastal plain were being inundated and fresh supplies of sediment were encountered. The clays were dispersed but the coarser constituents were pushed forward to become beaches. On all tropical coasts, the amounts of sand amassed in the coastal belts was especially huge. Beach volume increased as long as the rise of sea-level continued and the surplus sand was blown downwind to form coastal dunes. The beach-dune systems reached their greatest volume with the

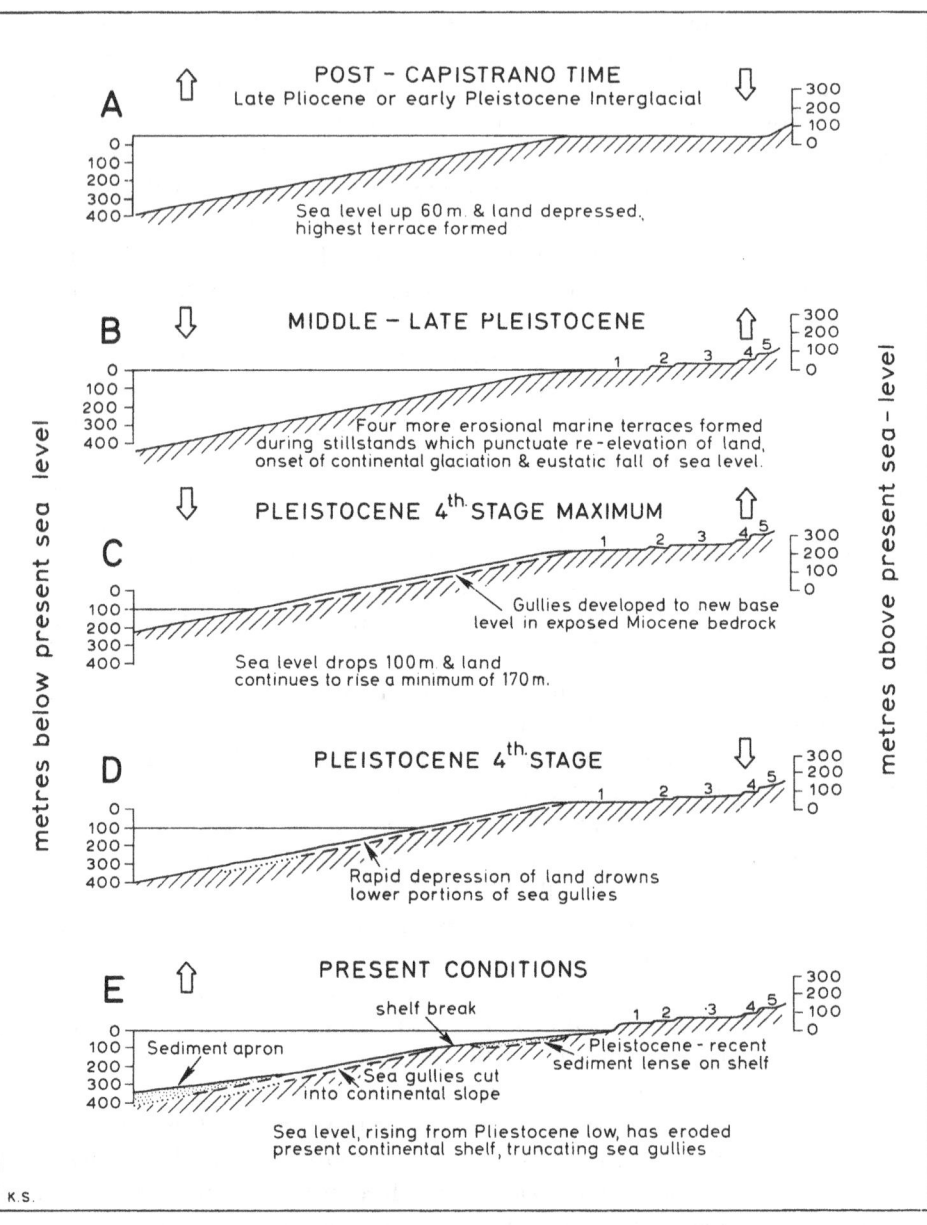

Figure V.21 Development of terraces and sea gullies during Pleistocene eustatic changes in sea-level on a tectonically active coast (*after Buffington and Moore, 1963*). The vertical arrows indicate the relative movements of sea or land. The diagram is based on conditions in the San Clemente locality, southern California.

approach of stillstand. Around 10 000 BP the rate of rise decreased markedly and in areas of rapid supply of river sediments, delta progradation became possible. As sea-level steadied about 5 000 BP new sediment supplies for the beach-dune systems were consequently no longer encountered, and marine processes brought about a net loss to the system due to reworking by storm waves and burrowing organisms. However, in areas where local conditions ensure continued and abundant supply of sand as on the east coast of Malaya, a succession of parallel ridges may result. If a berm survives a storm in such an environment a new one will form on its seaward side due to accretion during the following calmer conditions. At present beach ridges or their remnants are found as far as 6 km inland along the Endau river. Inland they reach heights of 8–9 m above present sea-level and diminish in height seawards. In Mexico, beach ridges adjoining the Grigalua river extend 40 km inland.

Fig V.22 illustrates one of the many points around the coasts of the British

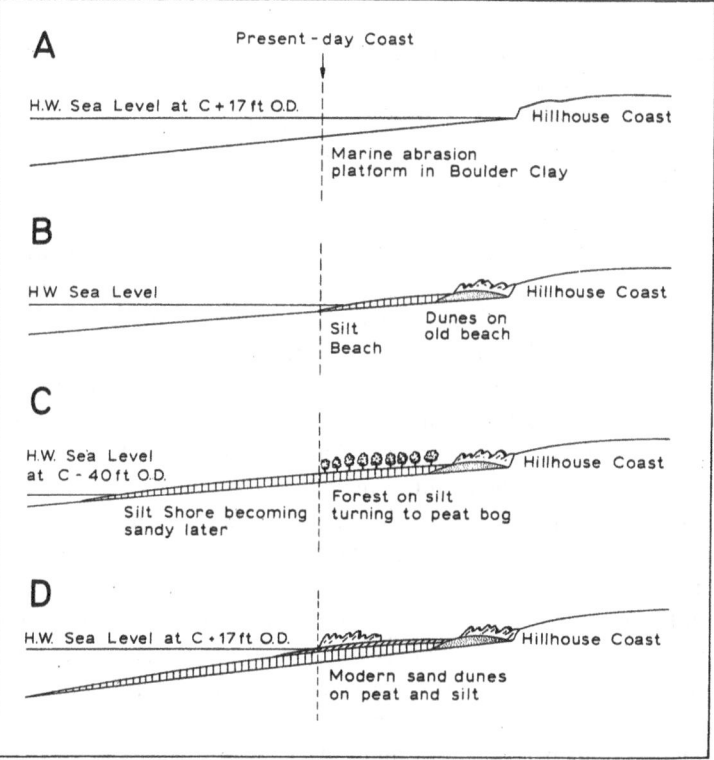

Figure V.22 Postglacial episodes on the sandy shore of south Lancashire (*from R. K. Gresswell, 1953,* Sandy shores in south Lancashire). At Stage B the dunes are made up of the Shirdley Hill Sand and the beach of Downholland silt.

Isles where there are records of postglacial changes in the forms of unconsolidated materials. R. K. Gresswell (1964) considered that, in early postglacial
times, the land-level on the east Lancashire coast, in relation to sea-level, was
at least 45 m higher than today. Marine transgression, presumably due to
eustatic sea-level rise exceeding isostatic land-level rise, culminated in the
formation of the Hillhouse beach, possibly 5000 years ago (fig. V.22.A). The
cliff foot associated with this feature is at 5 m OD suggesting a correlation
with similar features found between 0 and 10 m OD in many other coastal
areas of Britain. Eustatic rise then either diminished considerably or ceased
altogether, while either isostatic rise continued or local crustal warping
occurred. This slight relative elevation of the land favoured the deposition of
the Shirdley Hill Sand and the Downholland silt on the prograding shoreline
(fig. V.22.B), and forest and peat bog grew on the upper parts of the beach as
the coastline continued to retreat westward (fig. V.22.C). Finally land-level
fell in relation to sea-level and the erosion of the peat and other beds started
(fig. V.22.D). M. Schwartz's (1968) work, extending ideas of P. Bruun, has
shown how with other factors equal, within the last few millennia where some
approach to a balance between forces may be possible on a coastline of
deposition, the effect of a sea-level rise would be the landward displacement of
the beach profile, nearshore deposition compensating for the erosion of the
upper beach in such a way as to maintain water depth adjacent to the shore.
Factors not likely to remain equal include the amount of sediment supply
alongshore which is often the most important factor in determining whether
short-term stages of coastline evolution will be erosional or depositional.

Although the facts of late Quaternary times make the classical Johnsonian
theory linking the origin of barrier islands and related features to shorelines
of emergence unrealistic, there are some situations where accumulation and
emergence co-exist. The emergent beach at Gisborne, Poverty Bay in New
Zealand, comprises a belt of nearly 100 low, closely spaced and continuous
ridges about 5 km wide, rising from 4·5 m at the coast to 12 m inland and is
considered to be largely the result of tectonics. From the evidence of volcanic
ash falls warping appears to have taken place shortly before 1500 B.C. and
between A.D. 1000 and A.D. 1650. The effect on the shoreline plan of the
throwing-up of barrier beaches is to straighten its outline at the mouths of
outlets and low-lying indentations. If the sea has little tidal range and if the
climate is relatively dry, lagoons may develop. With marine oozes introduced
by incoming tides and the addition of river alluvium the filling up of lagoons
may proceed in clearly definable stages. Along the Pacific coast of Baja
California, Mexico, between 24° and 30° N, wind-blown sand is gradually
filling up many tidal lagoons. In this area the stages of lagoon and barrier
development have been studied in detail. At stage A, fig. V.23, the deepest
part of the lagoon was directly behind the barrier, so that the tidal currents

and resulting turbulence were strongest adjacent to the barrier. Deposition was confined to shallow marginal areas. The tidal flats expanded towards the barrier, but as strong and turbulent currents kept the channel clear of sediment it remained deep and stationary. Eventually, because of the distinctive condition of little addition of new sediment, a balance was approached as increased narrowing of the channels must have increased velocity and turbulence of the oscillating wedge of tidal water. Since the stage B barrier was deposited *c.* 1800 years ago, the barrier has prograded seaward approximately 1·6 km, perhaps due to changing climatic conditions leading to greater supply of sediment from adjacent land margins.

Another coastal environment where stages in coastline evolution are prominent is in river mouths where the term 'drowned valley' is a misnomer since nearly all coasts have been inevitably drowned in postglacial times. R. J. Russell (1967) suggests that a comparison between the north and south coasts of the Baltic well illustrates the major fallacy of the subdivision between coastlines of emergence and submergence. Instead, stages of alluviation in river mouths are recognizable, with the effects in the ria or classical 'drowned valley', either due to their depth or to the dearth of sediment supply or to diminished river discharge being least apparent. Many estuaries being progressively filled, contract in volume, depth and surface area until a depositional plain extends right to the coast often with an associated fringe of deposits strung out by marine processes (fig. V.24). Deltaic coasts are the other extreme where the scale of alluviation may be colossal; the volume of Recent

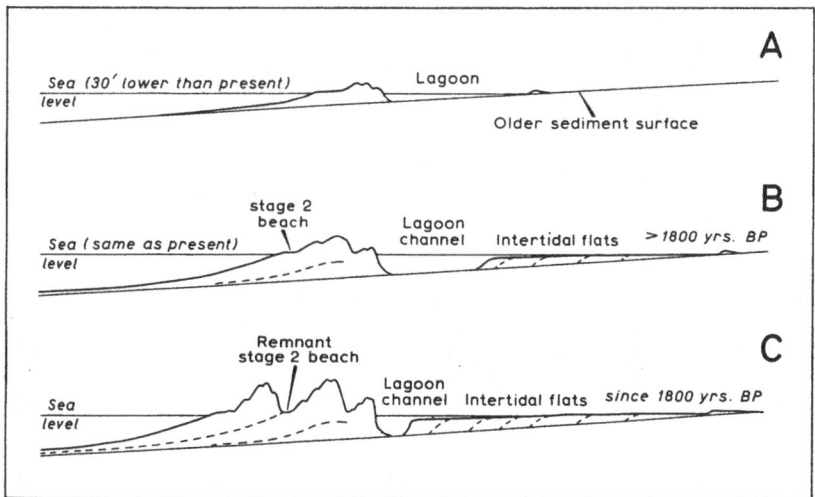

Figure V.23 Stages in the development of a coastal lagoon barrier (*from F. B. Phleger and G. C. Ewing, 1962,* Bull. geol. Soc. Am., *Vol. 73*), the Laguna Ojode Liebre, Baja California.

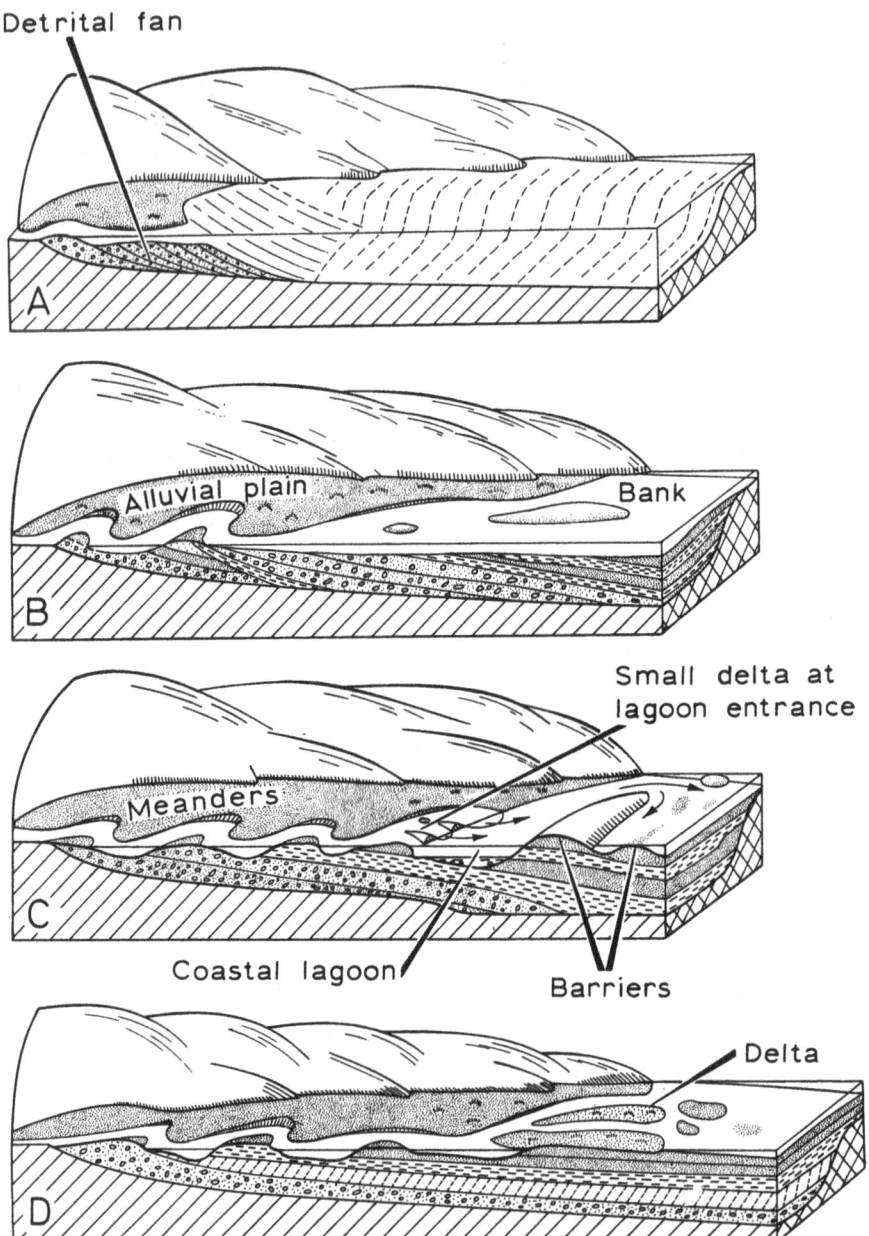

Figure V.24 Stages in the evolution of some estuaries and their transformation by valley infilling (*from F. Ottman, 1965*, Introduction à la géologie marine et littorale).

alluvial fill in the Lower Mississippi valley between Cape Givardeau, Missouri, and the end of the exposed valley walls in Louisiana is about 4200 km³.

In environments which favour their continued growth the rate of advance of some small features falls into recognizable stages, like Orford Ness which J. A. Steers (1939) described as extending 6 km southward since A.D. 1530. On the other hand, over spans of time, perhaps ranging from decades to millennia, certain self-arresting features may prevent unlimited extension of a depositional feature. Thus, as a spit narrows a river outlet, the current velocity and turbulence in the channel increases until at a critical point they inhibit deposition as a theoretical equilibrium state is approached. Frequently, however, outlets through a depositional barrier may move erratically with time. Breaches in spits open many times as on the Lake of Barberie which closes the Senegal delta, but are soon replaced if average conditions remain essentially the same. The history of Spurn Head, which curves 5·6 km across the mouth of the Humber has been examined in detail. G. de Boer (1964) suggests that episodes of destruction at intervals of about 240–250 years have punctuated considerable periods of persistence and growth. Rapid erosion of a wedge-shaped area of boulder clay at the root of the spit appears to be a factor leading to the periodic breaching near the mainland end of the spit. The breach operates as a flood-tide channel until deposition on the landward side leads to the growth of a new spit inside the remnants of the dismembered former spit. This mode of evolution may be particularly closely linked with the rapid retreat of the boulder clay cliffs at the root of the spit. In some situations the breaching of a spit may form a barrier island; in many instances the rate and persistence of sediment supply alongshore determines whether a depositional form is vulnerable to reworking. Where longshore drift is active on the fringe of an expanding delta, the ridges of previous spit-like developments may remain. These conspicuous forms are termed cherniers, as evergreen oaks flourished along those in the Mississippi delta. In south-west Louisiana, the system of cherniers is 100 km long, the regularity of spacing appearing to indicate that periods of erosion and retreat have alternated regularly with periods of accumulation and advance of shorelines.

The upward growth of coral reefs, in keeping pace with postglacial sea-level rise, is a significant enough stage in coastal evolution in itself. In addition, the sand cays on reefs might fall into an evolutionary scheme, although D. R. Stoddart's (1965) study describes cay types in terms of equilibria with the energy level of waves. R. F. Folk (1967) prefers to recognize an evolutionary sequence, beginning with the development of a sandbank or bar that hardly breaks the surface at high tide. Secondly, as the island grows to a certain size and permanence, a large area of dry sand remains beyond the reach of the highest tides, and during a third stage the island grows by

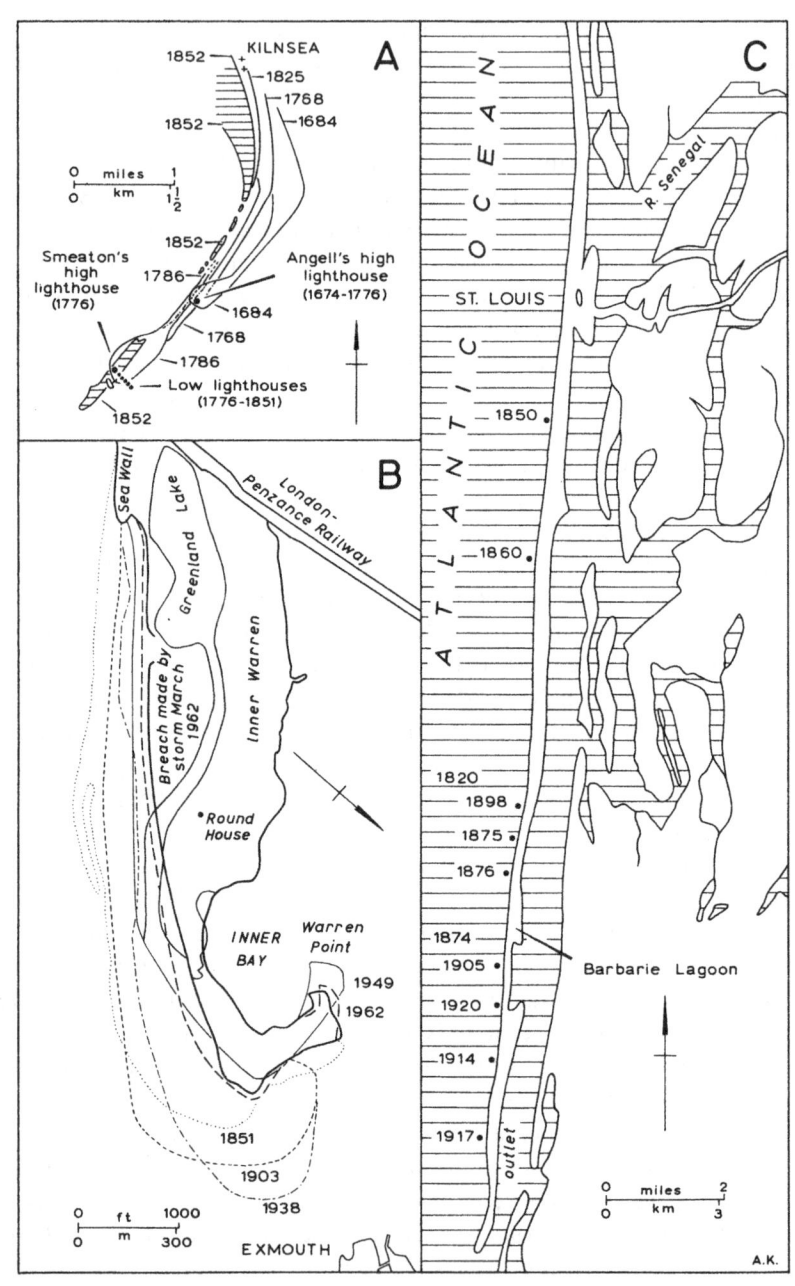

A

KILNSEA
1852
1825
1768
1684
1852

miles 1
km 1½

Smeaton's
high
lighthouse
(1776)
1852
1786
Angell's high
lighthouse
(1674-1776)
1684
1768
1786
Low lighthouses
(1776-1851)
1852

B

Sea Wall
Greenland Lake
London-Penzance Railway

Breach made by storm March 1962

Inner Warren

Round House

INNER BAY
Warren Point
1949
1962

1851
1903
1938

ft 1000
m 300

EXMOUTH

C

ATLANTIC OCEAN

R. Senegal

ST. LOUIS

1850

1860

1820
1898
1875
1876

1874
1905
1920

1914

1917

Barbarie Lagoon

outlet

miles 2
km 3

A.K.

accretion of beach or dune ridges, with the colonization by vegetation of the inner and older ridges. Both these steps are relatively ephemeral compared with the fourth stage when the island reaches sufficient width to allow vegetation to cover a central area out of reach of wave erosion. Building out of the lee-side, perhaps to several times the size of the island, in its third stage makes this possible and the growth of vegetation, the burrowing activities of crabs and nesting operations of birds gradually level out the older dunes in the central area.

Dune systems on coastal margins are usually intimately related to stages in the evolution of the adjacent depositional coastline. On many of the world's coasts, Recent dunes include two contrasting types. The older dunes are leached, oxidized, exhibit at least incipient soil profiles and are usually fixed by vegetation and are believed to have increased in volume as long as sea-level was rising. The newer system lying seaward has smaller volumes, continues to shift and appears inter-related with modern beaches and shore characteristics reflecting sand supply since sea-level steadied some millennia ago. The older dunes are subdued forms compared with the bolder outline of the more continuous coastal fringe of newer dunes.

G. Stages involving more than one Process and in the Evolution of the Land Surface as a whole

One of the reasons why stages in landform evolution are not often clear-cut is because more than one factor may be involved simultaneously. In some other situations the stages in landform evolution become those of one process succeeding a different formerly dominant one. Fluvioglacial phenomena are perhaps the most frequently recognized examples of such a combination. One of the most common developments is the ice-ponding of lakes and the effects on melt-water discharge of its periodic release. There may be one of two effects. First, ice-dammed lakes periodically, perhaps annually, empty rapidly due to water escaping beneath the ice. This self-emptying termed *jökulklaup* in Iceland, results when the height of the dammed lake, approaching nine-tenths that of the ice-dam, is sufficient to float the ice. The abruptness and volume of the escaping flow causes catastrophic floods downvalley from the glacier snout. A significant result is the progressive destruction of ice-deposited landforms as one of the stages of deglaciation. R. J. Price (1966)

Figure V.25 Development of spits.

A. Spurn Point, east Yorkshire, 1684–1852 (*from de Boer, 1964*).

B. Dawlish Warren, Devon, 1851–1962 (*from Kidson, 1964*).

C. Progressive extension across the mouth of the Senegal river (*from J. Larras, 1957*, Plages et côtes de sable).

describes how between 1948 and 1963 the Casement glacier, Alaska, retreated nearly 2 km north-east of a lake-covered esker system which melt-water streams began to destroy after the draining of the lake. A particularly common change is that the greater or more abrupt forms of melt water may trench the valley train of fluvioglacial sediments, thereby creating a terrace. However, after the abundant melt water of the early stage in deglaciation, the net longer term effect during glacier retreat is to decrease the competency of streams and to cause valley alluviation. Most significant are the spillways cut when melt water spills over a low point in a watershed boundary either as a lake spillway or as a sub-glacial chute. An example of the effects of such a catastrophic overflow is in the Hellemofiord area, at 67° N in Norway where a narrow canyon cuts 150–200 m down into the floor of a broad flat valley. Drainage patterns may also be modified substantially by diversions due to ice. In North Wales, streams like the Alyn system were diverted south by the barriers of the Welsh and Irish Sea ice. The spectacular diversion channel some 16 km west of Clinton, Iowa, in which Goose Lake lies, might have been first cut when the Illinoian glacier blocked the former Mississippi at Clinton, creating a lake in its valley which eroded the diversion channel as it spilled over the divide between the Maquoketa river and the Wapsipinicon river. The former course of the Allegheny river to the Erie basin was perhaps similarly diverted to the headstreams of the Ohio river. Eroded remnants of sands, silts, and clays once deposited in these glacial lakes provides evidence of their former occurrence, as in the north-trending valleys of the Illinois Ozarks. In north-west Europe, the northerly drainage was dammed by the Riss ice-sheet. As a result, a waterlogged area existed in front of the advancing ice whereas the southerly drainage in North America tended to keep the area in front of the ice relatively dry. The effect was less marked in the Würm glaciation as drainage was possible away from the ice towards the North Sea. However, it is possible, as in any situation where more than one agent is involved, that glacial diversions as an important stage in the development of a drainage system might be exaggerated. Abundance of melt water is only a brief phase in glaciations which in turn occupied only a fraction of Quaternary time. For instance, the diversion of the ancient River Teays to form the Ohio river, originally interpreted by some as one of the largest drainage diversions wrought by glaciation in North America, is according to analyses of the ancient channel sediments probably a preglacial event. On the other hand, the effect of glaciation on drainage patterns that might be caused by isostatic depressions of part of a land area might be considered. Diversion by this means may have been possible, for instance, where the lowest pass-point in the direction of a deep canyon between the south-west part of the mountain plateau area and Rombaksbotn, near Narvik, is only 7·5 m higher than that in the direction of the valley zone to the north-west.

In some areas glaciation, without introducing a diversion stage into the development of a drainage system has left a pattern of a river re-establishing its course. Drumlins, for instance, have significantly controlled the headward growth of tributaries to the Rock river in Wisconsin and limited the size of their watersheds, and have determined the winding course of the incised River Kelvin in Glasgow. The flooding by rising sea-levels of glacier-deposited landforms is a distinctive stage in the evolution of many lowlands. The Uppsala esker continues northward as an underwater ridge some 50 m high. Dogger Bank, rising to within 20 m of the sea surface, is a relic of a moraine which strikes from near the Yorkshire coast east towards the northern tip of Denmark. Westport Bay in the west of Ireland provides a classic example of the drowning of a drumlin field and many less extensive areas, as near Galway or at Berwick, on the mouth of the Tweed, show the stages of drumlin destruction by the sea, ending in a low mound of boulders.

A more unusual example of a glacial action inter-related with another process occurs in volcanically active areas in high latitudes, like Alaska. More than one-eighth of the surface area in Iceland is made up of palagonite breccias and tuffs, believed to be formed from basalt magma under considerable hydrostatic pressure, escaping upwards and sideways through the ice. The long rather flat-topped ridge of Dalsheidi is such a flow, occupying the axis of an ancient glaciated valley. Although the possible stages of evolution outlined in fig. V.26 describe an unusual example, similar stages in valley development are also observed in fluvially eroded valleys invaded by a lava stream.

Inevitably, the interdependence of stream and shore processes in their contribution to landform development has already been mentioned in several contexts, but it is perhaps worth recalling this point here. Just as it would be difficult to isolate stages in the evolution of streams from the changes at the coast affecting regional base-level, so are many stages in coastline evolution profoundly influenced by river activity. The significance of wind removal or accumulation of material in introducing distinctive stages into landform evolution in certain coastal, fluvial, glacial, and even slope environments is worth reiterating.

Stages in slope and stream evolution may also be interlinked, phases of accelerated slope weathering confining by colluvial infillings a channel in which the evacuation of debris becomes more difficult. Conversely with reduced colluvial supply, and consequently more rapid removal of bedload, the channel floors are incised and slope foots undercut. Of significance for broader contexts are theoretical schemes in which the adjustment between stream and slope, or lack of it, is of less importance than the evolution of their combined effects. Some of the reconstructions of the land surface as a whole are by necessity only very tentative suggestions. Studies which attempt the

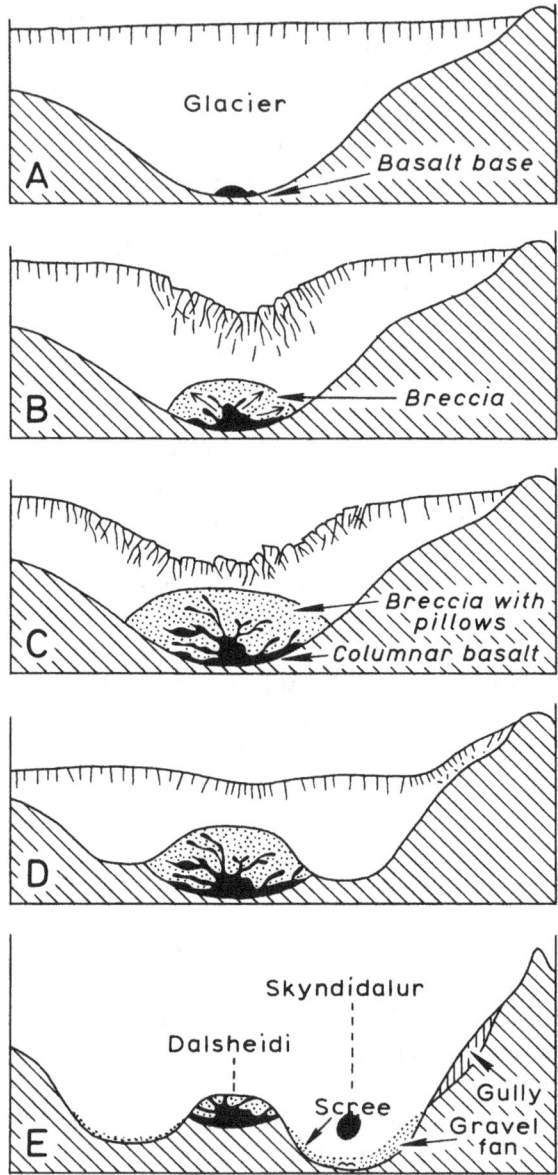

Figure V.26 The landform evolution of a sub-glacier lava flow (*from Walker and Blake, 1966*). Basalt may have flowed 35 km or more beneath the ancient Dalsheidi glacier in Iceland.

daunting task of reconstructing stages in the evolution of the land surface as a whole are often directed towards one of two objectives. The purpose may be largely geological, aimed at reconstructing the landform of a given period or periods in the past. Such studies of stratigraphical geomorphology often include the description of unconformities which have no direct significance in the present-day landscape. Alternatively the objective may be, largely by the use of relict fragments in the present land surface, the reconstruction of stages in the evolution of the present relief form.

An example of a study largely orientated towards reconstructions of stratigraphical geomorphology is J. C. Frye and A. B. Leonard's (1957) study of the Great Plains, at present a semi-arid near-treeless zone of west-central United States. In the Miocene, an erosional surface of gentle slopes and broad valleys developed on Mesozoic and Permian rocks. Locally minor tributaries were sharply incised but maximum relief probably did not exceed 75 m with deposition in lower parts (fig. V.27.A). Progressive reduction of relief proceeded throughout the Neogene by gradual engulfment of the erosional relief by the spreading of alluviation on to valley slopes. Bedrock areas were reduced to circum-alluviated hills and small discontinuous areas along former major divides (fig. V.27.B). Neogene deposition culminated in plains of low relief (fig. V.27.C) marked only by natural levees, channel scours, and small low swells where bedrock was not buried. A thick and widespread mantle of

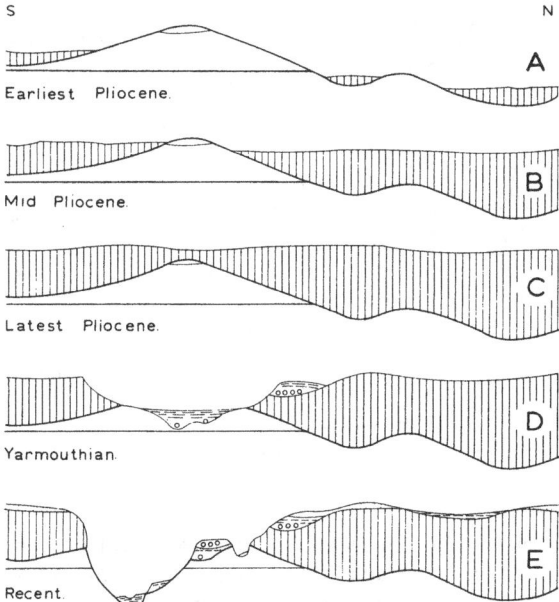

Figure V.27 Pliocene and Pleistocene evolution of the land surface in the central and southern High Plains of the United States (*from Frye and Leonard, 1957*).

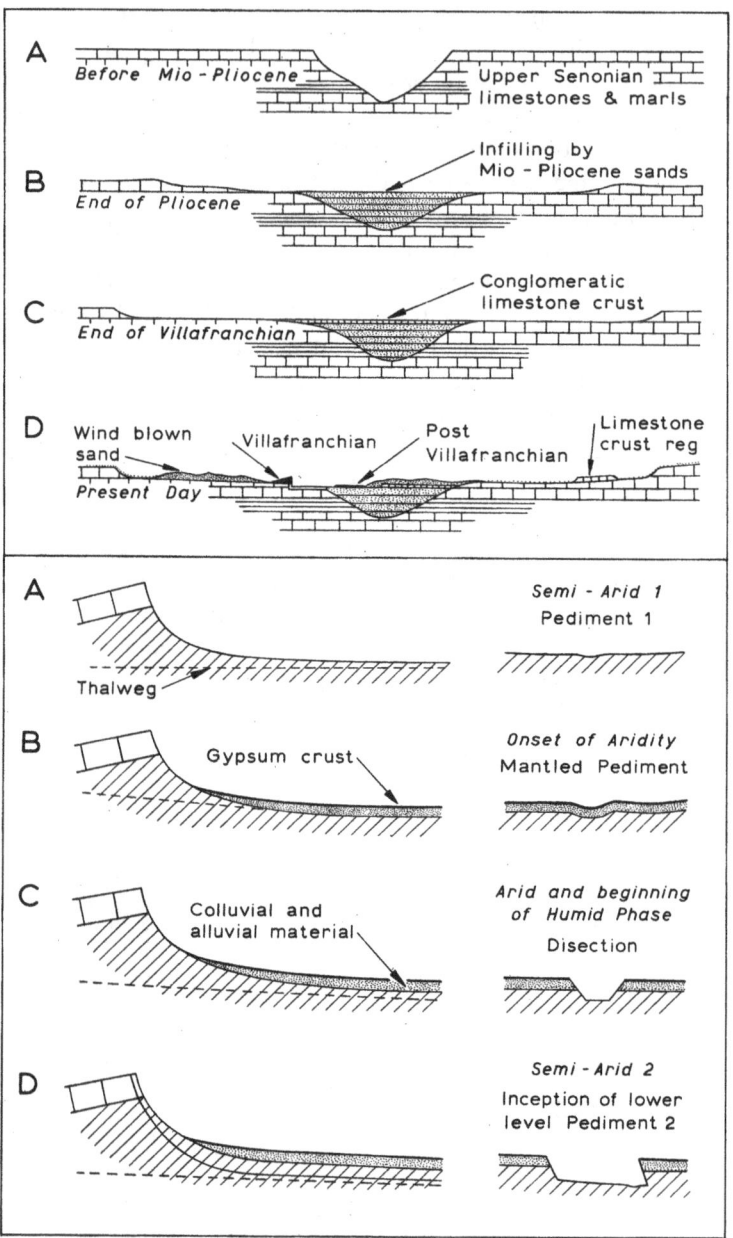

Figures V.28 and *V.29* Schemes of land-surface evolution in Tunisia (*from Coque, 1962*). (*Above*) a hamada in the extreme south. (*Below*) A slope and river channel in unresistant rocks.

dense calcium carbonate, locally encrusted on top developed over most of the region. At the beginning of the Pleistocene relief contrasts developed due to phases of entrenchment and associated alluvial terraces produced by sharp climatic changes from late Cenozoic times onward. The fall of the Pearlette ash in late Kansan times provides a precise datum for the next stage (fig. V.27.D). A feature of the present relief (fig. V.27.E) is that the incision of the valleys is sufficiently deep for the maintenance of perennial streams.

The most decisive stages in the evolution of relief near the desert in

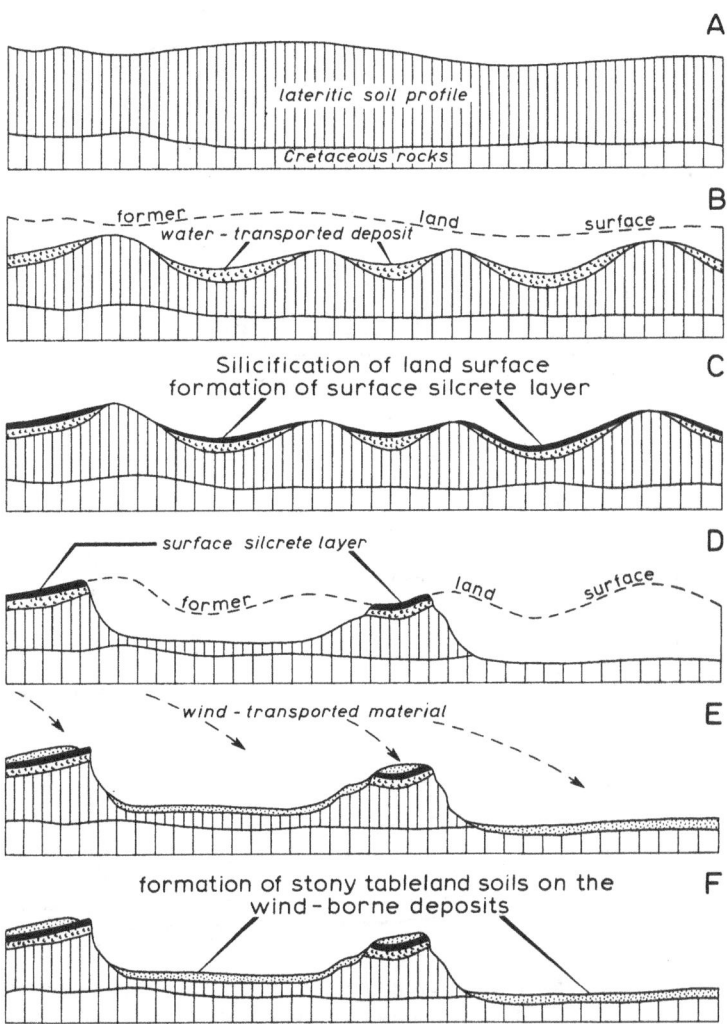

Figure V.30 Tertiary and earlier Quaternary pedological and geomorphological history in the south-east of the Australian arid zone (*from Jessup, 1961*).

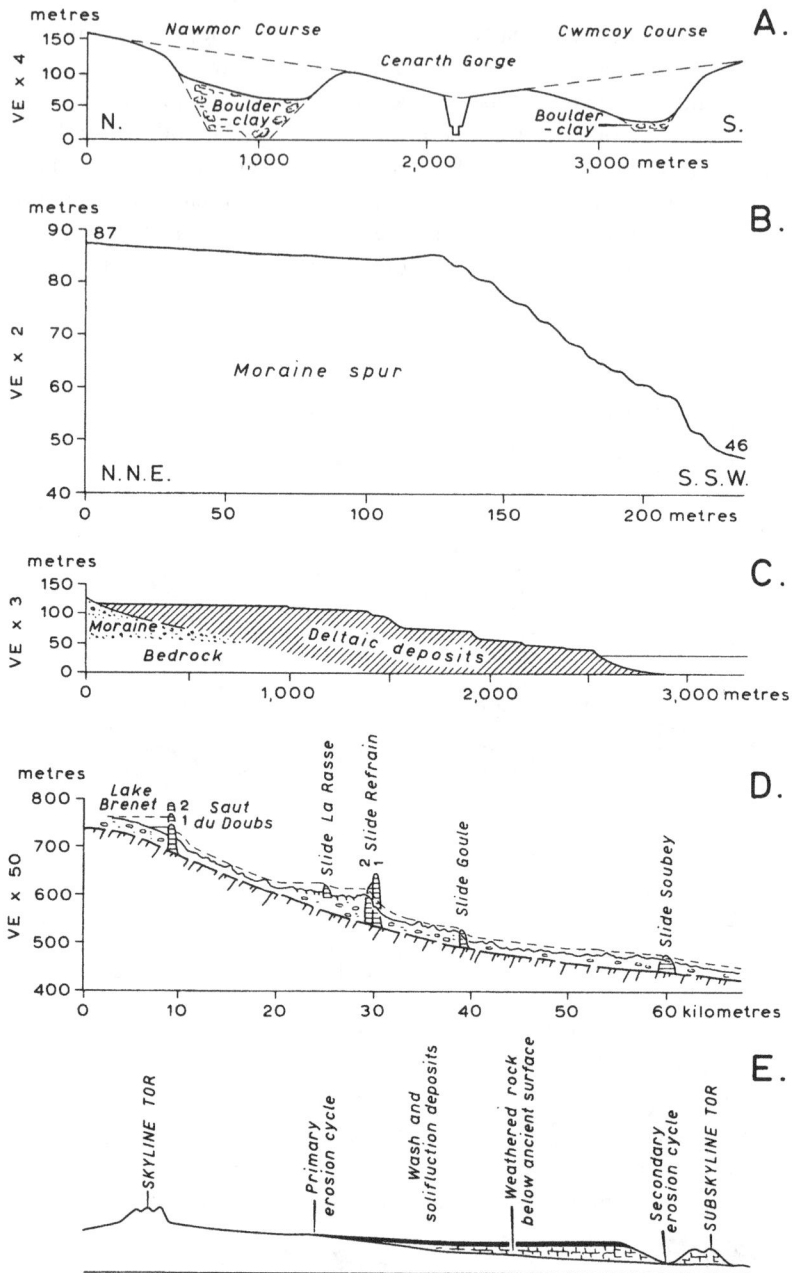

Figure V.31 Stages in landform evolution involving a sequence of effects due to different, consecutive processes.

A. A meandering river course replaced by a lower, shorter incised course following diversion due to deposition of boulder clay plugs (*from Jones, 1965*). The

Tunisia were pre-Quaternary. The same Quaternary climatic changes as led to stages in relief development farther north (fig. V.29) operated in the south but with effectiveness reduced to the extent that changes in pre-Quaternary landforms were negligible. The Mio–Pliocene deposits in the valleys cut down between the residual hills of the hamada surfaces puts the main period of dissection before the end of the Miocene, possibly as a response to large radius folding related to late-Atlas orogenic movements. Valley incision continued during the Oligocene and for most of the Miocene to depths of 55–60 m below the surrounding tableland, with slope-retreat broadening the valleys (fig. V.28.A). Subsequent deposition of several dozen metres of sediment has reduced the relief contrast somewhat and in many cases made the direction of flow of the formerly well-developed drainage net uncertain. In part the sediments came from the erosion of slopes adjacent to the rising level of the sedimentary fills (fig. V.28.B). The development of a conglomeratic crust cemented and indurated with calcium during the Villafranchian provides a detailed record of relief at the beginning of the Quaternary (fig. V.28.C). One feature is that the maximum depth of accumulation of debris throughout Quaternary time was only 5 m. Some slope development occurred in relation to local base-levels, but this was essentially a Villafranchian feature and there is no indication of exhumation of the buried valleys (fig. V.28.D).

On the northern fringe of the Sahara in Tunisia R. Coque (1962) envisages that, according to the amount of precipitation in a drainage basin, a main channel tends to develop sufficiently to evacuate the large quantities of

---

incision of the pre-existing meandering course of the River Teifi into the plateau can itself be recognized as an earlier stage, perhaps related to uplift in preglacial times.

B. A sequence of former shorelines notching a moraine spur at Elvegård in the Narvik district *(from R. Dahl, 1968,* Norsk geog. Tidsskr, *Vol. 22),* the result of isostatic rebound.

C. A sequence of former shorelines notching a glacial outwash fan, Gurreholms Elv, Kjove Land, East Greenland *(from D. E. Sugden and B. S. John, 1956,* Geog. Jour., *Vol. 131),* again the result of isostatic rebound.

D. Landslide dams providing instances of distinctive stages in the development of the long profile of a river in a mountainous region *(after A. Buxtorf, 1922,* Eclogae Geol. Helvetiae, *Vol. 16).* These conspicuous steps are in the longitudinal profile of the Doubs river in the French Alps.

E. An example of a land surface interpreted as the product of successive cycles of erosion *(from King, 1967).* The section shows the relation envisaged for many parts of Africa between skyline tors, interpreted as relics of an older denudation cycle, and sub-skyline tors, seen as youthful features of a new cycle.

water and debris occasionally carried on to lower slopes by rillwash and sheetwash during torrential rains (fig. V.29.A). The pattern of stages in the dissection of these surfaces, shown in fig. V.29 is similar regardless of regional or local base-levels; climatic changes appear to be the controlling factor. A phase of higher temperatures, reduced precipitation, and associated thinning of vegetation cover led to alluviation replacing lateral planation. Debris accumulations blocking channels encouraged lateral diversions and the consequent building-up of alluvial fans. This mantle of sediments thinned downslope, the steeper gradient facilitating, to some extent, the evacuation of small-sized debris (fig. V.29.B). The relatively thin depth of the alluvial mantles, being only a few metres thick at the most, introduced only a very slightly more pronounced concavity into the outline of the slopes and indicates that the periods of accumulation were short lived. However the porosity of the mantle accentuated the effect of the drier climate and gypsum crusts developed (fig. V.29.B). The reduction in sediment supply to channels, however, favoured some incision of the main drainage channels by the debris-free water of the occasional floods, a process accentuated at the beginning of the ensuing pluvial phase (fig. V.29.C). Subsequently the effect of greater and more frequent floods of the more humid phase was a reversion to lateral planation leading to the development of a second, lower pediment surface (fig. V.29.D).

According to R. W. Jessup's (1961) schematic interpretation of the evolution of relief in the south-eastern part of the Australian arid zone, the alternation of pluvial and non-pluvial periods during the Tertiary and Quaternary created a series of relief patterns unlike those postulated for Tunisia. Unlike many parts of Australia there are, in the south-east portion of the arid area no remnants of the actual land surface on which a deep lateritic soil developed (fig. V.30.A). Continental uplift and down-warping of the Lake Eyre basin in the Miocene led to the dissection of the post-laterite erosion surface (fig. V.30.B). Subsequent cementation of the upper part of the detrital deposits formed a silcrete capping (fig. V.30.C). The extensive younger erosional land surface developed into the lower part of the kaolinized materials when earth movements ceased (fig. V.30.D). As much of the area was a vast internal drainage basin, sedimentation continued from late Tertiary into the Quaternary. Intensified desiccation led to the deflation of lacustrine sediments which were deposited over a wide area and even on the higher tablelands. As in Tunisia, gypsum (fig. V.30.E), wind-blown from dried-up lakes, was laid down. Wind transportation ceased as the climate became more humid, stony tableland soils developed (fig. V.30.F) together with renewed but limited drainage incision and surface stripping.

Among workers in some humid mid-latitude areas, particularly in Britain, there was, until a decade ago, intense interest in the search for stages in the evolution of relief as a whole, studied by mapping erosion surface remnants and the interpretation of altitudinal ranges covering any discrete groupings that emerged. This pattern of investigation followed that of H. Baulig's pioneer work on the denudation chronology of the Massif Central (Baulig, 1928). The altitudinal ranges were regarded as an expression of phases of partial peneplanation before a relative fall in sea-level led to similar level-lings at lower altitudes, related to the new base-level. However, weathering crusts, distinctively weathered soils, dust falls, and other reference levels are usually absent, the evidence is largely morphological, was prepared before former sea-levels were indicated and dated by non-morphological evidence, and is based on theoretical concepts regarded by some geomorphologists as oversimplified. Reappraisals of the existing evidence of denudation chrono-logies, detailed replies to criticisms or to alternative interpretations and the accumulation of additional facts are few. It is therefore difficult to evaluate the present significance of these interpretations of denudation chronology in contemporary geomorphology; nonetheless the records of relatively flat seg-ments of land surface at certain discrete altitudinal ranges provide a most useful introductory generalization in the description of the relief of an area.

Appendix

Some simple methods of field measurement

Traditionally the younger student interested in landforms has been trained to evaluate arguments, to acquire patiently over years of field experience an eye for the country, to practise the arts of field sketching and photography, and to be content in the meantime with the interpretations of more experienced eyes. On reflection, it seems possible that the intellectual importance of debate in training the mind has, in geomorphology, obscured the importance of the basic step of science, that of collecting incontrovertible basic measurements. Equally obscured is the fact that a basic step in learning is the finding out of facts for oneself.

In the belief that some readers may wish to replace many of the examples scattered through this book with those of their own observation, it is considered that this book would be incomplete without some suggestions on how those with little experience and with limited means and time at their disposal either to do fieldwork or to search for the appropriate advanced manual or article on techniques, might go out into the field and return with some measurements or samples to be analysed.

A. Slope Measurement

During the last 20 years geomorphologists have come to regard hillslopes as the most important relief form, and to agree that the measurement of profiles in the field is a desirable stage in slope study. Slopes are usually measured with an Abney level, a conveniently carried pocket-sized instrument used in conjunction with ranging poles and a tape measure. With little expense and requiring no great skill an alternative is the construction and use of a slope pantometer, a simple device reflecting in concept the theme of the present appendix. Made from well-seasoned wood or with right-angle girders of light-weight alloy, the device consists of two uprights, each with a bolt near top and bottom (fig. App.1.A). One of the uprights has a large protractor scale attached, centred on the upper bolt. Two cross-pieces with holes exactly 1·5 m apart complete the parallelogram. One of the spirit bubbles in a builder's level makes it possible to set the uprights vertically for measuring the slope declivity. To facilitate the drawing of the results, tables of horizontal and vertical equivalents for slope angles are easily prepared, and the value for each angle in the profile sequence is read off and accumulated to give the points with which to trace the profile on graph paper. Representation of the results in the form of a subdivided histogram provides a compact summary of the relative frequency of measurements of a given angle, and by summing separately the frequencies for each of three equal sub-lengths of the profile also conveys simultaneously some impression of the slope shape (fig. App.1.B).

The procedure for measuring a series of slope angles down a profile is to place the protractor upright on the position occupied by the leading upright during the previous measurement. However, before taking measurements, three decisions on the position and length of the profile line are required to ensure some comparability between profiles. First, in dome-shaped hills with curved contours the two dimensions of the slope-profile will vary with down-slope changes in the plan shape and are perhaps best avoided in preliminary studies. Secondly, the actual alignment of the profile, always at right angles to the contours, might be drawn through a randomly located point designated *rn* in fig. App.1.C, or perhaps the lines might be located to answer a specific problem, such as the variation in slope form in relation to caprock thickness or proximity to stream channel. Thirdly, to avoid exaggerating the extent as well as the flatness that exists on a slope-profile summit such as *S* in fig. App.1.C, the inclination of the ridge crest, or the steepest gradient that exists in any direction away from the vicinity of the profile summit, should be measured. The summit area shown in fig. App.1.C, enlarged in frame D, is traversed by an inclined ridge crest lettered *MNO*, which an extension of the slope-profile line intersects at *N*. As the hypothetical contours are 200 yards apart with a 100-foot contour interval, the inclination of the ridge is about 9 degrees. In consequence declivities less than 9 degrees do not exist in this area. Therefore in the zone *XY* where the inclination along the slope profile line is between 6 and 12 degrees, trial measurements from different starting-points will eventually establish a slope length averaging 9 degrees which defines at its upper end the summit of the slope profile. In cases where there is a valley-floor gradient, the base of a slope, *B*, would be usefully defined by criteria similar to those which limit the summit end.

B. Soil Moisture Content

Innumerable aspects of the behaviour of erosional processes are closely related to a single factor. Dependent on this neglected factor, the water

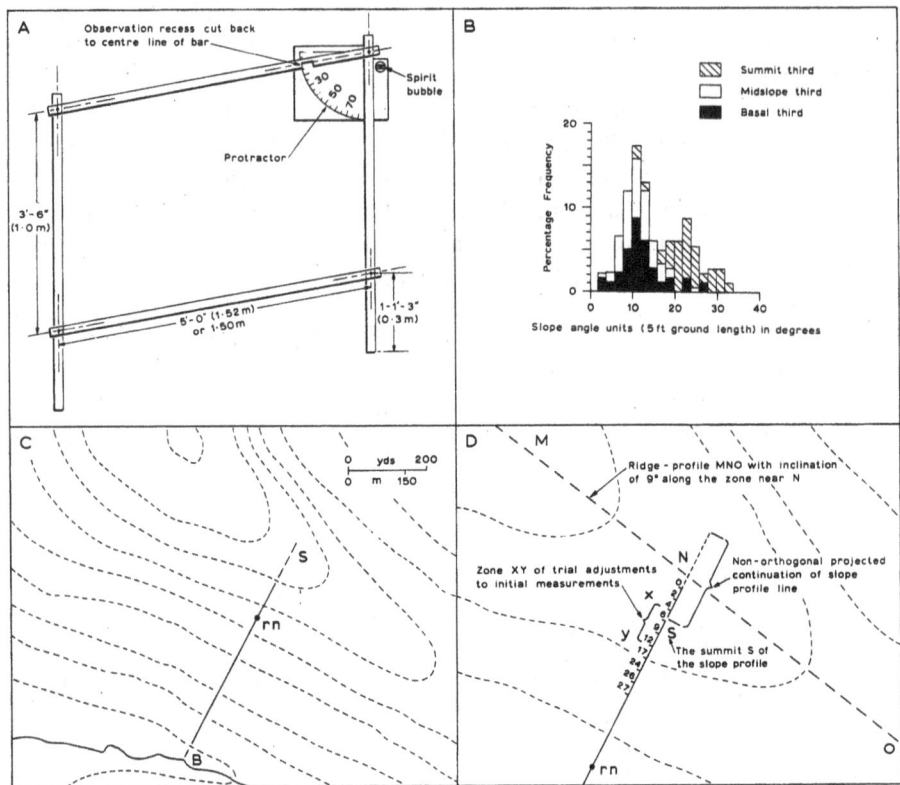

Figure App.1 Some procedures in the field study of hillslopes.

A. The slope pantometer, a simple device for the measurement of large numbers of unit-length slope angles *(from A. F. Pitty, 1968,* Jour. Geol., *Vol. 76).*

B. A subdivided histogram for representing slope-angle frequencies according to their position on the slope-profile *(from Pitty, 1969).* The example shows which angles are represented on a gritstone–shale slope, typical of many escarpments in the southern Pennines, and how the steepest angles are largely confined to the summit third of the profile.

C and D. The laying out of the slope profile orthogonal to contours and its delimitation *(from A. F. Pitty, 1966,* Zeit. f. Geomorph., *Vol. 10).* (C) The location is confined to an area where contours are straight and parallel in order to hold variations in a third-dimension constant. The line may be drawn through a grid reference point described by a randomly selected number, *rn.* S = summit, B = base. (D) The definition of the summit point on a slope profile where the summit is not flat.

content of the soil, is the mechanical behaviour of soil, such as elastic, plastic, or liquid states. These in turn influence the shrinkage, contraction, and shearing properties. The simple addition of water may, by altering the mode of failure, explain differences between landslide, mudflow, and sheetwash. Soil water will influence the proportion of precipitation which infiltrates into the soil and the amount which runs off on the surface; it will influence the effectiveness of freezing temperatures. In addition to physical conditions, most of the chemical and biological reactions of the soil are affected directly or indirectly by the moisture content of the soil, as is the vegetation cover.

Although many attempts have been made to express indirectly the total water content of the soil in terms of climatic data, and although few direct observations appear in the geomorphological literature, the measurement of soil-water content is straightforward.

Procedure

Although field collection of samples is involved, the drying and weighing is usually done in the laboratory. The main consideration is to avoid moisture loss from the soil sample between field and laboratory. Circular tins, approximately 6 cm in diameter and 1–2 cm deep, provide suitable sample containers provided that the lid is reasonably airtight. The collection of several samples from the same site at the same time, and their subsequent weighing one at a time at progressively longer intervals after the sampling time gives a measure of the efficiency of the containers. If the loss is measurable a correction factor can be worked out for use in subsequent calculations. Alternatively, a plastic bag packed with about 250 gm of soil and sealed tightly with a rubber band will, on return to the laboratory, provide a sub-sample which, if extracted from the centre of the bulk sample, resembles field moisture conditions closely.

In the laboratory a tin and wet soil are weighed, Wm. The lid is then removed from the sample tin, and tin and lid are placed in an oven. A temperature of 105°–110° C is maintained for 24 hours. The lid is replaced and the container placed in a desiccator where silica gel maintains a dry atmosphere, thus avoiding hygroscopic absorption of atmospheric humidity during cooling. When cool, the weight of the dried soil and tin is obtained, Wd. The tin, cleaned and dried, is then weighed, Wt. The moisture content, m, of the soil, expressed as a percentage of the dry soil, is given by

$$m = \frac{Wm - Wd}{Wd - Wt} \times 100 \text{ per cent}$$

It should be noted that as the result is expressed as a percentage of the dry soil, soil moisture contents from waterlogged sites may be as high as 200–

300 per cent. Also the calculation does not give a direct measure of the volume of water in the soil. This can be calculated only if the volume of the field sample is measured when collected. However, difficulties in extracting an undisturbed core of soil make this extra measurement unnecessary for many practical purposes.

C. Estimation of Stream-flow

The arguments for measuring stream-flow are compelling. Many authorities maintain that erosion by running water is still to be regarded as the dominant agent in landform sculpture, and detailed hydrological investigations have enabled geomorphologists, like G. H. Dury, to suggest new hypotheses concerning drainage evolution. Further, the volume of stream-flow is an essential item for the calculations of denudational loss which provide a vital aid in interpreting rates of change in the evolution of landforms. Not least, as water is one of man's most vital resources, no geomorphologist can equip himself with techniques to turn more readily to practical ends than those offered by some knowledge of stream-flow observation.

Fluctuations in stream-flow at a point are readily and accurately recorded by measuring the water height, or stage, in relation to a fixed datum. It can hardly be emphasized too strongly that one needs only a ruler to learn at first hand a great deal about the actual behaviour of streams. As an additional observation, the timing of float velocities provides a useful indication of the volume of stream-flow involved. While it is evident that floats necessarily provide only an approximation to the flow rate, specially constructed flumes with continuous recording equipment provide detail beyond the needs of most geomorphological investigations. In addition flumes are too expensive to form the basis of a dense network which could provide information on geographical contrasts in a small area. As an alternative, the current meter, being portable, has an advantage for a study of variations in the intensity of fluvial processes within a study area, but this instrument is expensive and is costly to maintain and to calibrate. The float method has, therefore, some advantages. L. B. Leopold, M. G. Wolman, and J. P. Miller (1964) point out that floats are particularly valuable during floods. In addition the lack of sophistication in the method can be offset by careful planning because uniformity of practice gives the highest degree of comparability between stream gaugings and also because careful selection of the reach for measurement is the most important factor if gauging of stream-flow is to be reliable.

Procedure

The channel should be open and straight, preferably with high, vertical banks to contain high water-flows, and with a natural or artificial constriction downstream which, by maintaining a smooth water surface upstream, establishes a clear relation between stage and discharge. There should be some permanent reference point to use as a datum for determining stage height. A stick, weighted with a nail at one end to make it float upright, provides a serviceable float.

Fig. App.2.A illustrates the measurement of the positions of 7 floats, in a hypothetical stream reach, as they cross the upstream line UV, and then again as they cross the downstream line DE. The distance between these two lines,

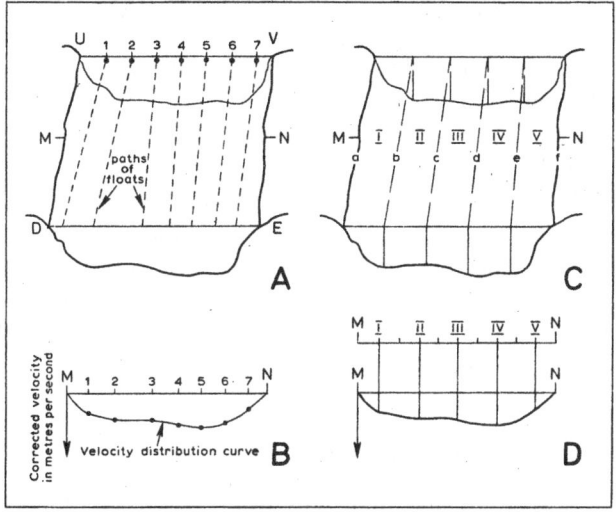

Figure App.2 Estimation of stream flow by floats (for explanation, see text).

in metres, is divided by the float velocity measured in seconds. The result is then multiplied by a correction factor, 0·85 to give a velocity representative of total depth of water flowing beneath the float. For a bed of canal-like smoothness a correction factor of 0·90 would be appropriate as would a factor of 0·80 for a very rough boulder-strewn channel. MN is a line across the stream, half-way between the lines UV and DE. The points at which the paths of the floats cross this line in fig. VI.2.A are shown in fig. App.2.B together with the corrected velocities which are joined to provide a velocity distribution curve.

The measured reach is now divided into 3 or more panels of equal width. In fig. App.2.C there are 5, indicated by Roman numerals I to V and separated by panel boundaries labelled a to f. The velocity, V, at the line MN, for the midpoint of each panel, is read off from the appropriate point on the velocity distribution curve (fig. App.2.D).

Stream depths at the panel boundaries are measured along both lines UV

and *DE*. The average of a pair of depth readings provides the values for the depth at the panel boundaries *a* to *f* along the mid-reach line, *MN*. The width, *w*, between panel boundaries along *MN* is also measured. The discharge, *q*, for the panels is then obtained,

$$q\,\mathrm{I} = w \times \frac{\text{depth } a + \text{depth } b}{2} \times V\,\mathrm{I}$$

$$q\,\mathrm{II} = w \times \frac{\text{depth } b + \text{depth } c}{2} \times V\,\mathrm{II}$$

$$q\,\mathrm{V} = w \times \frac{\text{depth } e + \text{depth} f}{2} \times V\,\mathrm{V}$$

The total discharge, *Q*, in cubic metres per second, is the sum of the individual panel discharges.

Ten measurements of discharge, if they cover a wide range of runoff conditions, may be sufficient to define a graph between stage height and discharge. With the aid of this graph, known as a rating curve, it is then possible to estimate the discharge volume from a reading of the stage alone. It is worth noting that the scatter of points about the rating curve is a helpful indication of the consistency of the discharge measurements, and that the shape of rating curves may vary from one station to another due to differences in channel configuration.

D. Particle-size Analysis of Sands and Pebbles

Particle-size analysis is a fundamental stage in the description of sediment transportation and deposition, and of depositional landforms. In areas where mechanical weathering is important particle-size analysis again provides basic data. As a lengthy pre-treatment is required to remove organic matter from soils and as the size analysis of clay and fine silts involves several stages, some initial familiarity with the usefulness of particle-size analysis is most readily obtained by studying problems involving only sandy or coarse silt sediments. The sieving of sands and coarse silt particles, free from organic matter, is straightforward.

Procedure

Take about 100 gm of sediment which have been thoroughly dried in an oven at a temperature not greater than 105°–110° C. Pass the sediment through a 2-mm sieve and weigh the coarser fraction before discarding it. When the fraction finer than 2 mm has been weighed, the weight of the coarser fraction can be recorded as a percentage of the total sample.

Arrange a nest of about five sieves with the mesh aperture size increasing upwards and with a collecting pan at the base. Pour the sub-sample of sediment finer than 2 mm on to the uppermost sieve and cover with the lid and shake the nest of sieves. After about 8 minutes further effort will not produce an appreciable increase in the degree of sorting obtained. In shaking, pressure should be applied to lid and base of the nest to maintain a tight fit between sieves. The basic motion in shaking should be rotary, to keep the particles moving round the sieves, combined with vertical jolts to facilitate the sorting process. The contents of each sieve are emptied on to a sheet of paper and then funnelled into a container for weighing, together with any grains wedged in the mesh of the sieve which can be dislodged by careful brushing. The percentage of the total weight retained at each size division is obtained,

$$\frac{Ws}{Wd} \times 100$$

where Ws is the weight of material retained at a given size grade and Wd is the original dry weight of sub-sample. The weight of material on the collecting pan is the percentage of the total weight finer than the mesh size of finest sieve. The percentage weight of material on the pan, when added to that of material retained on the finest sieve, gives the percentage of the total weight finer than the mesh size of the next-to-the-finest sieve size. In this way a cumulative percentage is worked out for material finer than each sieve size, the cumulative curve drawn up, and indices for describing median size and sorting interpolated, as shown in fig. III.4.

The approach for analysing the sizes of gravels and pebbles is similar in principle to sieving. The three main axes of a pebble – length, breadth, and depth – are defined at right-angles to each other and measured using either a ruler or preferably calipers. The pebble is weighed. For a sample of about fifty pebbles it is possible to calculate the cumulative percentage weight of pebbles smaller than a series of given length or breadth dimensions, or of a mean value of the three axes, and then to plot a cumulative curve. Ratios between the axes such as Cailleux's flatness index $\frac{l+b}{2d}$ can be calculated.

These results are plotted as a histogram, pebble analysis involves little expense compared with the cost of sieves and the principles of particle-size analysis and its results can be appreciated from a study of pebbles alone if the cost of sieves is prohibitive. However, it should not be beyond the ingenuity of some to find cheap, mass-produced items to provide a range of aperture sizes that would subdivide sand-sized particles. The actual sizes might be difficult to estimate accurately but a series of samples would be effectively ranked. The most important point in analysing sediments is to avoid wherever possible the collection, in one sample, of material which can be seen to be of more than one type. For instance, in stratified materials, sample parallel to the bedding; in rivers sample upstream from confluences where unlike materials become mixed.

E. Heavy Minerals

A complete heavy mineral analysis requires great expertise, specialized equipment, and a lot of time. However, in certain localities it is possible to study how one heavy mineral moves in relation to lighter particles by separating the magnetic component of a sand, if present, with a simple horse-shoe magnet. For example, using this method on barchan dunes in southern Peru, Hastenrath (1967) demonstrated a deficiency of the heavy mineral on the slip face of the dunes.

Table App.1 Deficiency of heavy minerals on the slip-face of barchan dunes in southern Peru, demonstrated by the use of a horse-shoe magnet (*Hastenrath, 1967*).

	PERCENTAGE MAGNETITE OF TOTAL SAMPLE IN THREE DUNES		
East horn	0·8	1·1	11·3
West horn	3·5	1·8	7·6
Windward face	1·5	3·1	3·4
Crest	1·8	2·4	1·8
Slipface	0·8	—	1·2

F. Soil pH

The measurement of pH is useful in weathering and in vegetation studies since it forms one of the most comprehensive single summaries of chemical conditions in the soil. For instance, on a granite catena in Rhodesia the pH of twenty-nine samples was highly correlated with base saturation ($r = 0.73$). In southern Scotland, soils developed from greywackes and shales on Broad Law have low pH values of 4·0–4·9 which correlate with the low exchangeable base content and degree of base saturation. In most situations where values higher than 8·5 occur, these usually indicate the presence of sodium carbonate. Values of pH which are low when compared with soils in different geomorphic situations or formed closer to bedrock, or from land surfaces of different ages, indicate the severity of leaching. For example, a surface horizon of a soil developed on granodiorite at Elberton, Georgia, had a pH of 4·8–5·2 compared with 8·7 at the surface of the unweathered bedrock. Changes downslope may reflect local changes in leaching intensity, or, as is particularly the case on chalk scarp slopes, the rapidity of renewal of unweathered fragments from a base-rich bedrock. Changes of pH in relation to time of exposure of depositional landforms are particularly noteworthy. Table App.2 illustrates the relationship between exposure time of glacial moraines and the pH of the surface soil. Although it must be stressed that pH readings can have no precise significance, the examples show that estimates of pH are well worth recording.

Table App.2 The relationship between pH and age of moraines (*from Stork, 1963, and from R. F. Chandler, 1942*, Soil Sci. Soc. Am. Proc., *Vol. 7*).

LOCALITY			
Storglaciar, Kebnekajse massif, N. Sweden		Mendenhall Glacier, near Juneau, Alaska	
Exposure in years	pH	Exposure in years	pH
0	8·1	15	5·4
5	7·7	90	5·1
10	7·8	250	4·3
15	6·5	approx. 1000	3·7
30	6·1	—	—
50	5·4	—	—

Procedure

To estimate soil pH by the Barium Sulphate method, stoppered test tubes 20 cm long and 1 cm internal diameter are used. The amount of soil placed in the test tube depends on the texture of soil, ranging from a depth of 1·5 cm for a clayey soil to 4 cm for a gravelly soil. The depth is then made up to 5 cm with an appropriate amount of barium sulphate. Distilled water is added to bring the height of the column up to 10 cm. This is then made up to 11·5 cm by the addition of Soil Indicator, a soil-testing reagent. The top end of the test tube is then stoppered with a second rubber bung and the suspension shaken vigorously. The test tube is then placed vertically in a rack, and when 2·5 cm of the supernatant liquid is clear, the colour of the liquid is compared with a range of standard colours on a special pH chart. Interpolation between the chart colour chips is not easy, and improved results are obtained if several samples are analysed at the same time and then ranked in order of increasing intensity of colour.

Among the factors which complicate the interpretation of pH observations is soil moisture. Waterlogged soils are commonly close to neutral but may, on drying, fall from 7 to 4 as a result of oxidation. High carbon dioxide concentrations may lower pH by as much as a unit compared with similar soils where the carbon dioxide concentration is low. At a given point variations of 0·5 pH units may occur during the course of a year.

The complexity of factors which might produce a given pH value makes most interpretations tentative, and very rarely can full use be made of really accurate measurements of pH of a field soil. In consequence little information is lost if measurements are made to the nearest 0·2 of a pH unit. Therefore more sophisticated methods than the Barium Sulphate method are unnecessary for most field study purposes.

G. Dissolved Solids

A comprehensive study of chemical weathering requires a great deal of expertise and equipment. However, the average rate of chemical denudation is very nearly that of any of the major constituents and calcium carbonate, which is quantitatively the most important in many drainage waters, is neither difficult nor expensive to determine accurately. Its study, in addition to providing first-hand insight into chemical weathering also can improve one's appreciation of the practical problems of maintaining water quality and of controlling pollution.

Procedure

Pipette 100 ml of a drainage water sample into a flask or dish, and add 1 ml of 4N sodium hydroxide solution. Add 1 calcium hardness indicator tablet which, when dissolved, colours the solution pink. Fill a burette with N/50 E.D.T.A. and add this dropwise to the sample. The end-point of this titration is reached when the sample solution becomes violet, and the addition of a further drop of N/50 E.D.T.A. produces no further colour change. If possible the sample should be agitated during the titration and the flask or dish should stand on a white background. Calcium carbonate content, or calcium hardness in non-limestone areas, in parts per million is given by

$$CaCO_3 = \frac{\text{E.D.T.A. used, in ml} \times 1000}{100}$$

The reagents are standard for the determination of water hardness and are therefore readily obtained.

H. Suspended Sediment

Although the measurement of suspended sediment is one of the most important techniques in measuring contemporary denudation rates, the technique when compared with those already described, requires greater skill and much more time. Also an expensive, highly sensitive balance is essential.

Procedure

The bung of the sampling bottle has two openings; a nozzle about 1 cm in diameter to point upstream and a tube through which air can be expelled as the bottle fills up. The bottle is raised and lowered at points across a stream. By maintaining a constant rate of lowering and raising the amounts of water entering the bottle in shallower or in slower moving parts of the stream will be proportional to the depth and velocity of flow. Accordingly, the bulk sample is representative of the amount of sediment moving through the cross-section.

In the laboratory the volume of the sample is carefully measured. It is then heated to evaporate some of the water. A two-piece funnel is assembled with a filter paper, of known weight and capable of retaining particles as small as 0·2 microns, placed on the filter disc of the lower part of the funnel. The upper part of the funnel is then clamped securely on top. A water-filter pump, attached to a tap, maintains a vacuum in the flask which supports the funnel. When all the water has passed through the funnel, the filter paper is removed, oven-dried, placed in a desiccator to cool, then weighed. The computed weight of sediment is divided by the weight of the sample. The result, multiplied by one million, gives the concentration of suspended sediment in parts per million.

Any small area with some relief offers scope for landform study. In fact with the contrast between north- and south-facing slopes differences equal to several degrees of latitude are on the doorstep. With contrasts between seasons, an indication of past climates is to some extent compressed into the present. If the student of geomorphology sets up a simple hypothesis, sees some advantage in making measurements to test it, and carefully collects field data by using any one of the above methods, or others not described here, his achievement will be substantial. Even if there is no apparent or comprehensible pattern, his accomplishment is still to have performed one of the most difficult tasks in landform study, that of collecting field measurements.

Selected Bibliography

General

BAKKER, J. B. 1960. Some observations in connection with recent Dutch investigations about granite weathering and slope development in different climates and climate changes, *Zeit. f. Geomorph.*, Suppbd 1, pp. 69–92.

BECKINSALE, R. P. 1966. *Land, air and ocean*, 4th rev. ed. (Butterworth, London), 448 pp.

BIRD, E. C. F. 1968. *Coasts* (Aust. Nat. Univ. Press, Canberra), 246 pp.

BIRD, J. B. 1967. *The physiography of Arctic Canada* (John Hopkins Press, Baltimore), 336 pp.

BIROT, P. 1960. *Le cycle d'érosion sous les différents climats* (Centro de Pesquisas de geografia do Brasil, Univ. of Brazil, Rio de Janeiro), 137 pp.; trans. by C. I. Jackson and K. M. Clayton, 1968, *The cycle of erosion in different climates* (Batsford, London), 144 pp.

BIROT, P. 1959. *Précis de géographie physique générale* (Armand Colin, Paris), 403 pp.; trans. by M. Ledésert, 1966, *General physical geography* (Harrap, London), 360 pp.

BIROT, P. 1965. *Géographie physique générale de la zone intertropicale* (Centre Doc. Univ., Paris), 290 pp.

BLACK, R. F. 1969. Geology, especially geomorphology, of northern Alaska, *Arctic*, Vol. 22, pp. 283–99.

BLOOM, A. L. 1969. *The surface of the earth* (Prentice-Hall, New York), 152 pp.

BUTZER, K. W. 1964. *Environment and archaeology* (Methuen, London), 524 pp.

BUTZER, K. W., HANSEN, C. L., and GLADFELTER, B. G. 1968. *Desert and river in Nubia. Geomorphology and pre-historic environments at the Aswan reservoir* (Univ. Wisconsin Press, Milwaukee), 562 pp.

CHARLESWORTH, J. K. 1957. *The Quaternary era* (Arnold, London), Vol. 1, 592 pp.; Vol. 2, pp. 593–1700.

CLAYTON, K. M. (Ed.). 1964. A bibliography of British geomorphology, *Brit. Geomorph. Research Group Occ. Paper*, No. 1, 211 pp.

COQUE, R. 1962. *La Tunisie présaharienne* (Armand Colin, Paris), 476 pp.

COTTON, C. A. 1960. *Geomorphology, an introduction to the study of landforms*, 7th rev. ed. (Whitcombe and Tombs, Christchurch), 505 pp.

DAVIES, J. L. 1969. *Landforms of cold climates* (Aust. Nat. Univ. Press, Canberra), 200 pp.

DAVIS, W. M. 1909. *Geographical essays* (Ginn, Boston), 777 pp.; republished 1954 (Dover, New York).

DERRUAU, M. 1963. *Précis de géomorphologie* (Masson, Paris), 393 pp.

DURY, G. H. 1959. *The face of the earth* (Penguin, Harmondsworth), 223 pp.

EASTERBROOK, D. J. 1969. *Principles of geomorphology* (McGraw-Hill, New York), 462 pp.

EMBLETON, C. and KING, C. A. M. 1968. *Glacial and periglacial geomorphology* (Arnold, London), 608 pp.

FAIRBRIDGE, R. W. (Ed.). 1968. *The encyclopedia of geomorphology* (Rheinhold, New York), 1295 pp.

FLINT, R. F. 1957. *Glacial and Pleistocene geology* (Wiley, New York), 553 pp.

GILLULY, A., WATERS, A. C., and WOODWARD, A. O. 1968. *Principles of geology* (Freeman, San Francisco), 687 pp.

GODARD, A. 1965. *Recherches de géomorphologie en Écosse du nord-ouest* (Cen. Nat. Rech. Sci., Paris), 701 pp.

GUILCHER, A. 1954. *Morphologie littorale et sous-marine* (Presses Univ. France, Paris), 215 pp.; trans. and rev. by B. W. Sparks and R. H. W. Kneese, 1958, *Coastal and submarine morphology* (Methuen, London), 274 pp.

HOLMES, A. 1965. *Principles of physical geology*, 5th rev. ed. (Nelson, London), 1288 pp.

HORTON, R. E. 1945. Erosional development of streams and their drainage basins: hydrophysical approach to quantitative morphology, *Bull. geol. Soc. Am.*, Vol. 56, pp. 275–370.

INSTITUTE OF GEOGRAPHY, U.S.S.R. ACADEMY OF SCIENCES. 1969. *The physical geography of China* (Praeger, New York), Vol. 1. 448 pp.; Vol. 2. 337 pp.

JENNINGS, J. N. and MABBUTT, J. A. (Eds). 1967. *Landform studies from Australia and New Guinea* (Cambridge Univ. Press, Cambridge), 434 pp.

KEEN, M. J. 1968. *An introduction to marine geology* (Pergamon, Oxford), 218 pp.

KING, C. A. M. 1959. *Beaches and coasts* (Arnold, London), 403 pp.

KING, L. C. 1967. *The morphology of the earth*, 2nd ed. (Oliver and Boyd, Edinburgh), 726 pp.

LAKE, P. 1958. *Physical geography*, 4th ed. (Ed. J. A. Steers) (Cambridge Univ. Press, Cambridge), 483 pp.

LINTON, D. L. 1951. Problems of Scottish scenery, *Scot. geog. Mag.*, Vol. 67, pp. 65–85.

LLIBOUTRY, L. 1964. *Traité de glaciologie* (Masson, Paris), Vol. 1, 427 pp.; Vol. 2, 1040 pp.

LONGWELL, C. R., FLINT, R. F., and SANDERS, J. E. 1969. *Physical geology* (Wiley, New York), 685 pp.

MACAR, P. (Ed.). 1966. *L'évolution des versants* (Univ. Liège et Acad. Roy. Belg., Liège), 384 pp.

MACHATSCHEK, F. 1968. *Geomorphologie*, 9th ed. (Teubner, Stuttgart), 209 pp.; trans. by D. J. Davies, 1969, *Geomorphology* (Oliver and Boyd, Edinburgh), 212 pp.

MARR, J. E. 1900. *The scientific study of scenery* (Methuen, London), 368 pp.; 4th rev. ed., 1912, 372 pp.

MATTSON, A. 1962. Morphologische Studien in Südschweden und auf Bornholm über nicht glaziale Formenwelt der Felsskulptur, *Lund Studies in geog.*, Ser. A, No. 20, 357 pp.

MAULL, O. 1958. *Handbuch der Geomorphologie* (Deuticke, Vienna), 600 pp.

MCGILL, J. T. 1960. *Selected bibliography of coastal geomorphology of the world* (Los Angeles), 50 pp.

MILLER, T. G. 1953. *Geology and scenery in Britain* (Batsford, London), 224 pp.

NICHOLS, R. L. 1960. Geomorphology of Marguerite Bay area, Palmer peninsula, Antarctica, *Bull. geol. Soc. Am.*, Vol. 71, pp. 1421–50.

NICHOLS, R. L. 1966. Geomorphology of Antarctica, *Am. geophys. Union, Antarctica Res. Ser.*, Vol. 8, pp. 1–46.

PENCK, W. 1924. *Die morphologische Analyse; ein Kapitel der physikalischen Geologie* (Engelhorns, Stuttgart), 283 pp.; trans. by H. Czech and K. C. Boswell, 1953, *Morphological analysis of land forms* (Macmillan, London), 429 pp.

PINCHEMEL, P. 1954. *Les plaines de craie du nord-ouest du Bassin Parisien et du sud-est du Bassin de Londres et leurs bordures* (Armand Colin, Paris), 502 pp.

RAPP, A. 1960. Development of mountain slopes in Kärkevagge and surroundings, northern Scandinavia, *Geografiska Annaler*, Vol. 42, pp. 65–187.

RANKAMA, K. (Ed.). 1965. *The Quaternary* (Interscience, New York), Vol. 1, 300 pp.; Vol. 2, 477 pp.

ROGNON, P. 1967. *Le massif de l'Atakor et ses bordures (Sahara centrale), étude géomorphologique* (Cent. Nat. Rech. Sci., Paris), 559 pp.

RUSSELL, R. J. 1967. *River plains and sea coasts* (Univ. of California Press, Berkeley), 173 pp.

SHARP, R. P. 1960. *Glaciers* (Oregon state system of higher education, Eugene), 78 pp.

SISSONS, J. B. 1967. *The evolution of Scotland's scenery* (Oliver and Boyd, Edinburgh), 259 pp.

SMALL, R. J. 1970. *The study of landforms* (Cambridge Univ. Press, Cambridge), 486 pp.

SPARKS, B. W. 1960. *Geomorphology* (Longmans, London), 371 pp.

STEERS, J. A. 1964. *The coastline of England and Wales*, 2nd ed. (Cambridge Univ. Press, Cambridge), 750 pp.

THORNBURY, W. D. 1969. *Principles of geomorphology*, 2nd ed. (Wiley, New York), 594 pp.

TRICART, J. 1963. *Géomorphologie des régions froides* (Presses Univ. France, Paris), 289 pp.; trans. by E. Watson, 1970, *Geomorphology of cold environments* (Macmillan, London), 320 pp.

TRICART, J. 1965. *Principes et méthodes de la géomorphologie* (Masson, Paris), 496 pp.

TRICART, J. 1965. *Le modelé des régions chaudes forêts et savanes* (Société d'édition d'enseignement supérieur, Paris), 322 pp.

TRICART, J. and CAILLEUX, A. 1960–61. *Le modelé des régions sèches* (Centre Doc. Univ., Paris), Vol. 1, 129 pp.; Vol. 2, 179 pp.

TRICART, J. and CAILLEUX, A. 1967. *Le modelé des régions périglaciaires* (Société d'édition d'enseignement supérieur, Paris), 512 pp.

WEBER, H. 1967. *Die Oberflächenformen des festen Lands* (Teubner, Leipzig), 367 pp.

WEST, R. G. 1963. Problems of the British Quaternary, *Proc. geol. Ass.*, Vol. 74, pp. 147–86.

WEST, R. G. 1968. *Pleistocene geology and biology, with especial reference to the British Isles* (Longmans, London), 377 pp.

WILHELMY, H. 1958. *Klimamorphologie der Massengesteine* (Westermann, Braunschweig), 238 pp.

WOLDSTEDT, P. 1954–65. *Das Eiszeitalter* (Enke, Stuttgart), Vol. 1, 374 pp.; Vol. 2, 438 pp.; Vol. 3, 328 pp.

WOOLDRIDGE, S. W. and MORGAN, R. S. 1959. *An outline of geomorphology: the physical basis of geography* (Longmans, London), 409 pp.

WRIGHT, H. E. and FREY, D. G. (Eds). 1965a. *The Quaternary of the United States* (Princeton Univ. Press, Princeton), 922 pp.

WRIGHT, H. E. and FREY, D. G. (Eds). 1965b. *International studies on the Quaternary* (Geol. Soc. Am., Special Paper No. 84), 565 pp.

WRIGHT, H. E. and OSBURN, W. H. (Eds). 1968. *Arctic and alpine environments* (Indiana Univ. Press, Bloomington), 308 pp.

ZEUNER, F. E. 1959. *The Pleistocene period*, 2nd ed. (Hutchinson, London), 447 pp.

Chapter I. Definitions, nature and basic postulates

A. Definitions

FENNEMAN, N. M. 1936. Cyclic and non-cyclic aspects of erosion, *Bull. geol. Soc. Am.*, Vol. 47, pp. 173–86.

HJULSTRÖM, F. 1935. Studies of the morphological activity of rivers as illustrated by the river Fyris, *Bull. geol. Inst. Uppsala*, No. 25, pp. 221–527.

LEOPOLD, L. B., WOLMAN, M. G., and MILLER, J. P. 1964. *Fluvial processes in geomorphology* (Freeman, San Francisco), 522 pp.

TRICART, J. 1965. *Principes et méthodes de la géomorphologie* (Masson, Paris), 496 pp.

WOOLDRIDGE, S. W. 1951. The role and relations of geomorphology, *in* L. D. Stamp and S. W. Wooldridge (Eds), *London essays in geography* (Longmans, London), pp. 19–31.

B. Nature of Geomorphology

1. Description and interpretation

ALEXANDER, C. S. 1966. A method of descriptive shore classification and mapping as applied to the northeast coast of Tanganyika, *Ann. Ass. Amer. Geogrs*, Vol. 56, pp. 128–40.

ARBER, M. A. 1949. Cliff profiles of Devon and Cornwall, *Geog. Jour.*, Vol. 114, pp. 191–6.

CAILLEUX, A. and TRICART, J. 1956. Le problème de la classification des faits géomorphologiques, *Ann. Géogr.*, Vol. 65, pp. 162–86.

COTTON, C. A. 1954. Deductive morphology and the genetic classification of coasts, *Sci. Monthly*, Vol. 78, pp. 163–81.

CURTIS, L. F., DOORNKAMP, J. C., and GREGORY, K. J. 1965. The description of relief in field studies of soils, *Jour. Soil Sci.*, Vol. 16, pp. 16–30.

DARBY, H. C. 1962. The problem of geographical description, *Trans. Inst. Brit. Geogrs*, No. 30, pp. 1–14.

DURY, G. H. 1963. Geographical description: an essay in criticism, *Aust. Geog.*, Vol. 9, pp. 67–78.

FOURNEAU, R. 1963. Essai de cartographie géomorphologique, *Rev. Belge Géog.*, Vol. 87, No. 3, pp. 1–7.

HAMILTON, R. A. *et al.* 1957. Surveying aneroids: their uses and limitations, *Geog. Jour.*, Vol. 123, pp. 481–98.

HORTON, R. E. 1932. Drainage basin characteristics, *Trans. Am. geophys. Union*, Vol. 13, pp. 350–61.

McCOY, R. M. 1969. Drainage network analysis with K-band radar imagery, *Geog. Rev.*, Vol. 59, pp. 493–512.

MILLER, V. C. and MILLER, C. F. 1961. *Photogeology* (McGraw-Hill, New York), 248 pp.

MORISAWA, M. 1958. Measurements of drainage-basin outline form, *Jour. Geol.*, Vol. 66, pp. 587–90.

MONKHOUSE, F. J. and WILKINSON, H. R. 1971. *Maps and diagrams*, 3rd rev. ed. (Methuen, London), 522 pp.

OLLIER, C. D. 1967. Landform description without stage names, *Aust. Geog. Studies*, Vol. 5, pp. 73–80.

PITTY, A. F. 1967. Some problems in selecting a ground-surface length for slope-angle measurement, *Rev. géomorph. dyn.*, Vol. 18, pp. 66–71.

REED, B., GALVIN, C. J., and MILLER, J. P. 1962. Some aspects of drumlin geometry, *Am. Jour. Sci.*, Vol. 260, pp. 200–10.

SAVIGEAR, R. A. G. 1965. A technique of morphological mapping, *Ann. Ass. Amer. Geogrs*, Vol. 55, pp. 514–38.

SPIRIDONOV, A. I. 1952. *Geomorfologicheskoe kartografirovanie* (Moscow); trans. by S. Kömmling, 1956, *Geomorphologische Kartographie* (VEB Deutscher Verlag der Wiss., Berlin), 160 pp.

TRICART, J., HIRSCH, A. R., and LE BOURDIEC, F. 1965. Présentation d'un extrait de carte géomorphologique détaillée, *Zeit. f. Geomorph.*, Vol. 9, pp. 133–65.

ZAKRZEWSKA, B. 1967. Trends and methods in land form geography, *Ann. Ass. Amer. Geogrs*, Vol. 57, pp. 128–65.

2. Process and form

ALLEN, J. R. L. 1968. *Current ripples: their relation to patterns of water and sediment motion* (North-Holland, Amsterdam), 433 pp.

BAGNOLD, R. A. 1941. *The physics of blown sand and desert dunes*, 3rd ed., 1960 (Methuen, London), 265 pp.

BRADLEY, W. H. 1966. Tropical lakes, copropel, and oil shale, *Bull. geol. Soc. Am.*, Vol. 77, pp. 1333–8.

WOOLDRIDGE, S. W. 1958. The trend of geomorphology, *Trans. Inst. Brit. Geogrs*, Vol. 25, pp. 29–35.

3. Artistic and scientific elements

AMERICAN GEOLOGICAL INSTITUTE. 1957. *Glossary of geology and related sciences* (Amer. geol. Inst., Washington), 325 pp.

BAULIG, H. 1956. *Vocabulaire Franco-Anglo-Allemand de géomorphologie*, (Les Belles Lettres, Paris), 229 pp.

de BOER, G. 1964. Spurn Head: its history and evolution, *Trans. Inst. Brit. Geogrs*, No. 34, pp. 71–89.

BRYAN, K. 1940. The retreat of slopes, *Ann. Ass. Amer. Geogrs*, Vol. 30, pp. 254–67.

CHORLEY, R. J. 1958. The shape of drumlins, *Jour. Glaciol.*, Vol. 3, pp. 339–44.

CHORLEY, R. J., DUNN, A. J., and BECKINSALE, R. P. 1964. *The history of the study of landforms* (Methuen, London), Vol. 1, 678 pp.

CRAIG, G. Y. 1969. Communication in geology, *Scot. Jour. Geol.*, Vol. 5, pp. 305–21.

DAVIES, G. L. 1969. *The earth in decay. A history of British geomorphology 1578–1878* (Macdonald, London), 390 pp.

DYLIKOWA, A. 1962. Notion et terme 'périglaciaire', *Biul. Peryglacjalny*, Vol. 11, pp. 149–64.

HAMELIN, L. E. and CLIBBON, P. 1962. Vocabulaire périglaciaire bilingue, *Cahier de géogr. de Québec*, Vol. 6, pp. 201–26.

HUTCHINGS, G. 1960. *Landscape drawing* (Methuen, London), 102 pp.

INTERNATIONAL FEDERATION FOR INFORMATION PROCESSING AND INTERNATIONAL COMPUTATION CENTRE. 1968. *IFIP–ICC vocabulary of information processing* (North-Holland, Amsterdam), 208 pp.

LeGRAND, H. E. 1960. Metaphor in geomorphic expression, *Jour. Geol.*, Vol. 68, pp. 576–9.

LEOPOLD, L. B., and SKIBITZKE, H. E. 1967. Observations on unmeasured rivers, *Geografiska Annaler*, Vol. 49A, pp. 247–55.

TERS, M. 1960. Méthodes et techniques modernes en géomorphologie, *Inf. géog.*, Vol. 24, pp. 156–65.

4. Qualitative and quantitative aspects

AHNERT, F. 1966. Zur Rolle der elektronischen Rechnenmaschine und des mathematischen Modells in der Geomorphologie, *Geog. Zeit.*, Vol. 54, pp. 118–33.

BAULIG, H. 1959. Morphométrie, *Ann. Géogr.*, Vol. 68, pp. 385–408.

CHORLEY, R. J. 1966. The application of statistical methods in geomorphology, *in* G. H. Dury (Ed.), *Essays in geomorphology* (Heinemann, London), pp. 275–387.

CREAGER, J. S., MCMANUS, D. A., and COLLIAS, E. E. 1962. Electronic data processing in sedimentary size analysis, *Jour. Sed. Petrol.*, Vol. 32, pp. 833–9.

DEVDARIANI, A. S. 1967. *Matematicheskiy analiz v geomorfologii* (Nedra Press, Moscow), 155 pp.

DOORNKAMP, J. C. (Ed.). 1969. The use of computers in geomorphological research, *Brit. Geomorph. Research Group, Occasional Paper No. 6*, 90 pp.

FERRAR, A. M. 1961. The depth of some lakes in Snowdonia, *Geog. Jour.*, Vol. 127, pp. 205–8.

KRUMBEIN, W. C. and MILLER, R. L. 1953. Design of experiments for statistical analysis of geological data, *Jour. Geol.*, Vol. 61, pp. 510–32.

LeGRAND, H. E. 1962. Perspective on problems of hydrogeology, *Bull. geol. Soc. Am.*, Vol. 73, pp. 1147–52.

MACKIN, J. H. 1963. Rational and empirical methods of investigation in geology, *in* C. C. Albritton (Ed.), *The fabric of geology* (Freeman, Cooper, Stanford), pp. 135–63.

MELTON, M. A. 1958. Correlation structure of morphometric properties of drainage systems and their controlling agents, *Jour. Geol.*, Vol. 66, pp. 442–60.

MORISAWA, M. 1964. Development of drainage systems on an upraised lake floor, *Am. Jour. Sci.*, Vol. 262, pp. 340–54.

PÉGUY, C. P. 1948. Introduction à l'emploi des méthodes statistiques en géographie physique, *Rev. géog. Alpine*, Vol. 36, pp. 1–103.

PINCHEMEL, P. 1950. L'étude des réseaux hydrographiques, *Bull. Ass. géog. Fr.*, Nos 208–9, pp. 72–80.

PITTY, A. F. 1968. Some comments on the scope of slope analysis based on frequency distributions, *Zeit. f. Geomorph.*, Vol. 12, pp. 350–55.

PITTY, A. F. 1969. A scheme of hillslope analysis. 1. Initial considerations and calculations, *Univ. Hull Occ. Papers in Geog.*, No. 9, 76 pp.

SCHEIDEGGER, A. E. 1960. Mathematical methods in geology, *Am. Jour. Sci.*, Vol. 258, pp. 218–21.

SCHUMM, S. A. 1960. The effect of sediment type on the shape and stratification of some modern fluvial deposits, *Am. Jour. Sci.*, Vol. 258, pp. 177–84.

SLAYMAKER, H. O. (Ed.). 1968. Morphometric analysis of maps, *Brit. Geomorph. Res. Group Occ. Papers*, No. 4, 67 pp.

SMITH, F. G. 1966. *Geological data processing: using* FORTRAN IV (Harper and Row, New York), 284 pp.

SPATE, O. H. K. 1960, Quantity and quality in geography, *Ann. Ass. Amer. Geogrs.* Vol. 50, pp. 377–94.

STRAHLER, A. N. 1954. Statistical analysis in geomorphic research, *Jour. Geol.*, Vol. 62, pp. 1–25.

SVENSSON, H. 1959. Is the cross-section of a glacial valley a parabola?, *Jour. Glaciol.*, Vol. 3, pp. 362–3.

TRICART, J. 1947. Sur quelques indices morphométriques, *C. r. Acad. Sci.*, Vol. 225, pp. 747–9.

TROEH, F. R. 1965. Landform equations fitted to contour maps, *Am. Jour. Sci.*, Vol. 263, pp. 616–27.

WOOD, W. F. and SNELL, J. B. 1957. The dispersion of geomorphic data around measures of central tendency, *Ann. Ass. Amer. Geogrs.* Vol. 47, pp. 184–5.

5. Laboratory and field work

BROWN, E. H. 1969. The teaching of fieldwork and the integration of physical geography, *in* R. U. Cooke and J. H. Johnson (Eds), *Trends in geography: an introductory survey* (Pergamon, Oxford), pp. 70–8.

BRUUN, P. 1966. Model geology: prototype and laboratory streams, *Bull. geol. Soc. Am.*, Vol. 77, pp. 959–74.

BRUUN, P. 1968. Model geology: prototype and laboratory streams: reply, *Bull. geol. Soc. Am.*, Vol. 79, pp. 395–8.

BUTZER, K. W. and CUERDA, J. 1962. Coastal stratigraphy of southern Mallorca and its implications for the Pleistocene chronology of the Mediterranean sea, *Jour. Geol.*, Vol. 70, pp. 398–416.

CAZALIS, P. 1961. Géomorphologie et processus expérimental, *Cahiers géog. Québec*, No. 9, pp. 33–50.

DERBYSHIRE, E. 1958. The identification and classification of glacial drainage channels from aerial photographs, *Geografiska Annaler*, Vol. 40, pp. 188–95.

HOOKE, R. Le B. 1968. Model geology: prototype and laboratory streams: discussion, *Bull. geol. Soc. Am.*, Vol. 79, pp. 391–4.

KELLER, W. D. 1963. Field work – our scientific birthright, *Jour. Geol. Ed.*, Vol. 11, pp. 119–23.

KING, C. A. M. 1966. *Techniques in geomorphology* (Arnold, London), 342 pp.

MCGEE, W. J. 1897. Sheetflood erosion, *Bull. geol. Soc. Am.*, Vol. 8, pp. 87–112.

MEIER, M. F. 1967. Why study glaciers, in the context of water resources?, *Trans. Am. Geophys. Union*, Vol. 48, pp. 798–802.

OGILVIE, A. G. 1936. The earth sculpture laboratory, *Geog. Jour.*, Vol. 87, pp. 145–9.

ØSTREM, G. 1965. Problems of dating ice-cored moraines, *Geografiska Annaler*, Vol. 47A, pp. 1–38.

POUQUET, J. 1969. Géomorphologie et ère spatiale, *Zeit. f. Geomorph.*, Vol. 13, pp. 414–71.

POWELL, J. W. 1875. *Exploration of the Colorado River of the West and its tributaries* (Govt. printing office, Washington), 285 pp.

SCHACHORI, A. and SEGINER, I. 1962. Sprinkling assembly for simulation of design storms as a means for erosion and runoff studies, *Bull. Ass. Internl Hydrol. Soc.*, Vol. 7, pp. 57–72.

STEWART, A. B. 1965. Soil in the field and in the laboratory, *Jour. Soil Sci.*, Vol. 16, pp. 171–82.

YOXALL, W. H. 1969. The relationship between falling base level and lateral erosion in experimental streams, *Bull. geol. Soc. Am.*, Vol. 80, pp. 1379–84.

WURM, A. 1935. Morphologische Analysen und Experiment, *Zeit. f. Geomorph.*, Vol. 10, pp. 1–24.

6. The role of geomorphology

BAKKER, J. P. 1959. Recherches néerlandaises de géomorphologie appliquée, *Rev. géomorph. dyn.*, Vol. 10, pp. 67–84.

HOWARD, A. D. 1967. Drainage analysis in geologic interpretation: a summation, *Bull. Amer. Ass. Petrol. Geols*, Vol. 51, pp. 2246–59.

KENNEDY, W. Q. 1962. Theoretical factors in geomorphological analysis, *Geol. Mag.*, Vol. 99, pp. 304–12.

KLIMASZEWSKI, M. 1960. Problèmes concernant la carte géomorphologique détaillée, son importance et pratique, *Przeglad Geogr.*, Vol. 32, pp. 459–85.

LeGRAND, H. E. 1962. Perspective on problems of hydrogeology, *Bull. geol. Soc. Am.*, Vol. 73, pp. 1147–52.

LINTON, D. L. 1968. The assessment of scenery as a natural resource, *Scot. geog. Mag.*, Vol. 84, pp. 219–38.

TRICART, J. 1962. *L'épiderme de la Terre. Esquisse d'une géomorphologie appliquée* (Masson, Paris), 167 pp.

TRICART, J. 1966. La place de la géomorphologie dans l'étude de la mise en valeur des deltas tropicaux, *in* UNESCO, *Scientific problems of the humid tropical zone deltas and their implications* (UNESCO, Paris), pp. 15–22.

WALTON, K. *et al.* 1966. The vertical displacement of shorelines in Highland Britain, *Trans. Inst. Brit. Geogrs*, No. 39, 145 pp.

ZVEDER, L. N. 1967. Reconstructing Early Jurassic plains of accumulation from occurrences of diamonds in present drainage network, *Internl Geol. Rev.*, Vol. 10 (1968), pp. 1362–6.

C. Basic Postulates

1. Catastrophism and uniformitarianism

DOBBIE, C. H. and WOLF, P. O. 1953. The Lynmouth flood of August 1952, *Proc. Inst. Civil Eng.*, Vol. 2, pp. 522–88.

DOEGLAS, D. J. 1959. Sedimentology of recent and old sediments: a comparison, *Geol. en Mijnbouw*, Vol. 9, pp. 228–30.

DOUGLAS, I. 1964. Intensity and periodicity in denudation processes with special reference to the removal of material in solution by rivers, *Zeit. f. Geomorph.*, Vol. 8, pp. 453–73.

GOULD, S. J. 1965. Is Uniformitarianism necessary?, *Am. Jour. Sci.*, Vol. 263, pp. 223–8.

GRETENER, P. E. 1967. Significance of the rare event in geology, *Bull. Amer. Ass. Petrol. Geols*, Vol. 51, pp. 2197–206.

HARRINGTON, J. W. 1967. The first, first principles of geology, *Am. Jour. Sci.*, Vol. 265, pp. 449–61.

HUME, J. D. and SCHALK, M. 1967. Shoreline processes near Barrow, Alaska: a comparison of the normal and catastrophic, *Arctic*, Vol. 20, pp. 86–103.

KIDSON, C. 1953. The Exmoor storm and the Lynmouth floods, *Geography*, Vol. 38, pp. 1–9.

MALDE, H. E. 1968. The catastrophic Late Pleistocene Bonneville flood in the Snake River plain, Idaho, *U.S. Geol. Surv. Prof. Paper 596*, 52 pp.

MILLER, D. J. 1960. Giant waves in Lituya Bay, Alaska, *U.S. Geol. Surv. Prof. Paper 354–C*, pp. 51–86.

NAIRN, A. E. M. 1965. Uniformitarianism and environment, *Palaeogeog., Palaeoclim., Palaeoecol.*, Vol. 1, pp. 5–11.

PLAYFAIR, J. 1802. *Illustrations of the Huttonian theory of the earth* (Creech, Edinburgh), 528 pp. (Facsimile reproduction by Univ. of Illinois Press, Urbana.)

SCHUMM, S. A. and CHORLEY, R. J. 1964. The fall of Threatening Rock, *Am. Jour. Sci.*, Vol. 262, pp. 1041–54.

STEARNS, H. T. 1962. Evidence of Lake Bonneville flood along Snake river below King Hill, Idaho, *Bull. geol. Soc. Am.*, Vol. 73, pp. 385–88.

WOLMAN, M. G., and MILLER, J. P. 1960. Magnitude and frequency of forces in geomorphic processes, *Jour. Geol.*, Vol. 68, pp. 54–74.

2. Stillstands and the mobility of earth structures

AXELROD, D. I. 1962. Post-Pliocene uplift of the Sierra Nevada, California, *Bull. geol. Soc. Am.*, Vol. 73, pp. 183–98.

BADEN-POWELL, D. F. W. 1963. Isostatic recovery in Scotland, *Nature*, Vol. 199, pp. 546–7.

BLOOM, A. L. 1963. Late-Pleistocene fluctuations of sea-level and postglacial crustal rebound in coastal Maine, *Am. Jour. Sci.*, Vol. 261, pp. 862–79.

BLOOM, A. L. 1967. Pleistocene shorelines: a new test of isostasy, *Bull. geol. Soc. Am.*, Vol. 78, pp. 1477–94.

BOURDIER, F. 1959. Origines et succès d'une théore géologique illusoire: l'eustatisme appliqué au terrasses alluviales, *Rev. géomorph. dyn.*, Vol. 10, pp. 16–29.

CHAMBERLIN, T. C. 1909. Diastrophism as the ultimate basis of correlation, *Jour. Geol.*, Vol. 17, pp. 685–93.

CHORLEY, R. J. 1963. Diastrophic background to twentieth-century geomorphological thought, *Bull. geol. Soc. Am.*, Vol. 74, pp. 953–70.

CHRISTENSEN, M. N. 1965. Late Cenozoic deformation in the central Coast Ranges of California, *Bull. geol. Soc. Am.*, Vol. 76, pp. 1105–24.

CHURCHILL, D. M. 1965. The displacement of deposits formed at sea-level, 6500 years ago in southern Britain, *Quaternaria*, Vol. 7, pp. 239–49.

DIXEY, F. 1938. Some observations on the physiographical development of central and southern Africa, *Trans. geol. Soc. S.Afr.*, Vol. 41, pp. 113–71.

FAIRBRIDGE, R. W., 1960. The changing level of the sea, *Sci. American*, Vol. 204, pp. 70–9.

FAIRBRIDGE, R. W. 1961. Eustatic changes in sea level, *in* L. H. Ahrens *et al.*, *Physics and chemistry of the earth*, Vol. 4 (Pergamon, New York), pp. 99–185.

FARRAND, W. R. 1962. Postglacial uplift in North America, *Am. Jour. Sci.*, Vol. 260, pp. 181–99.

DONN, W. L., FARRAND, W. R., and EWING, M. 1962. Pleistocene ice volumes and sea-level lowering, *Jour. Geol.*, Vol. 70, pp. 206–14.

FLINT, R. F. 1966. Comparison of interglacial marine stratigraphy in Virginia, Alaska, and Mediterranean areas, *Am. Jour. Sci.*, Vol. 264, pp. 673–84.

GILLULY, J. 1966. Orogeny and geochronology, *Am. Jour. Sci.*, Vol. 264, pp. 97–111.

GUILCHER, A. 1969. Pleistocene and Holocene sea level changes, *Earth Sci. Rev.*, Vol. 5, pp. 69–97.

HAILS, J. R. 1965. A critical review of sea-level changes in eastern Australia since the Last Glacial, *Aust. geog. Studies*, Vol. 3, pp. 63–78.

KING, L. C. 1956. A geomorphological comparison between eastern Brazil and Africa (central and southern), *Quart. Jour. geol. Soc.*, Vol. 112, pp. 445–74.

KUENEN, P. H. 1950. *Marine geology* (Wiley, New York), 568 pp.

KUENEN, P. H. 1955. Sea level and crustal warping, *Geol. Soc. Am. Spec. Paper 62*, pp. 193–204.

KUKKAMÄKI, T. J. (Ed.). 1964. Symposium on Recent crustal movements in Finland with bibliography, *Fennia*, Vol. 89, No. 2, 89 pp.

LIU KUANG-YEH. 1964. Neotectonic movement of the North China platform, *Internl Geol. Rev.*, Vol. 10 (1968), pp. 857–69.

McGINNIS, L.D. 1968. Glacial crustal bending, *Bull. geol. Soc. Am.*, Vol. 79, pp. 769–76.

MILLER, A. A. 1939. Attainable standards of accuracy in the determination of pre-glacial sea levels by physiographic method, *Jour. Geomorph.*, Vol. 2, pp. 95–115.

SHALER, N. S. 1894. Pleistocene distortions of the Atlantic seacoast, *Bull. geol. Soc. Am.*, Vol. 5, pp. 199–202.

SUESS, E. 1888. *Das Antlitz der Erde* (Tempsky, Vienna), Vol. 2; trans. by H. B. C. Sollas and W. J. Sollas, 1906, *The face of the earth* (Clarendon Press, Oxford), Vol. 2, 556 pp.

TANNER, W. F. 1968. Tertiary sea level symposium – introduction, *Palaeogeog., palaeoclim., palaeoecol.*, Vol. 5, pp. 7–14.

VERSTAPPEN, H. T. 1960. On the geomorphology of raised coral reefs and its tectonic significance, *Zeit. f. Geomorph.*, Vol. 4, pp. 1–28.

WEBB, S. D. and TESSMAN, N. 1968. A Pliocene vertebrate fauna from low elevation in Manatee County, Florida, *Am. Jour. Sci.*, Vol. 266, pp. 777–811.

WELLMAN, H. W. 1966. Active wrench faults of Iran, Afghanistan and Pakistan, *Geol. Rundschau*, Vol. 55, pp. 716–35.

3. The Cycle of Erosion

BAULIG, H. 1928. Les hauts niveaux d'érosion eustatique dans le Bassin de Paris, *Ann. Géogr.*, Vol. 37, pp. 289–305 and 385–406.

BAULIG, H. 1952. Surfaces d'aplanissement, *Ann. Géogr.*, Vol. 61, pp. 161–83 and 245–62.

BRETZ, J. H. 1962. Dynamic equilibrium and the Ozark land forms, *Am. Jour. Sci.*, Vol. 260, pp. 427–38.

BROWN, E. H. 1961. Britain and Appalachia: a study in the correlation and dating of planation surfaces, *Trans. Inst. Brit. Geogrs*, No. 29, pp. 91–100.

DAVIS, W. M. 1909. The geographical cycle, *in Geographical Essays* (Ginn, New York), 777 pp.; reprinted, 1954 (Dover, Boston), pp. 249–78.

FENNEMAN, N. M. 1936. Cyclic and non-cyclic aspects of erosion, *Bull. geol. Soc. Am.*, Vol. 47, pp. 173–86.

HEMPEL, L. 1958. Probleme der Oberflächenformung in Grossbritannien unter klimamorphologischer Fragestellung, *Pet. geog. Mitt.*, Vol. 102, pp. 13–27.

HILLS, E. S. 1961. Morphotectonics and the geomorphological sciences, with special reference to Australia, *Quart. Jour. geol. Soc.*, Vol. 117, pp. 77–89.

HODGSON, J. M., CATT, J. A. and WEIR, A. H. 1967. The origin and development of Clay-with-flints and associated soil horizons on the South Downs, *Jour. Soil Sci.*, Vol. 18, pp. 85–102.

LOUIS, H. 1967. Reliefumkehr durch Rumpfflächenbildung in Tanganyika, *Geografiska Annaler*, Vol. 49A, pp. 256–67.

MACAR, P. 1955. Appalachian and Ardennes levels of erosion compared, *Jour. Geol.*, Vol. 63, pp. 253–67.

NEEF, E. 1955. Zur genese das Formenbildes der Rumpfgebirge, *Pet. geog. Mitt.*, Vol. 101, pp. 183–92.

PITTY, A. F. 1968. The scale and significance of solutional loss from the limestone tract of the southern Pennines, *Proc. Geol. Ass.*, Vol. 79, pp. 153–77.

RICH, J. L. 1938. Recognition and significance of multiple erosion surfaces, *Bull. geol. Soc. Am.*, Vol. 49, pp. 1695–722.

SPARKS, B. W. 1949. The denudation chronology of the dip-slope of the South Downs, *Proc. Geol. Ass.*, Vol. 60, pp. 165–215.

TILLEY, P. 1968. Early challenges to Davis' concept of the cycle of erosion, *Prof. Geogr*, Vol. 20, pp. 265–9.

4. Morphoclimatic zones

BÜDEL, J. 1948. Die klima-morphologischen Zonen der Polarländer, *Erdkunde*, Vol. 2, pp. 22–53.

BÜDEL, J. 1951. Klima-morphologische Beobachtungen in Süditalien, *Erdkunde*, Vol. 5, pp. 73–6.

BÜDEL, J. 1963. Klima-genetische Geomorphologie, *Geol. Rundschau*, Vol. 15, pp. 269–85.

BÜDEL, J. 1969. Das System der klimagenetischen Geomorphologie, *Erdkunde*, Vol. 23, pp. 165–83.

CHORLEY, R. J. 1957. Climate and morphometry, *Jour. Geol.*, Vol. 65, pp. 628–38.

COTTON, C. A. 1947. *Climatic accidents in landscape-making* (Whitcombe and Tombs, Wellington), 353 pp.; facsimile reproduction, 1969 (Hafner, New York).

FRYE, J. C. 1959. Climate and Lester King's 'Uniformitarian nature of hillslopes', *Jour. Geol.*, Vol. 67, pp. 111–13.

GALLI-OLIVIER, C. 1969. Climate: a primary control of sedimentation in the Peru–Chile Trench, *Bull. geol. Soc. Am.*, Vol. 80, pp. 1849–52.

HOLMES, C. D. 1956. Geomorphic development in humid and arid regions, *Am. Jour. Sci.*, Vol. 253, pp. 377–90.

KING, L. C. 1957. The uniformitarian nature of hillslopes, *Trans. Edin. geol. Soc.*, Vol. 17, pp. 81–102.

LEHMANN, H. 1954. Das Karstphänomen in verschiedenen Klimazonen, *Erdkunde*, Vol. 8, pp. 112–22.

LOUIS, H. 1957. Rumpfflächenproblem, Erosionzylkus und Klimageomorphologie, *Pet. geog. Mitt.*, Ergänzungsband 262.

PENCK, A. 1914. The shifting of the climatic belts, *Scot. geog. Mag.*, Vol. 30, pp. 281–93.

SEUFFERT, O. 1969. Klimatische und nichtklimatische Faktoren der Fussflächen-entwicklung im Bereich der Gebirgsvorländer und Grabenregionen Sardiniens, *Geol. Rundschau*, Vol. 58, pp. 98–110.

TRICART, J. 1957. Comparison entre les conditions de façonnement des lits fluviaux en zone tempérée et zone intertropicale, *C. r. Acad. Sci.*, Vol. 245, pp. 555–7.

TRICART, J. 1953. Climat et géomorphologie, *Inf. Géog.*, Vol. 17, pp. 39–51.

TRICART, J. 1961. Les caractéristiques fondamentals du système morphologique des pays tropicaux humides, *Inf. Géog.*, Vol. 25, pp. 155–69.

TRICART, J. and CAILLEUX, A. 1965. *Introduction à la géomorphologie climatique* (S.E.D.E.S.), 307 pp.

TROLL, C. 1948. Der subnivale oder periglaziale Zyklus der Denudation, *Erdkunde*, Vol. 2, pp. 1–21.

VISHER, S. S. 1937. Regional contrasts in erosion in Indiana, with especial attention to the climatic factor in causation, *Bull. geol. Soc. Am.*, Vol. 48, pp. 897–930.

5. Structure, process and stage

BRETZ, J. H. 1962. Dynamic equilibrium and the Ozark land forms, *Am. Jour. Sci.*, Vol. 260, pp. 427–38.

CHOLLEY, A. 1950. Morphologie structurale et morphologie climatique, *Ann. Géogr.*, Vol. 59, pp. 321–35.

DYLIK, J. 1957. Dynamical geomorphology. Its nature and methods, *Bull. Soc. Sci. Lettres Lodz Cl. III*, No. 8, pp. 1–42.

HACK, J. T. 1960. Interpretation of erosional topography in humid temperate regions, *Am. Jour. Sci.*, Vol. 258A (Bradley volume), pp. 80–97.

HACK, J. T. 1966. Circular patterns and exfoliation in crystalline terrane, Grandfather Mountain area, North Carolina, *Bull. geol. Soc. Am.*, Vol. 77, pp. 975–86.

HEMPEL, L. 1959. Rezente und fossile Zertalungsformen im mediterranean Spanien, *Die Erde*, Vol. 90, pp. 38–59.

HILLS, E. S. 1961. Morphotectonics and the geomorphological sciences, with special reference to Australia, *Quart. Jour. geol. Soc.*, Vol. 117, pp. 77–89.

HOLLINGWORTH, S. E. 1962. The climatic factor in the geological record, *Quart. Jour. geol. Soc.*, Vol. 118, pp. 1–21.

KING, L. C. 1953. Canons of landscape evolution, *Bull. geol. Soc. Am.*, Vol. 64, pp. 721–62.

KITTS, D. B. 1963. Historical explanation in geology, *Jour. Geol.*, Vol. 71, pp. 297–313.

MELTON, F. A. 1959. Aerial photography and structural geomorphology, *Jour. Geol.*, Vol. 67, pp. 351–71.

NIINI, H. 1967. The dependence of relief on the structure and composition of the bedrock in Western Inari, Finnish Lapland, *Fennia*, Vol. 97, No. 2, 28 pp.

SYLVESTER-BRADLEY, P. C. 1967. Evolution versus entropy. *Proc. Geol. Ass.*, Vol. 78, pp. 137–47.

TUAN, YI-FU. 1962. Structure, climate and basin land forms in Arizona and New Mexico, *Ann. Ass. Amer. Geogrs*, Vol. 52, pp. 51–68.

6. The necessity for simplification of geomorphological complexity

ALLEN, P. 1964. Sedimentological models, *Jour. Sed. Petrol.*, Vol. 34, pp. 289–93.

CHORLEY, R. J. 1962. Geomorphology and General Systems Theory. *U.S. Geol. Surv. Prof. Paper 500-B*, 10 pp.

CHORLEY, R. J. and HAGGETT, P. (Eds.). 1967. *Models in geography* (Methuen, London), 816 pp.

CONACHER, A. J. 1969. Open systems and dynamic equilibrium in geomorphology: a comment. *Aust. geog. Studies*, Vol. 7, pp. 153–8.

HOLMES, C. D. 1964. Equilibrium in humid-climate physiographic processes. *Am. Jour. Sci.*, Vol. 262, pp. 436–45.

HOOKE, R. LeB. 1968. Steady-state relationships on arid-region alluvial fans in closed basins. *Am. Jour. Sci.*, Vol. 266, pp. 609–29.

HOWARD, A. D. 1965. Geomorphological systems – equilibrium and dynamics. *Am. Jour. Sci.*, Vol. 263, pp. 302–12.

KRUMBEIN, W. C. 1968. Statistical models in sedimentology. *Sedimentology*, Vol. 10, pp. 7–23.

MILLER, J. P. 1961. Solutes in small streams draining single rock types, Sangre de Cristo Range, New Mexico, *U.S. Geol. Surv. Water-supply Paper 1535-F*, 23 pp.

OLLIER, C. D. 1968. Open systems and dynamic equilibrium in geomorphology. *Aust. geog. Studies*, Vol. 6, pp. 167–70.

SMALLEY, I. J. and VITA-FINZI, C. 1969. The concept of system in the earth sciences, particularly geomorphology, *Bull. geol. Soc. Am.*, Vol. 80, pp. 1591–4.

WRIGHT, C. W. 1958. Order and disorder in nature, *Proc. Geol. Ass.*, Vol. 69, pp. 77–82.

D. Some Conclusions about Geomorphology

ALBRITTON, C. C. (Ed.). 1963. *The fabric of geology* (Freeman and Cooper, Stanford), 372 pp.

BAULIG, H. 1957. Les méthodes de la géomorphologie d'après M. Pierre Birot. *Ann. Géog.*, Vol. 66, pp. 97–124 and 221–36.

BAULIG, H. 1958. La leçon de Grove Karl Gilbert, *Ann. Géog.*, Vol. 67, pp. 289–307.

BIROT, P. 1958. Les tendances actuelles de la géomorphologie en France, *Zeit. f. Geomorph.*, Vol. 2, pp. 123–34.

DORALL, R. D. 1968. Geomorphology – the emergence of a science, *Geographica*, Vol. 4, pp. 24–8.

DYLIK, J. 1953. Caractères du développement de la géomorphologie moderne. *Bull. Soc. Sci. Lettres Lodz Cl. III*, No. 4, 40 pp.

FOLK, R. L. and FERM, J. C. 1966. A portrait of Paul D. Krynine. *Jour. Sed. Petrol.*, Vol. 36, pp. 851–63.

GELLERT, J. 1957. Systematik und Problemstellung der physischen Geographie. *Geog. Berichte*, Vol. 2, pp. 89–102.

HARRISON, J. M. 1968. Geological sciences in the world scientific community. *Quart. Jour. geol. Soc.*, Vol. 124, pp. 1–8.

MARTIN, L. 1950. William Morris Davis: investigator, teacher and leader in geomorphology. *Ann. Ass. Amer. Geogrs.*, Vol. 40, pp. 172–80.

MILLER, J. P. 1959. Geomorphology in North America. *Przeglad Geogr.*, Vol. 31, pp. 567–86.

PEEL, R. F. 1967. Geomorphology: trends and problems. *The Advancement of Science*, Vol. 24, pp. 205–16.

RUSSELL, R. J. 1949. Geographical geomorphology. *Ann. Ass. Amer. Geogrs.*, Vol. 39, pp. 1–11.

SCHUMM, S. A. and LICHTY, R. W. 1965. Time space, and causality in geomorphology. *Am. Jour. Sci.*, Vol. 263, pp. 110–19.

STEERS, J. A. 1960. Physiography: some reflections and trends. *Geography*, Vol. 45, pp. 1–15.

WOOLDRIDGE, S. W. 1958. The trend of geomorphology. *Trans. Inst. Brit. Geogrs.*, No. 25, pp. 29–35.

Although most geomorphologists would feel out of their depth in deep space, the following articles dealing with the surface features of the Moon and Mars are examples of sources that will be of general interest to many students of terrestrial landforms:

GILBERT, G. K. 1893. The Moon's face – a study of the origin of its features. *Bull. Phil. Soc. Washington*, Vol. 12, pp. 241–92.

MACKIN, J. H. 1969. Origin of lunar maria. *Bull. geol. Soc. Am.*, Vol. 80, pp. 735–48.

MARCUS, A. H. 1968. Martian craters: number density. *Science*, Vol. 160, pp. 1333–4.

QUAIDE, W. L. and OBERBECK, V. R. 1969. Geology of the Apollo landing sites. *Earth Sci. Rev.*, Vol. 5, pp. 255–78.

RENNILSON, T. T., DRAGG, D. L., MORRIS, E. C., SHOEMAKER, E. M., and TURKEVICH, A. 1966. Lunar surface topography *in* Surveyor I Mission report – Part II. Scientific data and results (*Jet Propulsion Laboratory Tech. Rep.* 32–1023, Pasadena), pp. 7–44.

RONCA, L. B. 1966. An introduction to the geology of the moon. *Proc. Geol. Ass.*, Vol. 77, pp. 101–25.

RONCA, L. B. 1969. Recent advances in lunar geology. *Proc. Geol. Ass.*, Vol. 80, pp. 365–78.

RONCA, L. B. 1970. An introduction to the geology of Mars, *Proc. Geol. Assoc.*, Vol. 81, pp. 111–28.

RONCA, L. B. and GREEN, R. R. 1970. Statistical geomorphology of the lunar surface, *Bull. geol. Soc. Am.*, Vol. 80, pp. 337–52.

SHTEYNBERG, G. S. 1965. Morphology of moon craters and cirques, compared to certain volcanic formations of Kamchatka, *Internl. Geol. Rev.*, Vol. 8 (1966), pp. 1440–50.

Chapter II. Landforms and structure

A. Geophysical Considerations

1. Earth movements and mountain building

BEMMELEN, R. W. VAN. 1968. On the origin and evolution of the earth's crust, *Geol. Rundschau*, Vol. 51, pp. 657–705.

BILLINGS, M. P. 1960. Diastrophism and mountain building, *Bull. geol. Soc. Am.*, Vol. 71, pp. 363–98.

ELDER, J. W. 1968. Convection: the key to dynamical geology, *Sci. Prog.*, Vol. 56, pp. 1–33.

GASKELL, T. F. 1970. *Physics of the earth* (Thames and Hudson, London), 216 pp.

GILLULY, J. 1963. The tectonic evolution of the western United States, *Quart. Jour. geol. Soc.*, Vol. 119, pp. 133–74.

HARLAND, W. B. 1969. The origin of continents and oceans, *Geol. Mag.*, Vol. 106, pp. 100–4.

HEEZEN, B. C., THORP, M., and EWING, M. 1959. The floors of the oceans. I. The North Atlantic, *Geol. Soc. Am. Special Paper*, 65, 122 pp.

HILLS, E. S. 1963. Outlines of structural geology (Methuen, London), 483 pp.

KUENEN, P. H. 1967. Geosynclinal sedimentation, *Geol. Rundschau*, Vol. 56, pp. 1–19.

LOVERING, J. F. 1958. The nature of the Mohorovičić discontinuity, *Trans. Am. geophys. Union*, Vol. 39, pp. 947–55.

MESCHERIKOV, J. A. 1959. Contemporary movements in the earth's crust, *Internl. geol. Rev.*, Vol. 1, pp. 40–51.

MILNES, A. G. 1969. On the orogenic history of the central Alps, *Jour. Geol.*, Vol. 77, pp. 108–12.

POLDERVAART, A. 1955. The crust of the earth. A symposium, *Geol. Soc. Am. Spec. Paper* 62, 767 pp.

RÜCKLIN, H. 1963. Die Entstehung des Gross-reliefs der Erde, *Geog. Zeit.*, Vol. 51, pp. 183–238.

SITTER, U. DE. 1954. Gravitational gliding tectonics, *Am. Jour. Sci.*, Vol. 252, pp. 371–95.

THOMPSON, G. A. and TALWANI, M. 1964. Crustal structure from Pacific Basin to Central Nevada, *Jour. geophys. Res.*, Vol. 69, pp. 4813–37.

TRÜMPY, R. 1960. Paleotectonic evolution of the central and western Alps, *Bull. geol. Soc. Am.*, Vol. 71, pp. 843–907.

WYLLIE, P. 1963. The Mohorovičić discontinuity and the orogenic cycle, *Trans. Amer. geophys. Union*, Vol. 44, pp. 1064–71.

WYLLIE, P. 1965. A modification of the geosyncline and tectogene-hypothesis, *Geol. Mag.*, Vol. 102, pp. 231–45.

2. Continental drift

BULLARD, E. C. 1964. Continental drift, *Quart. Jour. geol. Soc.*, Vol. 120, pp. 1–33.

GARLAND, G. C. 1966. Continental drift, *Roy. Soc. Canada Spec. Publ.*, No. 9, 140 pp.

HARLAND, W. B. 1967. Tectonic aspects of continental drift, *Sci. Progress*, Vol. 55, pp. 1–14.

HEEZEN, B. C. 1960. The rift in the ocean floor, *Sci. American*, Vol. 205, pp. 99–111.

HURLEY, P. M. 1968. The confirmation of continental drift, *Sci. American*, Vol. 218, pp. 52–64.

MAACK, R. 1969. *Kontinental drift und Geologie des südatlantischen Ozeans* (Gruyter, Berlin), 164 pp.

RAVEN, T. 1960. Alpine folding as related to continental drift, *Eclogae Geol. Helv.*, No. 1, pp. 161–8.

RUNCORN, S. K. (Ed.). 1962. *Continental drift* (Academic Press, New York), 338 pp.

SHERLOCK, R. L. 1934. Notes on the Amazon, *Geol. Mag.*, Vol. 71, pp. 112–16.

STONELEY, R. 1966. The Niger delta region in the light of the theory of continental drift, *Geol. Mag.*, Vol. 103, pp. 385–97.

TOIT, A. L. DU. 1937. *Our wandering continents* (Oliver and Boyd, Edinburgh), 366 pp. (reprinted 1957).

WEGENER, A. 1922. *Die Entstehung der Kontinente und Ozeane*, 3rd ed. (Vieweg, Braunschweig), 144 pp.; translated by J. G. A. Skerl, 1924, *The origin of continents and oceans* (Methuen, London), 212 pp.

3. Vulcanicity

ALSOP, L. E. and OLIVER, J. E. (Eds). 1969. Premonitory phenomena associated with several recent earthquakes and related problems, *Trans. Am. Geophys. Union*, Vol. 50, pp. 376–410.

BULLARD, F. M. 1962. *Volcanoes in history, in theory, in eruption* (Univ. of Texas Press, Austin), 441 pp.

BENIOFF, H. 1951. Earthquakes and rock creep, *Bull. Seismol. Soc. Am.*, Vol. 41, pp. 31–62.

BOUT, P. and DERRUAU, M. 1966. Recherches sur les volcans explosifs du Japon, *Mém. et Doc. Centre Rech. Doc. Cartog. géog. Ed. C.N.R.S.*, Vol. 10, No. 4, 89 pp.

BULLEN, K. E. 1954. *Seismology* (Methuen, London), 132 pp.

COATS, R. R., HAY, R. L., and ANDERSON, C. A. 1968. Studies in volcanology, *Geol. soc. Am. Mem. 116*, 678 pp.

COTTON, C. A. 1952. *Volcanoes as landscape forms* (Whitcombe and Tombs, Christchurch), 416 pp.; facsimile reproduction, 1969 (Hafner, New York).

DOLLAR, A. T. J. 1957. The Midlands earthquake of February 11, 1957, *Nature*, Vol. 179, pp. 507–10.

EATON, J. P. and MURATA, K. J. 1960. How volcanoes grow, *Science*, Vol. 132, pp. 925–38.

FISKE, R. S. and KOYANAGI, R. Y. 1968. The December 1965 eruption of Kilauea volcano, Hawaii, *U.S. Geol. Surv. Prof. Paper 607*, 21 pp.

GUTENBERG, B. and RICHTER, C. F. 1954. *Seismicity of the earth* (Princeton Univ. Press, Princeton), 310 pp.

JONES, J. G. 1968. Intraglacial volcanoes of the Laugarvatn region, south-west Iceland, *Quart. Jour. geol. Soc.*, Vol. 124, pp. 197–212.

KARNIK, V. 1969. *Seismicity of the European area* (Reidel, Dordrecht), Part I, 364 pp.

MACDONALD, G. A. 1953. Pahoehoe, Aa, and block lava, *Am. Jour. Sci.*, Vol. 251, pp. 169–91.

MCBIRNEY, A. R. 1959. Factors governing emplacement of volcanic necks, *Am. Jour. Sci.*, Vol. 257, pp. 431–48.

NEUMANN VAN PADANG, M. 1955. Present position regarding the catalogue of the active volcanoes of the world, *Bull. volcanologique* 17, 141–4.

RANDALL, B. A. O. 1959. Intrusive phenomena of the Whin Sill, east of the River North Tyne, *Geol. Mag.*, Vol. 96, pp. 385–92.

RICHTER, C. F. 1958. *Elementary seismology* (Freeman, San Francisco), 768 pp.

RITSEMA, A. R. 1969. Seismo-tectonic implications of a review of European earthquake mechanisms, *Geol. Rundschau*, Vol. 59, pp. 36–56.

RITTMAN, A. 1960. *Vulkane und ihre Tatigkeit* (Enke, Stuttgart), 336 pp.; trans. by E. A. Vincent, 1962, *Volcanoes and their activity* (Interscience, New York), 305 pp.

SMITH, R. L. 1960. Ash flows, *Bull. geol. Soc. Am.*, Vol. 71, pp. 795–842.

THORARLINSSON, S. 1968. Some problems of volcanism in Iceland, *Geol. Rundschau*, Vol. 57, pp. 1–20.

YAGI, K., KAWANO, Y., and AOKI, K. 1963. Types of quaternary activity in northeastern Japan, *Bull. volcanologique*, Vol. 26, pp. 223–35.

WENTWORTH, C. K. 1927. Estimates of marine and fluvial erosion in Hawaii, *Jour. Geol.*, Vol. 35, pp. 117–33.

WENTWORTH, C. K. 1954. The physical behaviour of basaltic lava flow, *Jour. Geol.*, Vol. 62, pp. 425–38.

WILLIAMS, H. 1954. Problems and progress in volcanology, *Quart. Jour. geol. Soc.*, Vol. 109, pp. 311–32.

4. Hydrothermal activity and meteorite craters

CASSIDY, W. A. *et al.* 1965. Meteorites and craters of Campo del Cielo, Argentina, *Science*, Vol. 149, pp. 1055–64.

ESCHER, B. G. 1955. Three caldera-shaped accidents: volcanic calderas, meteoric scars and lunar cirques, *Bull. volcanologique*, Vol. 16, pp. 55–70.

FLEISCHER, R. L., PRICE, P. B., and WOODS, R. T. 1969. A second tektite fall in Australia, *Earth and Planetary Sci. Letters*, Vol. 7, pp. 51–2.

HELLYER, B. 1969. Statistics of meteorite falls, *Earth and Planetary Sci. Letters*, Vol. 7, pp. 148–50.

KRINOV, E. L. 1961. The Kaalijarv meteorite craters on Saarema island, Estonian SSR, *Am. Jour. Sci.*, Vol. 259, pp. 430–40.

SPENCER, L. J. 1933. Meteorite craters as topographical features on the earth's surface, *Geog. Jour.*, Vol. 81, pp. 227–48.

RINEHART, J. S. 1969. Old Faithful geyser performance 1870 through 1966, *Bull. volcanologique*, Vol. 33, pp. 153–63.

WHITE, D. E. 1957. Thermal waters of volcanic origin, *Bull. geol. Soc. Am.*, Vol. 68, pp. 1637–58.

WHITE, D. E. 1967. Some principles of geyser activity, mainly from Steamboat Springs, Nevada, *Am. Jour. Sci.*, Vol. 265, pp. 641–84.

B. Geological Considerations

AUDEN, J. B. 1954. Drainage and fracture patterns in north-west Scotland, *Geol. Mag.*, Vol. 91, pp. 337–51.

BRADDOCK, W. A. and EICHER, D. L. 1962. Block-glide landslides in the Dakota group of the Front Range foothills, Colorado, *Bull. geol. Soc. Am.*, Vol. 73, pp. 317–24.

BRADLEY, W. C. 1963. Large scale exfoliation in massive sandstones of the Colorado plateau, *Bull. geol. Soc. Am.*, Vol. 74, pp. 519–28.

CAINE, N. 1967. The tors of Ben Lomond, Tasmania, *Zeit. f. Geomorph.*, Vol. 11, pp. 418–29.

CAMERMAN, C. 1953. La gelivité des matériaux pierreux, *Bull. Soc. Belge Géol.*, Vol. 62, pp. 17–34.

CHAPMAN, C. A. 1958. Control of jointing by topography, *Jour. Geol.*, Vol. 66, pp. 552–8.

CORBEL, J. 1957. Les Karsts du nord-ouest de l'Europe, *Inst. Études Rhodaniennes, Mém. et Doc.*, Lyon, 12, 541 pp.

EPSTEIN, J. B. 1967. Structural control of wind gaps and water gaps and of stream capture in the Stroudsburg Area, Pennsylvania and New Jersey, *U.S. Geol. Surv. Prof. Paper 550-B*, pp. B80–B86.

FLINT, R. F. 1963. Altitude, lithology and the Fall Zone in Connecticut, *Jour. Geol.*, Vol. 71, pp. 683–97.

GILBERT, G. K. 1895. Lake basins created by wind erosion, *Jour. Geol.*, Vol. 3, pp. 47–9.

HANCOCK, P. L. 1968. Joints and faults: the morphological aspects of their origins, *Proc. Geol. Ass.*, Vol. 79, pp. 141–51.

HARRISON, K. and THACKERAY, A. D. 1940. On the direction of certain valleys, *Geol. Mag.*, Vol. 77, pp. 82–8.

HILL, P. A. 1966. Joints: their initiation and propagation with respect to bedding, *Geol. Mag.*, Vol. 103, pp. 276–9.

HODGSON, R. A. 1961. Regional study of jointing in Comb Ridge–Navajo Mountain area, Arizona and Utah, *Bull. Amer. Ass. Petrol. Geols.*, Vol. 45, pp. 1–38.

HODGSON, R. A. 1965. Genetic and geometric relations between structures in basement and overlying sedimentary rocks, with examples from Colorado Plateau and Wyoming, *Bull. Amer. Ass. Petrol. Geols.*, Vol. 49, pp. 935–49.

HOLLINGWORTH, S. E., TAYLOR, J. H., and KELLAWAY, G. A. 1944. Large-scale superficial structures in the Northamptonshire ironstone field, *Quart. Jour. geol. Soc.*, Vol. 100, pp. 1–44.

KALTERHERBERG, J. VON and KÜHN-VELTEN, H. 1967. Klüfte und Talrichtungen im Turon des südöstlichen Münsterlandes, *Geol. Rundschau*, Vol. 56, pp. 726–48.

LIEBLING, R. S. and KERR, P. F. 1965. Observations on quickclay, *Bull. geol. Soc. Am.*, Vol. 76, pp. 853–78.

MARR, J. E. 1906. The influence of the geological structure of English Lakeland upon its present features – a study in physiography, *Quart. Jour. geol. Soc.*, Vol. 62, pp. lxvi–cxxviii.

MILLE, G. DE, SHOULDICE, J. R., and NELSON, H. W. 1964. Collapse structures related to evaporites of the Prairie formation, Saskatchewan, *Bull. geol. Soc. Am.*, Vol. 75, pp. 307–16.

NICKELSEN, R. P. and HOUGH, V. N. D. 1967. Jointing in the Appalachian plateau of Pennsylvania, *Bull. geol. Soc. Am.*, Vol. 78, pp. 609–30.

NICHOLSON, R. 1963. A note on the relation of rock fracture and fiord direction, *Geografiska Annaler*, Vol. 45, pp. 303–4.

O'BRIEN, C. A. E. 1957. Salt diapirism in south Persia, *Geologie en Mijnbouw* (The Hague), Vol. 19, pp. 357–76.

PLAFKER, G. 1964. Oriented lakes and lineaments in northeastern Bolivia, *Bull. geol. Soc. Am.*, Vol. 75, pp. 503–22.

PRICE, N. J. 1959. Mechanics of jointing in rocks, *Geol. Mag.*, Vol. 96, pp. 149–67.

RANDALL, B. A. O. 1961. On the relationship of valley and fiord directions to fracture pattern of Lyngen, Troms, northern Norway, *Geografiska Annaler*, Vol. 43, pp. 336–8.

ROBERTSON, E. C. 1955. Experimental study of the strength of rocks, *Bull. geol. Soc. Am.*, Vol. 66, pp. 1275–314.

SCHMITTHENNER, H. 1954. Die Regeln der morphologischen Gestaltung im Schichtstufenland, *Pet. geog. Mitt.*, Vol. 98, pp. 3–10.

SELWOOD, E. B. and COE, K. 1963. Large-scale terminal curvature affecting the cliffs west of Castletown Berehaven, west Cork, *Proc. Geol. Ass.*, Vol. 74, pp. 461–5.

SWEETING, M. M. 1955. The landforms of north-west Co. Clare, Ireland, *Trans. Inst. Brit. Geogrs*, No. 21, pp. 33–49.

WILSON, G. 1952. The influence of rock structures on coastline and cliff development around Tintagel, north Cornwall, *Proc. Geol. Ass.*, Vol. 63, pp. 20–48.

WISE, D. U. 1964. Microjointing in basement, Middle Rocky Mountains of Montana and Wyoming, *Bull. geol. Soc. Am.*, Vol. 75, pp. 287–306.

WRIGHT, M. D. 1964. Cementation and compaction of the Millstone Grit of the central Pennines, England, *Jour. Sed. Petrol.*, Vol. 34, pp. 756–60.

ZISCHINSKY, U. 1969. Über Bergzerreissung und Talzusschub, *Geol. Rundschau*, Vol. 58, pp. 974–83.

Chapter III

A and B. Selected References on Climate, Weather, and Hydrology

BARRY, R. G. and CHORLEY, R. J. 1968. *Atmosphere, weather and climate*, 2nd ed., 1971 (Methuen, London), 319 pp.

BECKINSALE, R. P. 1957. The nature of tropical rainfall, *Tropical Agriculture*, Vol. 34, pp. 76–98.

BRUCE, J. P. and CLARK, R. H. 1966. *Introduction to hydrometeorology* (Pergamon, Oxford), 319 pp.

FOURNIER, F. 1949. Les facteurs climatiques de l'érosion du sol, *Bull. Ass. géog. Fr.*, No. 202–3, pp. 97–103.

GEIGER, R. 1967. *Das Klima der bodennehen Luftschict* (Vieweg, Brunswick), 646 pp.; trans. by Scripta Technica Inc., 1965, *The climate near the ground* (Harvard Univ. Press, Cambridge, Mass.), 611 pp.

HUTCHINSON, G. E. 1957. *A treatise on limnology* (Wiley, New York), Vol. 1, 1015 pp.

JOHNSTON, G. H., and BROWN, R. J. E. 1961. Effect of a lake on distribution of permafrost in the Mackenzie River delta, *Nature*, Vol. 92, pp. 251–2.

LANGBEIN, W. B. and SCHUMM, S. A. 1958. Yield of sediment in relation to mean annual precipitation, *Trans. Am. geophys. Union*, Vol. 39, pp. 1076–84.

LINTON, D. L. 1959. River flow in Great Britain 1955–56, *Nature*, Vol. 183, pp. 714–16.

MEINZER, O. E., *et al.* 1942. *Hydrology* (McGraw-Hill, New York), 712 pp.

MUNN, R. E. 1966. *Descriptive micrometeorology* (Academic Press, New York), 245 pp.

SUTTON, O. G. 1960. *Understanding weather* (Penguin, Harmondsworth), 215 pp.

WILLIAMS, P. J. 1961. Climatic factors controlling the distribution of certain frozen ground phenomena, *Geografiska Annaler*, Vol. 43, pp. 339–47.

WARD, R. C. 1967. *Principles of hydrology* (McGraw-Hill, London), 402 pp.

B. Amounts and Motions of Water and Ice

BASCOM, W. N. 1959. Ocean waves, *Sci. American*, Vol. 201, pp. 74–84.

BASCOM, W. N. 1964. *Waves and beaches. The dynamics of the ocean surface* (Doubleday, New York), 267 pp.

DARBYSHIRE, J. 1955. An investigation of storm waves in the north Atlantic Ocean, *Proc. Roy. Soc. A*, Vol. 230, pp. 560–9.

DEFANT, A. 1953. *Ebbe und Flut des Meeres der Atmosphäre und der Erdfeste* (Springer-Verlag, Berlin); trans. by A. J. Pomerans, 1958, *Ebb and Flow* (Univ. of Michigan, Ann Arbor), 121 pp.

DRAPER, L. and DOBSON, P. J. 1965. Rip currents on a Cornish beach, *Nature*, Vol. 206, p. 1249.

HOYT, W. G. and LANGBEIN, W. B. 1955. *Floods* (Princeton Univ. Press, Princeton), 469 pp.

KAMB, B. 1964. Glacier geophysics, *Science*, Vol. 146, pp. 353–65.

KAMB, B. and LA CHAPELLE, E. 1964. Direct observation of the mechanism of glacier sliding over bedrock, *Jour. Glaciol.*, Vol. 5, pp. 159–72.

KIDSON, C. 1953. The Exmoor storm and the Lynmouth floods, *Geography*, Vol. 38, pp. 1–9.

LLIBOUTRY, L. 1965. How glaciers move, *New Scientist*, Vol. 28, pp. 734–6.

LLIBOUTRY, L. 1968. General theory of subglacial cavitation and sliding of temperate glaciers, *Jour. Glaciol.*, Vol. 7, pp. 21–58.

MATHEWS, W. H. 1964. Water pressure under a glacier, *Jour. Glaciol.*, Vol. 5, pp. 235–40.

MEIER, M. F. and POST, A. 1969. What are glacier surges?, *Canad. Jour. Earth Sci.*, Vol. 6, pp. 807–17.

MELLOR, M. 1959. Ice-flow in Antarctica, *Jour. Glaciol.*, Vol. 3, pp. 377–84.

MORISAWA, M. 1968. *Streams, their dynamics and morphology* (McGraw-Hill, New York), 175 pp.

NYE, J. F. 1952. The mechanics of glacier flow, *Jour. Glaciol.*, Vol. 2, pp. 81–93.

OSTENSO, N. A., SELLMANN, P. V., and PÉWÉ, T. L. 1965. The bottom topography of Gulkana glacier, Alaska Range, Alaska, *Jour. Glaciol.*, Vol. 5, pp. 651–60.

PATERSON, W. S. P. 1969. *The physics of glaciers* (Pergamon, Oxford), 250 pp.

SHARP, R. P. 1954. Glacier flow: a review, *Bull. geol. Soc. Am.*, Vol. 65, pp. 821–38.

STENBORG, T. 1969. Studies of the internal drainage of glaciers, *Geografiska Annaler*, Vol. 51A, pp. 13–41.

TRICKER, R. A. R. 1964. *Bores, breakers, waves and wakes* (Mills and Boon, London), 250 pp.

WEERTMAN, J. 1957. On the sliding of glaciers, *Jour. Glaciol.*, Vol. 3, pp. 33–8.

WEERTMAN, J. 1964. The theory of glacier sliding, *Jour. Glaciol.*, Vol. 5, pp. 287–303.

C. Mechanical and Frictional Forces

BAGNOLD, R. A. 1941. *The physics of blown sand and desert dunes*, reprinted, 1960 (Methuen, London), 265 pp.

CHORLEY, R. J. 1959. The geomorphic significance of some Oxford soils, *Am. Jour. Sci.*, Vol. 257, pp. 503–15.

CULLING, W. E. H. 1963. Soil creep and the development of hillside slopes, *Jour. Geol.*, Vol. 71, pp. 127–61.

KING, L. C. 1957. The uniformitarian nature of hillslopes, *Trans. Edin. geol. Soc.*, Vol. 17, pp. 81–102.

KIRKBY, M. J. 1967. Measurement and theory of soil creep, *Jour. Geol.*, Vol. 75, pp. 359–78.

SCHEIDEGGER, A. E. 1961. *Theoretical geomorphology* (Springer, Berlin), 333 pp.

SCHEIDEGGER, A. E. 1964. Some implications of statistical mechanics in geomorphology, *Bull. Internl. Ass. Sci. Hydrol.*, Vol. 9, pp. 12–16.

STRAHLER, A. N. 1952. Dynamic basis of geomorphology, *Bull. geol. Soc. Am.*, Vol. 63, pp. 923–38.

D. Sediments and Mechanical Characteristics of Soils

ANDEL, T. H. VAN. 1959. Reflections on the interpretation of heavy minerals, *Jour. Sed. Petrol.*, Vol. 29, pp. 153–63.

ANDERSON, H. W. 1951. Physical characteristics of soils related to erosion, *Jour. Soil and Water Conservation*, Vol. 6, pp. 129–33.

AVENARD, J. M. 1962. La solifluxion ou quelques méthodes de mécanique des sols appliquées au problème géomorphologique des versants (S.E.D.E.S., Paris), 164 pp.

BAKKER, J. P. 1957. Quelques aspects du problème des sédiments corrélatifs en climat tropical humid, *Zeit. f. Geomorph.*, Vol. 1, pp. 1–43.

BAVER, L. D. 1948. *Soil physics* (Wiley, New York), 398 pp.

BEAUJEU-GARNIER, J. 1955. Sur la présence de formation de type dit 'périglaciaire' en Algérie orientale. *C. r. Acad. Sci.*, Vol. 240, pp. 1246–8.

BERTHOIS, L. 1950. Méthode d'étude des galets. Application à l'étude de l'évolution des galets marins actuels, *Rev. géomorph. dyn.*, Vol. 1, pp. 199–225.

BERTHOIS, L. 1959. *Techniques d'analyses granulométriques* (Centre Doc. Univ., Paris), 64 pp.

BLACK, R. F. 1954. Permafrost: a review, *Bull. geol. Soc. Am.*, Vol. 65, pp. 839–56.

BREWER, R. 1964. *Fabric and mineral analysis of soils* (Wiley, New York), 470 pp.

BRYAN, R. B. 1968. The development, use and efficiency of indices of soil erodibility, *Geoderma*, Vol. 2, pp. 5–26.

CAILLEUX, A. and TAYLOR, G. 1954. *Cryopédologie: étude des sols gelés* (Hermann, Paris), 218 pp.

CAILLEUX, A. and TRICART, J. 1959. *Initiation à l'étude des sables et des galets* (Centre Doc. Univ., Paris), Vol. I, 376 pp.; Vol. II, 194 pp.; Vol. III, 202 pp.

CAPPER, L. and CASSIE, W. F. 1969. *The mechanics of engineering soils*, 5th ed. (Spon, London), 309 pp.

CLARK, M. J., LEWIN, J., and SMALL, R. J. 1967. The sarsen stones of the Marlborough Downs and their geomorphological implications, *Univ. Southampton, Research Series in Geog.*, No. 4, pp. 3–40.

CZEPPE, Z. 1959. Remarks on frost-heave, *Czasopismo Geogr.*, Vol. 30, pp. 195–202.

DOEGLAS, D. J. 1968. Grain-size indices, classification and environment, *Sedimentology*, Vol. 10, pp. 83–100.

DYLIK, J. 1963. Nouveaux problèmes du pergélisol pléistocene, *Acta Geogr. Lodz*, No. 17, 93 pp.

EASTERBROOK, D. J. 1964. Void ratios and bulk densities as means of identifying Pleistocene tills, *Bull. geol. Soc. Am.*, Vol. 75, pp. 745–50.

FOLK, R. L. 1962. Of skewnesses and sands, *Jour. Sed. Petrol.*, Vol. 32, pp. 145–6.

FOLK, R. L. and WARD, W. C. 1957. Brazos river bar: a study of the significance of grain-size parameters, *Jour. Sed. Petrol.*, Vol. 27, pp. 3–26.

FOLK, R. L. and WEAVER, C. E. 1952. A study of the texture and composition of chert, *Am. Jour. Sci.*, Vol. 250, pp. 498–510.

FRIEDMAN, G. M. 1961. Distinction between dune, beach and river sands from their textural characteristics, *Jour. Sed. Petrol.*, Vol. 31, pp. 514–29.

FRIEDMAN, G. M. 1967. Dynamic processes and statistical parameters compared for size frequency distribution of beach and river sands, *Jour. Sed. Petrol.*, Vol. 37, pp. 327–54.

GRIFFITHS, J. C. 1967. *Scientific method in analysis of sediments* (McGraw-Hill, New York), 508 pp.

HAILS, J. R. 1964. A reappraisal of the nature and occurrence of heavy minerals along parts of the east Australian coast, *Aust. Jour. Sci.*, Vol. 27, pp. 22–3.

HAILS, J. R. 1967. Significance of statistical parameters for distinguishing sedimentary environments in New South Wales, Australia, *Jour. Sed. Petrol.*, Vol. 37, pp. 1059–69.

HERDAN, G. 1960. *Small particle statistics,* 2nd rev. ed. (Butterworths, London), 418 pp.

INMAN, D. L. 1952. Measures for describing the size distribution of sediments, *Jour. Sed. Petrol,* Vol. 22, pp. 125–45.

IRANI, R. R. and CLAYTON, F. C. 1963. *Particle size: measurement, interpretation and application* (Wiley, New York), 165 pp.

JOHANSSON, C. E. 1963. Orientation of pebbles in running water. A laboratory study, *Geografiska Annaler,* Vol. 45, pp. 85–112.

JUNGERIUS, P. D. 1965. Some aspects of the geomorphological significance of soil textures, in Eastern Nigeria, *Zeitschrift für Geomorphologie,* Vol. 9, pp. 332–45.

KITTLEMAN, L. R. 1964. Application of Rosin's distribution in size frequency analysis of clastic rocks, *Jour. Sed. Petrol.,* Vol. 34, pp. 483–502.

KÖSTER, E. 1962. Möglichkeiten und Grenzen granulometrischer und morphometrischer Untersuchungsmethoden in der geographischen und geologischen Forschung, *Pet. geog. Mitt.,* Vol. 106, pp. 111–15.

KRINSLEY, D. H. and DONAHUE, J. 1968. Environmental interpretation of sand grain surface textures by electron microscopy, *Bull. geol. Soc. Am.,* Vol. 79, pp. 743–8.

KRINSLEY, D. H. and DONAHUE, J. 1968. Pebble surface textures, *Geol. Mag.,* Vol. 105, pp. 521–5.

KRINSLEY, D. H. and FUNNELL, B. M. 1965. Environmental history of sand grains from the Lower and Middle Pleistocene of Norfolk, England, *Quart. Jour. geol. Soc.,* Vol. 121, pp. 435–61.

KRUMBEIN, W. C. 1938. Size frequency distribution of sediments and the normal *phi* curve, *Jour. Sed. Petrol.,* Vol. 8, pp. 84–90.

LOVEDAY, J. 1962. Plateau deposits of the southern Chiltern Hills, *Proc. Geol. Ass.,* Vol. 73, pp. 83–102.

MCCAMMON, R. B. 1962. Efficiencies of percentile measurements for describing the mean size and sorting of sedimentary particles, *Jour. Geol.,* Vol. 70, pp. 453–65.

MCCANN, S. B. 1962. Some supposed 'raised beach' deposits at Corran, Loch Linnhe, and Loch Etive, *Geol. Mag.,* Vol. 99, pp. 131–42.

MELTON, M. A. 1965. The geomorphic and paleoclimatic significance of alluvial deposits in southern Arizona, *Jour. Geol.,* Vol. 73, pp. 1–38.

MOSS, A. J. 1962. The physical nature of common sandy and pebbly deposits. Part I, *Am. Jour. Sci.,* Vol. 260, pp. 337–73.

MOSS, A. J. 1963. The physical nature of common sandy and pebbly deposits. Part II, *Am. Jour. Sci.,* Vol. 261, pp. 297–343.

PAAS, W. 1962. Rezente und fossile Böden auf niederrheinischen Terrassen und deren Deckschichten, *Eiszeitalter und Gegenwart,* Vol. 12, pp. 165–230.

PENNY, L. F. and CATT, J. A. 1967. Stone orientation and other structural features of tills in East Yorkshire, *Geol. Mag.,* Vol. 104, pp. 344–60.

PETTIJOHN, F. J. 1957. *Sedimentary rocks* (Harper, New York), 718 pp.

PFEIFFER, H. 1965. Schwermineralseifen an der südlichen Ostsee-Küste, *Baltica*, Vol. 2, pp. 205–13.

RAGG, J. M. and BIBBY, J. S. 1966. Frost weathering and solifluction deposits in southern Scotland, *Geografiska Annaler*, Vol. 48A, pp. 12–23.

RICHTER, K. 1932. Die Bewegungsrichtung des Inlandeises, rekonstruiert aus den Kritzen und Längsachsen der Geschiebe, *Zeit. f. Geschiebeforschung*, Vol. 8, pp. 62–6.

SANDFORD, K. S. 1965. Notes on the gravels of the Upper Thames floodplain between Lechlade and Dorchester, *Proc. Geol. Ass.*, Vol. 76, pp. 61–75.

SHEPARD, F. P. and YOUNG, R. 1961. Distinguishing between beach and dune sands, *Jour. Sed. Petrol.*, Vol. 31, pp. 196–214.

SKEMPTON, A. W. 1953. Soil mechanics in relation to geology, *Proc. Yorks. geol. Soc.*, Vol. 29, pp. 33–62.

THAMES, J. L. and URSIC, S. J. 1960. Runoff as a function of moisture storage capacity, *Jour. geophys. Res.*, Vol. 2, pp. 651–4.

THEBAULT, J. 1968. Contribution à l'étude des galets, *Rev. géomorph. dyn.*, Vol. 18, pp. 49–72.

TERZAGHI, K. 1965. *Theoretical soil mechanics* (Wiley, New York), 510 pp.

VISHER, G. S. 1969. Grain size distributions and depositional processes, *Jour. Sed. Petrol.*, Vol. 39, pp. 1074–106.

VISTELIUS, A. B. 1958. *Strukturnye diagrammy* (Izd. Akad. Nauk S.S.S.R., Moscow), 157 pp.; trans. by R. Baker, 1966, *Structural diagrams* (Pergamon, Oxford), 178 pp.

WARD, W. H. 1945. The stability of natural slopes, *Geog. Jour.*, Vol. 105, pp. 170–97.

WAUGH, B. 1965. A preliminary electron microscope study of the development of authigenic silica in the Penrith sandstone, *Proc. Yorks. geol. Soc.*, Vol. 35, pp. 59–69.

WEBSTER, R. 1966. The drifts in the Vale of White Horse of north Berkshire and Wiltshire, *Proc. Geol. Ass.*, Vol. 77, pp. 255–62.

WENTWORTH, C. K. 1922. A field study of the shapes of river pebbles, *U.S. Geol. Surv. Bull. 730*, pp. 103–15.

E. Geochemical Considerations

BUTLER, J. R. 1957. The geochemistry and mineralogy of rock weathering, II, the Nordmarka area, *Geochemica, Cosmochemica Acta*, Vol. 6, pp. 268–81.

CORNWALL, I. W. 1958. *Soils for the archaeologist* (Dent, London), 230 pp.

FIELDES, M. and SWINDALE, L. D. 1954. Chemical weathering of silicates in soil formation, *N.Z. Jour. Sci. Tech. B*, Vol. 36, pp. 140–54.

FREDERICKSON, A. F. 1951. Mechanism of weathering, *Bull. geol. Soc. Am.*, Vol. 62, pp. 221–32.

GRIM, R. E. 1968. *Clay mineralogy*, 2nd ed. (McGraw-Hill, New York), 596 pp.

HARRISS, R. C. and ADAMS, J. A. S. 1966. Geochemical and mineralogical studies on the weathering of granitic rocks, *Am. Jour. Sci.*, Vol. 264, pp. 146–73.

JAEGER, J. L. 1957. *La géochimie* (Presses Univ. France, Paris), 119 pp.

KELLER, W. D. 1957. *The principles of chemical weathering* (Lucas, Columbia), 111 pp.

KELLER, W. D. and REESMANN, A. L. 1963. Dissolved products of artificially pulverised silicate minerals and rocks: Part II, *Jour. Sed. Petrol.*, Vol. 33, pp. 426–37.

LOUGHNAN, F. C. 1962. Some considerations in the weathering of the silicate minerals, *Jour. Sed. Petrol.*, Vol. 32, pp. 284–90.

MASON, B. H. 1966. *Principles of geochemistry*, 3rd ed. (Wiley, New York), 329 pp.

PETTIJOHN, F. J. 1941. Persistence of heavy minerals and geologic age, *Jour. Geol.*, Vol. 49, pp. 610–25.

RONDEUA, A. 1958. Géomorphologie et géochimie, *C. r. Soc. géol. Fr.*, pp. 288–90.

SIEVER, R. 1962. Silica solubility 0–200°C., and the diagenesis of siliceous sediments, *Jour. Geol.*, Vol. 70, pp. 127–50.

SMITH, W. W. 1962. Weathering of some Scottish basic igneous rocks with reference to soil formation, *Jour. Soil Sci.*, Vol. 13, pp. 202–15.

TODD, T. W. 1968. Paleoclimatology and the relative stability of felspar minerals under atmospheric conditions, *Jour. Sed. Petrol.*, Vol. 38, pp. 832–44.

WOLFF, R. G. 1967. Weathering of Woodstock granite, near Baltimore, Maryland, *Am. Jour. Sci.*, Vol. 265, pp. 106–17.

F. Biological Activity

ARANDA, J. M., and COUTTS, J. R. H. 1963. Micrometeorological observations in an afforested area in Aberdeenshire: rainfall characteristics, *Jour. Soil Sci.*, Vol. 14, pp. 124–33.

ATTIWEL, P. M. 1968. Loss of elements from decomposing litter, *Ecology*, Vol. 49, pp. 142–5.

BAY, C. E., WUNNECHE, G. W., and HAYS, O. E. 1952. Frost penetration into soils as influenced by depth of snow, vegetation cover, and air temperatures, *Trans. Am. geophys. Union*, Vol. 33, pp. 541–6.

BOULAINE, J. 1961. Sur le rôle de la végétation dans le formation des carapaces calcaires méditerranéenes, *C. r. Acad. Sci.*, Vol. 253, pp. 2568–70.

BOYER, P. 1959. De l'influence des termites de la zone intertropicale sur la configuration de certains sols, *Rev. géomorph. dyn.*, Vol. 10, pp. 41–4.

BUTUZOVA, O. V. 1962. Role of the root system of trees in the formation of microrelief, *Soviet Soil Sci.*, Vol. 4, pp. 364–72.

CAROZZI, A. V. 1967. Recent calcite-cemented sandstone generated by the equatorial tree *Iroko* (*Chlorophora excelsa*), Daloa, Ivory Coast, *Jour. Sed. Petrol.*, Vol. 37, pp. 597–600.

CHEMIN, E. 1921. Action corrosives des racines sur le marbre, *C. r. Acad. Sci.*, Vol. 173, pp. 1014–16.

CUNNINGHAM, R. K. 1963. The effect of clearing a tropical forest soil, *Jour. Soil Sci.*, Vol. 14, pp. 334–45.

DAHLSKOG, S. 1966. Sedimentation and vegetation in a Lapland mountain delta, *Geografiska Annaler*, Vol. 48A, pp. 86–101.

ERHART, E. H. 1956. *La genèse des sols en tant que phénomène géologique. Esquisse d'une theorie géologique et géochimique. Biostasie et rhexistasie* (Masson, Paris), 90 pp.

ERHART, E. H. 1951. Sur l'importance des phénomènes biologiques dans la formation des cuirasses ferrugineuses en zone tropicale, *C. r. Acad. Sci.*, Vol. 233, pp. 804–6.

EVANS, J. W. 1968. The role of *Penitella penita* (Conrad 1837) (Family Pholadidae) as eroders along the Pacific coast of North America, *Ecology*, Vol. 49, pp. 156–9.

EVERITT, B. L. 1968. Use of the cottonwood in an investigation of the Recent history of a flood plain, *Am. Jour. Sci.*, Vol. 266, pp. 417–39.

HACK, J. T. and GOODLETT, J. C. 1960. Geomorphology and forest ecology of a mountain region in the Central Appalachians, *U.S. Geol. Surv. Prof. Paper 347*, 66 pp.

HEALY, T. R. 1968. Bioerosion on shore platforms in the Waitemata formation, Auckland, *Earth Sci. Jour.*, Vol. 2, pp. 26–37.

HOPKINS, B. 1960. Rainfall interception by a tropical forest in Uganda, *East Afr. Agric. Jour.*, Vol. 25, pp. 255–8.

JENNY, H., GESSEL, S. P., and BINGHAM, F. T. 1949. Comparative study of decomposition rates of organic matter in temperate and tropical regions, *Soil Sci.*, Vol. 68, pp. 419–32.

JONES, R. J. 1965. Aspects of the biological weathering of limestone pavements, *Proc. Geol. Ass.*, Vol. 76, pp. 421–33.

KELLER, W. D. and FREDERICKSON, A. 1952. Role of plants and colloidal acids in the mechanism of weathering, *Am. Jour. Sci.*, Vol. 250, pp. 594–609.

KIDSON, C. 1959. The uses and limitations of vegetation in shore stabilization, *Geography*, Vol. 44, pp. 241–50.

KRUMBEIN, W. E. 1969. Über den Einfluss der Mikroflora auf die exogene Dynamik (Verwitterung und Krustenbildung), *Geol. Rundschau*, Vol. 58, pp. 333–63.

LOVERING, T. S. 1959. Geological significance of accumulator plants in rock weathering, *Bull. geol. Soc. Am.*, Vol. 70, pp. 781–800.

LULL, H. W. and PIERCE, R. S. 1959. Frost and forest soil, *Internl. Ass. Sci. Hydrol.*, No. 48, pp. 40–8.

LUTZ, H. J. 1960. Movement of rocks by uprooting of forest trees, *Am. Jour. Sci.*, Vol. 258, pp. 752–6.

LUX, H. 1964. Die biologischen Grundlagen der Strandhaferpflanzung und Silbergrasansaat im Dünenbau. (Ihre Untersuchung über die Möglichkeiten biologischer Dünenbau – und Dünenbefestigungs – massnahmen inden Sylter Dünengebieten bei vorhaudener und fehlender Sandablagerung), *Angewandte Pflanzensoziologie*, Vol. 20, pp. 5–53.

LYNTS, G. W. 1966. Relationship of sediment-size distribution to ecologic factors in Buttonwood Sound, Florida Bay, *Jour. Sed. Petrol.*, Vol. 36, pp. 66–74.

MCLEAN, R. F. 1967. Measurements of beachrock erosion by some tropical marine gastropods, *Bull. Marine Sci.*, Vol. 17, pp. 551–61.

MILLMAN, A. P. 1957. Biogeochemical investigations in areas of copper–tin mineralization in south-west England, *Geochim. Cosmochim. Acta*, Vol. 12, pp. 85–93.

MORTENSEN, H. 1964. Eine einfache Methode der Messung der Hangabtragung unter Wald und einige bisher damit gewonnenen Ergebnisse, *Zeit. f. Geomorph.*, Vol. 8, pp. 212–22.

MORTLAND, M. M., LAWTON, K., and UEHARA, G. 1956. Alteration of biotite to vermiculite by plant growth, *Soil Sci.*, Vol. 82, pp. 477–81.

NESTEROFF, W. D. and MÉLIÈRES, F. 1967. L'érosion littorale du pays de Caux, *Bull. Soc. géol. Fr.*, Ser. 7, Vol. 9, pp. 159–69.

NEUMANN, A. C. 1966. Observations on coastal erosion in Bermuda and measurements of the boring sponge *Cliona lamp, Limnol. Oceanog.*, Vol. 11, pp. 92–108.

PITTY, A. F. 1966. An approach to the study of karst water, illustrated by results from Poole's Cavern, Buxton, *Univ. Hull Occ. Papers in Geog.*, No. 5, 70 pp.

RAHN, P. H. 1969. The relationship between forested slopes and angles of repose for sand and gravel, *Bull. geol. Soc. Am.*, Vol. 80, pp. 2123–8.

RANSON, G. 1959. Érosion biologique des calcaires côtiers et autres calcaires d'origine animale, *C. r. Acad. Sci.*, Vol. 249, pp. 438–40.

RAUP, H. M. 1951. Vegetation and cryoplanation, *Ohio Jour. Sci.*, Vol. 51, pp. 105–16.

RHOADS, D. C. 1963. Rates of sediment reworking by *Yoldia limatula* in Buzzards Bay, Massachusetts, and Long Island Sound, *Jour. Sed. Petrol.*, Vol. 33, pp. 723–7.

SIEVER, R. 1962. Silica solubility, 2000°C, and the diagenesis of siliceous sediments, *Jour. Geol.*, Vol. 70, pp. 127–50.

VANN, J. H. 1959. Landform–vegetation relationships in the Atrato delta, *Ann. Ass. Amer. Geogrs*, Vol. 49, pp. 345–60.

WEBLEY, D. M., HENDERSON, M. E. K., and TAYLOR, I. F. 1963. The microbiology of rocks and weathered stones, *Jour. Soil Sci.*, Vol. 14, pp. 102–12.

G. Human Activity

DEMEK, J. 1969. Beschleunigung der geomorphologischen Prozesse durch die Wirkung des Menschen, *Geol. Rundschau*, Vol. 58, pp. 111–21.

DOUGLAS, I. 1967. Man, vegetation, and the sediment yield of rivers, *Nature*, Vol. 215, pp. 925–8.

JENNINGS, J. N. 1965. Man as a geological agent, *Aust. Jour. Sci.*, Vol. 28, pp. 150–6.

KIDSON, C. 1961. The Norfolk Broads, *Nature*, Vol. 192, pp. 314–15.

LAMBERT, J. M., JENNINGS, J. N., SMITH, C. T., GREEN, C., and HUTCHINSON, J. N. 1960. The making of the Broads, *Roy. Geog. Soc. Research Mem.*, No. 3, 153 pp.

MEADE, R. H. 1969. Errors in using modern stream-load data to estimate natural rates of denudation, *Bull. geol. Soc. Am.*, Vol. 80, pp. 1265–74.

NELSON, J. G. 1966. Man and geomorphic process in the Chemung River valley, New York and Pennsylvania, *Ann. Ass. Amer. Geogrs*, Vol. 56, pp. 24–32.

SCHULTZE, J. H. 1951–2. Über das Verhältnis zwischen Denudation und Boden Erosion, *Die Erde*, Vol. 3, pp. 220–33.

THOMAS, W. L. (Ed.). 1967. *Man's role in changing the face of the earth*, 7th impr. (Univ. Chicago Press, Chicago), 1193 pp.

TICHY, F. 1960. Die vom Menschen gestaltete Erde, *Die Erde*, Vol. 91, pp. 220–33.

WALLWORK, K. L. 1956. Subsidence in the mid-Cheshire industrial areas, *Geog. Jour.*, Vol. 122, pp. 40–53.

WILKINSON, H. R. 1963. Man and the natural environment, *Univ. Hull Occ. Papers in Geog.*, No. 1, 35 pp.

Chapter IV. Inter-relationships between processes and landforms

Some general references

AXELSSON, V. 1967. The Laitaure delta. A study of deltaic morphology and processes, *Geografiska Annaler*, Vol. 49A, pp. 1–127.

BALCHIN, W. G. V. and PYE, N. 1955. Piedmont profiles in the arid cycle, *Proc. Geol. Ass.*, Vol. 66, pp. 167–81.

BRESSAU, S. 1957. Abrasion, Transport und Sedimentation in der Beltsee, *Die Küste*, Vol. 6, pp. 64–102.

BUTZER, K. W. 1965. Desert landforms at the Kurkur Oasis, Egypt, *Ann. Ass. Amer. Geogrs*, Vol. 55, pp. 578–91.

CAVAILLE, A. 1953. L'érosion actuelle en Quercy, *Rev. géomorph. dyn.*, Vol. 2, pp. 57–74.

COLBY, B. R. 1963. Fluvial sediments – a summary of source, transportation, deposition, and measurement of discharge, *U.S. Geol. Surv. Bull. 1181*, pp. A1–A47.

CORBEL, J. 1959. Érosion en terrain calcaire (vitesse d'érosion et morphologie), *Ann. Géogr.*, Vol. 68, pp. 97–120.

CORBEL, J. 1961. Morphologie périglaciaire dans l'Arctique, *Ann. Géogr.*, Vol. 70, pp. 1–24.

CORBEL, J. 1964. L'érosion terrestre, étude quantitative (méthodes-techniques – résultats), *Ann. Géogr.*, Vol. 73, pp. 385–412.

COTTON, C. A. 1961. The theory of savanna planation, *Geography*, Vol. 46, pp. 89–101.

COTTON, C. A. 1962, Plains and inselbergs of the humid tropics, *Trans. Roy. Soc. N.Z.*, Vol. 88, pp. 269–77.

DAVIDSON, J. 1963. Littoral processes and morphology on Scania flat-coasts, *Lund Studies in Geog.* Ser. A, No. 23.

DENNY, C. S. 1967. Fans and pediments, *Am. Jour. Sci.*, Vol. 265, pp. 81–105.

DRESCH, J. 1957. Pédiments et glacis d'érosion, pédiplains et inselbergs, *Inf. géog.*, Vol. 21, pp. 183–96.

DUCHAUFOUR, P. 1960. *Précis de pédologie* (Cen. Doc. Univ., Paris), 478 pp.

EARDLEY, A. J. 1966. Rates of denudation in the high plateaus of southwestern Utah, *Bull. geol. Soc. Am.*, Vol. 77, pp. 777–80.

FITZPATRICK, E. A. 1958. An introduction to the periglacial geomorphology of Scotland, *Scot. geog. Mag.*, Vol. 74, pp. 28–36.

FLINT, R. F. 1952. The ice age in the North American Arctic, *Arctic*, Vol. 5, pp. 135–52.

FOX, C. 1950. Land erosion, *Endeavour*, Vol. 9, pp. 178–82.

GALLOWAY, R. W. 1961. Periglacial phenomena in Scotland, *Geografiska Annaler*, Vol. 43, pp. 348–53.

GEIKIE, A. 1868. On denudation now in progress, *Geol. Mag.*, Vol. 5, pp. 249–54.

GILLULY, J. 1964. Atlantic sediments, erosion rates and the evolution of the Continental Shelf: some speculations, *Bull. geol. Soc. Am.*, Vol. 75, pp. 483–92.

GILLULY, J., REED, J. C., and CADY, W. M. 1970. Sedimentary volumes and their significance, *Bull. geol. Soc. Am.*, Vol. 81, pp. 353–76.

HADLEY, R. F. 1967. Pediments and pediment-forming processes, *Jour. geol. Ed.*, Vol. 15, pp. 83–9.

HAMELIN, L. E. and COOK, F. A. 1967. *Le périglaciaire par l'image (illustrated glossary of periglacial phenomena)* (Presses Univ. Laval, Québec), 237 pp.

HODGSON, W. A. 1966. Coastal processes around the Otago peninsula, *N.Z. Jour. Geol. Geophys.*, Vol. 9, pp. 76–90.

HÖGBOM, B. 1914. Über die geologische Bedeutung des Frostes, *Bull. geol. Inst. Univ. Uppsala*, No. 12, pp. 257–389.

INMAN, D. L. and BAGNOLD, R. A. 1963. Beach and nearshore processes. II. Littoral processes *in* M. N. Hill (Ed.). *The sea* (Interscience, New York), Vol. 3, pp. 529–53.

KOON, D. 1955. Cliff retreat in the south west United States, *Am. Jour. Sci.*, Vol. 253, pp. 44–53.

LLIBOUTRY, L. 1961. Phénomènes cryonivaux dans les Andes de Santiago (Chile), *Biul. Peryglacjalny*, Vol. 10, pp. 209–24.

LOUIS, H. 1964. Über Rumpflächen und Talbildung in den wechselfeuchten Tropen besonders nach Studien in Tanganyika, *Zeit. f. Geomorph.*, Vol. 8, pp. 43*–70*.

MAMMERICKX, J. 1964. Quantitative observations on pediments in the Mojave and Sonoran deserts (southwestern United States), *Am. Jour. Sci.*, Vol. 262, pp. 417–35.

MENSCHING, H. VON. 1969. Bergfussflächen und des System der Flächenbildung in den ariden Subtropen und Troppen, *Geol. Rundschau*, Vol. 58, pp. 62–82.

MOHR, E. C. J. and VAN BAREN, F. A. 1954. *Tropical soils; a critical study of soil genesis as related to climate, rock and vegetation* (van Hoeve, The Hague), 498 pp.

MORTENSEN, H. 1949. Rumpffläche – Stufenlandschaft – Alternierende Abtragung, *Pet. geog. Mitt.*, Vol. 93, pp. 1–14.

NOMENYI, F. 1952. Annotated and illustrated bibliographic material on the morphology of rivers, *Bull. geol. Soc. Am.*, Vol. 63, pp. 595–644.

NORRMAN, J. O. 1964. Lake Vättern. Investigations on shore and bottom morphology, *Geografiska Annaler*, Vol. 46, pp. 1–238.

OLLIER, C. D. 1969. *Weathering* (Oliver and Boyd, Edinburgh), 304 pp.

PASSARGE, S. 1929. Das Problem der Inselberglandschaften, *Zeit. f. Geomorph.*, Vol. 4, pp. 109–22.

PÉCSI, M. 1963. Periglaziale Erscheinungen in Ungern, *Pet. geog. Mit.*, Vol. 107, pp. 161–82.

PEEL, R. F. 1966. The landscape in aridity, *Trans. Inst. Brit. Geogrs*, No. 38, pp. 1–23.

PUGH, J. C. 1956. Fringing pediments and marginal depressions in the inselberg landscape of Nigeria, *Trans. Inst. Brit. Geogrs*, No. 22, pp. 15–31.

RAHN, P. H. 1966. Inselbergs and nickpoints in southwestern Arizona, *Zeit. f. Geomorph.*, Vol. 10, pp. 217–25.

RAHN, P. H. 1967. Sheetfloods, streamfloods and the formation of pediments, *Ann. Ass. Amer. Geogrs*, Vol. 57, pp. 593–604.

RITTER, D. F. 1967. Rates of denudation, *Jour. geol. Ed.*, Vol. 15, pp. 154–9.

RUSSELL, R. J. 1967. Aspects of coastal morphology, *Geografiska Annaler*, Vol. 49A, pp. 299–309.

RUXTON, B. P. and BERRY, L. 1961. Notes on faceted slopes, rock fans and domes on granite in the east-central Sudan, *Am. Jour. Sci.*, Vol. 259, pp. 194–206.

RUXTON, B. P. and MCDOUGALL, I. 1967. Denudation rates in northeast Papua from potassium-argon dating of lavas, *Am. Jour. Sci.*, Vol. 265, pp. 545–61.

SHARP, R. P. 1966. Kelso dunes, Mojave desert, California, *Bull. geol. Soc. Am.*, Vol. 77, pp. 1045–74.

SPARKS, B. W. 1962. Rates of operation of geomorphological processes, *Geography*, Vol. 47, pp. 145–53.

SUNDBORG, Å. 1956. The river Klarälven, a study of fluvial processes, *Geografiska Annaler*, Vol. 38, pp. 127–316.

SUNDBORG, Å. 1967. Some aspects on fluvial sediments and fluvial morphology. I. General views and graphic methods, *Geografiska Annaler*, Vol. 49A, pp. 333–43.

TABER, S. 1943. Perennially frozen ground in Alaska; its origin and history, *Bull. geol. Soc. Am.*, Vol. 54, pp. 1433–548.

TE PUNGA, M. 1956. Altiplanation terraces in southern England, *Biul. Peryglacjalny*, Vol. 4, pp. 331–8.

TE PUNGA, M. 1957. Periglaciation in southern England, *Tijds. Kon. Nederl. Aardrijsh Gen.*, Vol. 74 , pp. 401–12.

TRICART, J., and VOGT, H. 1967. Quelques aspects du transport des alluvions grossières et du façonnement des lits fluviaux, *Geografiska Annaler*, Vol. 49A, pp. 351–66.

TROLL, C. 1944. Strukturböden, Solifluktion und Frostklimate der Erde, *Geol. Rundschau*, Vol. 34, pp. 545–694.

TWIDALE, C. R. 1967. Origin of the piedmont angle as evidenced in South Australia, *Jour. Geol.*, Vol. 75, pp. 393–411.

WILHELMY, H. 1958. *Klimamorphologie der Massengesteine* (Westermann, Braunschweig), 238 pp.

YOUNG, A. 1958. A record of the rate of erosion on Millstone Grit, *Proc. Yorks. geol. Soc.*, Vol. 31, pp. 149–56.

A. Rock Breakdown

1. Erosion

BAKKER, J. P. 1965. A forgotten factor in the interpretation of glacial stairways, *Zeit. f. Geomorph.*, Vol. 9, pp. 18–34.

BALLARD, A. J. and SORENSEN, F. H. 1968. Preglacial structure of Georges Basin and Northeast Channel, Gulf of Maine, *Bull. Amer. Ass. Petrol. Geols*, Vol. 52, pp. 494–500.

BATTLE, W. R. B., and LEWIS, W. V. 1951. Temperature observations in bergschrunds and their relationship to cirque erosion, *Jour. Geol.*, Vol. 59, pp. 537–45.

BRADLEY, W. C. 1958. Submarine abrasion and wave-cut platforms, *Bull. geol. Soc. Am.*, Vol. 69, pp. 967–74.

COTTON, C. A. 1968. Marine cliffing according to Darwin's theory, *Trans. Roy. Soc. N.Z.*, Vol. 6, pp. 187–208.

CRARY, A. P. 1966. Mechanism for fiord formation indicated by studies of an ice-covered inlet, *Bull. geol. Soc. Am.*, Vol. 77, pp. 911–30.

CRICKMAY, C. H. 1960. Lateral activity in a river of northwestern Canada, *Jour. Geol.*, Vol. 68, pp. 377–91.

CURTIS, R. H., and EVERNDEN, J. F. 1958. Rate of marine planation suggested by K/A dates, *Bull. geol. Soc. Am.*, Vol. 69, p. 1680.

DAHL, R. 1965. Plastically sculptured detail forms on rock surfaces in northern Nordland, Norway, *Geografiska Annaler*, Vol. 47A, pp. 83–140.

DAVIES, G. L. 1960. Platforms developed in the boulder clay of the coastal margins of counties Wicklow and Wexford, *Irish Geog.*, Vol. 4, pp. 107–16.

FAIRCHILD, H. L. 1905. Ice erosion theory a fallacy, *Bull. geol. Soc. Am.*, Vol. 16, pp. 13–74.

GEIKIE, J. 1878. On the glacial phenomena of the Long Island, or Outer Hebrides, *Quart. Jour. geol. Soc.*, Vol. 34, pp. 819–66.

GJESSING, J. 1966. Some effects of ice erosion on the development of Norwegian valleys and fiords, *Norsk Geog. Tidsskr.*, Vol. 20, pp. 273–99.

GJESSING, J. 1967. Potholes in connection with plastic scouring forms, *Geografiska Annaler*, Vol. 49A, pp. 178–87.

GRAWE, O. R. 1936. Ice as an agent of rock weathering: a discussion, *Jour. Geol.*, Vol. 44, pp. 173–82.

HARLAND, W. B. 1957. Exfoliation joints and ice action, *Jour. Glaciol.*, Vol. 3, pp. 8–12.

HAYNES, V. M. 1968. The influence of glacial erosion and rock structure on corries in Scotland, *Geografiska Annaler*, Vol. 50A, pp. 221–33.

HOLTEDAHL, H. 1967. Notes on the formation of fjords and fjord-valleys, *Geografiska Annaler*, Vol. 49A, pp. 188–203.

KLIMASZEWSKI, M. 1964. On the effect of the preglacial relief on the course and magnitude of glacial erosion in the Tatra Mountains, *Geogr. Polonica*, Vol. 2, pp. 116–21.

LEWIS, W. V. 1948. Valley steps and glacial valley erosion, *Trans. Inst. Brit. Geogrs*, Vol. 13 (1947), pp. 19–44.

LEWIS, W. V. 1954. Pressure release and glacial erosion, *Jour. Glaciol.*, Vol. 2, pp. 417–22.

LEWIS, W. V. (Ed.), 1960. Norwegian cirque glaciers, *Roy. Geog. Soc. Research Ser.*, No. 4, 104 pp.

LEIGHTON, M. M. and BROPHY, J. A. 1961. Illinoian glaciation in Illinois, *Jour. Geol.*, Vol. 69, pp. 1–31.

LEOPOLD, L. B., EMMETT, W. W., and MYRICK, R. M. 1966. Channel and hillslope processes in a semi-arid area, New Mexico, *U.S. Geol. Surv. Prof. Paper 352-G*, pp. 193–253.

LINTON, D. L. 1963. The forms of glacial erosion, *Trans. Inst. Brit. Geogrs*, No. 33, pp. 1–28.

PEEL, R. F. 1951. A study of two Northumbrian spillways, *Trans. Inst. Brit. Geogrs*, No. 15 (1949), pp. 73–89.

ROBINSON, A. H. W., 1949. Deep clefts in the Inner Sound of Raasay, *Scot. Geog. Mag.*, Vol. 65, pp. 20–5.

RUSSELL, R. J. 1963. Recent recession of tropical cliffy coasts, *Science*, Vol. 139, pp. 9–15.

SCHUMM, S. A. 1962. Erosion on miniature pediments in Badlands National Monument, south Dakota, *Bull. geol. Soc. Am.*, Vol. 73, pp. 719–24.

SCHWARTZ, M. L. 1968. The scale of shore erosion, *Jour. Geol.*, Vol. 76, pp. 508–17.

SHARP, R. P. 1940. Geomorphology of the Ruby–East Humboldt range, Nevada, *Bull. geol. Soc. Am.*, Vol. 51, pp. 337–72.

SHEPPARD, T. 1912. *The lost towns of the Yorkshire coast and other chapters bearing upon the geography of the district* (Brown, London), 328 pp.

SISSONS, J. B. 1958. Sub-glacial stream erosion in southern Northumberland, *Scot. geog. Mag.*, Vol. 74, pp. 163–74.

SUGDEN, D. E. 1968. The selectivity of glacial erosion in the Cairngorm mountains, Scotland, *Trans. Inst. Brit. Geogrs.* No. 45, pp. 79–92.

SUGDEN, D. E. 1969. The age and form of corries in the Cairngorms, *Scot. geog. Mag.*, Vol. 85, pp. 34–46.

THOMPSON, H. R. 1950. Some corries of north-west Sutherland, *Proc. Geol. Ass.*, Vol. 87, pp. 145–55.

TIETZE, W. 1961. Über die Erosion von unter Eis fliessenden Wasser, *Mainzer geog. Studien Festschr. Panzer* (Westermann, Braunschweig), pp. 125–42.

TSCHANG, T. L. 1957. Potholes in the river beds of north Taiwan, *Erdkunde*, Vol. 11, pp. 296–303.

2. Frost action

ARNAUD, R. J. ST. and WHITESIDE, E. P. 1963. Physical breakdown in relation to soil development, *Jour. Soil Sci.*, Vol. 14, pp. 267–81.

BONIFAY, E. 1955. Le rôle du gel dans la fissuration des galets de roche calcaire, *C. r. Acad. Sci.*, Vol. 240, pp. 896–8.

BOYÉ, M. 1952. Gélivation et cryoturbation dans le massif du Mont-Perdu (Pyrénées Centrales), *Pirineos*, Vol. 23, pp. 5–29.

CAILLEUX, A. 1943. Fissuration de la craie par le gel, *Bull. Soc. géol. Fr.*, 5th ser., Vol. 13, pp. 511–20.

CZEPPE, Z. 1964. Exfoliation in a periglacial climate, *Geog. Polonica*, Vol. 2, pp. 5–10.

DYLIK, J., and KLATKA, T. 1952. Recherches microscopiques sur la désintégration périglaciaire, *Bull. Soc. Sci. Lettres Lodz*, Cl. III, Vol. 3, pp. 1–12.

SCHMID, J. 1955. *Der Bodenfrost als morphologischer Faktor* (Hüthig, Heidelburg), 144 pp.

WIMAN, S. 1963. A preliminary study of experimental frost weathering, *Geografiska Annaler*, Vol. 45, pp. 113–21.

3. Other mechanical processes of rock breakdown

BEAUMONT, P. 1968. Salt weathering on the margin of the Great Kavir, Iran, *Bull. geol. Soc. Am.*, Vol. 79, pp. 1683–4.

BIROT, P. 1954. Désagrégation des roches cristallines sous l'action des sels. *C. r. Acad. Sci.*, Vol. 238, p. 1145.

BLACKWELDER, E. 1933. The insolation hypothesis of rock weathering, *Am. Jour. Sci.*, Vol. 226, pp. 97–113.

BLANCK, E. and PASSARGE, S. 1925. Die chemische Verwitterung in der ägyptischen Wüste (Univ. Abhandl. aus dem Gebiet der Auslandskunde, Hamburg), Vol. 17, 110 pp.

KUENEN, P. 1960. Experimental abrasion 4: Eolian action, *Jour. Geol.*, Vol. 68, pp. 427–49.

MORTENSEN, H. 1933. Die 'Salzsprengung' und ihre Bedeutung für die regionalklimatische Gliederung der Wüsten, *Pet. geog. Mitt.*, Vol. 79, pp. 130–5.

OLLIER, C. D. 1963. Insolation weathering: examples from central Australia, *Am. Jour. Sci.*, Vol. 261, pp. 376–81.

PASSARGE, S. 1924. Das Problem der Skulptur Inselberglandschaften, *Pet. geog. Mitt.*, Vol. 70, pp. 66–70 and 117–20.

PEDRO, G. 1957. Nouvelles recherches sur l'influence des sels dans la désagrégation des roches, *C. r. Acad. Sci.*, Vol. 244, pp. 2822–4.

SHARP, R. P. 1964. Wind-driven sand in Coachella Valley, California. *Bull. geol. Soc. Am.*, Vol. 75, pp. 785–804.

4. Chemical weathering

ALEXANDER, F. E. S. 1959. Observations on tropical weathering: a study of the movement of iron, aluminium and silicon in weathering rocks at Singapore, *Quart. Jour. geol. Soc.*, Vol. 115, pp. 123–44.

BALL, D. F. 1964. Deepweathering profile on the Island of Rhum, Inverness-shire, *Scot. geog. Mag.*, Vol. 80, pp. 22–7.

BARTON, D. C. 1938. The disintegration and exfoliation of granite in Egypt, *Jour. Geol.*, Vol. 46, pp. 109–11.

BIROT, P. 1947. Résultats de quelques experiences sur la désagrégation des roches cristallins, *C. r. Acad. Sci.*, Vol. 225, pp. 745–7.

BIROT, P. 1962. *Contribution a l'étude de la désagrégation des roches* (Cen. Doc. Univ., Paris), 232 pp.

BUNTING, B. T. 1961. The role of seepage moisture in soil formation, slope development, and stream initiation, *Am. Jour. Sci.*, Vol. 259, pp. 503–18.

DAHL, R. 1967. Post-glacial micro-weathering of bedrock surfaces in the Narvik district of Norway, *Geografiska Annaler*, Vol. 49A, pp. 155–66.

DOORNKAMP, J. C. 1968. The role of inselbergs in the geomorphology of southern Uganda, *Trans. Inst. Brit. Geogrs*, No. 44, pp. 151–62.

EMERY, K. O. 1960. Weathering of the Great Pyramid, *Jour. Sed. Petrol.*, Vol. 30, pp. 140–3.

FALCONER, J. D. 1911. *The geology and geography of northern Nigeria* (Macmillan, London), 295 pp.

FITZPATRICK, E. A. 1963. Deeply weathered rock in Scotland: its occurrence, age and contribution to the soils, *Jour. Soil Sci.*, Vol. 14, pp. 33–43.

GLINKA, K. D. 1914. *Die Typen der Bodenbildung, ihre klassifikation und geographische Verbreitung* (Gebrüder Borntraeger, Berlin), 365 pp.; trans. by C. F. Marbut, 1927, *The great soil groups of the world and their development* (Edwards Bros., Ann Arbor), 235 pp.

GOODCHILD, J. G. 1890. On the weathering of limestones, *Geol. Mag.*, Vol. 17, pp. 463–6.

GUILCHER, A. and PONT, P. 1957. Étude expérimentale de la corrosion littorale des calcaire, *Bull. Ass. géog. Fr.*, No. 265–6, pp. 48–62.

HAY, R. L. 1959. Origin and weathering of late Pleistocene ash deposits on St Vincent, B.W.I., *Jour. Geol.*, Vol. 67, pp. 65–87.

HAY, R. L. 1960. Rate of clay formation and mineral alteration in a 4000-year-old volcanic ash soil on St Vincent, B.W.I., *Am. Jour. Sci.*, Vol. 258, pp. 354–68.

HENDRICKS, D. M. and WHITTIG, L. D. 1968. Andesite weathering, *Jour. Soil Sci.*, Vol. 19, pp. 135–47.

HILL, D. E. and TEDROW, J. C. F. 1961. Weathering and soil formation in the Arctic environment, *Am. Jour. Sci.*, Vol. 259, pp. 84–101.

KLINGE, H. 1965. Podzol soils in the Amazon basin, *Jour. Soil Sci.*, Vol. 16, pp. 95–103.

NICHOLS, R. L. 1963. Geologic features demonstrating aridity of McMurdo Sound area, Antarctica, *Am. Jour. Sci.*, Vol. 261, pp. 20–31.

NOSSIN, J. J. and LEVELT, W. M. 1967. Igneous rock weathering on Singapore Island, *Zeit. f. Geomorph.*, Vol. 11, pp. 14–35.

NYE, P. H. 1954. Some soil-forming processes in the humid tropics. I: A field study of a catena in West African forest, *Jour. Soil Sci.*, Vol. 5, pp. 7–21.

OLLIER, C. D. 1965. Some features of granite weathering in Australia, *Zeit. f. Geomorph.*, Vol. 9, pp. 285–304.

PHEMISTER, T. C. and SIMPSON, S. 1949. Pleistocene deep weathering in north-east Scotland, *Nature*, Vol. 164, pp. 318–19.

REICHE, P. 1950. A survey of weathering processes and products, *Univ. New Mexico Publ. in Geol.*, No. 1, 87 pp.

RUXTON, B. P. 1958. Weathering and surface erosion in granite at the piedmont angle, Balos, Sudan, *Geol. Mag.*, Vol. 95, pp. 353–77.

SCHUNKE, E. 1969. Die Schichtstufenhänge des Leine–Weser–Berglandes–Methoden und Ergebnisse ihrer Untersuchung, *Geol. Rundschau*, Vol. 58, pp. 446–64.

SWEETING, M. M., GROOM, G. E., WILLIAMS, V. H., PIGOTT, C. D., SMITH, D. I., and WARWICK, G. T. 1965. Denudation in limestone regions; a symposium, *Geog. Jour.*, Vol. 131, pp. 24–56.

5. Weathering forms

BÖGLI, A. 1951. Probleme der Karrenbildung, *Geog. Helvetica*, Vol. 6, pp. 191–204.

BÖGLI, A. 1960. Kalklösung und Karrenbildung, *Zeit. f. Geomorph.*, Suppbd 2, pp. 4–21.

BRAJNIKOV, B. 1953. Les 'pains de sucre' du Brésil sont-ils enracinés, *C. r. géol. Soc. Fr.*, pp. 267–9.

CALKIN, P. and CAILLEUX, A. 1962. A quantitative study of cavernous weathering (taffonis) and its application to glacial chronology in Victoria Valley, Antarctica, *Zeit. f. Geomorph.*, Vol. 6, pp. 317–24.

DAHL, R. 1966. Block fields, weathering pits and tor-like forms in the Narvik mountains, Nordland, Norway, *Geografiska Annaler*, Vol. 48A, pp. 55–85.

DEMEK, J. 1964. Castle koppies and tors in the Bohemian Highlands (Czechoslovakia), *Biul. Peryglacjalny*, Vol. 14, pp. 195–216.

DRAGOVICH, D. 1969. The origin of cavernous surfaces (tafoni) in granitic rocks of southern South Australia, *Zeit. f. Geomorph.*, Vol. 13, pp. 163–81.

EAKIN, H. M. 1916. The Yukon–Koyukuk region, Alaska, *U.S. Geol. Surv. Bull. 631*, 88 pp.

FEININGER, T. 1969. Pseudokarst on quartz diorite, *Zeit. f. Geomorph.*, Vol. 13, pp. 287–96.

FENELON, P. (Ed.). 1968. *Phénomènes karstiques*, Cen. Doc. Cartog. Géog. Mem. et Doc., 392 pp.

GERBER, E. 1959. Form und Bildung alpiner Talböden, *Geog. Helvetica*, Vol. 14, pp. 117–238.

GERSTENHAUER, A. 1960. Der tropische Kegelkarst in Tabasco (Mexico), *Zeit. f. Geomorph.*, Suppbd 2, pp. 22–48.

GODARD, A. 1966. Morphologie des solces et des massifs anciens: les 'tors' et le probleme de leur origine, *Rev. géog. de l'Est*, Vol. 6, pp. 153–70.

GUILCHER, A. 1958. Coastal corrosion forms in limestone around the Bay of Biscay, *Scot. geog. Mag.*, Vol. 74, pp. 137–49.

HEDGES, J. 1969. Opferkessel, *Zeit. f. Geomorph.*, Vol. 13, pp. 22–55.

HSI-LIN, T. 1961. The pseudokarren and exfoliation forms of granite on Pulan Ubin, Singapore, *Zeit. f. Geomorph.*, Vol. 5, pp. 302–12.

LeGRAND, H. 1952. Solution depressions in diorite in North Carolina, *Am. Jour. Sci.*, Vol. 250, pp. 566–85.

LOUIS, H. 1963. Über Sockelfläche und Hüllfläche des Reliefs, *Zeit. f. Geomorph.*, Vol. 7, pp. 353–66.

LINTON, D. L. 1955. The problem of tors, *Geog. Jour.*, Vol. 121, pp. 470–87.

MABBUTT, J. 1952. A study of granite relief from south-west Africa, *Geol. Mag.*, Vol. 89, pp. 87–96.

PALMER, H. S. 1927. Lapiés in Hawaiian basalts, *Geog. Rev.*, Vol. 17, pp. 627–31.

PALMER, J. 1956. Tor formation at the Bridestones in north-east Yorkshire, and its significance in relation to problems of valley-side development and regional glaciation, *Trans. Inst. Brit. Geogrs*, Vol. 22, pp. 55–72.

PALMER, J. and NEILSON, R. A. 1962. The origin of granite tors on Dartmoor, Devonshire, *Proc. Yorks. geol. Soc.*, Vol. 33, pp. 315–40.

SMITH, J. F. and ALBRITTON, C. C. 1941. Solutional effects on limestone as a function of slope, *Bull. geol. Soc. Am.*, Vol. 52, pp. 61–78.

SWEETING, M. M. 1958. The karstlands of Jamaica, *Geog. Jour.*, Vol. 124, pp. 184–99.

THOMAS, M. F. 1966. Some geomorphological implications of deep weathering patterns in crystalline rocks in Nigeria, *Trans. Inst. Brit. Geogrs*, Vol. 40, pp. 173–93.

THOMAS, M. F. 1967. A bornhardt dome in the plains near Oyo, western Nigeria, *Zeit. f. Geomorph.*, Vol. 11, pp. 239–61.

THORP, M. B. 1967. Closed basins in younger granite massifs, northern Nigeria, *Zeit. f. Geomorph.*, Vol. 11, pp. 459–80.

TROMBE, F. 1952. *Traité de spéléologie* (Payot, Paris), 367 pp.

TWIDALE, C. R. 1964. A contribution to the study of domed inselbergs, *Trans. Inst. Brit. Geogrs*, No. 34, pp. 91–113.

WALL, J. R. D. and WILFORD, G. E. 1966. A comparison of small-scale features on microgranodiorite and limestone in west Sarawak, Malaysia, *Zeit. f. Geomorph.*, Vol. 10, pp. 462–8.

WARWICK, G. T. 1962. *In* C. H. D. Cullingford (Ed.), *British Caving: an introduction to speleology* (Routledge, London), Chaps 1–5, pp. 11–217.

WHITE, L. S. 1949. Processes of erosion on steep slopes of Oahu (Hawaii), *Am. Jour. Sci.*, Vol. 247, pp. 168–86.

WILLIAMS, P. W. 1966. Limestone pavements with special reference to western Ireland, *Trans. Inst. Brit. Geogrs*, No. 40, pp. 155–72.

ZWITTKOVITS, F. 1969. Alters- und Hohengliederung der Karren in den Nördlichen Kalkalpen, *Geol. Rundschau*, Vol. 58, pp. 378–95.

B. Transportation

1. Transportation of particles in a fluid

BAGNOLD, R. A. 1937. The transport of sand by wind, *Geog. Jour.*, Vol. 89, pp. 409–38.

BEADNELL, H. J. L. 1910. The sand dunes of the Libyan desert, *Geog. Jour.*, Vol. 35, pp. 379–95.

BEHRE, C. H. 1926. Sand flotation in nature, *Science*, Vol. 63, pp. 405–8.

CHERRY, J. A. 1966. Sand movement along equilibrium beaches north of San Francisco, *Jour. Sed. Petrol.*, Vol. 36, pp. 341–57.

DAVIDSON, C. 1888. Note on the movement of scree material, *Quart. Jour. geol. Soc.*, Vol. 44, pp. 232–8.

DOWNING, B. H. 1968. Subsurface erosion as a geomorphological agent in Natal, *Trans. geol. Soc. S. Africa*, Vol. 121, pp. 131–4.

DURAND, R. 1951. Transport hydraulique de graviers et galets en conduite, *La Houille Blanche*, Spec. No. B, pp. 609–19.

EMERY, K. O. 1955. Transportation of rocks by driftwood, *Jour. Sed. Petrol.*, Vol. 25, pp. 51–7.

ENGELN, O. D. VON. 1918. Transportation of debris by ice-bergs, *Jour. Geol.*, Vol. 26, pp. 74–81.

FRÉCAUT, R. 1966. Les transports solides de fond des cours d'eau: techniques et possibilités des méthodes de mesure directe, *Rev. géog. de l'Est*, Vol. 6, pp. 321–8.

GARDNER, J. 1969. Observations on surficial talus movement, *Zeit. f. Geomorph.*, Vol. 13, pp. 317–23.

GILBERT, G. K. 1914. The transportation of debris by running water, *U.S. Geol. Surv. Prof. Paper* 86, 263 pp.

GIRESSE, P. 1965. Exemples de transport côtier sur le littoral catalan espagnol. Mécanismes de tirage, *Cahiers Océanog.*, Vol. 17, pp. 99–106.

GÖHREN, H. 1966. Beobachtungen über den Einfluss des Oberwassers auf die Sandbewegung in der Aussenweser, *Die Küste*, Vol. 14, pp. 157–69.

HARDY, J. R. 1964. The movement of beach material and wave action near Blakeney Point, Norfolk, *Trans. Inst. Brit. Geogrs*, Vol. 34, pp. 53–69.

JOHNSON, J. W. 1956. Dynamics of nearshore sediment movement, *Bull. Amer. Ass. Petrol. Geols*, Vol. 40, pp. 2211–32.

JOLLIFFE, I. P. 1964. An experimental design to compare the relative rates of movement of different sizes of beach pebbles, *Proc. Geol. Ass.*, Vol. 75, pp. 67–87.

JOPLING, A. V. 1966. Some principles and techniques used in reconstructing the hydraulic parameters of a paleo-flow regime, *Jour. Sed. Petrol.*, Vol. 36, pp. 5–49.

KENT, P. E. 1966. The transportation mechanism in catastrophic falls, *Jour. Geol.*, Vol. 74, pp. 79–82.

KIDSON, C. and CARR, A. P. 1959. The movement of shingle over the sea bed close inshore, *Geog. Jour.*, Vol. 125, pp. 380–9.

MEADE, R. H. 1969. Landward transport of bottom sediments, *Jour. Sed. Petrol.*, Vol. 39, pp. 222–34.

MCGEE, W. J. 1897. Sheetflood erosion, *Bull. geol. Soc. Am.*, Vol. 8, pp. 87–112.

MELAND, N. and NORRMAN, J. O. 1969. Transport velocities of individual size fractions in heterogeneous bedload, *Geografiska Annaler*, Vol. 51A, pp. 127–44.

PHILLIPS, A. W. 1963. Tracer experiments at Spurn Head, Yorkshire, England, *Shore and Beach*, Vol. 31, pp. 30–5.

SHARP, R. P. 1964. Wind-driven sand in Coachella valley, California, *Bull. geol. Soc. Am.*, Vol. 75, pp. 785–804.

SHREVE, R. L. 1968. Leakage and fluidization in air-layer lubricated avalanches, *Bull. geol. Soc. Am.*, Vol. 79, pp. 653–8.

STAN, C. and GHENOVICI, A. 1964. River and sea transport in the Braila–Sulina sector of the Danube, *Rev. Roum. Géol. Géophys. Géog., Sér. Géog.*, Vol. 8, pp. 45–50.

STRIDE, A. H. 1963. Current-swept sea floors near the southern half of Great Britain, *Quart. Jour. geol. Soc.*, Vol. 119, pp. 175–99.

WOOD, J. 1825. Remarks on the moving of rocks by ice, *Am. Jour. Sci.*, Vol. 9, pp. 144–5.

2. Amounts of sediment transport

ANDERSON, H. W. 1954. Suspended sediment discharge as related to stream-flow, topography, soil and land use, *Trans. Am. geophys. Union*, Vol. 35, pp. 268–81.

BERTHOIS, L. and BARBIER, M. 1954. Apports sedimentaires en suspension dans La Loire (année 1953), *Bull. Soc. Géol. Fr.*, Sér. 6, Vol. 20, pp. 237–42.

BOON, J. D. 1969. Quantitative analysis of beach sand movement, *Sedimentology*, Vol. 13, pp. 85–103.

BIRKELAND, P. W. 1968. Mean velocity and boulder transport during Tahoe age floods of the Truckee river, California–Nevada, *Bull. geol. Soc. Am.*, Vol. 79, pp. 137–42.

BOGARDI, J. 1951. Mesure du debit solide des rivieres en Hongrie, *La Houille Blanche*, Vol. 6, pp. 108–26.

COUTTS, J. R. H., KANDIL, M. F., NOWLAND, T. L., and TINSLEY, J. 1968. Use of radioactive ^{59}Fe for tracing soil particle movement, *Jour. Soil Sci.*, Vol. 19, pp. 311–41.

DEPETRIS, P. T. and GRIFFEN, J. J. 1968. Suspended load in the Rio de la Plata drainage basin, *Sedimentology*, Vol. 11, pp. 53–60.

DIACONU, O. 1969. Resultats de l'étude de l'écoulement des alluvions en suspension des rivieres de la Roumanie, *Bull. Internl Ass. Sci. Hydrol.*, Vol. 14, pp. 51–89.

ELLISON, W.D. 1945. Some effects of raindrops and surface flow on soil erosion and infiltration, *Trans. Am. geophys. Union*, Vol. 26, pp. 415–29.

FOURNIER, F. 1960. *Climat et érosion*, (Presses Univ. France, Paris), 201 pp.

FOURNIER, F. 1969. Transports solides effectués par les cours d'eau, *Bull. Internl Ass. Sci. Hydrol.*, Vol. 14, pp. 7–49.

INGLE, J. C. 1966. *The movement of beach sand* (Elsevier, Amsterdam), 221 pp.

KUHLMAN, H. 1958. Quantitative measurements of aeolian sand transport, *Geog. Tidsskr.*, Vol. 57, pp. 51–74.

MICHAUD, J. and CAILLEUX, A. 1950. Vitesses des movements du sol au Cambeyron, *C. r. Acad. Sci.*, Vol. 230, pp. 314–15.

MIN TIEH, T. 1941. Soil erosion in China, *Geog. Rev.*, Vol. 31, pp. 570–90.

ØSTREM, G. 1965. Problems of dating ice-cored moraines, *Geografiska Annaler*, Vol. 47A, pp. 1–38.

PARDE, M. 1953. La turbiditié des rivières et ses facteurs géographiques, *Rev. Géog. alpine*, Vol. 41, pp. 399–421.

POUQUET, J. 1951. *L'érosion* (Presse Univ. France, Paris), 126 pp.

SCHOFIELD, J. C. 1967. Sand movement at Mangatawhiri spit and Little Omaha bay, *N.Z. Jour. Geol. Geophys.*, Vol. 10, pp. 697–721.

SCHUMM, S. A. 1956. The role of creep and rainwash on the retreat of badland slopes, *Am. Jour. Sci.*, Vol. 254, pp. 693–706.

SCHUMM, S. A. 1967. Rates of surficial rock creep on hillslopes in western Colorado, *Science*, Vol. 155, pp. 560–1.

SELBY, M. J. 1966. Methods of measuring soil creep, *Jour. Hydrol (N.Z.)*, Vol. 5, pp. 54–63.

TODD, O. J. and ELIASSEN, S. 1940. The Yellow River problem, *Trans. Amer. Soc. Civ. Eng.*, Vol. 105, pp. 346–453.

U.S. DEPT. OF AGRICULTURE. 1965. Proceedings of the Federal Inter-Agency sedimentation conference, 1963; *U.S. Dept. Agric. Misc. Publ.* No. 970, 933 pp.

VOLLBRECHT, K. 1966. The relationship between wind records, energy of longshore drift, and energy balance off the coast of a restricted body of water, as applied to the Baltic, *Marine Geology*, Vol. 4, pp. 119–47.

WHETTEN, J. T., KELLEY, J. C., and HANSON, L. G. 1969. Characteristics of Columbia River sediment and sediment transport, *Jour. Sed. Petrol.*, Vol. 39, pp. 1149–66.

WISCHMEIER, W. H. and SMITH, D. D. 1958. Rainfall energy and its relationship to soil loss, *Trans. Am. geophys. Union*, Vol. 39, pp. 285–91.

YOUNG, A. 1960. Soil movement by denudational processes on slopes, *Nature*, Vol. 187, pp. 220–2.

3. Mass movement

ANDERSSON, J. G. 1906. Solifluxion, a component of subaerial denudation, *Jour. Geol.*, Vol. 14, pp. 91–112.

BLACKWELDER, E. 1928. Mudflows as a geologic agent in semi-arid mountains, *Bull. geol. Soc. Am.*, Vol. 39, pp. 465–83.

COLLIN, A. 1846. *Recherches expérimentales sur les glissements spontanés des terrains argileux,* (Carilian-Goeury and Dalmont, Paris), trans. by W. R. Schriever *et al.*, 1956, *Landslides in clay* (Univ. of Toronto Press, Toronto), 160 pp.

COORAY, P. G. 1958. Earthslips and related phenomena in the Kandy district, Ceylon, *The Ceylon Geogr.*, Vol. 12, pp. 75–90.

DHONAU, T. J. and DHONAU, N. B. 1963. Glacial structures on the north Norfolk coast, *Proc. Geol. Ass.*, Vol. 74, pp. 433–9.

DYLIK, J. 1967. Solifluxion, congelifluxion and related slope processes, *Geografiska Annaler*, Vol. 49A, pp. 167–77.

ECKEL, E. B. (Ed.). 1958. Landslides and engineering practice, *National Research Council: Highway Research Board Spec. Rep. 29*, 232 pp.

FAIRBRIDGE, R. W. 1950. Landslide patterns on oceanic volcanoes and atolls, *Geog. Jour.*, Vol. 115, pp. 84–8.

GERLACH, T. 1959. Needle ice and its role in the displacement of the cover of waste material in the Tatra mountains, *Przeglad Geogr.*, Vol. 31, pp. 603–5.

GROVE, A. T. 1953. Account of a mudflow on Bredon Hill, Worcestershire, April 1951, *Proc. Geol. Ass.*, Vol. 64, pp. 10–13.

HOLLINGWORTH, S. E. 1934. Some solifluction phenomena in the northern Lake District, *Proc. Geol. Ass.*, Vol. 45, pp. 167–88.

JAHN, A. 1967. Some features of mass movement on Spitzbergen slopes, *Geografiska Annaler*, Vol. 49A, pp. 213–25.

JOHNSON, R. H. 1965. A study of the Charlesworth landslides near Glossop, *Trans. Inst. Brit. Geogrs*, No. 37, pp. 111–26.

JONES, F. O., EMBODY, D. R., and PETERSON, W. L. 1961. Landslides along the Columbia River valley, northeastern Washington, *U.S. Geol. Surv. Prof. Paper 367*, 98 pp.

KAYSER, B. 1963. L'érosion par franes en Lucanie, *Mediterranée*, Vol. 4, pp. 93–100.

KUPSCH, W. O. 1962. Ice-thrust ridges in western Canada, *Jour. Geol.*, Vol. 70, pp. 582–94.

MACAR, P. 1947. Les chutes de l'Inkrisi (Congo occidental) et leurs divers modes d'érosion, *Ann. Soc. géol. Belg.*, Vol. 82, pp. 38–51.

MCLEAN, R. F. and DAVIDSON, C. F. 1968. The role of mass movement in shore platform development along the Gisborne coastline, New Zealand, *Earth Sci. Jour.*, Vol. 2, pp. 15–25.

PRIOR, D. R., STEPHENS, N., and ARCHER, D. R. 1968. Composite mudflows on the Antrim coast of north-east Ireland, *Geografiska Annaler*, Vol. 50A, pp. 65–78.

RICE, R. M., CORBETT, E. S., and BAILEY, R. G. 1969. Soil slippage related to vegetation, topography and soil in southern California, *Water Resources Research*, Vol. 5, pp. 647–59.

RUDBERG, S. 1958. Some observations concerning mass movement on slopes in Sweden, *Geol. Foren. I. Stock. Körh.*, Vol. 80, pp. 114–25.

RUTTEN, M. G. 1960. Ice-pushed ridges, permafrost and drainage, *Am. Jour. Sci.*, Vol. 258, pp. 293–7.

SCHMID, J. 1955. *Der Bodenfrost als morphologischer Faktor* (Hüthig, Heidelberg), 144 pp.

SHARPE, C. F. S. 1938. *Landslides and related phenomena; a study of mass-movements of soil and rock* (Columbia Univ. Press, New York), 137 pp.

URBANEK, J. 1968. Slide classification, *Geograficky Casopis*, Vol. 20, pp. 221–36.

WATSON, R. A. and WRIGHT, H. E. 1963. Landslides on the east flank of the Chuska Mountains, northwestern New Mexico, *Am. Jour. Sci.*, Vol. 261, pp. 525–48.

WAHRHAFTIG, C. and COX, A. 1959. Rock glaciers in the Alaska Range, *Bull. geol. Soc. Am.*, Vol. 70, pp. 383–436.

WILLIAMS, P. J. 1957. Some investigations into solifluction features in Norway, *Geog. Jour.*, Vol. 123, pp. 42–58.

WILLIAMS, P. J. 1959. An investigation into processes occurring in solifluction, *Am. Jour. Sci.*, Vol. 257, pp. 481–90.

ZARUBA, Q. and MENCL, V. 1969. *Landslides and their control* (Elsevier, Amsterdam), 205 pp.

4. Transportation of dissolved material

DAVIS, S. N. 1964. Silica in streams and ground water, *Am. Jour. Sci.*, Vol. 262, pp. 870–91.

DOUGLAS, I. 1969. The efficiency of humid tropical denudation systems, *Trans. Inst. Brit. Geogrs*, No. 46, pp. 1–19.

EK, C. 1964. Note sur les eaux de fonte des glaciers de la Haute Maurienne. Leur action sur les carbonate, *Rev. Belge Géog.*, Vol. 88, pp. 127–56.

EK, C. 1966. Faible agressivité des eaux de fonte des glaciers: l'exemple de la Marmolada (Dolomites), *Ann. Soc. géol. Belg.*, Vol. 89, pp. 177–88.

GARRELS, R. M. 1965. Silica: role in the buffering of natural waters, *Science*, Vol. 148, p. 69.

GORHAM, E. 1961. Factors influencing supply of major ions to inland waters, with special reference to the atmosphere, *Bull. geol. Soc. Am.*, Vol. 72, pp. 795–840.

HEM, J. D. 1959. Study and interpretation of the chemical characteristics of natural waters, *U.S. Geol. Surv. Water-supply Paper 1473*, 269 pp.

KRAUSKOPF, K. B. 1956. Dissolution and precipitation of silica at low temperatures, *Geochim. Cosmochim. Acta*, Vol. 10, pp. 1–26.

LIVINGSTONE, D. A. 1963. The chemical composition of rivers and lakes, *U.S. Geol. Surv. Prof. Paper 440-G*, 64 pp.

MACKENZIE, F. T. and GARRELS, R. M. 1966. Chemical mass balance between rivers and oceans, *Am. Jour. Sci.*, Vol. 264, pp. 507–25.

MOORE, E. S. and MAYNARD, J. E. 1929. Solution, transportation and precipitation of iron and silica, *Econ. Geol.*, Vol. 24, pp. 272–303, 365–402, 506–27.

ROUGERIE, G. 1961. Étude comparative de l'évacuation de la silice en milieux cristallins tropical humide et tempéré humide, premiers resultats, *Ann. Géog.*, Vol. 70, pp. 45–50.

C. Deposition

General

ALLEN, J. R. L. 1965. A review of the origin and characteristics of recent alluvial sediments, *Sedimentology*, Vol. 5, pp. 89–191.

JOHANSSON, C. E. 1965. Structural studies of sedimentary deposits, *Lund Studies in Geog.*, No. 42A, 61 pp.

NORRIS, R. M. 1956. Crescentic beach cusps and barchan dunes, *Bull. Amer. Ass. Petrol. Geols*, Vol. 40, pp. 1681–6.

Slope deposits

ANDREWS, J. T. 1961. The development of scree slopes in the English Lake District and central Quebec–Labrador, *Cahiers Géog. Québec*, No. 10, pp. 219–30.

BENEDICT, J. B. 1966. Radiocarbon dates from a stone-banked terrace in the Colorado Rocky Mountains, *Geografiska Annaler*, Vol. 48A, pp. 24–31.

BLAGBROUGH, J. W. and BREED, W. J. 1967. Protalus ramparts on Navajo Mountain, southern Utah, *Am. Jour. Sci.*, Vol. 265, pp. 759–72.

BURKALOW, A. VAN. 1945. Angle of repose and angle of sliding friction, an experimental study, *Bull. geol. Soc. Am.*, Vol. 56, pp. 669–708.

CAINE, N. 1967. The texture of talus in Tasmania, *Jour. Sed. Petrol.*, Vol. 37, pp. 796–803.

DYLIK, J 1960. Rhythmically stratified slope waste deposits, *Biul. Peryglacjalny*, Vol. 8, pp. 31–41.

MUDGE, M. R. 1965. Rockfall-avalanche and rockslide-avalanche deposits at Sawtooth Ridge, Montana, *Bull. geol. Soc. Am.*, Vol. 76, pp. 1003–14.

PARIZEK, E. J. and WOODRUFF, J. F. 1957. Description and origin of stone layers in soils of the southeastern States, *Jour. Geol.*, Vol. 65, pp. 24–34.

PÉCSI, M. 1967. Relationship between slope geomorphology and Quaternary slope sedimentation, *Acta Geol. Acad. Sci. Hungaricae*, Vol. 11, pp. 307–21.

PÉCSI, M. 1969. Genetic classification of slope sediments, *Biul. Peryglacjalny*, Vol. 18, pp. 15–27.

PIPPAN, T. 1969. Studies on grus and block deposits on mountain slopes in Austria, *Biul. Peryglacjalny*, Vol. 18, pp. 29–42.

TINKLER, K. J. 1966. Slope profiles and scree in the Eglwyseg valley, North Wales, *Geog. Jour.*, Vol. 132, pp. 379–85.

TIVY, J. 1962. An investigation of certain slope deposits in the Lowther Hills, Southern Uplands of Scotland, *Trans. Inst. Brit. Geogrs*, No. 30, pp. 59–74.

WATSON, E. 1969. The slope deposits in the Nant Iago valley, near Cader Idris, Wales, *Biul. Peryglacjalny*, Vol. 18, pp. 95–113.

Mudflows and alluvial fans

ANSTEY, R. L. 1966. A comparison of alluvial fans in West Pakistan, *Pakistan Geog. Rev.*, Vol. 21, pp. 14–20.

BEATY, C. B. 1961. Boulder deposits in Flint Creek valley, western Montana, *Bull. geol. Soc. Am.*, Vol. 72. pp. 1015–20.

BEATY, C. B. 1963. Origin of alluvial fans, White Mountains, California and Nevada, *Ann. Ass. Amer. Geogrs*, Vol. 53, pp. 516–35.

BEATY, C. B. 1970. Age and estimated rate of accumulation of an alluvial fan, White Mountains, California, U.S.A., *Am. Jour. Sci.*, Vol. 268, pp. 50–77.

BLISSENBACH, E. 1954. Geology of alluvial fans in semiarid regions, *Bull. geol. Soc. Am.*, Vol. 65, pp. 175–90.

BULL, W. B. 1963. Alluvial-fan deposits in Western Fresno County, California, *Jour. Geol.*, Vol. 71, pp. 243–41.

BULL, W. B. 1968. Alluvial fans, *Jour. geol. Ed.*, Vol. 16, pp. 101–6.

BLUCK, B. J. 1964. Sedimentation of an alluvial fan in southern Nevada, *Jour. Sed. Petrol.*, Vol. 34, pp. 395–400.

Clastic deposition in rivers

BAGNOLD, R. A. 1968. Deposition in the process of hydraulic transport, *Sedimentology*, Vol. 10, pp. 45–56.

FENNEMAN, N. M. 1906. Floodplains produced without floods, *Bull. Amer. geog. Soc.*, Vol. 38, pp. 89–91.

HICKIN, E. J. 1969. A newly identified process of point bar formation in natural streams, *Am. Jour. Sci.*, Vol. 267, pp. 999–1010.

KRIGSTRÖM, A. 1962. Geomorphological studies of sandur plains and their braided rivers in Iceland, *Geografiska Annaler*, Vol. 44, pp. 328–46.

RASID, H. 1966. Morphology of the Jamuna flood plain, *Oriental Geogr.*, Vol. 10, pp. 57–72.

WILLIAMS, G. E. 1966. Planar cross-stratification formed by the lateral migration of shallow streams, *Jour. Sed. Petrol.*, Vol. 36, pp. 742–6.

WILLIAMS, P. F. and RUST, B. R. 1969. The sedimentology of a braided river, *Jour. Sed. Petrol.*, Vol. 39, pp. 649–79.

Chemical deposition in streams

CAILLEUX, A. 1965. Quaternary secondary chemical deposition in France, *Geol. Soc. Am. Spec. Paper 84*, pp. 125–39.

EK, C. and PISSART, A. 1965. Dépôt de carbonate de calcium par congélation et teneur en bicarbonate des eaux résiduelles, *C. r. Acad. Sci.*, Vol. 260, pp. 929–32.

GAMS, I. 1963. Nekatere znacilnosti Krke in njenih pritokov, *Geografski Zbornik*, Vol. 8, pp. 92–110; English summary. Some characteristics of the river Krka and its affluents, pp. 108–9.

GREGORY, J. W. 1911. Constructive waterfalls, *Scot. geog. Mag.*, Vol. 27, pp. 537–46.

SENCU, V. 1967. Morphologie und Enstehung des Steinsalzkarstes bei Slănic-Prahova, *Rev. Roumanie géol. géophys. géog.*, Vol. 11, pp. 49–65.

Deltas

ANDEL, T. H. VAN. 1967. The Orinoco delta, *Jour. Sed. Petrol.*, Vol. 37, pp. 297–310.

HAY, T. 1926. Delta formation in the English lakes, *Geol. Mag.*, Vol. 63, pp. 292–301.

JOPLING, A. V. 1965. Hydraulic factors controlling the shape of laminae in laboratory deltas, *Jour. Sed. Petrol.*, Vol. 35, pp. 777–91.

MOORE, D. 1966. Deltaic sedimentation, *Earth Sci. Rev.*, Vol. 1, pp. 87–104.

SHIRLEY, M. L. and RAGSDALE, J. A. (Eds). 1966. *Deltas in their geologic framework* (Houston Geol. Soc., Houston), 251 pp.

STRAATEN, L. M. J. U. VAN. 1960. Some recent advances in the study of deltaic sedimentation, *Liverpool and Manchester geol. Jour.*, Vol. 2, pp. 411–42.

STRAATEN, L. M. J. U. VAN. 1964. *Deltaic and shallow marine deposits*, (Elsevier, Amsterdam).

Estuaries

AHNERT, F. 1960. Estuarine meanders in the Chesapeake Bay area, *Geol. Rev.*, Vol. 50, pp. 390–401.

BERGDAHL, A. 1960. Glacifluvial estuaries on the Närke Plain, *Lund Studies in Geog.*, Ser. A, No. 13.

BERTHOIS, L. 1960. Étude dynamique de la sédimenatation dans la Loire, *Cahiers Océanogr.*, Vol. 12, pp. 631–57.

BERTHOIS, L. 1965. *Techniques d'études estuariennes* (Centre Doc. Univ., Paris), 144 pp.

EVANS, G. 1965. Intertidal flat sediments and their environments of deposition in the Wash, *Quart. Jour. Geol. Soc.*, Vol. 121, pp. 209–45.

GUILCHER, A. 1953. Mesures de la vitesse de sédimentation et d'érosion dans des estuaires breton, *C. r. Acad. Sci.*, Vol. 237, pp. 1345–7.

LAUF, G. H. (Ed.). 1967. Estuaries, *Am. Ass. Adv. Sci.*, Publ. 83, 757 pp.

NOTA, D. J. G. and LORING, D. H. 1964. Recent depositional conditions in the St Lawrence River and Gulf; a reconnaissance survey, *Marine Geol.*, Vol. 2, pp. 198–235.

PRICE, W. A. and KENDRICK, M. 1963. Field model investigation into the reasons for siltation in the Mersey estuary, *Proc. Instn. Civ. Engrs*, Vol. 24, pp. 273–518.

ROBINSON, A. H. W. 1960. Ebb-flood channel systems in sandy bays and estuaries, *Geography*, Vol. 45, pp. 183–99.

Coastal deposits

ALLEN, J. R. L. 1965. Coastal geomorphology of eastern Nigeria: beach ridge barrier islands and vegetated tidal flats, *Geol. en Mijnbouw*, Vol. 44, pp. 1–21.

BLUCK, B. J. 1967. Sedimentation of beach gravels: examples from South Wales, *Jour. Sed. Petrol.*, Vol. 37, pp. 128–56.

ELLIOTT, E. L. 1958. Sandspits of the Otago coast, *N.Z. Geogr*, Vol. 14, pp. 65–74.

EVANS, O. F. 1942. The origin of spits, bars and related structures, *Jour. Geol*, Vol. 50, pp. 846–65.

GUILCHER, A. and KING, C. A. M. 1961. Spits, tombolos and tidal marshes in Connemara and west Kerry, *Proc. Roy. Irish Acad.*, Vol. 61B, pp. 283–338.

HEY, R. W. 1967. Sections in the beach-plain deposits of Dungeness, Kent, *Geol. Mag.*, Vol. 104, pp. 361–70.

KIDSON, C. 1963. The growth of sand and shingle spits across estuaries, *Zeit. f. Geomorph.*, Vol. 7, pp. 1–22.

KING, C. A. M. and WILLIAMS, W. W. 1949. The formation and movement of sand bars by wave action, *Geog. Jour.*, Vol. 113, pp. 70–85.

MACNEIL, F. S. 1954. Organic reefs and banks and associated detrital sediments, *Am. Jour. Sci.*, Vol. 252, pp. 385–401.

NICHOLS, R. L. 1961. Characteristics of beaches formed in polar climates, *Am. Jour. Sci.*, Vol. 259, pp. 694–708.

PUTSY, N. P. 1965. Beach ridge development in Tabasco, Mexico, *Ann. Ass. Amer. Geogrs*, Vol. 55, pp. 112–24.

SCHOLL, D. W. 1963. Sedimentation in modern coastal swamps, southwestern Florida, *Bull. Amer. Ass. Petrol. Geols*, Vol. 47, pp. 1581–603.

SILVESTER, R. 1965. Coral reefs, atolls and guyots, *Nature*, Vol. 207, pp. 681–8.

STEERS, J. A. 1937. The Coral Islands and associated features of the Great Barrier Reefs, *Geog. Jour.*, Vol. 89, pp. 1–28 and 119–46.

STODDART, D. R. 1965. The shape of atolls, *Marine geology*, Vol. 3, pp. 369–83.

TANNER, W. F. 1961. Offshore shoals in area of energy deficit, *Jour. Sed. Petrol.*, Vol. 31, pp. 87–95.

THOM, B. G. 1964. Origin of sand beach ridges, *Aust. Jour. Sci.*, Vol. 26, pp. 351–2.

WIENS, H. J. 1959. Atoll development and morphology, *Ann. Ass. Amer. Geogrs*, Vol. 49, pp. 35–54.

ZENKOVITCH, V. P. 1959. On the genesis of cuspate spits along lagoon shores, *Jour. Geol.*, Vol. 67, pp. 269–77.

Eolian deposits

BAGNOLD, R. A. 1951. Sand formations in southern Arabia, *Geog. Jour.*, Vol. 177, pp. 78–87.

BEHEIRY, S. A. 1967. Sand forms in the Coachella valley, southern California, *Ann. Ass. Amer. Geogrs*, Vol. 57, pp. 25–48.

COOPER, W. S. 1958. Coastal sand dunes of Oregon and Washington, *Geol. Soc. Am.*, Mem. 72, 169 pp.

FINKEL, H. J. 1959. The barchans of southern Peru, *Jour. Geol.*, Vol. 67, pp. 614–47.

FLINT, R. F. and BOND, G. 1968. Pleistocene sand ridges and pans in western Rhodesia, *Bull. geol. Soc. Am.*, Vol. 79, pp. 299–314.

HASTENRATH, S. L. 1967. The barchans of the Arequipa region, southern Peru, *Zeit. f. Geomorph.*, Vol. 11, pp. 300–31.

INMAN, D. L., EWING, G. C., and CORLISS, J. B. 1966. Coastal sand dunes of Guerrero Negro, Baja California, Mexico, *Bull. geol. Soc. Am.*, Vol. 77, pp. 787–802.

JENNINGS, J. N. 1967. Cliff top dunes, *Aust. geog. Studies*, Vol. 5, pp. 40–9.

MADIGAN, C. T. 1936. The Australian sand-ridge deserts, *Geog. Rev.*, Vol. 26, pp. 205–27.

McKEE, E. D. and TIBBITTS, G. C. 1964. Primary structures of a seif dune and associated deposits in Libya, *Jour. Sed. Petrol.*, Vol. 34, pp. 5–17.

PERRIN, R. M. S. 1956. Nature of 'Chalk Heath' soils, *Nature*, Vol. 178, p. 31.

SHARP, R. P. 1966. Kelso dunes, Mojave desert, California, *Bull. geol. Soc. Am.*, Vol. 77, pp. 1045–74.

SIMONETT, D. S. 1960. Development and grading of dunes in western Kansas, *Ann. Ass. Amer. Geogrs*, Vol. 50, pp. 216–41.

VERSTAPPEN, H. T. 1968. On the origin of longitudinal (seif) dunes, *Zeit. f. Geomorph.*, Vol. 12, pp. 200–20.

Glacial deposits

ANDREWS, J. T., and SMITHSON, B. B. 1966. Till fabrics of the cross-valley moraines of north-central Baffin Island, Northwest Territories, Canada, *Bull. geol. Soc. Am.*, Vol. 77, pp. 271–90.

CHARLESWORTH, J. K. 1929. The South Wales end moraine, *Quart. Jour. Geol. Soc.*, Vol. 95, pp. 335–58.

FALCONER, G., IVES, J., LØKEN, O., and ANDREWS, J. T. 1965. Major end moraines in eastern and central Arctic Canada, *Geogr. Bull.*, No. 7, pp. 137–53.

GELLERT, J. F. 1966. Morphologie der Eisrandzonen der letzten skandinavischen Vereisung in Mittel- und Osteuropa, *Geog. Berichte*, Vol. 11, pp. 99–121.

GRAVENOR, C. P. 1953. The origin of drumlins, *Am. Jour. Sci.*, Vol. 251, pp. 674–81.

GRAVENOR, C. P. and KUPSCH, W. O. 1959. Ice-disintegration features in western Canada, *Jour. Geol.*, Vol. 67, pp. 48–64.

HARRIS, S. A. 1967. Origin of part of the Guelph drumlin field and the Galt and Paris moraines, Ontario: a reinterpretation, *Can. Geogr*, Vol. 11, pp. 16–34.

HOLMSEN, G. 1963. Glacial deposits in southeastern Norway, *Am. Jour. Sci.*, Vol. 261, pp. 880–9.

IVES, J. D. 1967. Glacial terminal and lateral features in northeast Baffin Island: illustrations with descriptive notes, *Geog. Bull.*, Vol. 9, pp. 62–70 and 106–14.

KIRBY, R. P. 1969. Variation in glacial deposition in a sub-glacial environment: an example from Midlothian, *Scot. Jour. Geol.*, Vol. 5, pp. 49–53.

MCKENZIE, G. D. 1969. Observations on a collapsing kame terrace in Glacier Bay National Monument, south-eastern Alaska, *Jour. Glaciol.*, Vol. 8, pp. 413–25.

ORME, A. R. 1967. Drumlins and the Weichsel glaciation of Connemara, *Irish Geog.*, Vol. 5, pp. 262–74.

ØSTREM, G. 1964. Ice-cored moraines in Scandinavia, *Geografiska Annaler*, Vol. 46, pp. 282–337.

PRICE, R. J. 1966. Eskers near the Casement glacier, Alaska, *Geografiska Annaler*, Vol. 48A, pp. 111–25.

PRICE, R. J. 1969. Moraines, sandar, kames and eskers near Breidamerkurjökull, Iceland, *Trans. Inst. Brit. Geogrs*, No. 46, pp. 17–43.

WEISSE, R. 1968. Endmöranen oder Oser örtlich von Glöwen, *Geog. Berichte*, Vol. 13, pp. 277–91.

D. Inter-relationships between Rock Breakdown, Transport, and Deposition

1–3. Examples of studies examining time lags and the influence both of source area and of rate of supply on further sedimentary processes and characteristics.

ANDERSON, H. W. 1957. Relating sediment yields to watershed variables, *Trans. Am. geophys. Union*, Vol. 38, pp. 921–4.

CULLEN, D. J. 1966. Fluviatile run-off as a factor in the primary dispersal of submarine gravels, *Sedimentology*, Vol. 7, pp. 191–201.

DAVIS, G. H. 1961. Geologic control of mineral composition of stream waters of the eastern slope of the southern Coast Ranges, California, *U.S. Geol. Surv. Water-supply Paper 1535-B*, 30 pp.

HEIDEL, S. G. 1956. The progressive lag of sediment concentration with flood waves, *Trans. Am. geophys. Union*, Vol. 47, pp. 56–66.

ICHIKAWA, M. 1958. On the debris supply from mountain slopes and its relation to river bed deposition, *Sci. Rep. Tokyo, Kyoiku Daigaku*, No. 49, 24 pp.

MANNER, S. B. 1958. Factors affecting sediment delivery rates in the Red Hills physiographic area, *Trans. Am. geophys. Union*, Vol. 39, pp. 669–75.

4. Covariance between phenomena

ANDRIESSE, J. P. 1969. A study of the environment and characteristics of tropical podzols in Sarawak (East Malaysia), *Geoderma*, Vol. 2, pp. 201–27.

BASCOM, W. N. 1951. The relationship between sand size and beach face slope, *Trans. Am. geophys. Union*, Vol. 32, pp. 866–74.

BETTENAY, E. 1962. The salt lake systems and their associated aeolian features in the semi-arid region of Western Australia, *Jour. Soil Sci.*, Vol. 13, pp. 10–17.

BLUCK, B. J. 1969. Particle rounding in beach gravels, *Geol. Mag.*, Vol. 106, pp. 1–14.

CAMELS, A. and CAILLEUX, A. 1966. Variabilité des galets le long d'une même plage, *C. r. Soc. géol. Fr.*, pp. 84–5.

CARR, A. P. 1969. Size grading along a pebble beach: Chesil beach, England, *Jour. Sed. Petrol.*, Vol. 39, pp. 297–310.

EYLES, R. J. 1968. Stream net ratios in west Malaysia, *Bull. geol. Soc. Am.*, Vol. 79, pp. 701–12.

FURLEY, P. A. 1968. Soil formation and slope development. 2. The relationships between soil formation and gradient angle in the Oxford area, *Zeit. f. Geomorph.*, Vol. 12, pp. 24–42.

HOLMES, C. D. 1960. Evolution of till-stone shapes, central New York, *Bull. geol. Soc. Am.*, Vol. 71, pp. 1645–60.

KNEBEL, H. J., KELLEY, J. C., and WHETTEN, J. T. 1968. Clay minerals of the Columbia river: a qualitative, quantitative and statistical evaluation, *Jour. Sed. Petrol.*, Vol. 38, pp. 600–11.

KUENEN, P. 1956. Experimental abrasion of pebbles. 2: rolling by current, *Jour. Geol.*, Vol. 64, pp. 336–68.

MACKNEY, D. and BURNHAM, C. P. 1964. A preliminary study of some slope soils in Wales and the Welsh borderland, *Jour. Soil Sci.*, Vol. 15, pp. 319–30.

MELTON, M. A. 1958. Correlation structure of morphometric properties of drainage systems and their controlling agents, *Jour. Geol.*, Vol. 66, pp. 442–60.

MELTON, M. A. 1958. Geometric properties of mature drainage systems and their representation in an E_4 space, *Jour. Geol.*, Vol. 66, pp. 35–56.

MELTON, M. A. 1965. Debris-covered hillslopes of the southern Arizona desert – consideration of their stability and sediment contribution, *Jour. Geol.*, Vol. 73, pp. 715–29.

PITTMAN, E. D. 1969. Destruction of plagioclase twins by stream transport, *Jour. Sed. Petrol.*, Vol. 39, pp. 1432–7.

PLUMLEY, W. J. 1948. Black Hills terrace gravels: a study in sediment transport, *Jour. Geol.*, Vol. 56, pp. 526–77.

RAGG, J. M. and BIBBY, J. S. 1966. Frost weathering and solifluction products in southern Scotland, *Geografiska Annaler*, Vol. 48A, pp. 12–23.

ROUGERIE, G. 1951. A propos de l'étude morphoscopique des galets équatoriaux, *C. r. Soc. géol. Fr.*, pp. 80–2.

SCHIEDEGGER, A. E. 1966. Horton's law of stream numbers, *Water Resources Research*, Vol. 4, pp. 651–5.

SMART, J. S. 1969. Topological properties of channel networks, *Bull. geol. Soc. Am.*, Vol. 80, pp. 1757–74.

SNEED, E. D. and FOLK, R. L. 1958. Pebbles in the Lower Colorado river, Texas: a study in particle morphogenesis, *Jour. Geol.*, Vol. 66, pp. 114–50.

STRAW, A. 1968. Late Pleistocene glacial erosion along the Niagara escarpment of southern Ontario, *Bull. geol. Soc. Am.*, Vol. 79, pp. 889–910.

WAGGONER, P. E. and BINGHAM, C. 1961. Depth of loess and distance from source, *Soil Sci.*, Vol. 92, pp. 396–401.

WALKER, P. H. 1964. Sedimentary properties and processes on a sandstone hillside, *Jour. Sed. Petrol.*, Vol. 34, pp. 328–34.

WATSON, J. P. 1964. A soil catena on granite in Southern Rhodesia. I. Field observations, *Jour. Soil. Sci.*, Vol. 15, pp. 238–57.

WINTERER, E. L. and VON DER BORCH, C. C. 1968. Striated pebbles in a mudflow deposit, South Australia, *Palaeogeog., Palaeoclim., Palaeoecol.*, Vol. 5, pp. 205–11.

WOLDENBERG, M. J. 1969. Spatial order in fluvial systems: Horton's laws derived from mixed hexagonal hierarchies of drainage basin areas, *Bull. geol. Soc. Am.*, Vol. 80, pp. 97–112.

5. Reworking of sediments in streams

CARLSTON, C. W. 1965. The relation of free meander geometry to stream discharge and its geomorphic implications, *Am. Jour. Sci.*, Vol. 263, pp. 864–85.

HORLOCK, J. H. 1955. Erosion in meanders, *Nature*, Vol. 176, pp. 1034.

LANGBEIN, W. B. and LEOPOLD, L. B. 1964. Quasi-equilibrium states in channel morphology, *Am. Jour. Sci.*, Vol. 262, pp. 782–94.

LEOPOLD, L. B. and WOLMAN, M. G. 1957. River channel patterns: braided, meandering and straight, *U.S. Geol. Surv. Prof. Paper 282-B*, pp. 39–85.

LEOPOLD, L. B. and WOLMAN, M. G. 1960. River meanders, *Bull. geol. Soc. Am.*, Vol. 71, pp. 769–94.

MØLLER, J. T. 1963. Accumulation and abrasion in a tidal area. Cartographic methods and results, *Geog. Tidsskr.*, Vol. 62, pp. 56–79.

SCHUMM, S. A. 1960. The effect of sediment type on the shape and stratification of some modern fluvial deposits, *Am. Jour. Sci.*, Vol. 258, pp. 177–84.

SCHUMM, S. A. 1963. Sinuosity of alluvial rivers on the Great Plains, *Bull. geol. Soc. Am.*, Vol. 74, pp. 1089–1100.

WERTZ, J. B. 1966. The flood cycle of ephemeral mountain streams in the southwestern United States, *Ann. Ass. Amer. Geogrs*, Vol. 56, pp. 598–633.

6. Dynamic depositional landforms

DIETZ, R. S. 1963. Wave-base, marine profile of equilibrium, and wave-built terraces: a critical appraisal, *Bull. geol. Soc. Am.*, Vol. 74, pp. 971–90.

DOLAN, R. 1966. Beach changes on the Outer Banks of North Carolina, *Ann. Ass. Amer. Geogrs*, Vol. 56, pp. 699–711.

HARMS, J. C. 1969. Hydraulic significance of some sand ripples, *Bull. geol. Soc. Am.*, Vol. 80, pp. 363–96.

HOOKE, R. LeB. 1967. Processes on arid-region alluvial fans, *Jour. Geol.*, Vol. 75, pp. 438–60.

LANGBEIN, W. B. and LEOPOLD, L. B. 1968. River channel bars and dunes – theory of kinematic waves, *U.S. Geol. Surv. Prof. Paper 422-L*, 20 pp.

LETTAU, K. and LETTAU, H. 1969. Bulk transport of sand by the barchans of the Pampa de La Toya in southern Peru, *Zeit. f. Geomorph.*, Vol. 13, pp. 182–95.

MILLER, R. L. and ZEIGLER, J. M. 1958. A model relating dynamics and sediment pattern in equilibrium in the region of shoaling waves, breaker zone, and foreshore, *Jour. Geol.*, Vol. 66, pp. 417–41.

PALMER, H. R. 1834. Observations on the motion of shingle beaches, *Phil. Trans.*, Vol. 124, pp. 567–76.

PRICE, W. A. 1963. Patterns of flow and channeling in tidal inlets, *Jour. Sed. Petrol.*, Vol. 33, pp. 279–90.

SHARP, R. P. 1963. Wind ripples, *Jour. Geol.*, Vol. 71, pp. 617–36.

TRICART, J. and MAINGUET, M. 1965. Caractéristiques granulométriques de quelques sables éolien du désert péruvien, aspects de la dynamique des barkanes, *Rev. géomorph. dyn.*, Vol. 15, pp. 110–21.

7. Reworking of sediments into distinctive patterns

BALL, D. F. 1967. Stone pavements in soils of Caernarvonshire, North Wales, *Jour. Soil. Sci.*, Vol. 18, pp. 103–8.

BALL, D. F. and GOODIER, R. 1968. Large sorted stone-stripes in the Rhinog Mountains, North Wales, *Geografiska Annaler*, Vol. 50A, pp. 54–9.

CAINE, T. N. 1963. The origin of sorted stripes in the Lake District, northern England, *Geografiska Annaler*, Vol. 45, pp. 172–9.

CORTE, A. 1963. Particle sorting by repeated freezing and thawing, *Science*, Vol. 142, pp. 499–501.

COSTIN, A. B., THOM, B., WIMBUSH, D. J., and STUIVER, M. 1967. Nonsorted steps in the Mt. Kosciusko area, Australia, *Bull. geol. Soc. Am.*, Vol. 78, pp. 979–92.

DZULYNSKI, S. 1963. Polygonal structures in experiments and their bearing upon some periglacial phenomena, *Bull. Acad. Polonaise Sci.*, Vol. 11, pp. 145–50.

FLEMMING, N. C. 1964. Tank experiments on the sorting of beach material during cusp formation, *Jour. Sed. Petrol.*, Vol. 34, pp. 112–22.

HAY, T. 1936. Stone stripes, *Geog. Jour.*, Vol. 87, pp. 47–50.

HOPKINS, D. M. and SIGAFOOS, R. S. 1954. Discussion: role of frost thrusting in the formation of tussocks, *Am. Jour. Sci.*, Vol. 252, pp. 55–9.

JOHNSSON, G. 1960. Cryoturbation at Zaragoza, northern Spain, *Zeit. f. Geomorph.*, Vol. 4, pp. 74–80.

KLATKA, T. 1961. Problemes des sols striés de la partie septentrionale de la presqu'île de Sörkapp (Spitzbergen), *Biul. Peryglacjalny*, Vol. 10, pp. 291–320.

KLATKA, T. 1962. Champs de pierre de Lysogory, origine et âge, *Acta. geog. Lodz*, No. 12, 129 pp.

KUENEN, P. H. 1948. The formation of beach cusps, *Jour. Geol.*, Vol. 56, pp. 34–40.

LACHENBRUCH, A. H. 1962. Mechanics of thermal contraction cracks and ice-wedge polygons in permafrost, *Geol. Soc. Am. Spec. Paper*, No. 70, 69 pp.

MACKAY, J. R. 1962. Pingos of the Pleistocene Mackenzie delta area, *Geog. Bull.*, No. 18, pp. 21–63.

MÜLLER, F. 1959. Beobachtungen uber Pingos in Östgronlands und in der Kanadischen Arktis, *Medd. Grøland Dañm.*, Vol. 153, 127 pp.

NEAL, J. T., LANGER, A. M., and KERR, P. F. 1968. Giant desiccation polygons of Great Basin playas, *Bull. geol. Soc. Am.*, Vol. 79, pp. 69–90.

PHILBERTH, K. 1964. Recherches sur les sols polygonaux et striés, *Biul. Peryglacjalny*, Vol. 13, pp. 99–198.

POSER, H. 1933. Das Problem des Strukturbodens, *Geol. Rundschau*, Vol. 24, pp. 105–21.

RUSSELL, R. J. and MCINTIRE, W. G. 1965. Beach cusps, *Bull. geol. Soc. Am.*, Vol. 76, pp. 307–20.

SCHREIBER, K. F. 1969. Beobachtungen über die Entstehung von 'Buckelweiden' auf der Hochflächen des schweizer Jura, *Erdkunde*, Vol. 23, pp. 280–90.

SVENSSON, H. 1964. Traces of pingo-like frost mounds, *Lund studies in Geog. series A*, No. 30, pp. 93–106.

SVENSSON, H. *et al.* 1967. Polygonal ground and solifluction features, *Lund studies in Geog. series A*, No. 40, 67 pp.

TABER, S. 1929. Frost heaving, *Jour. Geol.*, Vol. 37, pp. 428–61.

TALLIS, J. H. and KERSHAW, K. A. 1959. Stability of stone polygons in North Wales, *Nature*, Vol. 183, pp. 485–6.

WASHBURN, A. L. 1956. Classification of patterned ground and review of suggested origins, *Bull. geol. Soc. Am.*, Vol. 67, pp. 823–65.

VERGER, F. 1964. Mottureaux et gilgais, *Ann. Géog.*, Vol. 73, pp. 413–30.

8. Recombination of weathered chemical elements

BASSETT, H. 1954. Silicification of rocks by surface waters, *Am. Jour. Sci.*, Vol. 252, pp. 733–5.

BRÜCKNER, W. 1955. The mantle-rock ('laterite') of the Gold Coast and its origin, *Geol. Rundschau*, Vol. 43, pp. 307–27.

CHARLES, G. 1949. Sur la formation de la carapice zonaire en Algérié, *C. r. Acad. Sci.*, Vol. 228, pp. 261–3.

DEMANGEOT, J. 1961. Pseudo-cuestas de la zone intertropicale, *Bull. Ass. Géog. Fr.*, No. 296–7, pp. 2–17.

DROSTE, J. B. 1956. Alteration of clay minerals by weathering in Wisconsin tills, *Bull. geol. Soc. Am.*, Vol. 67, pp. 911–18.

ENGEL, C. G. and SHARP, R. P. 1958. Chemical data on desert varnish, *Bull. geol. Soc. Am.*, Vol. 69, pp. 487–518.

FREISE, E. W. 1936. Bodenverkrustung in Brasilien, *Zeit. f. Geomorph.*, Vol. 9, pp. 233–48.

GUILCHER, A. 1961. Le 'beach-rock' ou grès de plage, *Ann. Géog.*, Vol. 70, pp. 113–25.

KELLER, W. D. 1964. Processes of origin and alteration of clay minerals, *in Soil clay mineralogy – a symposium* (Chapel Hill, Univ. North Carolina Press), pp. 3–76.

LAMOTTE, M. and ROUGERIE, G. 1962. Les apports allochtones dans la genèse des cuirasses ferrugineuses, *Rev. géomorph. dyn.*, Vol. 13, pp. 145–60.

LANGFORD-SMITH, T. and DURY, G. H. 1965. Distribution, character and attitude of the duricrust in the northwest of New South Wales and the adjacent areas of Queensland, *Am. Jour. Sci.*, Vol. 263, pp. 170–90.

LARSEN, G. and CHILINGAR, G. V. 1967. *Diagenesis in sediments* (Elsevier, Amsterdam), 551 pp.

MAUD, R. R. 1965. Laterite and lateritic soil in coastal Natal, South Africa, *Jour. Soil Sci.*, Vol. 16, pp. 60–72.

MOSELEY, F. 1965. Plateau calcrete, calcreted gravels, cemented dunes and related deposits of the Maallegh-Bomba region of Libya, *Zeit. f. Geomorph.*, Vol. 9, pp. 166–85.

MOSS, R. P. 1965. Slope development and soil morphology in a part of south-west Nigeria, *Jour. Soil Sci.*, Vol. 16, pp. 192–209.

PRIDER, R. T. 1966. The lateritized surface of Western Australia, *Aust. Jour. Sci.*, Vol. 28, pp. 443–51.

ROUGERIE, G. 1959. Latéritisation et pédogenèse intertropicale, *Inf. géog.*, Vol. 23, pp. 199–206.

STEPHENS, C. G. 1966. Origin of silicretes of central Australia, *Nature*, Vol. 209, p. 496.

STODDART, D. R. and CANN, J. R. 1965. Nature and origin of beachrock, *Jour. Sed. Petrol.*, Vol. 35, pp. 243–7.

TAYLOR, J. C. M., and ILLING, L. V. 1969. Holocene intertidal calcium carbonate cementation, Qatar, Persian Gulf, *Sedimentology*, Vol. 12, pp. 69–107.

YAALON, D. H. 1967. Factors affecting the lithification of eolianite and interpretation of its environmental significance in the coastal plain of Israel, *Jour. Sed. Petrol.*, Vol. 37, pp. 1189–99.

9. Influence of weathered material on subsequent weathering

ACKERMANN, E. 1962. Butserstein-Zeugen vorzeitlicher Grundwasserschwankungen, *Zeit. f. Geomorph.*, Vol. 6, pp. 148–82.

BÜDEL, J. 1957. Die 'Doppelten Einebnungsflächen' in den feuchten Troppen, *Zeit. f. Geomorph.*, Vol. 1, pp. 201–28.

GILLULY, J. 1937. Physiography of the Ajo region, Arizona, *Bull. geol. Soc. Am.*, Vol. 48, pp. 323–48.

MABBUTT, J. A. 1966. Mantle-controlled planation of pediments, *Am. Jour. Sci.*, Vol. 264, pp. 78–91.

MORTENSEN, H. 1927. *Der Formenschatz der nordchilenischen Wüste* (Weidmann, Berlin), 191 pp.

WAHRHAFTIG, C. 1965. Stepped topography of the southern Sierra Nevada, *Bull. geol. Soc. Am.*, Vol. 76, pp. 1165–90.

YOUNG, R. G. 1964. Fracturing of sandstone cobbles in caliche-cemented terrace gravels, *Jour. Sed. Petrol.*, Vol. 34, pp. 887–9.

10. The concept of quasi-equilibrium

CARLSTON, C. W. 1969. Downstream variations in the hydraulic geometry of streams: special emphasis on mean velocity, *Am. Jour. Sci.*, Vol. 267, pp. 499–509.

CARLSTON, C. W. 1969. Longitudinal slope characteristics of rivers of the midcontinent and the Atlantic East Gulf slopes, *Bull. Internl Ass. Sci. Hydrol.*, Vol. 14, pp. 21–31.

DURY, G. H. 1963. Rivers in geographical teaching, *Geography*, Vol. 48, pp. 18–30.

HACK, J. T. 1956. Studies of longitudinal stream profiles in Virginia and Maryland, *U.S. Geol. Surv. Prof. Paper 294-B*, pp. 45–97.

LEOPOLD, L. B. and LANGBEIN, W. B. 1962. The concept of entropy in landscape evolution, *U.S. Geol. Surv. Prof. Paper 500-A*, 20 pp.

MACKIN, J. H. 1948. Concept of the graded river, *Bull. geol. Soc. Am.*, Vol. 59, pp. 463–512.

SCHUMM, S. A. and LICHTY, R. W. 1965. Time, space, and causality in geomorphology, *Am. Jour. Sci.*, Vol. 263, pp. 110–19.

SOUCHEZ, R. 1961. Pente d'équilibre de creusement et force tractrice, *Bull. Soc. Roy. Belge géog.*, Nos I–IV, pp. 145–58.

E. Inter-relationships between Form and Process

ALLEN, J. R. L. 1965. Scour marks in snow, *Jour. Sed. Petrol.*, Vol. 35, pp. 331–8.

BRUNET, P. 1956. Deux processus d'érosion en haute montagne, *Rev. géomorph. dyn.*, Vol. 7, pp. 143–7.

CAILLEUX, A. 1948. Le ruissellement en pays tempéré non-montagneux, *Ann. Géog.*, Vol. 57, pp. 21–39.

CALLENDER, G. S. 1951. The effect of the altitude of the firn area of a glacier's response to temperature variations, *Jour. Glaciol.*, Vol. 1, pp. 573–6.

FLINT, R. F. 1943. Growth of North American ice sheet during the Wisconsin age, *Bull. geol. Soc. Am.*, Vol. 54, pp. 325–62.

JENNINGS, J. N. and BIK, M. J. 1962. Karst morphology in Australian New Guinea, *Nature*, Vol. 194, pp. 1036–8.

RATHJENS, C. 1951. Der Hochkarst im System der klimatischen morphologie, *Erdkunde*, Vol. 5, pp. 310–15.

SCHOELLER, H. 1950. Les variation de la teneur en gas carbonique des eaux souterraines en fonction de l'altitude, *C. r. Acad. Sci.*, Vol. 230, pp. 560–1.

SPREEN, W. C. 1947. A determination of the effect of topography on precipitation, *Trans. Amer. geophys. Union*, Vol. 28, pp. 285–90.

F. The Significance of Changes in Weather and in Climate

1. Local changes due to differing aspects

BALCHIN, W. G. V. and PYE, N. 1947. A microclimatological investigation of Bath and the surrounding districts, *Quart. Jour. Roy. Met. Soc.*, Vol. 73, pp. 297–319.

BEATY, C. B. 1962. Asymmetry of stream patterns and topography in the Bitterroot Range, Montana, *Jour. Geol.*, Vol. 70, pp. 347–54.

CAILLEUX, A. and CALKIN, P. 1963. Orientation of hollows in cavernously weathered boulders in Antarctica, *Biul. Peryglacjalny*, Vol. 12, pp. 147–50.

CARSON, C. E. and HUSSEY, K. M. 1962. The oriented lakes of arctic Alaska, *Jour. Geol.*, Vol. 70, pp. 417–39.

FLINT, R. F. and FIDALGO, F. 1964. Glacial geology of the east flank of the Argentine Andes between latitude 39° 10′ S. and latitude 41° 20′ S., *Bull. geol. Soc. Am.*, Vol. 75, pp. 335–52.

GARNETT, A. 1937. Insolation and relief, *Trans. Inst. Brit. Geogrs*, No. 5, 71 pp.

GILBERT, G. K. 1904. Systematic asymmetry of crest lines in the High Sierra of California, *Jour. Geol.*, Vol. 12, pp. 579–88.

HEMBREE, C. H. and RAINWATER, F. H. 1961. Chemical degradation on opposite flanks of the Wind River Range, Wyoming, *U.S. Geol. Surv. Water-supply Paper 1535-E*, 9 pp.

JENNINGS, J. N. 1957. On the orientation of parabolic or U-dunes, *Geog. Jour.*, Vol. 123, pp. 474–80.

LANDSBERG, S. Y. 1956. The orientation of dunes in Britain and Denmark in relation to the wind, *Geog. Jour.*, Vol. 122, pp. 176–89.

LEVERETT, F. 1929. Pleistocene glaciations of the northern hemisphere, *Bull. geol. Soc. Am.*, Vol. 40, pp. 745–60.

LINTON, D. L. 1959. Morphological contrasts of eastern and western Scotland, *in* R. Miller and J. W. Watson (Eds), *Geographical essays in memory of Alan G. Ogilvie*, pp. 16–45.

MAARLEVELD, G. C. 1960. Wind directions and coversands in the Netherlands, *Biul. Peryglacjalny*, Vol. 8, pp. 49–58.

MABBUTT, J. A., WOODING, R. A., and JENNINGS, J. N. 1969. The asymmetry of Australian desert sand ridges, *Aust. Jour. Sci.*, Vol. 32, pp. 159–60.

MELTON, M. A. 1960. Intravalley variation in slope angles related to microclimate and erosional environment, *Bull. geol. Soc. Am.*, Vol. 71, pp. 133–44.

ØSTREM, G., BRIDGE, C. W., RENNIE, W. F. 1967. Glacio-hydrology, discharge and sediment transport in the Decade Glacier area, Baffin Island, N.W.T., *Geografiska Annaler*, Vol. 49A, pp. 268–82.

SEDDON, B. 1957. Late-glacial cwm glaciers in Wales, *Jour. Glaciol.*, Vol. 3, pp. 94–9.

SOONS, J. M. and RANIER, J. N. 1968. Micro-climate and erosion processes in the southern Alps, New Zealand, *Geografiska Annaler*, Vol. 50A, pp. 1–15.

SPEITZER, H. 1960. Hangformung und Asymmetrie der Bergrücken in den Alpen und im Taurus, *Zeit. f. Geomorph.*, Suppbd 1, pp. 211–36.

STOKES, W. L. 1964. Incised, wind-aligned stream patterns of the Colorado plateau, *Am. Jour. Sci.*, Vol. 262, pp. 808–16.

TEMPLE, P. H. 1965. Some aspects of cirque distribution in the west-central Lake District, northern England, *Geografiska Annaler*, Vol. 47A, pp. 185–93.

THORP, J. 1934. The asymmetry of the Pepino Hills of Puerto Rico in relation to the trade winds, *Jour. Geol.*, Vol. 42, pp. 537–45.

WHITE, E. M. 1961. Drainage alignment in western South Dakota, *Am. Jour. Sci.*, Vol. 259, pp. 207–10.

WHITE, W. A. 1966. Drainage asymmetry and the Carolina capes, *Bull. geol. Soc. Am.*, Vol. 77, pp. 223–40.

WILLIAMS, M. A. J. 1969. Prediction of rainsplash erosion in the seasonally wet tropics, *Nature*, Vol. 222, pp. 763–5.

2 and 3. Changes in the effectiveness of processes with time and examples of past climates

ARNBORG, L., WALKER, H. J., and PEIPPO, J. 1967. Suspended load in the Colville River, Alaska, 1962, *Geografiska Annaler*, Vol. 49A, pp. 131–44.

AXELROD, D. I. and BAILEY, H. P. 1969. Paleotemperature analysis of Tertiary floras, *Palaeogeog., palaeoclim., palaeoecol.*, Vol. 6, pp. 163–95.

BACHMANN, F. 1966. *Fossile Strukturböden und Eiskeile* (Kunz-Druck, Zürich), 177 pp.

BECKINSALE, R. P. 1965. Climatic change: a critique of modern theories, *in* J. B. Whittow and P. D. Wood (Eds), *Essays in geography for Austin Miller* (Univ. of Reading, Reading), pp. 1–38.

BOWEN, R. N. C. 1966. *Palaeotemperature analysis* (Elsevier, Amsterdam), 265 pp.

BROECKER, W. S., EWING, M., and HEEZEN, B. C. 1960. Evidence for an abrupt change in climate close to 11,000 years ago, *Am. Jour. Sci.*, Vol. 258, pp. 429–48.

BÜDEL, J. 1951. Die Klimazonen des Eiszeitalters, *Eiszeitalter und Gegenwart*, Vol. 1, pp. 16–26; trans. 1959, Climatic zones of the Pleistocene, *Internl Geol. Rev.*, Vol. 1, pp. 72–9.

BÜDEL, J. 1963. Die pliozänen und quartären Pluvialzeiten der Sahara, *Eiszeitalter und Gegenwart*, Vol. 14, pp. 161–87.

COETZEE, J. A. 1964. Evidence for a considerable depression of the vegetation belts during the Upper Pleistocene on the East African mountains, *Nature*, Vol. 204, pp. 564–6.

COOPE, G. R. 1962. A Pleistocene coeopterous fauna with arctic affinities from Fladbury, Worcestershire, *Quart. Jour. Geol. Soc.*, Vol. 118, pp. 103–23.

COTTON, C. A. 1958. Alternating Pleistocene morphogenetic systems, *Geol. Mag.*, Vol. 95, pp. 125–36.

CROLL, J. 1875. *Climate and time and their geological relations* (E. Stanford, London), 577 pp.

CURRY, R. R. 1966. Glaciation about 3,000,000 years ago in Sierra Nevada, *Science*, Vol. 154, pp. 770–1.

CURRY, R. R. 1969. Holocene climatic and glacial history of the central Sierra Nevada, California, *Geol. Soc. Am. Spec. Paper 123*, pp. 1–47.

CZEPPE, Z. 1960. Annual course and the morphological effect of the vertical frost movements of soil at Hornsund, Vestspitzbergen, *Bull. Acad. Polon. Sci. Sér. Sci. géol. géog.*, No. 2, pp. 145–8.

CZEPPE, Z. 1968. The annual rhythm of morphogenetic processes in Spitzbergen, *Geog. Polonica*, Vol. 14, pp. 657–65.

DENTON, G. H. and ARMSTRONG, R. L. 1969. Miocene–Pliocene glaciations in southern Alaska, *Am. Jour. Sci.*, Vol. 267, pp. 1121–42.

DIMBLEBY, G. W. 1952. Pleistocene ice-wedges in north-east Yorkshire, *Jour. Soil Sci.*, Vol. 5, pp. 1–19.

DRESCH, J. 1954. Formes et limites climatiques et paléoclimatiques en Afrique du Nord, *Ann. Géog.*, Vol. 58, pp. 56–9.

DURY, G. H. 1954. Weather, climate and river erosion in the Ice Age, *Science News*, Vol. 33, pp. 65–88.

DYLIK, J. 1960. Rhythmically stratified slope waste deposits, *Biul. Peryglacjalny*, Vol. 8, pp. 31–41.

EMILIANI, C. 1955. Pleistocene temperatures, *Jour. Geol.*, Vol. 63, pp. 538–78.

EWING, M. and DONN, W. L. 1956. A theory of ice ages, *Science*, Vol. 123, pp. 1061–6.

FLINT, R. F. 1959. Pleistocene climates in eastern and southern Africa, *Bull. geol. Soc. Am.*, Vol. 70, pp. 343–74.

FLINT, R. F. and BRANDTNER, F. 1961. Climatic changes since the Last Interglacial, *Am. Jour. Sci.*, Vol. 259, pp. 321–8.

FRENZEL, B. 1967. *Die Klimaschwankungen des Eiszeitalters* (Vieweg, Braunschweig), 296 pp.

FRYE, J. C. 1962. Comparison between Pleistocene deep-sea temperatures and glacial and interglacial episodes, *Bull. geol. Soc. Am.*, Vol. 73, pp. 263–6.

GENTILLI, J. 1961. Quaternary climates of the Australian region, *Ann. New York Acad. Sci.*, Vol. 95, pp. 465–501.

GRIPP, K. 1964. Winter-Phänomena am Meeresstrand, *Zeit. f. Geomorph.*, Vol. 8, pp. 326–31.

GROVE, A. T. and WARREN, A. 1968. Quaternary landforms and climate on the south side of the Sahara, *Geog. Jour.*, Vol. 134, pp. 194–208.

GUILLIEN, Y. 1955. La couverture végétale de l'Europe pléistocène, *Ann. Géog.*, Vol. 64, pp. 241–76.

HEY, R. W. 1963. Pleistocene screes in Cyrenaica (Libya), *Eiszeitalter und Gegenwart*, Vol. 14, pp. 77–84.

HOINKES, H. C. 1968. Glacier variation and weather, *Jour. Glaciol.*, Vol. 7, pp. 3–20.

HOMMERIL, P., and LARSONNEUR, C. 1963. Quelques effets morphologiques du gel intense de l'hiver 1963 sur le littoral bas-normand, *Cahiers Océanog.*, Vol. 15, pp. 638–50.

HOUTMAN, T. J. 1965. Winter hydrological conditions of coastal waters south of Kaikoura peninsula, *N.Z. Jour. Geol. Geophys.*, Vol. 8, pp. 807–19.

JOHNSSON, G. 1959. True and false ice-wedges in southern Sweden, *Geografiska Annaler*, Vol. 41, pp. 15–33.

474 Introduction to Geomorphology

KAISER, K. 1960. Klimazeugen des periglazialen Dauerfrostbodens in Mittel- und Westeuropa, *Eiszeitalter und Gegenwart*, Vol. 11, pp. 121–41.

LADURIE, E. LE R. 1967. *Histoire du climat depuis l'an mil* (Flammarion, Paris), 379 pp.

LÜTIG, G. 1965. Interglacial and interstadial periods, *Jour. Geol.*, Vol. 73, pp. 579–91.

MILANKOVITCH, M. 1938. Astrononische Mittel zur Erforschung der erdgeschichtlichen Klimate, *Hardbuch der Geophysik*, Vol. 9, pp. 593–698.

MOBERLY, R., BAVER, L. D., and MORRISON, A. 1965. Source and variation of Hawaiian littoral sand, *Jour. Sed. Petrol.*, Vol. 35, pp. 589–98.

NAIRN, A. E. M. (Ed.). 1961. *Descriptive palaeoclimatology* (Interscience, London), 380 pp.

NAIRN, A. E. M. (Ed.). 1964. *Problems in palaeoclimatology* (Interscience, London), 705 pp.

OLDFIELD, F. 1960. Late Quaternary changes in climate, vegetation and sea-level in lowland Lonsdale, *Trans. Inst. Brit. Geogrs*, Vol. 28, pp. 99–117.

OTVOS, E. G. 1965. Sedimentation-erosion cycles of single tidal periods on Long Island Sound beaches, *Jour. Sed. Petrol.*, Vol. 35, pp. 604–9.

PÉWÉ, T. L., CHURCH, R. E., and ANDRESEN, M. J. 1969. Origin and paleoclimatic significance of large-scale patterned ground in the Donnelly Dome area, Alaska, *Geol. Soc. Am. Special Paper 103*, 87 pp.

PÉWÉ, T. L. (Ed.). 1969. *The periglacial environment: past and present* (McGill–Queens University Press, Montreal), 487 pp.

POSER, H. 1948. Boden- und Klimaverhältnisse in Mittel- und Westeuropa während der Würmeiszeit, *Erdkunde*, Vol. 2, pp. 53–68.

PRENTICE, J. E. and MORRIS, P. G. 1959. Cemented screes in the Manifold valley, north Staffordshire, *East Midld Geogr.*, Vol. 2, pp. 16–20.

REEVES, C. C. 1965. Pleistocene climate of the Llano Estacado, *Jour. Geol.*, Vol. 73, pp. 181–8.

REEVES, C. C. 1968. *Introduction to Palaeolimnology* (Elsevier, Amsterdam), 228 pp.

REID, C. 1887. On the origin of the dry chalk valleys and of the Coombe rock, *Quart. Jour. geol. Soc.*, Vol. 43, pp. 364–73.

SCHWARZBACH, M. 1961. *Das Klima der Vorzeit* (Enke, Stuttgart), 275 pp.; trans. by R. O. Muir, 1963, *Climates of the past* (Van Nostrand, London), 328 pp.

ŠEGOTA, T. 1967. Paleotemperature changes in the Upper and Middle Pleistocene, *Eiszeitalter und Gegenwart*, Vol. 18, pp. 127–41.

SHOTTON, F. W. 1960. Large-scale patterned ground in the valley of the Worcestershire Avon, *Geol. Mag.*, Vol. 97, pp. 404–8.

STARKEL, L. 1966. Post-glacial climate and the moulding of European relief, *Roy. Met. Soc. Symp. on World climate 8000 to 0 BC*, pp. 15–33.

STRAHLER, A. N. 1966. Tidal cycles of changes in an equilibrium beach, Sandy Hook, New Jersey, *Jour. Geol.*, Vol. 74, pp. 247–68.

TABER, B. A. 1950. Intensive frost action along lake shores, *Am. Jour. Sci.*, Vol. 248, pp. 784–93.

TUCKFIELD, C. G. 1964. Gully erosion in the New Forest, Hampshire, *Am. Jour. Sci.*, Vol. 262, pp. 795–807.

UNESCO. 1963. Changes of climate, *Arid Zone Research*, Vol. 20 (UNESCO, Paris), 488 pp.

WALKER, D. and WEST, R. G. (Eds). 1970. *Studies in the vegetational history of the British Isles* (Cambridge Univ. Press, Cambridge), 266 pp.

WATERS, R. S. 1961. Involutions and ice-wedges in Devon, *Nature*, Vol. 189, pp. 389–90.

WATSON, E. 1965. Periglacial structures in the Aberystwyth region of central Wales, *Proc. Geol. Soc.*, Vol. 76, pp. 443–62.

WATT, A. S., PERRIN, R. M. S., and WEST, R. G. 1966. Patterned ground in Breckland: structure and composition, *Jour. Ecol.*, Vol. 54, pp. 239–58.

WEAVER, C. E. 1967. Variability of a river clay suite, *Jour. Sed. Petrol.*, Vol. 37, pp. 971–4.

WOLMAN, M. G. 1959. Factors influencing erosion of a cohesive river bank, *Am. Jour. Sci.*, Vol. 257, pp. 204–16.

WRIGHT, H. E. 1961. Late Pleistocene climates of Europe: a review, *Bull. geol. Soc. Am.*, Vol. 257, pp. 204–16.

YEHLE, L. A. 1954. Soil tongues and their confusion with certain indicators of periglacial climate, *Am. Jour. Sci.*, Vol. 252, pp. 532–46.

ZEIGLER, J. M., and TUTTLE, S. D. 1961. Beach changes based on daily measurements of four Cape Cod beaches, *Jour. Geol.*, Vol. 69, pp. 583–99.

Chapter V. Landforms and time

A. Dating

ANDREWS, J. T. and WEBBER, P. J. 1964. A lichenometrical study of the north-western margin of the Barnes Ice Cap: a geomorphological technique, *Geogr. Bull.*, No. 22, pp. 80–104.

AVER, V. 1963. Late glacial and postglacial shoreline displacements in South America as established by tephrachronology, compared with displacements of the Baltic shorelines, *Fennia*, Vol. 89, No. 1, pp. 51–5.

AXELROD, D. I. 1966. Potassium-argon ages of some western Tertiary floras, *Am. Jour. Sci.*, Vol. 264, pp. 497–506.

BESCHEL, R. 1958. Lichenometrical studies in West Greenland, *Arctic*, Vol. 11, p. 254.

BLACKBURN, G. 1966. Radiocarbon dates relating to soil development, coastline changes, and volcanic ash deposition in south-east South Australia, *Aust. Jour. Sci.*, Vol. 29, pp. 50–2.

BOWEN, D. Q. 1966. Dating Pleistocene events in south-west Wales, *Nature*, Vol. 211, pp. 475–6.

COX, A., DALRYMPLE, G. B., and DOELL, R. R. 1967. Reversals of the Earth's magnetic field, *Sci. American*, Vol. 216, pp. 44–54.

DAVIS, M. B. 1963. On the theory of pollen analysis, *Am. Jour. Sci.*, Vol. 261, pp. 897–912.

DIMBLEBY, G. W. 1961. Transported material in the soil profile, *Jour. Soil Sci.*, Vol. 12, pp. 12–22.

EMILIANI, C. 1955. Pleistocene temperatures, *Jour. Geol.*, Vol. 63, pp. 538–78.

EVERNDEN, J. F., CURTIS, G. H., and KISTLER, R. 1950. Potassium-argon dating of Pleistocene volcanics, *Quaternaria*, Vol. 4, pp. 13–17.

EVERNDEN, J. F. and JAMES, G. T. 1964. Potassium-argon dates and the Tertiary floras of North America, *Am. Jour. Sci.*, Vol. 262, pp. 945–74.

FAEGRI, K. and IVERSEN, J. 1966. *Textbook of modern pollen analysis*; 2nd rev. ed. (Munksgaard, Copenhagen), 237 pp.

GEER, G. DE. 1912. A geochronology of the last 12,000 years, *C. r. XI Internl. Geol. Congress (Stockholm)*, Vol. 1, pp. 241–53.

GEER, G. DE. 1951. Conclusions from C_{14} and de Geer's chronology, Dani-Gotiglacial, with datings, *Geol. Fören Förhandl.*, Vol. 73, pp. 557–71.

GODWIN, H. 1941. Pollen analysis and Quaternary geology, *Proc. Geol. Ass.*, Vol. 52, pp. 328–61.

GODWIN, H. 1960. Radiocarbon dating and Quaternary history in Britain, *Proc. Roy. Soc. B*, Vol. 153, pp. 287–320.

HAMILTON, E. I. 1965. *Applied geochronology* (Academic Press, London), 267 pp.

HEATH, J. P. 1960. Repeated avalanches at Chaos Jumbles, Lassen Volcanic National Park, *Am. Jour. Sci.*, Vol. 258, pp. 744–51.

HOLMES, A. 1960. A revised geological time scale, *Trans. Edin. Geol. Soc.*, Vol. 17, pp. 183–216.

IVES, R. L. 1962. Dating of the 1746 eruption of Tres Virgenes volcano, Baja California del Sur, Mexico, *Bull. geol. Soc. Am.*, Vol. 73, pp. 647–8.

KAIZUKA, S. 1958. Tephrochronological studies in Japan, *Erdkunde*, Vol. 12, pp. 253–70.

KAYE, C. A. 1964. The upper limit of barnacles as an index of sea-level change on the New England coast during the past 100 years, *Jour. Geol.*, Vol. 72, pp. 580–600.

KELLEY, T. E. and BAKER, C. H. 1966. Color variations within glacial till, east-central North Dakota – a preliminary investigation, *Jour. Sed. Petrol.*, Vol. 36, pp. 75–80.

LANDIM, P. M. B. and FRAKES, L. A. 1968. Distinction between tills and other diamictons based on textural characteristics, *Jour. Sed. Petrol.*, Vol. 38, pp. 1213–23.

LIBBY, W. F. 1965. *Radiocarbon dating*, 2nd ed. (Univ. Chicago Press, Chicago), 175 pp.

MERRITT, R. S. and MULLER, E. H. 1959. Depth of leaching in relation to carbonate content of till in central New York State, *Am. Jour. Sci.*, Vol. 257, pp. 465–80.

MOORBATH, S. 1962. Lead isotope abundance studies on mineral occurrences in the British Isles and their geological significance, *Phil. Trans. Roy. Soc.*, Vol. 254A, pp. 295–360.

MÜNNICH, K. O. 1960. Die C_{14} Methode, *Geol. Rundschau*, Vol. 49, pp. 237–44.

PULLAR, W. A., 1967. Uses of volcanic ash beds in geomorphology, *Earth Sci. Jour.*, Vol. 1, pp. 164–77.

RONCA, L. B. and ZELLER, E. J. 1965. Thermoluminescence as a function of climate and temperature, *Am. Jour. Sci.*, Vol. 263, pp. 416–28.

ROSHOLT, J. N., EMILIANI, C., GEISS, J., KOCZY, F. F., and WANGERSKY, P. J. 1961. Absolute dating of deep-sea cores by the Pa^{231}/Th^{230} method, *Jour. Geol.*, Vol. 69, pp. 162–85.

SHACKLETON, N. 1967. Oxygen isotope analyses and Pleistocene temperatures reassessed, *Nature*, Vol. 215, pp. 15–17.

SHARP, R. P. 1960. Pleistocene glaciation in the Trinity Alps of northern California, *Am. Jour. Sci.*, Vol. 250, pp. 305–40.

SHARP, R. P. 1969. Semiquantitative differentiation of glacial moraines near Convict Lake, Sierra Nevada, California, *Jour. Geol.*, Vol. 77, pp. 68–91.

SHOTTON, F. W. 1966. The problems and contributions of methods of absolute dating within the Pleistocene period, *Quart. Jour. geol. Soc.*, Vol. 122, pp. 357–83.

SIMONSON, R. W. 1954. Identification and interpretation of buried soils, *Am. Jour. Sci.*, Vol. 252, pp. 705–32.

STORK, A. 1963. Plant immigration in front of retreating glaciers, with examples from the Kebnekaise area, northern Sweden, *Geografiska Annaler*, Vol. 45, pp. 1–22.

WAMBEKE, A. R. VAN. 1962. Criteria for classifying tropical soils by age, *Jour. Soil. Sci.*, Vol. 13, pp. 124–32.

ZEUNER, F. E. 1958. *Dating the past*, 4th ed., rev. and enl. (Methuen, London), 516 pp.

B. The Initial Form

AMBROSE, J. W. 1964. The exhumed paleoplains of the pre-Cambrian shield of North America, *Am. Jour. Sci.*, Vol. 262, pp. 817–57.

BOYÉ, M., MOULLINE, M., PRATVIEL, L., and VIGUIER, C. 1968. Relations entre la forme des cours inférieurs de la Garonne et de la Dordogne et les topographies souterraines des terrains, *Rev. géomorph. dyn.*, Vol. 18, pp. 83–91.

BURK, C. A. 1968. Buried ridges within continental margins, *Trans. New York Acad. Sci.*, Vol. 30, pp. 397–409.

CARSOLA, A. J. 1954. Microrelief on the Arctic Sea floor, *Bull. Amer. Ass. Petrol. Geols*, Vol. 28, pp. 1587–601.

DAY, A. A. 1959. The continental margin between Brittany and Ireland, *Deep-Sea Research*, Vol. 5, pp. 249–65.

DURRANCE, E. M. 1969. The buried channels of the Exe, *Geol. Mag.*, Vol. 106, pp. 174–89.

FISHER, R. V. 1964. Resurrected Oligocene hills, eastern Oregon, *Am. Jour. Sci.*, Vol. 262, pp. 713–25.

HAWKINS, A. B. 1962. The buried channel of the Bristol Avon, *Geol. Mag.*, Vol. 99, pp. 369–74.

HINSCHBERGER, F. 1964. La repartition des fonds sous-marins dans le vestibule du Goulet de Brest, *C. r. Acad. Sci.*, Vol. 258, pp. 6497–9.

HOLTEDAHL, H. 1958. Some remarks on geomorphology of continental shelves off Norway, Labrador, and southeast Alaska, *Jour. Geol.*, Vol. 66, pp. 461–71.

KLIMASZEWSKI, M. 1960. On the influence of preglacial relief on the extension and development of glaciation and deglaciation of mountainous regions, *Przeglad Geogr.*, Vol. 32 (Supp.), pp. 41–9.

MALLORY, J. K. 1964. Topography of the seabed off the coast of southern Africa, *S. Afr. Jour. Sci.*, Vol. 60, pp. 105–9.

MARINESCU, A. and SELATIU, O. 1965. An underwater valley in front of the Rumanian shore of the Black Sea, *Rev. Roum. Géol. Géophys. Géog., Sér. Géog.*, Vol. 9, pp. 77–80.

MARTIN, H. 1969. Paläomorphologische Formelemente in den Landschaften Südwest-Afrikas, *Geol. Rundschau*, Vol. 58, pp. 121–8.

MARTIN, R. 1966. Paleogeomorphology and its applications to exploration for oil and gas (with examples from Western Canada), *Bull. Amer. Ass. Petrol. Geols*, Vol. 50, pp. 2277–311.

MENARD, H. W. and DIETZ, R. S. 1952. Mendocino submarine escarpment, *Jour. Geol.*, Vol. 60, pp. 266–78.

MCMANUS, J. 1967. Pre-glacial diversion of the Tay drainage through the Perth gap, *Scot. Geog. Mag.*, Vol. 83, pp. 138–9.

SHEPARD, F. P. 1967. *The earth beneath the sea*, rev. ed. (John Hopkins, Baltimore), 242 pp.

SHEPARD, F. P. and DILL, R. F. 1966. *Submarine canyons and other sea valleys* (Rand McNally, Chicago), 381 pp.

SOONS, J. M. 1968. Raised submarine canyons: a discussion of some New Zealand examples, *Ann. Ass. Amer. Geogrs*, Vol. 58, pp. 606–13.

WALSH, P. T. 1966. Cretaceous outliers in south-west Ireland and their implications for Cretaceous palaeogeography, *Quart. Jour. Geol. Soc.*, Vol. 122, pp. 63–84.

WARNER, M. M. 1965. Cementation as a clue to structure, drainage patterns, permeability, and other factors, *Jour. Sed. Petrol.*, Vol. 35, pp. 797–804.

WILLIAMS, G. J. 1968. The buried channel and superficial deposits of the Lower Usk, and their correlation with similar features in the Lower Severn, *Proc. Geol. Ass.*, Vol. 79, pp. 325–48.

WINSLOW, J. H. 1966. Raised submarine canyons: an exploratory hypothesis, *Ann. Ass. Amer. Geogrs*, Vol. 56, pp. 634–72.

C. Stages in Slope Evolution

AGAR, R. 1960. Post-glacial erosion of the north Yorkshire coast from the Tees estuary to Ravenscar, *Proc. Yorks. geol. Soc.*, Vol. 32, pp. 409–28.

AHNERT, F. 1960. The influence of Pleistocene climates upon the morphology of cuesta scarps on the Colorado plateau, *Ann. Ass. Amer. Geogrs*, Vol. 50, pp. 139–56.

BAKKER, J. P. and LE HEUX, J. W. N. 1952. A remarkable new geomorphological law, *Proc. K. ned. Akad. Wet.*, Vol. 55B, pp. 399–410 and 554–71.

BECKETT, P. H. T. 1968. Soil formation and slope development. 1. A new look at Walther Penck's *Aufbereitung* concept, *Zeit. f. Geomorph.*, Vol. 12, pp. 1–24.

BRYAN, K. 1922. Erosion and sedimentation in the Papago County, Arizona, *U.S. Geol. Surv. Bull. 730*, pp. 19–90.

CARTER, C. S. and CHORLEY, R. J. 1961. Early slope development in an expanding stream system, *Geol. Mag.*, Vol. 98, pp. 117–30.

COLEMAN, A. M. and BALCHIN, W. G. V. 1959. The origin and development of surface depressions in the Mendip Hills, *Proc. Geol. Ass.*, Vol. 70, pp. 291–309.

DURAND, J. H. 1959. Les sols rouges et les croûtes en Algérie, *Service et Sci.*, Algiers, No. 7, 188 pp.

JAHN, A. 1968. Denudational balance of slopes, *Geog. Polonica*, Vol. 13, pp. 9–29.

KERNEY, M. P., BROWN, E. H., and CHANDLER, T. J. 1964. The Late-glacial and Post-glacial history of the Chalk escarpment near Brook, Kent, *Phil. Trans. Roy. Soc. B*, Vol. 248, pp. 135–204.

MACKAY, J. R. 1966. Segregated epigenetic ice and slumps in permafrost, Mackenzie delta area, N.W.T., *Geog. Bull.*, Vol. 8, pp. 59–80.

MELTON, M. A. 1965. The geomorphic and paleoclimatic significance of alluvial deposits in southern Arizona, *Jour. Geol.*, Vol. 73, pp. 1–38.

PENCK, W. 1924. *Die morphologische Analyse; ein Kapitel der physikalischen Geologie* (Engelhorns, Stuttgart), 283 pp.; trans. by H. Czech and K. C. Boswell, 1953, *Morphological analysis of land forms* (Macmillan, London), 429 pp.

RUXTON, B. P. and BERRY, L. 1961. Notes on faceted slopes, rock fans and domes on granite in the east-central Sudan, *Am. Jour. Sci.*, Vol. 259, pp. 194–206.

SAVIGEAR, R. A. G. 1952. Some observations on slope development in South Wales, *Trans. Inst. Brit. Geogrs*, No. 18, pp. 31–51.

SCHEIDEGGER, A. E. 1961. Mathematical models of slope development, *Bull. geol. Soc. Am.*, Vol. 72, pp. 37–50.

SCHUMM, S. A. 1956. Evolution of drainage systems and slopes in badlands at Perth Amboy, New Jersey, *Bull. geol. Soc. Am.*, Vol. 67, pp. 597–646.

SCHUMM, S. A. 1966. The development and evolution of hillslopes, *Jour. geol. Ed.*, Vol. 14, pp. 98–104.

TRICART, J. 1957. Mise au point: l'évolution des versants, *Inf. geog.*, Vol. 21, pp. 108–16.

TWIDALE, C. R. 1967. Origin of the piedmont angle as evidenced in South Australia, *Jour. Geol.*, Vol. 75, pp. 393–411.

WALKER, P. H. 1962. Soil layers on hillslopes: a study at Nowra, N.S.W., Australia, *Jour. Soil Sci.*, Vol. 13, pp. 167–77.

WATSON, R. A. and WRIGHT, H. E. 1963. Landslides on the east flank of the Chuska Mountains, northwestern New Mexico, *Am. Jour. Sci.*, Vol. 261, pp. 525–48.

D. Stages in Drainage System Evolution

AMBROSE, J. W. 1964. Exhumed paleoplains of the Precambrian shield of North America, *Am. Jour. Sci.*, Vol. 262, pp. 817–57.

BEATY, C. B. 1961. Topographic effects of faulting: Death Valley, California, *Ann. Ass. Amer. Geogrs*, Vol. 51, pp. 234–40.

BECKINSALE, R. P. and RICHARDSON, L. 1964. Recent findings in the physical development of the Lower Severn valley, *Geog. Jour.*, Vol. 130, pp. 87–105.

BEETE, J. B. 1862. On the mode of formation of some of the river-valleys in the south of Ireland, *Quart. Jour. geol. Soc.*, Vol. 18, pp. 378–403.

BLANC, A. 1958. Répertoire bibliographique critique des études de relief karstique en Yougoslavie depuis Jovan Cvijič, *Cen. Doc. Cartog. Géog. Mém. et Doc.*, Vol. 6, pp. 135–223.

BONYTHON, C. W. and MASON, B. 1953. The filling and drying of Lake Eyre, *Geog. Jour.*, Vol. 119, pp. 321–30.

BOWLER, J. M. and HARFORD, L. B. 1965. Quaternary tectonics and the evolution of the riverine plain near Echuca, Victoria, *Jour. geol. Soc. Aust.*, Vol. 13, pp. 339–54.

BROWN, E. H. 1952. The river Ystwyth, Cardiganshire: a geomorphological analysis, *Proc. Geol. Ass.*, Vol. 63, pp. 244–69.

BRUSH, L. M. and WOLMAN, M. G. 1960. Knickpoint behaviour in noncohesive material: a laboratory study, *Bull. geol. Soc. Am.*, Vol. 71, pp. 59–74.

CHANG, P. 1964. Relationship of block faulting to stream development on both flanks of Chin-Ling mountain regions *Intnl Geol. Revs*, Vol. 10 (1968), pp. 1428–39.

DURY, G. H. 1954. Contribution to a general theory of meandering valleys, *Am. Jour. Sci.*, Vol. 252, pp. 193–224.

DURY, G. H. 1959. A contribution to the geomorphology of Central Donegal, *Proc. Geol. Ass.*, Vol. 70, pp. 1–27.

DURY, G. H. 1960. Misfit streams: problems in interpretation, discharge and distribution, *Geog. Rev.*, Vol. 50, pp. 219–42.

DURY, G. H. 1964. Principles of underfit streams, *U.S. Geol. Surv. Prof. Paper 452A*, 67 pp.

EK, C. 1961. Conduits souterrains en relation avec les terrasses fluviales, *Ann. Soc. géol. Belg.*, Vol. 84, pp. 313–40.

GAMS, I. 1962. Slepe doline v Sloveniji; *Geografski Zbornik*, Vol. 7, pp. 265–306; English summary, Blind valleys in Slovenia, pp. 301–4.

GEORGE, P. 1948. Quelques formes karstiques de la Croatie occidentale et de la Slovanie méridionale (Yougoslavie), *Ann. Géog.*, Vol. 57, pp. 298–307.

HOLMES, D. A. 1968. The recent history of the Indus, *Geog. Jour.*, Vol. 134, pp. 367–82.

HJULSTROM, F. 1949. Climatic changes and river patterns, *Geografiska Annaler*, Vol. 31, pp. 83–9.

JONES, O. T. 1924. The Upper Towy drainage system, *Quart. Jour. geol. Soc.*, Vol. 80, pp. 568–609.

JONES, O. T. 1965. The glacial and post-glacial history of the lower Teifi valley, *Quart. Jour. geol. Soc.*, Vol. 121, pp. 247–81.

LENSEN, G. J. 1968. Analysis of progressive fault displacement during downcutting at the Branch River terraces, South Island, New Zealand, *Bull. geol. Soc. Am.*, Vol. 79, pp. 545–56.

LEOPOLD, L. B. and MILLER, J. P. 1954. A post-glacial chronology for some alluvial valleys in Wyoming, *U.S. Geol. Surv. Water-supply Paper 1261*, 90 pp.

LUMSDEN, G. I. and DAVIES, A. 1965. The buried channel of the River Nith and its marked change in level across the Southern Upland fault, *Scot. Jour. Geol.*, Vol. 1, pp. 134–43.

MILLER, A. A. 1935. The entrenched meanders of the Herefordshire Wye, *Geog. Jour.*, Vol. 75, pp. 160–78.

PITTY, A. F. 1965. A study of some escarpment gaps in the southern Pennines, *Trans. Inst. Brit. Geogrs*, No. 37, pp. 127–45.

PITTY, A. F. 1966. Landform studies in the Peak District of Derbyshire, unpubl. D.Phil. thesis, Univ. of Oxford, 378 pp.

REEVES, C. C. 1966. Pluvial lake basins of west Texas, *Jour. Geol.*, Vol. 74, pp. 269–91.

REEVES, C. C. 1968. *Introduction to paleolimnology* (Elsevier, Amsterdam), 228 pp.

RHODHAMEL, E. C. and CARLSTON, C. W. 1963. Geologic history of the Teays valley in West Virginia, *Bull. geol. Soc. Am.*, Vol. 74, pp. 251–74.

RICE, R. J. 1957. Some aspects of the glacial and postglacial history of the lower Goyt valley, *Proc. Geol. Ass.*, Vol. 68, pp. 217–27.

RUPPEL, E. T. 1967. Late Cenozoic drainage reversal, east-central Idaho, and its relation to possible undiscovered placer deposits, *Econ. Geol.*, Vol. 62, pp. 648–63.

SARKAR, S. K. and BASUMALLICK, S. 1968. Morphology, structure and evolution of a channel island in the Barakar river, Barakar, West Bengal, *Jour. Sed. Petrol.*, Vol. 38, pp. 747–54.

STRAW, A. 1963. The Quaternary evolution of the lower and middle Trent, *East Midld Geogr*, Vol. 3, pp. 171–89.

STRICKLIN, F. L. 1961. Degradational stream deposits of the Brazos river, central Texas, *Bull. geol. Soc. Am.*, Vol. 72, pp. 19–36.

ŠERKO, A. and MICHLER, I., trans. by N. Kuret, 1958, *La grotte de Postojna et les autres curiosites du Karst*; 2nd enl. ed. (Zavod Postojnska jama, Ljubljana), 191 pp.

TROLL, C. 1954. Über Alter und Gildung von Tal Mandern, *Erdkunde*, Vol. 8, pp. 286–302.

TUAN, Y. F. 1966. New Mexican gullies; a critical review and some recent observations, *Ann. Ass. Amer. Geogrs*, Vol. 56, pp. 573–97.

VARESCHI, V. 1963. Die Gabelteilung des Orinoco, *Pet. geog. Mitt.*, Vol. 107, pp. 241–8.

VITA-FINZI, C. 1969. Late Quaternary alluvial chronology of Iran, *Geol. Rundschau*, Vol. 58, pp. 951–73.

VITA-FINZI, C. 1969. *The Mediterranean valleys. Geological changes in historical time* (Cambridge Univ. Press, Cambridge), 140 pp.

WHITE, W. B. and SCHMIDT, V. A. 1966. Hydrology of a karst area in east-central West Virginia, *Water Res. Research*, Vol. 2, pp. 549–60.

E. Stages in Glaciated Areas

BIRKELAND, P. W. 1964. Pleistocene glaciation of the northern Sierra Nevada, north of Lake Tahoe, California, *Jour. Geol.*, Vol. 72, pp. 810–25.

BORNS, H. W. and GOLDTHWAIT, R. P. 1966. Late Pleistocene fluctuations of Kaskawulsh glacier, southwestern Yukon Territory, Canada, *Am. Jour. Sci.*, Vol. 264, pp. 600–19.

CHARLESWORTH, J. K. 1926. The readvance marginal kame-moraine of the south of Scotland and some later stages of retreat, *Trans. roy. Soc. Edinb.*, Vol. 55, pp. 25–50.

CHARLESWORTH, J. K. 1929. The South Wales end-moraine, *Quart. Jour. geol. Soc.*, Vol. 85, pp. 335–58.

DENTON, G. H. and STUIVER, M. 1966. Neoglacial chronology, northeastern St. Elias Mountains, Canada, *Am. Jour. Sci.*, Vol. 264, pp. 577–99.

DOWNIE, C. 1964. Glaciations of Mount Kilimanjaro, northeast Tanganyika, *Bull. geol. Soc. Am.*, Vol. 75, pp. 1–16.

DURY, G. H. 1953. A glacial breach in the north-western Highlands, *Scot. geog. Mag.*, Vol. 69, pp. 106–17.

DURY, G. H. 1957. A glacially breached watershed in Donegal, *Irish Geog.*, Vol. 3, pp. 80–171.

GROVE, J. M. 1966. The Little Ice Age in the massif of Mont Blanc, *Trans. Inst. Brit. Geogrs*, Vol. 40, pp. 129–43.

JENNESS, S. E. 1960. Late Pleistocene glaciation of eastern Newfoundland, *Bull. geol. Soc. Am.*, Vol. 71, pp. 161–80.

JOHN, B. S. 1967. Further evidence for a Middle Würm interstadial and a Main Würm glaciation of south-west Wales, *Geol. Mag.*, Vol. 104, pp. 630–3.

LEIGHTON, M. M. and BROPHY, J. A. 1966. Farmdale glaciation in northern Illinois and southern Wisconsin, *Jour. Geol.*, Vol. 74, pp. 478–99.

LINTON, D. L. 1951. Watershed breaching by ice in Scotland, *Trans. Inst. Brit. Geogrs*, Vol. 15 (1949), pp. 1–16.

LINTON, D. L. 1949 and 1951. Some Scottish river captures re-examined, *Scot. Geog. Mag.*, Vol. 65, pp. 123–31, and Vol. 67, pp. 31–44.

MCLENNAN, A. G. 1969. The last glaciation and deglaciation of central Lanarkshire, *Scot. Jour. Geol.*, Vol. 5, pp. 248–68.

MCKENZIE, G. D. 1969. Observations on a collapsing kame terrace in Glacier Bay National Monument, south-eastern Alaska, *Jour. Glaciol.*, Vol. 8, pp. 413–25.

MERCER, J. H. 1968. Variations of some Patagonian glaciers since the Late-Glacial, *Am. Jour. Sci.*, Vol. 266, pp. 91–109.

PENNY, L. F. 1964. A review of the Last Glaciation in Great Britain, *Proc. Yorks. geol. Soc.*, Vol. 34, pp. 387–411.

PORTER, S. C. 1964. Late Pleistocene glacial chronology on north-central Brooks Range, Alaska, *Am. Jour. Sci.*, Vol. 262, pp. 446–60.

PORTER, S. C. and DENTON, G. H. 1967. Chronology of neoglaciation in the North American Cordillera, *Am. Jour. Sci.*, Vol. 265, pp. 177–210.

SISSONS, J. B. 1967. Glacial stages and radiocarbon dates in Scotland, *Scot. Jour. Geol.*, Vol. 3, pp. 375–81.

SOONS, J. M. 1963. The glacial sequence in part of the Rakaia valley, Canterbury, New Zealand, *N.Z. Jour. Geol. Geophys.*, Vol. 6, pp. 735–56.

SYNGE, F. M. 1964. Some problems concerned with the glacial succession in south-east Ireland, *Irish Geog.*, Vol. 5, pp. 73–82.

SZUPRYCZYNSKI, J. 1968. Glaciations in the Spitzbergen area, *Geog. Polonica*, Vol. 14, pp. 175–83.

TOTTEN, S. M. 1969. Overridden recessional moraines of north-central Ohio, *Bull. geol. Soc. Am.*, Vol. 80, pp. 1931–46.

VIRKKALA, K. 1961. On the glacial geology of the Mämeenlinna region, southern Finland, *Bull. Comm. géol. Finlande*, No. 196, pp. 215–41.

YATES, E. M. and MOSELEY, F. 1967. A contribution to the glacial geomorphology of the Cheshire Plain, *Trans. Inst. Brit. Geogrs*, Vol. 42, pp. 107–25.

F. Stages in Coastline Evolution

ANDREWS, J. T. 1967. Problems in the analysis of the vertical displacement of shorelines, *Geog. Bull.*, Vol. 9, pp. 71–4.

BUFFINGTON, E. C. and MOORE, D. G. 1963. Geophysical evidence of the origin of gullied submarine slopes, San Clemente, California, *Jour. Geol.*, Vol. 71, pp. 356–70.

COTTON, C. A. 1963. Levels of planation of marine benches, *Zeit. f. Geomorph.*, Vol. 7, pp. 97–111.

DRISCOLL, E. M. and HOPLEY, D. 1968. Coastal development in a part of tropical Queensland, Australia, *Jour. Tropical Geog.*, Vol. 26, pp. 17–28.

FLEMMING, N. C. 1965. Form and relation to present sea level of Pleistocene marine erosion features, *Jour. Geol.*, Vol. 73, pp. 799–811.

FLINN, D. 1969. On the development of coastal profiles in the north of Scotland, Orkney and Shetland, *Scot. Jour. Geol.*, Vol. 5, pp. 393–9.

FOLK, R. L. 1967. Sand cays of Alacran reef, Yucutan, Mexico; morphology, *Jour. Geol.*, Vol. 75, pp. 412–37.

GLENIE, R. C., SCHOFIELD, J. C., and WARD, W. T. 1968. Tertiary sea levels in Australia and New Zealand, *Palaeogeog., Palaeoclim., Palaeoecol.*, Vol. 5, pp. 141–63.

GRESSWELL, R. K. 1964. The origin of the Mersey and Dee estuaries, *Geol. Jour.*, Vol. 4, pp. 77–86.

GUILCHER, A. 1949. Aspects et problèmes morphologiques du massif de Devon-Cornwall, *Rev. géog. alpine*, Vol. 37, pp. 689–717.

JOHNSON, D. W. 1919. *Shore processes and shoreline development* (J. Wiley, New York), 584 pp.

KANEKO, S. 1966. Rising promontories associated with a subsidiary coast and sea-floor in south-western Japan, *Trans. Roy. Soc. N.Z., Geol.*, Vol. 4, pp. 211–28.

KIDSON, C. 1952. Dawlish Warren: a study of the evolution of the sand spits across the mouth of the River Exe in Devon, *Trans. Inst. Brit. Geogrs.*, Vol. 16 (1950), pp. 67–80.

KING, W. B. R. 1954. The geological history of the English Channel, *Quart. Jour. geol. Soc.*, Vol. 110, pp. 77–101.

KLIEWE, H. 1963. Fortschritte und Problem beider Erforschung der Küstenentwicklung an der südlichen Ostsee, *Baltica*, Vol. 1, pp. 116–36.

KOFOED, J. W. and GORSLINE, D. S. 1966. Sediments of the Choptank River, Maryland, *South eastern geology*, Vol. 7, pp. 65–82.

LEWIS, W. V. 1938. Evolution of shoreline curves, *Proc. Geol. Ass.*, Vol. 49, pp. 107–27.

MACNEIL, F. S. 1954. The shape of atolls: an inheritance from subaerial erosion forms, *Am. Jour. Sci.*, Vol. 252, pp. 402–27.

MITCHELL, G. F. 1960. The Pleistocene history of the Irish Sea, *Adv. Sci.*, Vol. 68, pp. 313–25.

NAKAGAWA, H. 1965. Pleistocene sea levels along the Pacific coast of Japan, *Sci. Rep. Tohoku Univ., Ser. 2 (Geol.)*, Vol. 37, pp. 451–62.

ORME, A. 1962. Abandoned and composite seacliffs in Britain and Ireland, *Irish Geography*, Vol. 4, pp. 279–91.

PONS, L. J., JELGERSMA, S., WIGGERS, A. J., and DE JONG, J. D. 1963. Evolution of the Netherlands coastal areas during the Holocene; *Verh. Kon. Ned. Geol. Mijnbk. Gen. Geol.*, Ser. 21–2, pp. 197–208.

ROBINSON, A. H. W. 1966. Residual currents in relation to shoreline evolution of the East Anglian coast, *Marine Geology*, Vol. 4, pp. 57–84.

RUSSELL, R. J. 1967. Aspects of coastal morphology, *Geografiska Annaler*, Vol. 49A, pp. 299–309.

SCHLANGER, S. O. *et al.* 1963. Subsurface geology of Eniwetok Atoll, *U.S. Geol. Surv., Prof. Paper 260-b*, 75 pp.

SCHWARTZ, M. L. 1968. The scale of shore erosion, *Jour. Geol.*, Vol. 76, pp. 508–17.

SHIER, D. E. 1969. Vermetid reefs and coastal development in the Ten Thousands Islands, southwest Florida, *Bull. geol. Soc. Am.*, Vol. 80, pp. 485–508.

SISSONS, J. B. 1963. Scottish raised shoreline heights with particular reference to the Forth valley, *Geografiska Annaler*, Vol. 45, pp. 180–5.

STEERS, J. A. *et al.* 1939. Recent coastal changes in south-eastern England: a discussion, *Geog. Jour.*, Vol. 93, pp. 399–419 and 491–512.

STEPHENS, N. 1958. The evolution of the coastline of north-east Ireland, *Adv. Sci.*, Vol. 56, pp. 389–91.

STODDART, D. R. 1965. British Honduras cays and the low wooded island problem, *Trans. Inst. Brit. Geogrs*, Vol. 36, pp. 131–47.

TANNER, W. F. 1962. Reorientation of convex shores, *Am. Jour. Sci.*, Vol. 260, pp. 37–43.

VARJO, U. 1964. Über finnische Küsten und ihre Entstehung, *Fennia*, Vol. 91, No. 2, 104 pp.

VERSTAPPEN, H. T. 1954. The influence of climatic changes on the formation of coral islands, *Am. Jour. Sci.*, Vol. 252, pp. 428–35.

WILLIAMS, W. W. 1956. An east coast survey: some recent changes in the coast of East Anglia, *Geog. Jour.*, Vol. 122, pp. 317–34.

WILLIAMS, W. W. 1960. *Coastal changes* (Routledge and Kegan Paul, London), 220 pp.

ZENKOVICH, V. P. 1962. *Osnovii ucheniva o razvitsi toskich beregov*; (Academy of Sciences, Moscow), 710 pp.; trans. by O. G. Fry and edited by J. A. Steers, 1967, *Processes of coastal development* (Oliver and Boyd), 738 pp.

G. Stages involving more than one process and in the Evolution of the Landsurface as a whole

BALCHIN, W. G. V. 1952. The erosion surfaces of Exmoor and adjacent areas, *Geog. Jour.*, Vol. 118, pp. 453–76.

BATTISTINI, R. 1966. Le littoral du paléokarst de la presqu'île, de Narinda au Nord de Majunga (Madagascar): un example de côté contraposée, *Bull. Ass. géog. Fr.*, Nos. 346–7, pp. 41–9.

BAULIG, H. 1928. *Le plateau central de la France et sa bordure Méditerranéenne, étude morphologique* (Armand Colin, Paris), 589 pp.

BROWN, E. H. 1960. *The relief and drainage of Wales* (Univ. Wales, Press), 186 pp.

BRUNSDEN, D., KIDSON, C., ORME, A. R., and WATERS, R. S. 1964. Denudation chronology of parts of south-western England, *Field Studies*, Vol. 2, pp. 115–32.

CLAYTON, K. M. 1953. The glacial chronology of part of the middle Trent basin, *Proc. Geol. Soc.*, Vol. 64, pp. 198–207.

DIXEY, F. 1956. Erosion surfaces in Africa, *Trans. geol. Soc. S. Afr.*, Vol. 59, pp. 1–16.

DURY, G. H. 1955. Diversion of drainage by ice, *Science News*, Vol. 38, pp. 48–71.

EMBLETON, C. 1964. Sub-glacial drainage and supposed ice-dammed lakes in north-east Wales, *Proc. Geol. Ass.*, Vol. 75, pp. 31–8.

EVERARD, C. E. 1957. Erosion platforms on the borders of the Hampshire basin, *Trans. Inst. Brit. Geogrs*, No. 22, pp. 33–46.

FRYE, J. C. and LEONARD, A. B. 1957. Ecological interpretation of Pliocene and Pleistocene Stratigraphy in the Great Plains region, *Am. Jour. Sci.*, Vol. 255, pp. 1–11.

GEORGE, T. N. 1955. British Tertiary landscape evolution, *Sci. Prog.*, Vol. 43, pp. 291–307.

GEORGE, T. N. 1966. Geomorphic evolution in Hebridean Scotland, *Scot. Jour. Geol.*, Vol. 2, pp. 1–34.

GREGORY, K. J. 1965. Proglacial lake Eskdale after sixty years, *Tràns. Inst. Brit. Geogrs*, Vol. 36, pp. 149–62.

HOUGH, J. L. 1966. Correlation of glacial lake stages in the Huron-Erie and Michigan basins, *Jour. Geol.*, Vol. 74, pp. 62–77.

IVES, J. D. 1960. Former ice-dammed lakes and the deglaciation of the middle reaches of the George River, Labrador–Ungava, *Geog. Bull.*, No. 14, pp. 44–88.

JAMIESON, T. F. 1865. On the history of the last geological changes in Scotland, *Quart. Jour. geol. Soc.*, Vol. 21, pp. 161–203.

JESSUP, R. W. 1961. A Tertiary-Quaternary pedological chronology for the south-eastern portion of the Australia arid zone, *Jour. Soil Sci.*, Vol. 12, pp. 199–213.

JOHNSON, R. H. and RICE, R. J. 1961. The geomorphology of the south-west Pennine upland, *Proc. Geol. Ass.*, Vol. 72, pp. 21–32.

JOHNSON, R. H. 1965. The glacial geomorphology of the west Pennine slopes from Cliviger to Congleton, *in* J. B. Whittow and P. D. Wood (Eds), *Essays in geography for Austin Miller* (Univ. of Reading, Reading), pp. 58–93.

KIRBY, R. P. 1969. Morphometric analysis of glaciofluvial terraces in the Esk basin, Midlothian, *Trans. Inst. Brit. Geogrs*, No. 48, pp. 1–18.

LEWIS, P. F. 1960. Linear topography in the southwestern Palouse, Washington–Oregon, *Ann. Ass. Amer. Geogrs*, Vol. 50, pp. 98–111.

MANNERFELT, C. M. 1945. Nagra glacialmorfologiska formelement, *Geografiska Annaler*, Vol. 27, pp. 3–239.

MACAR, P. 1965. Apercu synthétique sur l'évolution géomorphologique de l'Ardenne, *La Géographie*, Vol. 17, No. 1, 11 pp.

MITCHELL, G. F. 1963. Morainic ridges on the floor of the Irish Sea, *Irish Geog.*, Vol. 4, pp. 335–44.

NICHOLS, R. L. 1968. Coastal geomorphology, McMurdo Sound, Antarctica, *Jour. Glaciol.*, Vol. 7, pp. 449–78.

QUENNELL, A. M. 1958. The structural and geomorphic evolution of the Dead Sea rift, *Quart. Jour. geol. Soc.*, Vol. 114, pp. 1–24.

ROBINSON, A. H. W. 1968. The submerged glacial landscape off the Lincolnshire coast, *Trans. Inst. Brit. Geogrs*, No. 44, pp. 119–32.

SERET, G. 1965. La succession des épisodes fluviatiles périglaciaires et fluvioglaciaires à l'aval des glaciers, *Zeit. f. Geomorph.*, Vol. 9, pp. 305–20.

STRAW, A. 1961. Drifts, meltwater channels and ice-margins in the Lincolnshire Wolds, *Trans. Inst. Brit. Geogrs*, Vol. 9, pp. 115–28.

STRIDE, A. H. 1959. On the origin of the Dogger Bank, in the North Sea, *Geol. Mag.*, Vol. 96, pp. 33–44.

SWEETING, M. M. 1963. Erosion cycles and limestone caverns in the Ingleborough district, *Geog. Jour.*, Vol. 115, pp. 63–78.

TOMLINSON, M. E. 1963. The Pleistocene chronology of the Midlands, *Proc. Geol. Ass.*, Vol. 74, pp. 187–202.

TODD, T. W. 1968. Dynamic diversion: influence of longshore current-tidal flow interaction on chernier and barrier island plains, *Jour. Sed. Petrol.*, Vol. 38, pp. 734–46.

WALKER, G. P. L. and BLAKE, D. H. 1966. The formation of a palagonite breccia mass beneath a valley glacier in Iceland, *Quart. Jour. geol. Soc.*, Vol. 122, pp. 45–61.

WOLFE, P. E. 1964. Late Cenozoic uplift and exhumed Rocky Mountains of central western Montana, *Bull. geol. Soc. Am.*, Vol. 75, pp. 493–502.

WOOLDRIDGE, S. W. 1950. The upland plains of Britain: their origin and geographical significance, *Adv. Sci.*, Vol. 7, pp. 162–75.

WOOLDRIDGE, S. W. and LINTON, D. L. 1955. *Structure, surface and drainage of south-east England* (Philip, London), 176 pp.

ZAKHAROV, Y. F. 1966. The age of relief in northern, subpolar and polar Urals, *Intnl. Geol. Rev.*, Vol. 9 (1967), pp. 1025–7.

Subject Index

Geographical Index

(excluding bibliography)

Author Index

(including bibliography)